Springer Transactions in Civil and Environmental Engineering

Springer Transactions in Civil and Environmental Engineering (STICEE) publishes the latest developments in Civil and Environmental Engineering. The intent is to cover all the main branches of Civil and Environmental Engineering, both theoretical and applied, including, but not limited to: Structural Mechanics, Steel Structures, Concrete Structures, Reinforced Cement Concrete, Civil Engineering Materials, Soil Mechanics, Ground Improvement, Geotechnical Engineering, Foundation Engineering, Earthquake Engineering, Structural Health and Monitoring, Water Resources Engineering, Engineering Hydrology, Solid Waste Engineering, Environmental Engineering, Wastewater Management, Transportation Engineering, Sustainable Civil Infrastructure, Fluid Mechanics, Pavement Engineering, Soil Dynamics, Rock Mechanics, Timber Engineering, Hazardous Waste Disposal Instrumentation and Monitoring, Construction Management, Civil Engineering Construction, Surveying and GIS Strength of Materials (Mechanics of Materials), Environmental Geotechnics, Concrete Engineering, Timber Structures.

Within the scopes of the series are monographs, professional books, graduate and undergraduate textbooks, edited volumes and handbooks devoted to the above subject areas.

More information about this series at http://www.springer.com/series/13593

S. M. Shiva Nagendra · Uwe Schlink ·
Andrea Müller · Mukesh Khare
Editors

Urban Air Quality Monitoring, Modelling and Human Exposure Assessment

Springer

Editors
S. M. Shiva Nagendra
Department of Civil Engineering
Indian Institute of Technology Madras
Chennai, Tamil Nadu, India

Uwe Schlink
Department of Urban
and Environmental Sociology
Helmholtz Centre for Environmental
Research UFZ
Leipzig, Germany

Andrea Müller
Molecular Systems Biology
Helmholtz Centre for Environmental
Research UFZ
Leipzig, Germany

Mukesh Khare
Department of Civil Engineering
Indian Institute of Technology Delhi
New Delhi, Delhi, India

ISSN 2363-7633 ISSN 2363-7641 (electronic)
Springer Transactions in Civil and Environmental Engineering
ISBN 978-981-15-5510-7 ISBN 978-981-15-5511-4 (eBook)
https://doi.org/10.1007/978-981-15-5511-4

This Springer imprint is published by the registered company Springer Nature Singapore Pte Ltd.
The registered company address is: 152 Beach Road, #21-01/04 Gateway East, Singapore 189721, Singapore

Preface

Urban air pollution poses a significant threat to human health, environment and the quality of life of people throughout the world. It also significantly contributes to climate change. The climate change has the potential to significantly modify the local and regional air quality through changes in chemical reaction rates, vertical mixing of pollutants (boundary layer height), and transport of pollutant. It is predicted that climate changes could increase ground level harmful air pollutants concentration in urban areas.

Poor air quality will largely affect human health, vegetation (crop yield) and material damage (economic loss). In cities, children and elderly people are more vulnerable to both climate and air quality changes. It is reported that climate factors such as temperature, humidity, and wind velocity and wind direction have considerable impact on chronic respiratory diseases. Asthma is also correlated with the changes in air temperature and humidity. The World Health Organization (WHO) has estimated that in developing countries, increasing urban air pollution (UAP) has resulted in more than 2 million deaths per annum along with various cases of respiratory illnesses (WHO, 2014). One of the major sources of UAP is the road transport sector. Besides, domestic, commercial and industrial activities also contribute to UAP.

Air pollution in urban areas involves multiple processes: the generation of pollutants at and their release from a source; their transport and transformation and removal from the atmosphere; and their effects on human beings, visibility, materials, and ecosystems. Therefore, modelling, monitoring and management of urban air quality are essential methods to achieve clean air and comfortable urban conditions in a sustainable manner.

This book dedicated to the topics focus on Air Quality Management in Cities and shares the practical experience of the contributing authors. This book is divided into two parts. Part I has seven chapters that discuss several topics related to urban air quality management. Chapter 1 provides introduction to urban air pollution including types of air pollutants and their emissions sources, air quality trends in developed and developing countries, current air pollution issues followed by air quality management options. Chapter 2 covers broadly on air quality monitoring techniques. This chapter starts with basics of air pollution measurement, air quality monitoring networks, measurement of gaseous pollutants and measurement of particulate pollutants followed by data quality control and quality assurance. Chapter 3 explains topics related to air quality modelling. This chapter starts with basic concepts on mathematical model formulation, followed by air pollution meteorology, air quality modelling techniques, air pollutants dispersion models, source apportionment models, statistical distributions models, limitations of air quality models, application models in local and urban air quality management. Chapter 4 details the principles of air quality management, air quality management frame work, air quality standards and legislations, air quality management practices in developed countries, air quality management practices in developing countries and problems associated with air quality management. Chapter 5 covers topics related to indoor air quality, particularly covering basics of indoor air pollution, sources of indoor air pollutants, indoor air quality and human comfort, indoor air quality modelling and indoor air management. Chapter 6 provides an overview on personal exposure and health risk assessment. In this chapter topics such as air pollution and chemical toxicity, air pollutants and health risks, personal exposure to air pollutants, health risks assessment and case studies are emphasized in greater detail. The last chapter (Chap. 7) describes the smart sensors for air quality management with special focus on air quality measuring sensors, types of smart air quality sensors, data acquisition, smart-sensors for mobile and personal monitoring, air quality maps and limitations of smart air quality sensors. Part II presents twenty three case studies related to the concepts introduced in Part I.

This book is intended for students, academicians, scientists, engineers and researchers to know more about air quality management. We wish to thank the contributing authors and to acknowledge the help of the eminent members of the scientific advisory, committee of the IICAQM conferences and all reviewers in peer reviewing the submitted book chapter contributions.

Chennai, India — Prof. Dr. S. M. Shiva Nagendra
Leipzig, Germany — Prof. Dr. Uwe Schlink
Leipzig, Germany — Dr. Andrea Müller
New Delhi, India — Prof. Dr. Mukesh Khare

Contents

Editors and Contributors

About the Editors

Dr. S. M. Shiva Nagendra is presently working as a Professor in the Department of Civil Engineering, Indian Institute of Technology Madras, India. He obtained his Ph.D. degree in Civil Engineering from IIT Delhi, India. Dr. Nagendra has more than 20 years' research experience mainly in the field of air quality monitoring, modelling, management, indoor air quality, personal exposure, control technologies and policy implications. His groups research is a mixture of filed and laboratory experiments, micro, local and urban scale air quality characterization, modelling and simulation. He has published more than 75 research publications in international and national peer reviewed journals, one reference book, and over 100 papers in peer reviewed conference proceedings. He is a member of several scientific committees, societies and technical institutions. He is the founding chairman of Indian International Conference on Air Quality Management (IICAQM) series and serves as the Vice President of Society for Indoor Environment.

Dr. Uwe Schlink is a Senior Researcher at the Department of Urban and Environmental Sociology, Helmholtz Centre for Environmental Research and Professor at University of Leipzig, Institute of Meteorology. He is leading the air quality and urban climate research group in the department. His research areas include urban climate adaptation, thermal comfort, urban air quality; personal exposure and vulnerability with extreme environmental situations in urban areas; statistical modelling and Bayesian inference. He has published around 75 articles in peer-reviewed journals and presented more than 100 conference papers. He is a member of several scientific committees and societies.

Dr. Andrea Müller is a Senior Scientist in the Department Molecular System Biology at the Helmholtz Centre for Environmental Research in Leipzig, Germany. Her research interests lie in air quality and their impact on human health. Her special research topics are related to: Exposure analyses (indoor and outdoor) of bio aerosols (pollen, mold, bacteria, allergens); Impact of environmental pollutant compounds (air particles) on cell systems; Toxicology and mode of action of single and mixture compounds; Development of methods for analyses of PAHs in particulate matter; Statistical analysis of association between exposure and health outcome; Modeling of distribution of air contaminants.

Dr. Mukesh Khare is presently working as a Professor in Department of Civil Engineering, Indian Institute of Technology Delhi, India. He is a Fellow of Institution of Engineers India and Fellow of Wessex Institute of Great Britain. He obtained his Ph.D. degree in Faculty of Engineering from Newcastle University, UK and has managed a range of environmental projects throughout his professional career. With a specialization in air quality modelling, Prof. Khare's experience has covered research and development studies, teaching, consulting, modelling, editorial activities. In addition, Prof. Khare has authored more than 200 research publications primarily for peer reviewed journals and conference proceedings. He is a member of High Level Task Force (HLTF), under the Prime Minister Office, Government of India.

Contributors

Patrizio Arrigo CNR Institute for Macromolecular Studies, Genoa, Italy

Amit Bafana CSIR NEERI Environmental Health Division, Nagpur, India

Tirthankar Banerjee DST-Mahamana Centre of Excellence in Climate Change Research, Institute of Environment and Sustainable Development, Banaras Hindu University, Varanasi, India

Ross Beardsley Ramboll, Novato and San Francisco, CA, USA

Margaret Bell School of Civil Engineering and Geosciences, Newcastle University, Newcastle upon Tyne, UK

V. S. Chithra Department of Civil Engineering, Mar Baselios Christian College of Engineering and Technology, Kuttikkanam, Kerala, India

Ashok Deshpande Berkeley Initiative Soft Computing (BISC)—Special Interest, Group (SIG)—Environmental Management Systems (EMS), University of California, Berkeley, CA, USA;
College of Engineering, Pune, India

S. Devaanandan Department of Civil Engineering, Indian Institute of Technology Madras, Chennai, India;
Puducherry Pollution Control Committee (PPCC), Puducherry, India

Saravana S. Devi CSIR NEERI Environmental Health Division, Nagpur, India

D. M. M. S. Dissanayaka Department of Transport and Logistics Management, Faculty of Engineering, University of Moratuwa, Moratuwa, Sri Lanka

M. Diya Department of Civil Engineering, Indian Institute of Technology Madras, Chennai, India

W. W. A. S. Fernando Department of Transport and Logistics Management, Faculty of Engineering, University of Moratuwa, Moratuwa, Sri Lanka

Rajyalakshmi Garaga Department of Civil Engineering, Indian Institute of Technology Guwahati, Guwahati, India

S. Geetha Rajalakshmi Engineering College, Chennai, Tamil Nadu, India

S. K. Goyal CSIR-National Environmental Engineering Research Institute, Delhi Zonal Centre, Naraina Industrial Area, New Delhi, India

Sunil Gulia Indian Institute of Technology, Hauz Khas, New Delhi, India;
CSIR-National Environmental Engineering Research Institute, Delhi Zonal Centre, Naraina Industrial Area, New Delhi, India

Gaurav Gupta National Institute of Technical Teachers' Training and Research, Chandigarh, India

Pardeep Kumar Gupta Punjab Engineering College, Chandigarh, India

N. Jagadisha Department of EEE, S.J.C.E, Mysuru, India

Abba Elizabeth Joseph Xavier University, Bhubaneswar, Harirajpur, India

Mukesh Khare Department of Civil Engineering, The Indian Institute of Technology, Hauz Khas, New Delhi, India

Soma Sekhara Rao Kolluru School of Environmental Science and Engineering, Indian Institute of Technology Kharagpur, Kharagpur, India

Sri Harsha Kota Department of Civil Engineering, Indian Institute of Technology Guwahati, Guwahati, India

Kannan Krishnamurthi CSIR NEERI Environmental Health Division, Nagpur, India

Rakesh Kumar CSIR-National Environmental Engineering Research Institute, Nagpur, India;
National Environmental Engineering Research Institute, Mumbai Zonal Laboratory, Mumbai, India

Sonal Kumari Department of Chemistry, Faculty of Science, Dayalbagh Educational Institute, Dayalbagh, Agra, India

Shailendra Kumar Kushawaha Department of Chemistry, Chhatrapati Shivaji Institute of Technology, Durg, C.G, India

Anita Lakhani Department of Chemistry, Faculty of Science, Dayalbagh Educational Institute, Dayalbagh, Agra, India

Shari Libicki Ramboll, Novato and San Francisco, CA, USA

K. S. Lokesh Department of Environmental Engineering, Sri Jayachamarajendra College of Engineering, Mysuru, Karnataka, India

Selvakumar Madhavan Rajalakshmi Engineering College, Chennai, Tamil Nadu, India

K. Maharaj Kumari Department of Chemistry, Faculty of Science, Dayalbagh Educational Institute, Dayalbagh, Agra, India

N. Manojkumar School of Civil Engineering, Vellore Institute of Technology (VIT), Vellore, TN, India

Jyothi S. Menon Public Health Foundation of India, Gurugram, India

Andrea Müller Department Molecular Systems Biology, Helmholtz Centre for Environmental Research UFZ, Leipzig, Germany

M. Muthuraj Department of Bioengineering, National Institute of Technology Agartala, Agartala, India

Sumita Nair Department of Applied Chemistry, Bhilai Institute of Technology, Durg, C.G, India

Abhilash T. Nair Department of Civil Engineering, Indian Institute of Technology Madras, Chennai, India;
Department of Applied Sciences and Humanities, National Institute of Foundry and Forge Technology, Ranchi, India

Anil Namdeo School of Civil Engineering and Geosciences, Newcastle University, Newcastle upon Tyne, UK

Pravin K. Naoghare CSIR NEERI Environmental Health Division, Nagpur, India

Sunil Ojha Ramboll India, Gurgaon, India

Asim Kumar Pal Department of Environmental Science and Engineering, IIT (ISM), Dhanbad, India

T. Panda MSB140A, Biochemical Engineering Laboratory, Department of Chemical Engineering, Indian Institute of Technology Madras, Chennai, India

Aditya Kumar Patra Department of Mining Engineering, Indian Institute of Technology Kharagpur, Kharagpur, India

Shamsh Pervez School of Studies in Chemistry, Pt. Ravishankar Shukla University, Raipur, C.G, India

Yasmeen F. Pervez Department of Chemistry, Chhatrapati Shivaji Institute of Technology, Durg, C.G, India

Anju Elizbath Peter Department of Civil Engineering, Indian Institute of Technology Madras, Chennai, India

R. Rajagopala Department of EEE, S.J.C.E, Mysuru, India

R. Ravi Krishna Department of Chemical Engineering, Indian Institute of Technology Madras, Chennai, India

Dinesh Sah Department of Chemistry, Faculty of Science, Dayalbagh Educational Institute, Dayalbagh, Agra, India

S. Sangeetha Department of EEE, CMR College of Engineering & Technology, Hyderabad, India

Saraswati Department of Civil Engineering, Indian Institute of Technology Madras, Chennai, India

Uwe Schlink Department of Urban and Environmental Sociology, Helmholtz Centre for Environmental Research UFZ, Leipzig, Germany

A. Seenivasan Department of Biotechnology, National Institute of Technology Andhra Pradesh, Tadepalligudem, Andhra Pradesh, India

B. S. Shilpa Department of Civil Engineering, Vidyavardhaka College of Engineering, Mysuru, Karnataka, India

Shipra Electricity Supply Sub Division Motipur, North Bihar Power Distribution Company Ltd., Muzaffarpur, Bihar, India

S. M. Shiva Nagendra Department of Civil Engineering, Indian Institute of Technology Madras, Chennai, Tamil Nadu, India

Nandita Singh Institute of Environment and Sustainable Development, Banaras Hindu University, Varanasi, India

Hemant Sood National Institute of Technical Teachers' Training and Research, Chandigarh, India

G. Srikanth Department of ECE, CMR Technical Campus, Hyderabad, India

B. Srimuruganandam School of Civil Engineering, Vellore Institute of Technology (VIT), Vellore, TN, India

A. D. Srinivasan Department of EEE, S.J.C.E, Mysuru, India

Seema Unnikrishnan National Institutes of Industrial Engineering, Mumbai, India

P. Venkatakrishnan Department of ECE, CMR Technical Campus, Hyderabad, India

Nidhi Verma Department of Chemistry, Faculty of Science, Dayalbagh Educational Institute, Dayalbagh, Agra, India

Puneet Kumar Verma Department of Chemistry, Faculty of Science, Dayalbagh Educational Institute, Dayalbagh, Agra, India

Prince Vijay Jawaharlal College of Engineering and Technology, Palakkad, Kerala, India

Krish Vijayaraghavan Ramboll, Novato and San Francisco, CA, USA

S. Vivek Balachander Xavier University, Bhubaneswar, Harirajpur, India

Jyoti Yadav Department of Computer Science, Savitribai Phule Pune University, Pune, India

List of Figures

List of Tables

Part I
Foundations of Air Pollution Monitoring, Modelling, Health Risk and Control Engineering

Chapter 1
Introduction to Urban Air Pollution

S. M. Shiva Nagendra, Mukesh Khare, Uwe Schlink, and Anju Elizbath Peter

1.1 Urban Air Quality

Air pollution refers to the release of pollutants into the air that are detrimental to human health and the planet as a whole. Its effects can range from higher disease risks to rising temperatures. The ambient air quality in many cities of developing countries are found to be at alarming levels due to excessive emissions added by rapid urbanization. An estimation of the world's major cities average PM_{10} and $PM_{2.5}$ concentration levels undertaken a few years ago revealed that they were 71 $\mu g/m^3$ and 23 $\mu g/m^3$, respectively (WHO 2016). Several studies (Anderson et al. 2004; Ostro and World Health Organization 2004) have reported commensurable escalation in daily death per 10 $\mu g/m^3$ increase in PM at some of the North American and Western European regions.

S. M. Shiva Nagendra (✉) · A. E. Peter
Department of Civil Engineering, Indian Institute of Technology Madras, Chennai, Tamil Nadu 600036, India
e-mail: snagendra@iitm.ac.in

M. Khare
Department of Civil Engineering, Indian Institute of Technology Delhi, New Delhi 110016, India
e-mail: mukeshk@civil.iitd.ernet.in

U. Schlink
Department of Urban and Environmental Sociology, Helmholtz Centre for Environmental Research UFZ, 04318 Leipzig, Germany
e-mail: uwe.schlink@ufz.de

S. M. Shiva Nagendra et al. (eds.), *Urban Air Quality Monitoring, Modelling and Human Exposure Assessment*, Springer Transactions in Civil and Environmental Engineering,
https://doi.org/10.1007/978-981-15-5511-4_1

1.2 Types of Air Pollutants and Their Emissions Sources

1.2.1 Types of Air Pollutants

The solid particles, liquid droplets or gases emanating from natural or man-made sources could be pollutants. There is presence of two kinds of pollutants which can be either primary or secondary. Generally, the direct emission of material from an activity make it primary pollutants, such as ash from a volcanic eruption, the carbon monoxide gas from a motor vehicle exhaust or sulphur dioxide released from factories. The reaction or interaction of primary pollutants results in the formation of secondary pollutants which are not emanated directly from the source. The ground-level ozone is an example of a secondary pollutant which is one of the important components of photochemical smog.

The allowable concentrations of the substances in ambient air are defined by criteria pollutants on the basis of national air quality standards (NAQS). Carbon monoxide, lead, nitrogen dioxide, ozone, particles and sulphur dioxide are considered as criteria air pollutants.

1.2.2 Types of Sources

There are four main types of air pollution sources (Fig. 1.1):

- **mobile sources**—such as cars, buses, planes, trucks and trains

Fig. 1.1 Mobile, stationary, area and natural sources all emit pollution into the air

- **stationary sources**—such as power plants, oil refineries, industrial facilities and factories
- **area sources**—such as agricultural areas, cities and wood burning fireplaces
- **natural sources**—such as wind-blown dust, wildfires and volcanoes.

According to the Environmental Protection Agency (EPA), the primary mobile source which includes the automobile contributes large amounts of air pollution. The stationary sources also are renowned as point sources that discharge huge amounts of pollution from a particular site. The numerous smaller sources of pollution individually are not so significant, together make area sources. The current problem of air pollution contributed by all these pollution sources except natural sources can be critical occasionally.

1.3 Air Quality Trends in Developed and Developing Countries

The death rate from air pollution is higher in countries that have a higher level of pollution. There is also an important regional divide: most European, North American and Latin American countries cluster near the origin at low pollution levels and low death rates. Nearly all countries with either a high death rate or high $PM_{2.5}$ concentration (or both) are in Africa or Asia.

Countries such as Qatar, Saudi Arabia, Oman, Kuwait and the UAE have a comparably lower risk of premature death, despite high levels of pollution. They do, however, have a significantly higher GDP per capita than their neighbours. It may therefore be expected that overall health, well-being and healthcare/medical standards in these nations significantly reduce the risk of mortality from respiratory illness. The influence of GDP and quality of life is also likely to explain the other extreme, where countries such as Afghanistan and the Central African Republic have a very high death rate, despite having mid-range pollution levels. With low GDP levels, overall health and healthcare quality is likely to increase the burden of pollution-related disease.

Drier conditions here mean the source of pollutants which may be notably different from elsewhere: much of the local air pollution could be derived from dust and sand particles as opposed to anthropogenic sources of industry and agriculture. If proximal sand and dust sources have less severe impacts on health, this may explain the strong outliers from this world region. Most of the world's richest countries have death rates in the range of 10--20 per 100,000. At the lowest incomes, many countries also have relatively low death rates.

The highest death rates from outdoor pollution occur in the middle of the income distribution: most notably countries with large industrialization. The span of death rates for any given level of income, however, also indicates that local contexts, policies, urban and transport systems, and healthcare can also have a substantial impact on air pollution death rates.

In 2014, about 92% of the world population occurred in places with subordinate ambient air quality as anticipated by the World Health Organization (WHO). Furthermore, the major cause of 7 million deaths worldwide in 2012 was due to air pollutants (Kuehn 2014) as reported by WHO. The increasing air pollution degrading the urban air quality, becoming a major environmental and public health concern worldwide and linked to stroke, cancer, infertility, respiratory and cardiovascular diseases, cognitive decline and other adverse medical conditions. The large number of death (approximately 90%) related to air pollution occurring in low- and middle-income countries which are present mainly was in South East Asia and Western Pacific regions.

Air pollution is acknowledged as a global public health issue. The 6.7 million deaths all over the world in 2016 was due to the air pollution as recorded by the global burden of diseases, injuries and risk factors (GBD) study. Moreover, the number of deaths is increased by 2050 without any action. The lancet commission enunciated that the biggest burden of illness will occur in the low- and middle-income countries as the most of deaths due to pollution occurring in these regions.

1.4 Current Air Pollution Issues

1.4.1 Climate Change Impacts on Public Health and Welfare

The risks to public health and the environment from climate change are substantial and far-reaching, expected to lead to more intense hurricanes and storms, heavier and more frequent flooding, increased drought and more severe wildfires---events that can cause deaths, injuries and billions of dollars of damage to property and the nation's infrastructure.

Carbon dioxide and other greenhouse gas pollution leads to more frequent and intense heat waves that increase mortality, especially among the poor and elderly. The climate change impact on public health includes increase in personal exposure to ground-level ozone pollution, heat stress and spread of waterborne illnesses, airborne allergens and other infectious diseases.

Other effects of greenhouse gas pollution noted in the scientific literature include acidification of ocean, upsurge in sea level and enhanced weather disturbances, outrage to forests and agriculture, species eradication and deprivation of ecosystem. Past studies documented the increase in extreme weather and climate events in recent decades, caused significant damage and disruption to human well-being, infrastructure, agriculture, biodiversity and ecosystem. Those most vulnerable to climate-related health effects---such as children, the elderly and the poor and future generations---face disproportionate risks. Recent studies also find that certain communities including low-income communities are disproportionately affected by certain climate-change-related impacts including heat waves, degraded air quality and extreme weather events, which are associated with increased deaths, illnesses and economic challenges.

1.4.2 Global Warming

Global warming is an environmental phenomenon caused by natural and anthropogenic air pollution. It refers to rising air and ocean temperatures around the world. This temperature rise is at least partially resulting of enhanced amount of greenhouse gases in the atmosphere. Greenhouse gases trap heat energy in the atmosphere.

Carbon dioxide (CO_2) is a greenhouse gas that has had the biggest effect on global warming. Carbon dioxide is emitted into the atmosphere by burning fossil fuels (coal, gasoline and natural gas). Humans have come to rely on fossil fuels to power cars and planes, heat homes and run factories. Other greenhouse gases emitted by natural and artificial sources also include methane, nitrous oxide and fluorinated gases. Methane is a major emission from coal plants and agricultural processes. Nitrous oxide is a common emission from industrial factories, agriculture and the burning of fossil fuels in cars. Fluorinated gases, such as hydrofluorocarbons, are emitted by industry. Fluorinated gases are often used instead of gases such as chlorofluorocarbons (CFCs). CFCs have been outlawed in many places because they deplete the ozone layer. Worldwide, many countries have taken steps to reduce or limit greenhouse gas emissions to combat global warming. The Kyoto Protocol, first upholded in Kyoto, Japan, in 1997, is an agreement between 183 countries that they will work to reduce their carbon dioxide emissions.

1.4.3 Long-Range Transport (LRT) and Transboundary Driven Air Pollution

Long-range transport (LRT) impinges air quality from far away sources. The Convention on Long-range Transboundary Air Pollution was accorded in 1979. This is intended as organized international gadget which has set forth for worthwhile alliance between 49 parties in Europe. In long-range transport, the air pollutants are carried away anywhere around the world from one topographical area and affects the far-flung and rearing sites. The pollutant sources in an area affect the nearby or distant area having no pollution due to long-range transport. The 181 major issues of air pollution transport were evidenced by Rodhe et al. (1972) by way of air mass trajectory calculations and reported the incidence of acid rain in Northern Europe as a result of presence of industrial sources in the south and west. The Indian subcontinent is bearing heavy particulate pollution due to the dust blowing from Thar Desert. Begum et al. (2011) also observed effect on quality of air in South Asia due to sandstorm coming from Oman and other Middle Eastern regions. Badarinath et al. (2010) observed dust aerosols over the Arabian Sea and Indian region due to long-range transport. The long-range transport of air pollution and its effect on chemistry of rain water was also recorded at sites in Himalayan region by Kulshrestha and Kumar (2014). The western air-masses carrying sulphate and nitrate components cause perennial detrimental effect on ecosystem of Himalayan region. The problem of transboundary air pollution transport is a serious matter of concern between and within the countries. The critical prevailing pollution problem in New Delhi can be an example of transboundary air pollution (Chandrappa and Kulshrestha 2015).

The buses running on diesel were liable for the existing air pollution of Delhi city in 1990s, and the residents of Delhi were extricated by replacement of diesel buses with CNG after supreme court order which improved the air quality of Delhi. But the winters of 2015 observed intense level of NO_2, SO_2, $PM_{2.5}$ and PM_{10} violating the NAAQS limit which were never observed in recent preceding winters. The transboundary air pollution could be responsible for corresponding unusual increase in air pollutant from recently developed brick kilns in adjoining states like Punjab, Haryana and Uttar Pradesh (Chandrappa and Kulshrestha 2015). Further, the heavy-duty trucks running on diesel infringing from neighbouring states endowed indicative amount of carbon and sulphur oxides and increased air pollution in Delhi (Gupta et al. 2015).

1.4.4 Atmospheric Brown Cloud

The atmospheric brown clouds (ABCs) consist of airborne particles, sub-micron size aerosols like soot and dust. The pollution moving over continents as well as ocean bodies results in formation of transcontinental and transoceanic plumes. The polluted air in ABCs absorbs and scatters incoming sunlight and result in global climate change. The atmospheric brown cloud composed of sulphate and nitrate aerosol which reflects solar radiations back into space and cool the atmosphere. But black carbon particles absorb sunlights and cause heating effect. The sulphate aerosol is the result of burning of fossil fuels, nitrate aerosols are produced due to emissions from vehicles, and black carbon is mainly emitted due to incomplete combustion of diesel, coal, gasoline and solid fuels. Both type of aerosols present in ABCs decrease the solar radiation arriving at the earth's surface and result in dimming effect. The nucleation of additional cloud droplets by aerosol enhanced the dimming effect. The evaporation of water from the earth's surface decreases due to the dimming effect which results in slowing down of the hydrological cycle. The lower rainfall and reduced monsoon circulation was also observed over the continents due to ABCs dimming effect as reported by some studies. The warming of atmosphere by ABCs at elevated levels from 2 to 6 km results in retreating of glaciers as well as snow packs in the Hindu Kush-Himalaya-Tibetan glaciers as observed by Ramanathan and Feng (2009).

1.5 Air Quality Management Options

Air quality management has two important outlooks---'decisive' and 'discreet'. The pollution is reduced at all extent of pollution problem from local to global scale in the former case. The long-term plan of 5, 10 or 15 years were made to accomplish the objectives for improving the air quality in case of decisive approach. In discreet approach, the inevitable hazards were prevented by making plan on a short notice for control of short-term pollution episodes. These short-term episodes were mostly of short duration of 34 h to 5 or 6 days.

The air pollution caused by vehicular emissions results in detrimental environmental and health effects from last two decades. The administrators, public and many lawmakers have taken a intense pursuit in infringing air quality. The several laws were adopted to maintain and protect the environment from vehicular emissions by the government. The Air (Prevention and Control of Pollution) Act, 1981, Environment (Protection) Act, 1988, the Motor Vehicles Act, 1988, and Central Motor Vehicle Rules, 1989, are some of the imperative laws which are inacted. The important preventive actions were taken by distressed authorities from time to time to control the air pollution as advised by supreme court. However, due to urbanization, absence of adequate public transport system, supply of poor-quality fuel to the consumer and traffic profusion have resulted into the degradation of local air quality, especially near traffic interchange and at busy straight roads in urban centres.

1.5.1 Ambient Air Quality Criteria

Air quality criteria are cause--effect relationships observed experimentally, epidemiologically, or in the field of exposure to various ambient levels of specific pollutants. Ambient air quality norms are intended in many countries to safeguard public health and welfare. In some countries, such as India and USA, they are manifested in legislations; while in others such as Australia and the United Kingdom, they are tended to be used as guidelines. The World Health Organization has formulated long-term air quality goals which are adopted widely.

Table 1.1 indicates that the criteria tend to be stated in two approach; as average levels and as extreme values to be outreached only a small magnitude of the time per year. In later case, it is chiefly the annual maximum, the second highest and 98 percentile concentrations which are of interest. Thus, the basic form of air pollution data or pollution prediction requested for air quality management is the yearly dispensation of concentrations at fixed averaging times and locations of interest throughout an air quality control region.

The goal of air quality management is to maintain a quality of air that protects human health and welfare, also includes the protection of animals, plants, ecosystems, etc. Government policy is the foundation for air quality management. The successful air quality program could not be maintained in absence of a proper strategy and competent legislation. A proper strategy indicates to policies in several areas including energy, transport, development, planning and the environment. Emission inventories, concentration measurements and dispersion models are major tools for

Table 1.1 World Health Organization recommended long-term air quality goals

Sl. no.	Pollutant	Averaging time	Limiting level (μg/m^3)
1	Sulphur oxides[a]	Annual mean 98% of concentration below[b]	60 200
2	Suspended particulates[a]	Annual mean 98% of concentration below[b]	4 120
3	Carbon monoxide	8-h average 1-h maximum	10,000 40,000
4	Photochemical oxidants	8-h average 1-h maximum	60 120
5	Nitrogen dioxide	1-h average not to be exceeded more than once a month	200–340

[a]Values for sulphur oxides and suspended particulates apply only in conjunction with one another
[b]The permissible 2% of observations over this limit may not fall on consecutive days

air quality assessment and are interdependent in application of air quality management. The complete schemes of relevant interrelationships in air quality management are shown in Fig. 1.2.

The approach needed for air pollution management in metropolitan establishment is complex situation which demands case to case accession as size and activities of

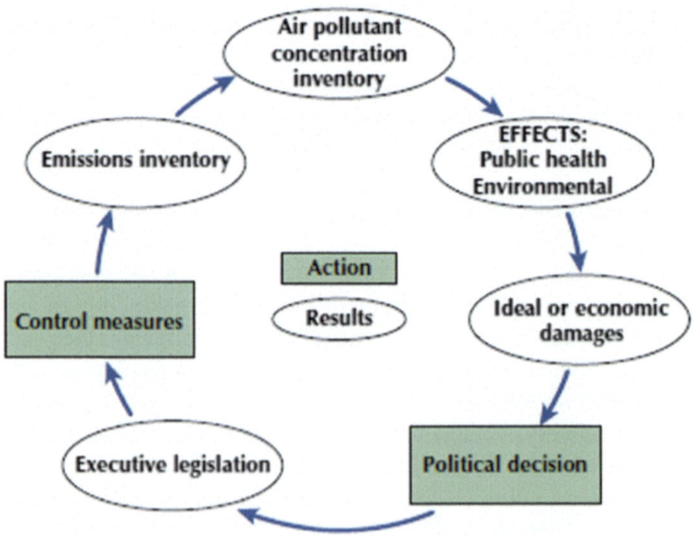

Fig. 1.2 Air quality management

urban area varies widely. The provenance and management approach for air pollution in urban area are specified to be done.

References

Anderson HR, Atkinson RW, Peacock J, Marston L, Konstantinou K, World Health Organization (2004) Meta-analysis of time-series studies and panel studies of particulate matter (PM) and ozone (O_3): report of a WHO task group (No. EUR/04/5046026). WHO Regional Office for Europe, Copenhagen

Badarinath KVS, Kharol SK, Kaskaoutis DG, Sharma AR, Ramaswamy V, Kambezidis HD (2010) Long-range transport of dust aerosols over the Arabian Sea and Indian region—a case study using satellite data and ground-based measurements. Glob Planet Change 72:164

Begum A, Biswas K, Pandit G et al (2011) Long range transport of soil dust and smoke pollution in the South Asian region. Atmos Pollut Res 2:151–157

Chandrappa R, Kulshrestha UC (2015) Sustainable air pollution management: theory and practice. https://doi.org/10.1007/978-3-319-21596-9

Gupta GP, Singh S, Kumar B, Kulshrestha UC (2015) Industrial dust sulphate and its effects on biochemical and morphological characteristics of Morus (Morus Alba) plant in NCR Delhi. Environ Monit Assess. https://doi.org/10.1007/s10661-015-4301-4

Kuehn BM (2014) WHO: more than 7 million air pollution deaths each year. JAMA 311:1486

Kulshrestha UC, Kumar B (2014) Airmass trajectories and long range transport of pollutants: review of wet deposition scenario in South Asia. Adv Meteorol 2014:596041. https://doi.org/10.1155/2014/596041

Ostro B, World Health Organization (2004) Outdoor air pollution: assessing the environmental burden of disease at national and local levels

Ramanathan V, Feng Y (2009) Air pollution, greenhouse gases and climate change: global and regional perspectives. Atmos Environ 43(2009):37–50

Rodhe H, Persson C, Åkesson O (1972) An investigation into regional transport of soot and sulfate aerosols. Atmos Environ (1967) 6(9), 675–693

WHO (2016) Urban ambient air pollution database. Available online: http://www.who.int/phe/health_topics/outdoorair/databases/who-aap-database-may 2016.xlsx (accessed on 4 July 2020)

Chapter 2
Air Quality Monitoring and Techniques

R. Ravi Krishna, S. M. Shiva Nagendra, Saraswati, and M. Diya

2.1 Air Quality Monitoring

The information on air quality of an area could be provided by air quality monitoring. The existing policies and their implementation could also be evaluated by monitoring of air quality. The planning, design and the establishment of monitoring network for fulfilling air quality objectives is one of the important components of any air quality monitoring programme (Sivertsen 2008). The air quality monitoring network has been designed primarily to ensure effective regulatory compliance in most of the developing countries. The 600 monitoring stations are present throughout in different cities, towns, states and union territories for establishing national ambient air quality monitoring (NAAQM) network in India. Additionally, CPCB (2014) recorded own monitoring station operated by state pollution control boards. The Indian Institute of Tropical Meteorology (IITM), Pune, also started monitoring of criteria air pollutants over Delhi recently with the direction of Ministry of Earth Sciences, Government of India (IITM 2014).

R. Ravi Krishna (✉)
Department of Chemical Engineering, Indian Institute of Technology Madras, Chennai 600036, India
e-mail: rrk@iitm.ac.in

S. M. Shiva Nagendra · Saraswati · M. Diya
Department of Civil Engineering, Indian Institute of Technology Madras, Chennai, Tamil Nadu 600036, India
e-mail: snagendra@iitm.ac.in

S. M. Shiva Nagendra et al. (eds.), *Urban Air Quality Monitoring, Modelling and Human Exposure Assessment*, Springer Transactions in Civil and Environmental Engineering,
https://doi.org/10.1007/978-981-15-5511-4_2

2.2 Design of Air Quality Monitoring Network

Design of ambient air quality monitoring station acts as the preliminary step in identifying the sources and the solutions for critical air pollution. Indian Standard (IS) 2000 Part 14 indicates that sampling of atmosphere helps us to determine the current status of air quality; to assess the potential health effects and other damages to the property; to develop and validate air pollution models; to reduce the pollution levels from the key contributors; and to forecast air quality for the people, so that precautionary steps could be taken before the severe impacts on health of human.

The air pollutants measurement of the ambient surrounding can be accomplished with a variety of objectives which define the scope, methodology and analysis of the data collected. One of the most commonly used methods to collect air pollutant is the time-averaged method using a high or low volume sampler. Time-averaged measurements are useful in evaluating general trends in ambient air quality in a given region or a season and for comparing trends between regions. Lower timescale measurements are useful in evaluating specific events in an environment. To accomplish this, there are other instruments that use indirect methods to estimate particulate and gaseous pollutants. The following discussion looks at the different types of the particulate and gaseous monitoring techniques.

2.3 Measurement of Gaseous Pollutants

2.3.1 Grab Sampling

Certain gases are present in the environment at concentration levels that are above the detection limit of the analytical instrument. In such cases, 'grab' samples can be collected and introduced in a suitable analytical instrument directly without much processing. As the name suggests, grab sampling involves the 'grabbing' of a certain volume of ambient air. The small volume of air samples is collected by the simplest way with the help of special containers within a second or minutes. Gas chromatography (GC) techniques can be used for analysis of the air samples. Plastic bags and stainless-steel or glass containers are mainly used for the sampling. The bags made of polymer are simple to use. Saran, Mylar, Scotchpak (aluminium foil, a laminate of polyethylene and Teflon) and Tedlar are some of the materials which are suitable for the sampling. The quantity of air for the sampling varies from 10 to 100 L. However, there are chances of contamination while filling the bags as the pumping of air sample is required for the sampling of air. The glass (or stainless-steel) containers can also be a simple method of sampling of air. The contamination or surface losses can be avoided by careful pre-treatment and conditioning of canisters before using them. The larger volume of air samples could be collected by pressurising the air samples as container volume is very low. The analytes (in the form of per cent or a ppm level in air matrix) present in compounds could be determined by using whole air sampling.

A pressure/vacuum gauge installed in the canister can indicate the presence of a gas sample. In a Tedlar bag, the shape of the bag is sufficient to indicate the presence of an air sample.

The low concentration of analytes makes them near or below the limit of detection of the analytical instrument. In such cases, the analyte in the environment is accumulated in a medium and transferred to the analytical instrument after some processing. The most commonly used method of analyte concentration is using an adsorbent tube. A pump (also known as a low volume sampler since the flow rate is up to 10 L/min) is used to process a certain volume of ambient air through a tube filled with a certain amount of adsorbent. The analyte adsorbed on the sorbent can be desorbed using a solvent and introducing the solvent into a suitable analytical instrument. The solvent extract can be processed further by concentration and injected into an analytical instrument. Alternatively, the sorbent tube can be heated very rapidly and the vapours that are flash-desorbed can be swept by an inert gas to the analytical instrument through a gas-sampling valve.

2.3.2 Analytical Techniques

This section describes the analytical techniques for some of the gaseous and vapour phase components in ambient air.

Sulphur Dioxide The technique that is most inexpensive for the analysis of SO_2 in ambient air is the modified West–Gaeke method (*US EPA Manual Reference Method*: 40 CFR Part 50, Appendix A). This involves the use of an impinger with an absorbing solution that draws in ambient air and absorbs the SO_2. The sampling rate is usually around 1 L/min, and the suggested sampling interval is 4 h. The absorbing solution for SO_2 is 35 mL of 0.04 M potassium tetrachloro mercurate (TCM). After the sampling, the absorbing solution is carefully transferred to another container and transported to the laboratory and is treated with 1 mL of 0.6% sulphamic acid, 2 mL of 0.2% formaldehyde solution and 2 mL pararosaniline solution and diluted with distilled water. The addition of the pararosaniline aids in the formation of a complex that has a distinct colour. The intensity of the colour is measured using a UV-Visible spectrophotometer, and the absorbance of 560 nm wavelength is measured. The absorbance is calibrated with the amount of sulphur in the solution. The external calibration is performed using sodium sulphite.

The alternative to this method is a continuous monitoring technique that uses a technique of UV fluorescence. This relies on the fluorescent emission of SO_2 at 220–240 nm. This is a much more expensive instrument, but can monitor ambient SO_2 continuously. This method is in use at all the continuous monitoring stations operated by the Central Pollution Control Board.

Nitrogen Oxides An impinger technique similar to the one used in the modified West–Gaeke method is used for the analysis of nitrogen oxides (*US EPA Manual*

Reference Method: EQN-1277-026). The absorbing solution for NO is an aqueous solution of 35 ml of distilled water containing 4 g of sodium hydroxide and 1 g of sodium arsenite. After the sampling, the absorbing solution is carefully transferred to another container and transported to the laboratory and is treated with 1 ml of hydrogen peroxide solution, 10 ml of sulphanilamide solution and 1.4 ml of N-(1-Naphthyl)-ethylenediamine Di-hydrochloride (NEDA) solution. The intensity of the colour that is formed in the reagent mixture is measured using a UV-Visible spectrophotometer, and the absorbance of 540 nm wavelength is measured. The external calibration is performed using sodium nitrite.

The alternative to this method is a continuous monitoring technique that uses a technique of chemiluminiscence. The chemiluminiscence detection technique is based on the principle of reaction of ozone (O_3) with nitric oxide (NO) which results in the formation of electronically excited nitrogen dioxide (NO_2*) and oxygen. The formed excited NO_2 molecules revert back to NO_2 immediately and results in light emission which is directly proportional to the NO concentration in the sample. The photomultiplier tube is used for the measurement of the intensity of light emission. The NOx (NO + NO_2) analysis is done by the NO_2 to NO converter. This is also in use at all the continuous monitoring stations operated by the Central Pollution Control Board (CPCB).

Carbon Monoxide Carbon monoxide (CO) is monitored in ambient air using a non-dispersive infrared analyser. There are several options of either handheld or static samplers. This is also a part of the instrumentation used in the continuous monitoring program of the CPCB.

Ozone Ozone is measured using UV absorbance in a wavelength of 254 nm. This is a continuous monitoring instrument.

Volatile and Semi-volatile Organic Compounds

Gas Chromatography The predominant technique for the analysis of all the VOCs and SVOCs is gas chromatography and various detectors that are applicable along with it. Gas chromatography (GC) is a technique that is used to separate components of a mixture using the relative affinity of the different analytes between mobile and stationary phase. The stationary phase is a chemical compound that is bound to a solid phase and is present as a column. The mobile phase in GC is a gas, typically an inert gas such as nitrogen, helium or argon. Figure 2.1 shows the schematic of a typical chromatography system.

The mobile phase (also known as the carrier gas) is supplied from a cylinder through a column that is placed in an oven. The carrier gas flow rate and the temperature of the oven are adjustable. The mixture of analytes such as the one in ambient air is injected into the column as a finite pulse. The sample injector can be heated up to 300 °C, which can facilitate the injection of liquid samples as well. The liquid solvent and the analytes vaporise in the injector and are carried into the column by the carrier gas. The analytes of the mixture travel through the column through a

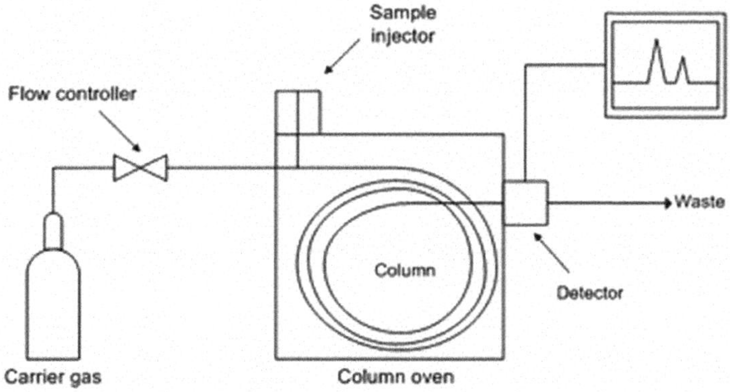

Fig. 2.1 Schematic of a gas chromatography system. *Image source* Wikipedia

series of adsorption and desorption steps. The analytes are separated in the process and exit at the end of the column, where they are sent to a detector. The output is a plot of the intensity measured by the detector as a function of time, and this plot (or the output) of a chromatography system is known as a chromatogram. Figure 2.2 shows a schematic of a sample chromatogram. An analyte that has no retention on the column comes out earlier. The retention of an analyte on the column is related to the partition constant of the analyte between the carrier gas and the stationary phase. GC columns are either packed columns of capillary columns. In the current practice of chromatography, there are a number of stationary phases compositions available for specific groups of chemicals. The typical maximum temperature that a stationary phase can withstand without degradation is around 300 °C.

Increasing the temperature or the flow rate in a GC analysis results in the decrease of the retention time. The parameters of temperature can be dynamically changed to manage the separation of a large number of analytes in the mixture of the extract. The

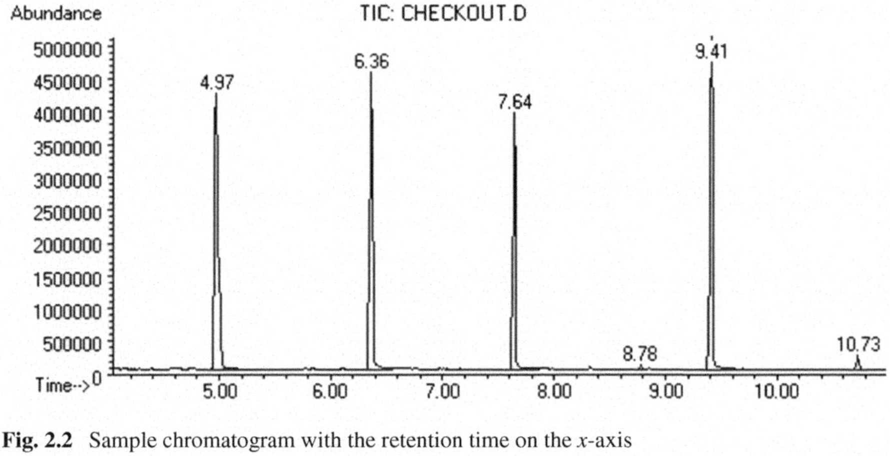

Fig. 2.2 Sample chromatogram with the retention time on the *x*-axis

efficiency of separation is the clear resolution of peaks without overlap. This usually happens with a combination of factors including the stationary phase composition, oven temperature and carrier gas flow rate. All these parameters along with a few others pertaining to the detectors along with the proposed run time of a given analysis are compiled and prescribed as a standard method. The run time is, however, under the controller of the user of the method and can be adjusted depending on the objectives of the analysis and the compounds of interest. Typically, trial and error runs are needed with standards of the analytes of interest to confirm the retention times and fix the parameters for the method.

A variety of detectors are available for use along with the GC. The most commonly used detectors are the following:

Flame Ionisation Detector (FID) The most commonly used detector is predominantly a hydrocarbon detector. A flame induced by the supply of hydrogen gas and air aids in the combustion of the analyte between two elements of the detector. Corresponding to the extent of combustion, an electrical signal is generated, which is calibrated with either mass or concentration of the analyte that was measured. This is a very sensitive detector for hydrocarbons and is extensively used. The FID is used in a number of hydrocarbon analysers even without chromatography (to get an estimate of what is commonly known as TPH or total petroleum hydrocarbons).

Thermal Conductivity Detector (TCD) This is known as a universal detector. This measures the thermal conductivity of an analyte in comparison with the thermal conductivity of a reference gas. Almost all compounds can be measured by this technique. However, the sensitivity of this technique is not very high, which implies that the detection limit is relatively high and is therefore not commonly used for environmental analysis.

Electron Capture Detector (ECD) This is a selective detector used for the analysis of chlorinated organic compounds. This is a very sensitive detector with a very high signal-to-noise ratio and is commonly used for the quantitative analysis of chlorinated organic compounds such as chlorinated pesticides, polychlorinated biphenyls and chlorinated solvents.

Mass Spectrometer (MS) The above three detectors have one significant disadvantage. The detectors cannot identify the analyte as they sense and measure the analyte eluting from the column. This poses a challenge in terms of accurately making a qualitative mapping of the peaks in the chromatogram with an identity of the analyte. The identification is necessary to prepare a calibration chart required for the quantitative analysis of the identified analytes. In non-selective detectors such as the FID, typically known standards must be injected individually to obtain the retention time under a set of column and flow conditions. The retention time then becomes the reference for identification of a peak in an unknown sample. Using the retention time alone as an indicator for the identification of a peak has significant uncertainties. However, with the careful choice of column and pre-sample cleanup procedures, the co-elution of other compounds may be minimised. In order to avoid this uncertainty, a different

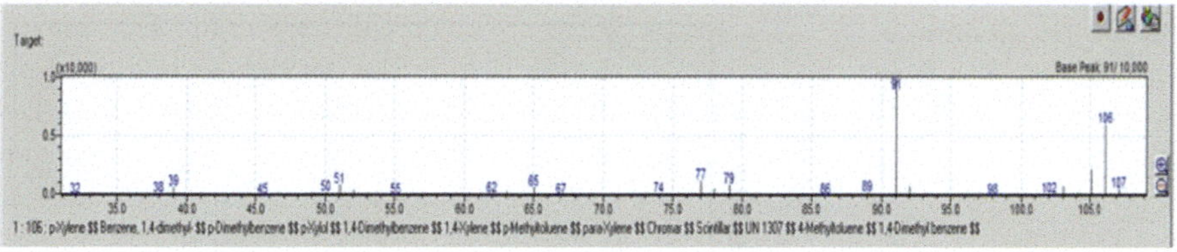

Fig. 2.3 Mass spectra of benzene

kind of detector called as a mass spectrometer (MS) is used. This is based on the principle of mass spectrometry, where the analyte entering the detector is ionised and then fragmented, and the mass/charge ratio of the fragments is analysed. Each analyte passing through the detector will give rise to a mass spectrum. A library of reference mass spectra is available, and comparison with the unknown sample spectrum will provide a quality match giving options of the closest possible chemicals that the unknown peak could be. This information is very useful in giving the analyst a better idea of what is present in the sample. Most current instruments with the GC-MS have the option of mass spectral libraries up to more than 300,000 compounds of environmental interest.

The GC-MS can be operated in a scan mode, where the chromatogram is created by scanning the entire mass range required. This is usually required for the acquisition of mass spectra and comparison with libraries. However, since considerable scanning resources are used, the sensitivity of the instrument is lower in this mode due to a low signal-to-noise ratio. In order to increase the signal-to-noise ratio and to increase the sensitivity, the GC-MS is operated in a selected ion monitoring (SIM) mode, where selected mass/charge ratios are probed. This results in a very drastic increase in sensitivity. However, this result in the truncation of mass spectral information and library searches cannot be done in this mode. This is a quantitation mode and must be used only after the identity of an analyte at a particular retention time that is already established (Fig. 2.3).

Liquid Chromatography For the GC analysis, the analyte should be sufficiently volatile. Since we are interested in the analysis of gas-phase chemicals, this is not an issue since the analysis is of vapour-phase chemicals. However, in certain protocols, the adsorbent that is used for the collection of the air sample demands the use of a non-volatile solvent. Under these conditions, it is preferred to us a liquid chromatography system (LC). The LC is the older of the chromatographic techniques. The components of this are the same as in the GC, except that the mobile phase is a liquid. The liquid is usually a solvent that the analyte is completely soluble in. The column is a packed column with a suitable phase composition. The choice of detectors that are used at the end of the column are as follows:

UV-Visible Spectrophotometer This is the most commonly used detector and is a very simple detector. The detector measures the absorbance of an organic molecule at different incident wavelengths of light in the UV or visible range. The transmission

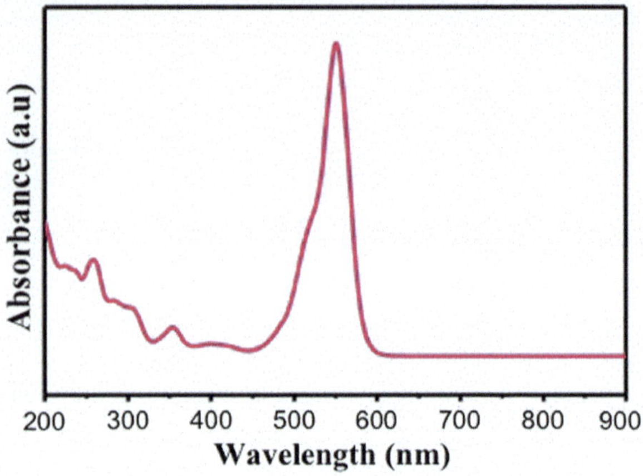

Fig. 2.4 UV absorbance spectrum

of light is measured and converted to an absorbance value. The absorbance is a function of the mass or the concentration of the analyte of interest. All compounds that absorb light display an absorbance spectrum as shown in Fig. 2.4. The quantitative calibration is usually done using the wavelength where the absorbance is the highest. A diode array of detectors (known as a diode array detector or a photo diode array) is sometimes used to obtain the entire spectra at every point of the chromatogram just like the GC-MS.

Fluorescence Detector This is a very specific detector with very high signal-to-noise ratio and therefore very high sensitivity. This relies on the fluorescing property of several organic molecules.

Mass Selective Detector (MS) This is a recent development (in the last few decades) and is deemed useful for the same reasons for which the GC-MS is useful. The instrument, however, requires an interface where the liquid sample can be aerosolised for entry into the mass spectrometer which is under vacuum.

2.4 Measurement of Particulate Pollutants

Various techniques have evolved over the period of time for the measurement of air quality related to the particulate matter (PM) measurements. Before we go to the actual techniques, let us try to understand the meaning of particulate matter and how they are different from the aerosol particles. By definition, the PM is solid or liquid particles collected on the filter paper from the atmospheric air for the physical, chemical and biological analysis to infer their characteristics. This term is slightly different from the term used aerosol particles. Thus, the aerosol particles are defined as suspension of liquid or solid particles in the air or any gaseous medium, and in our case, it is atmosphere. To understand the source characteristics and other properties

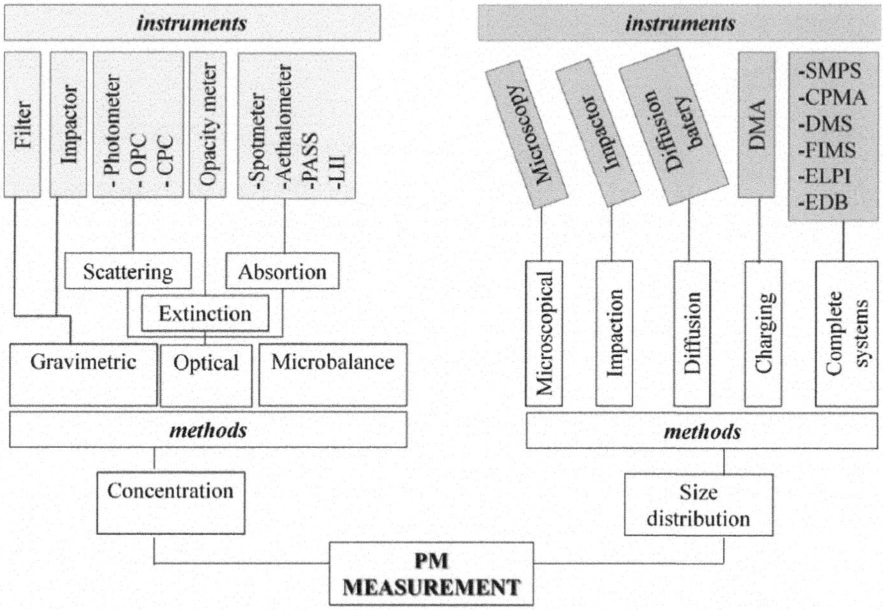

Fig. 2.5 Various instrumentations and methods adopted for the measurements of atmospheric aerosol particles. The methods and instruments depending upon the application of the techniques can be used for both the number size and mass estimation. Adopted from Simões Amaral et al. (2015)

to estimate the impact of aerosol on climate and ecosystem health, the accurate characterisation is very important (Fig. 2.5).

The measurement of the atmospheric aerosols can be broadly categorised in two distinct way (i) mass measurements, which is generally preferred when estimating the impact of aerosols on human health, and (ii) number and/or size measurements where the impact of aerosols on the climate is being investigated.

Optical Particle Counter (OPC) OPC is a stationary instrument which has been designed for the continuous measurement of airborne dust and its particle size distribution. It consists of 31 size channels with size varies from 250 nm to 34 μm and operates with $1.2\,\mathrm{l\,min^{-1}}$ flow rate. The principle of measurement here is the measurement of scattering light of the single particles, where a semiconductor laser serves as light source with wavelength in visible range at 655 nm. There is a aerosol inlet through which the sample air is traverse into the measuring cell. The light scattering in the measuring cell helps in the detection of particles in the sample air. The counting of scattering light pulse of every single particle is done, and the particle size is associated with particular intensity of its scattering light signal. The schematic diagram of measuring principle is shown in Fig. 2.6. The wavelength used in detecting the particles puts the lower limit in detecting and separating the particle size in these types of optical instruments. However, for higher particle size range (from 250 nm

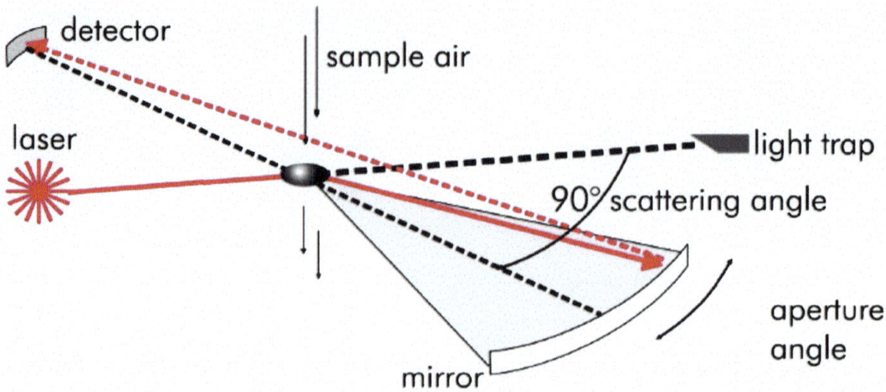

Fig. 2.6 Operating principle of optical particle counter (OPC) component is that the pulse from the emitted laser is used for counting the particles, whereas scattering of the light due to particle up to 90° is used to infer the size in given 31 channels

to coarse mode), these types of instruments are proven to be very efficient. The time resolution of this instrument is generally very high where a full scan from 250 nm to 34 μm can be carried out in 1–2 min. The laser beam is focused to a flat elliptical strip by means of illumination optics. The sample air is focused aerodynamically and then led as particle flow through the inner area of the measuring volume. When doing environmental measurements, the particle concentration of the sample air is normally so low, that statistically seen only one particle is in the measuring volume. Measuring at particle sources, technical particulate matters, or working places, very high particle concentrations can appear which require a previous dilution of the sample air. But since the entire sampling volume of 1.2 l min^{-1} is analysed, the dust monitor reasonably good counting statistics.

The scattering light emitted by every particle is being detected by a second optics under a scattering angle of 90° and then directed onto a receiver diode via a wide-angle mirror. The detector signal will be classified into size channels after amplification captive to its intensity. The particle size is proportional to the intensity of the detected scattering light signal, whereas the scattering light intensity is also being influenced by the particles refractive index, particle shape and orientation of the particle within the measuring volume. Positioning the detector into a 90° direction enables minimising the influence of the aerosol particles refractive index for determining the particle size.

Condensation Particle Counters (CPC) Optical particle counters are not able to measure particles much below 50 nm. To count smaller particles, condensation particle counters (CPC) are used and can count all particles down to 2 nm. In these counters, particles are detected and counted by laser scattering in a very similar way to a standard optical particle counter. In a CPC, particles are first grown by condensation of a volatile liquid present in the system to a size of 10–12 μm thus allowing easy detection. In a supersaturated vapour, suspended particles act as nuclei for vapour condensation and may grow to form droplets. This process is referred to as heterogeneous nucleation. Above a critical supersaturation, droplets may form by homogeneous nucleation where clusters of vapour molecules nucleate droplets. Supersaturation is defined as the actual vapour mixing ratio divided by the saturation vapour mixing ratio at that temperature and is generally expressed as a percentage. The critical supersaturation for homogeneous nucleation is typically around 300%. Droplets may form heterogeneously at supersaturations of less than 1%. This is the process by which clouds form in the atmosphere. The size of aerosol particles which will act as condensation nuclei is a function of the supersaturation it experiences. The smallest size at which condensation will occur at a particular supersaturation is referred to as the Kelvin diameter. This size dependence is because saturation vapour pressure is defined relative to a flat liquid surface. Vapour molecules are able to escape more easily from a curved surface, thus the saturation vapour pressure is increased with decreasing radius of curvature. The Kelvin diameter is the diameter at which a pure liquid droplet is in equilibrium with the vapour phase at that supersaturation.

In a CPC, supersaturation is carefully controlled to a high value, typically 100–200%, but below the critical supersaturation, so that droplets do not form in particle free air. There are a number of techniques by which such supersaturations may be achieved, including adiabatic expansion, flow mixing and diffusional thermal cooling. Most commercially available CPCs use the diffusional thermal cooling method.

In a counter using the diffusional thermal cooling method, particle-laden air passes through a heated porous block or wick in contact with the working fluid, which is usually butanol, and becomes saturated. In some models in order to ensure that all particles experience the highest possible supersaturation which occurs along the centre of the saturator, a sheath flow is used to constrain the particles. Here, the inlet flow is split into sample and sheath flows, the sheath flow is filtered and passes through the saturator wick. The sample flow is then injected into the centre of the laminar sheath flow through a capillary. On exiting the saturator, the saturated air is cooled in the condenser, becoming supersaturated and causing particles above a certain size to grow into drops. These drops are then focused through a nozzle, and pass through the laser beam and are counted. The minimum particle size which can be detected by a CPC is determined by the supersaturation achieved, which in turn depends on the temperature difference between saturator and condenser. In practice not all particles experience the same supersaturation, and hence, the efficiency of the counter gradually decreases at smaller sizes, and the lower cut-off size is defined as the size at which 50% of particles present are counted. The more uniform the supersaturation, the sharper the cut-off. Particle count is generally determined by counting individual pulses of scattered light. Relatively recently CPCs have been developed which use water as the working fluid. In these instruments, supersaturation is achieved by passing the sample through a wet tube, during which it becomes saturated, and the temperature of the sample reaches equilibrium with the walls. It then passes through a heated region of the wet tube, which raises the vapour pressure, as the rate of water vapour diffusion is greater than that of thermal diffusion, the sample becomes supersaturated. In other aspects, water CPCs operate in a similar manner to butanol CPCs described above. While CPCs do not provide any information on the original size of particles counted, they are often used with size-selecting devices such as differential mobility analysers, for example in the differential mobility particle sizer, so that size distribution information can be built up. Additionally, several CPCs with different lower cuts can be used to provide a course, but fast response size distribution of small particles. The response time of most commercial CPCs is typically less, which is not fast enough for most eddy covariance applications, though some newer models have 10 Hz data output. However, most also output raw pulses from the detection system, which with appropriate external processing allow eddy covariance flux measurements to be made. Again, using multiple CPCs, fluxes of ultrafine particles in specific size ranges can be measured. This is especially useful for studies of new particle formation processes and urban emissions. The Centre for Atmospheric Science currently owns twelve CPCs manufactured by TSI Inc., including several new water-based CPCs, one of which has been specially modified for aircraft use.

Gravimetric Method Particualte matter is collected on filter papers and then weighed to obtain the PM concentrations. The filters are weighed before and after the sampling, and the difference of weight of filter paper provides the mass of particles in the gravimetric method. The off-line measurement of PM mass concentrations with gravimetric sampling was suggested in Nussbaumer et al. (2008). All the granular modes like nucleation, accumulation and coarse modes of particles are collected on the filter in this method if there is absence of impactor or cyclone to eliminate larger particles. The presence of a pre-cyclone helps in cutting out particles higher than the desired cutoff and helps determining the concentration of PM_{10} and $PM_{2.5}$ as reported by Nussbaumer et al. (2008). Nussbaumer et al. (2008) confirmed the chemical analysis of collected particles on the filters. The mass concentrations of filters could be altered if the filters are not properly handled and conditioned. Therefore, the filters should be put in controlled conditions of temperature and relative humidity as also suggested by Nussbaumer et al. (2008) and Giechaskiel et al. (2014). Gravimetric analysis may also be applied with cascade impactors, that are used to obtain size distribution of ambient PM. For more detailed morphological analysis, the other techniques like scanning electron microscopy (SEM) and transmission electron microscopy (TEM) are necessary for the thorough investigation.

High Volume Sampler (HVS) Among the primary PM samplers used in the gravimetric method, the high volume sampler is one of the most commonly used instruments used. High volume air samplers typically sample more than 1500 cubic metres (m^3) of air through a pre-weighed filter over a particular period. Filter papers are then reweighed so the particulate loading of the air (in mg/m^3) can be calculated. The paper contents can also be examined and analysed. The PM_{10} head separates particles smaller than 10 μm from other airborne particulates. Particulates that are greater than 10 μm settle out in the impaction chamber are removed during cleaning or routine maintenance. The particles smaller than 10 μm are carried upward by the air flow and settle in a laboratory-supplied quartz fibre filter. The working principle of the instrument is PM_{10} and TSPM which are measured by passing air at flow rate of about 1 lpm through high efficiency cyclone which retains the dust particles greater than 10-μm size and allows only fines (less than 10-μm particles) to reach the glass microfibre filter where these particles are retained. The instrument provides instantaneous flow rate and the period of operation (on time) for calculation of air volume passed through the filter. Amount of particulates collected is determined by measuring the change in weight of the cyclone cup and filter paper.

Figure 2.7 shows the working principle of high volume sampler. The air sample is drawn into the instrument by a suction device which is primarily passed through a cyclone separator where the particle larger than 10 μm is separated. The smaller particle then passes through a pre-weighted filter paper which then takes back to the laboratory for weight measurement. The measurement of repairable particulate matter can be done accurately done with HVS, and TSPM can also be assessed by collection of dust retained in the cyclone cup.

Optical Methods The basic principal of detection, is based on the scattering in the optical detection methods, where a light beam lighted the aerosol particles, where light is irradiated in all the directions. Giechaskiel et al. (2014) explicated the simultaneous transformation of this light in other energy forms like absorption. The scattering and absorption help in the calculation of extinction of the light as mentioned by these authors. The real-time particle concentration measured by optical instruments which can be based on the principles of scattering, absorption and light extinction.

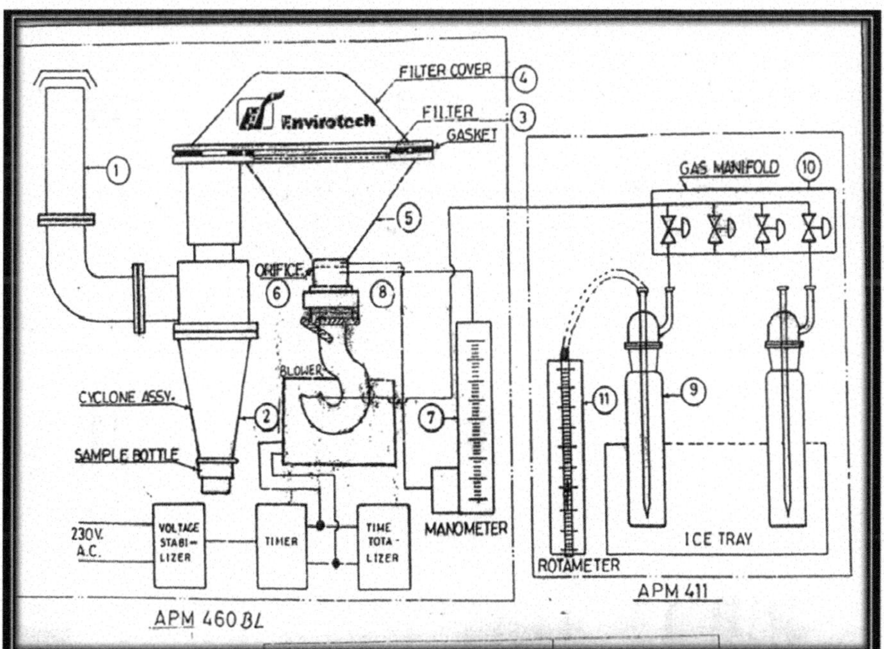

Fig. 2.7 Working principle of high volume sampler for the measurement of PM_{10}

Light Scattering There are two types of light scattering, one is light dispersal by single particles and other is dispersal by group of particles. Dispersion photometer measured the intensity of scattered light through many angles coming from group of particles as observed by Giechaskiel et al. (2014). The photometer detector is present in the scattering photometer for measuring the scattered light. The scattering light from group of particles existing in the optical detection volume was measured by the scattering photometer as explicated by Hinds (1999). The visible light (600 nm) is used by almost all commercial light scattering instruments at measuring angles of 90°, 45° or less than 30°. The respirable aerosol monitor (RAM) is one of the scattering photometers as specified by Vincent (2007). In the original version, the pump helps in the aspiration of aerosol, and then, the respirable aerosol fraction is separated by a cyclone. The scattering of light occurs at an angle of 45°–90° in the optical sensor zone, and there is photodiode for detecting the aerosol. There is a new automated digital version of the instrument, the DataRam 4, having ability of data registry. The concentration and particle average size as well as temperature and relative humidity are measured by this instrument. The DataRam 4 is used by Costa et al. (2012) in the laboratory and field experiments to do sampling of $PM_{2.5}$. The sampling in this compact instrument is done continuously with data storage as according to the author. The scattering photometer fabricated in University of California named as University of California Berkeley Particle and Temperature Sensors (UCB-PATS) is used by Chowdhury et al. (2013). The light dispersion occurs in photoelectric sensors chamber, and loss of ions by particles occurs in suspension in ionisation chamber. These two chambers are combined in the UCB-PATS having smoke detector technology which helps in proper measurements of fine particles as according to narration by the author. The light intensity scattering at an angle of 45° is measured by the light emitting diode (LED) and a photodiode which are present in the light dispersion chamber. The DustTrak is another scattering photometer developed by Prado et al. (2012) and Chowdhury et al. (2013). The dust tract is a portable light dispersion photometer which measured the particle mass in real time as stated by Chowdhury et al. (2013). The isolation of optical chamber having aerosol helps in cleanness of chamber and low maintenance of equipment which improves the measurement. The optical particle counter (OPC) is mostly used in case of light scattering by single particles as alluded by Giechaskiel et al. (2014). A diode laser light source is used by OPCs which lighted the particles present in the sample at a particular angle. The light scattering by the particles is measured by the photodetector. The counting and measurement of particles could be done at the same time based on the intensity of the flash. The OPCs and scattering photometer are identical to each other except the smaller size of optical detection volume in OPCs which result in lighting of one particle at a time as according to Giechaskiel et al. (2014). The photodetector detects the scattered light in the form of electric pulse. The calibration curve is used to ascertain the size of particle from the electric pulse height. The particle should be above 100 nm diameter for detecting by the photodetector. Jiang and Bell (2008) evaluated the concentration of $PM_{2.5}$ by using the SidePak Personal Aerosol Monitor with model AM510 (TSI, Shoreview, MN, USA). The collection chamber draws the aerosol sample continuously. The aerosol flow is intercepted by a laser. The scattering light from particles collected on 90° lens which concentrate it towards a photodetector. The aerosol mass concentration is proportional to the voltage converted from light by the detection circuit. The concentration of small particles is measured by the light scattering counters also known as condensation particle counters (CPCs). The small particles scattered insufficient light which is not detected by the conventional optical counters. The CPCs are identical to optical particle counters when the condensation process from a working fluid in the CPCs result in increase in size of small particles as narrated by Giechaskiel et al. (2014). The particle is evaluated with each flash of light while the enlarged particles are exposed to the focal point of a laser beam. The minimum size of particles detected in the CPCs is determined by the magnitude of supersaturation. The condensation particle counter is used by Leskinen et al. (2014) and Torvela et al. (2014) for sampling of particles from burning wood chips.

Light Absorption The black carbon (BC) concentration is measured by the equipment based on light absorption principle. Giechaskiel et al. (2014) observed that BC is a positive radiative agent and has large contribution to climate change due to strong absorption of light in the atmosphere. The direct and indirect effects of aerosols to the climate are concorded with absorption of light by aerosols which is a very dubious parameters and very difficult to be measured as observed by Lack et al. (2006). There are different techniques for measuring based on absorption of aerosol as specified by Giechaskiel et al. (2014). These are (i) the extinction and scattering difference method, (ii) the methods based on the measuring of the attenuation of light due to the PM collected on a filter, (iii) photoacoustic spectroscopy method and (iv) laser-induced incandescence (LII) method. The particle heating takes place for measuring the BC in the last two methods where heating is result of light absorption by the particles. (i) Spotmeters also called as reflectometers or smoke filter meters as light reflected over filters. Giechaskiel et al. (2014) elucidated that the difference in the light reflection by exposed and non-exposed spot on the filters is the measurement of concentration of particles where exhaust gas is first go through the filter for the filtration. (ii) BC concentration is also measured by aethalometer. The quartz fibre filters are used for the collection of PM in aethalometers where attenuation in absorption of light in the filter is measured as reported by Giechaskiel et al. (2014) and Krecl et al. (2007). The light-absorbing carbon (LAC) was measured mostly by the well-known optical instrument, i.e., aethalometer, as according to Krecl et al. (2007) besides the particle soot absorption photometer (PSAP). However, the conventional aethalometers are mostly used for environment monitoring and not for transitory emission tests as having several minutes of time resolution as according to Giechaskiel et al. (2014). The model having time resolution of 1–10 s came into the market in recent times. Both PSAP and aethalometer are based on the principle of light absorption by carbonated aerosols as affirmed by Krecl et al. (2007) where concentration is observed by measuring the attenuation of transmitted light through particles collected on the filter. The mass concentration of LAC in PM_1 was determined by aethalometer series 8100 (Magee Scientific, Berkeley, CA, USA) in an experiment by Krecl et al. (2007). The 880 nm LED and a 525 nm LED were used for lighting the aerosol present on the filter using the aethalometer and the PSAP, respectively. The BC and aerosol absorption coefficient mass concentrations were measured using a seven-wavelength aethalometer in a study by Gong et al. (2015). (iii) Photoacoustic soot sensor (PASS)—the heating of light-absorbing particles occurred due to absorption of amplitude-modulated light. Giechaskiel et al. (2014) reported that the acoustic pressure waves generated by the surrounding gas due to heat produced from the particles are enrolled by a microphone. The light-absorbing particles <300 nm, and larger particles (>300 nm) concentration is proportional to the enrolled signal. The aerosol absorption in 532 nm was measured by very acute method developed by Lack et al. (2006) having an excellent response time by using photoacoustic absorption spectroscopy. (iv) Laser-induced incandescence (LII)—the photomultiplier tube is used for measuring the particle decomposition which is the result of heating of particles below the carbon sublimation temperature by short laser pulse as per Giechaskiel et al. (2014). The number and average size of particles and soot volume is derived by analysing the incandescence intensity and decomposition rate as according to Santoro and Shaddix (2002). (c) The light extinction in aerosols is measured (i) Cavity ring down (CRD) is a system developed by Mellon et al. (2011) and Pettersson et al. (2004) in the laboratory. (ii) Opacity meter is another instrument that measure light extinction. Öztürk (2015) used the equipment mostly for the measurement of particles in the diesel engines. The atomisation and drying lead to generation of aerosol in the CRD experiment developed by Pettersson et al. (2004) where differential mobility analyser (DMA) selects the size of dried particles. There is a counter (TSI, model 3022A) for counting the selected particles in the CRD cell exit (TSI, model 3022A). The light fraction transmitted from an exhaust volume is measured by opacity meters as explained by Giechaskiel et al. (2014). The adsorption and scattering results in light extinction, i.e., opacity, which is the difference of incident and transmitted light and gives the particle concentration.

Bioaerosol Online Measurements Ultraviolet aerodynamic particle sizer (UVAPS: TSI Inc). The UVAPS is a reformed aerodynamic particle sizer (APS) with an additional UV laser to induce fluorescence in the biological particles for their real-time detection. It provides information on number concentration and size distribution of both total and fluorescent biological aerosol particles by means of measuring the aerodynamic diameter, side-scattering intensity and fluorescence intensity of each particle.

The primary operation of UVAPS in detecting the biological aerosol particles is based on laser-induced fluorescence produced by specific fluorophores (riboflavin, NADH, NAD(P) H, etc.) present naturally in the biological particles (Hairston et al. 1997; Brosseau et al. 2000; Agranovski et al. 2003a, b, 2004). These are the co-enzymes that are mostly linked with cellular metabolism (Pöhlker et al. 2012) and thus give a measure of viable biological particles in the atmosphere (Huffman et al. 2010; Hairston et al. 1997).

Although UVAPS detect and count the biological aerosol particles based on their intrinsic fluorescence, there are certain limitations associations with the instrument. The UVAPS cannot identify the type of bioaerosol nor differentiate them based on their varying fluorescent intensity. Biological particles with very weak fluorescence (such as specific spores, damaged cells, etc.) as well as smaller particles (since fluorophore abundance is a function of particle size) escape the fluorescence detection limit of UVAPS and will not be counted (Agranovski et al. 2003a, b, 2004; Kanani and Klapa 2007). Also, certain non-biological aerosol particles such as soot, polyaromatic hydrocarbon (PAH) and some secondary organic aerosols (SOA) may also exhibit fluorescence in the same detection limit and may interfere with the FBAP counts (Huffman et al. 2010). Since the concentrations of these non-biological fluorescent particles are generally low at size range >1 μm, and their contribution to the fluorescent particle count would not be significant (Huffman et al. 2010). A detailed description of the instrument is given below.

UVAPS Operation The UVAPS is a high-performance particle spectrometer that measures the aerodynamic diameter, the scattered light intensity and the fluorescence of airborne particles. The instrument measures the aerodynamic diameter (D_a) of the aerosol particles in the size range of 0.54–19.81 μm, by means of the time of flight technique that measures the velocity of each particle between two laser beams He–Ne red lasers; ($\lambda = 633$ nm) in an accelerating airflow through a nozzle. The scattered light from the particle is detected by avalanche photodiode (APD) module and is converted into electric signal, which has two crests (Fig. 2.8). Using a calibration table based on latex spheres, each time of flight is converted to aerodynamic particle size and is recorded in 52 channels.

Once the size of particle is estimated, the fluorescence is measured for the same particle by exiting it using an ultraviolet Nd:YAG laser ($\lambda = 355$ nm) and detecting the emissions within the range of 420–575 nm. Each particle illuminated by UV generated light which is detected with the help of a photomultiplier tube (PMT). The total fluorescence which is spectrally unresolved is reported for each individual particle as the fluorescence intensity increased in the channels. The data recorded in the first channel is considered to be non-fluorescent while in the following channels

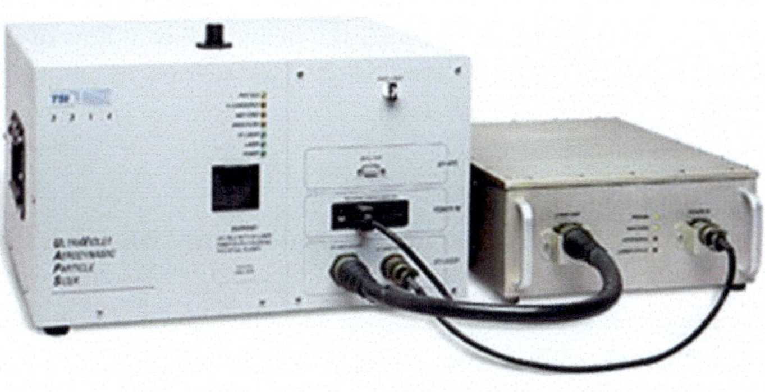

Fig. 2.8 A pictorial representation of UVAPS as obtained from TSI UVAPS manual. The unit on the left hand side shows the instrument panel, whereas the unit on the right had side depicts the dedicated power unit used to supply the high-voltage power to the UV laser in the instrument

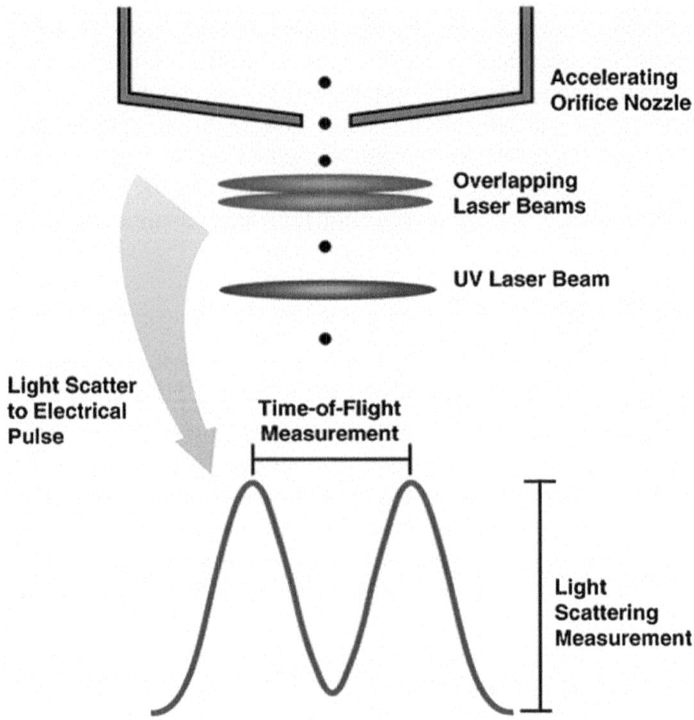

Fig. 2.9 Double-crested signal from particle passing through overlapping beams. Figure as obtained from TSI manual

have an increasing order of fluorescence intensity. Figure 2.9 shows the schematic representation of flow operating principle of UVAPS.

There is 5 min interval (where full diameter range scan was for 285 s and back-scanning was for 15 s) in the measuring process of UVAPS where air sample is drawn with a volumetric flow rate of 5 L min^{-1} (lpm) at ambient temperature and pressure. The equipment has splitted sample flow of 1.0 lpm and a sheath flow of the 4.0 lpm to confine the particle flow through the centre of air stream, by utilising pressure difference feedback control (Fig. 2.10).

2.5 Data Quality Assurance and Quality Control (QA/QC)

This is a very important aspect of all analysis since the data that is reported can be used to make important public policy decisions and therefore can be challenged in court. The veracity and the confidence of the analysis results must be documented and reported. A large fraction of the QA/QC aspects are statistical quantities related

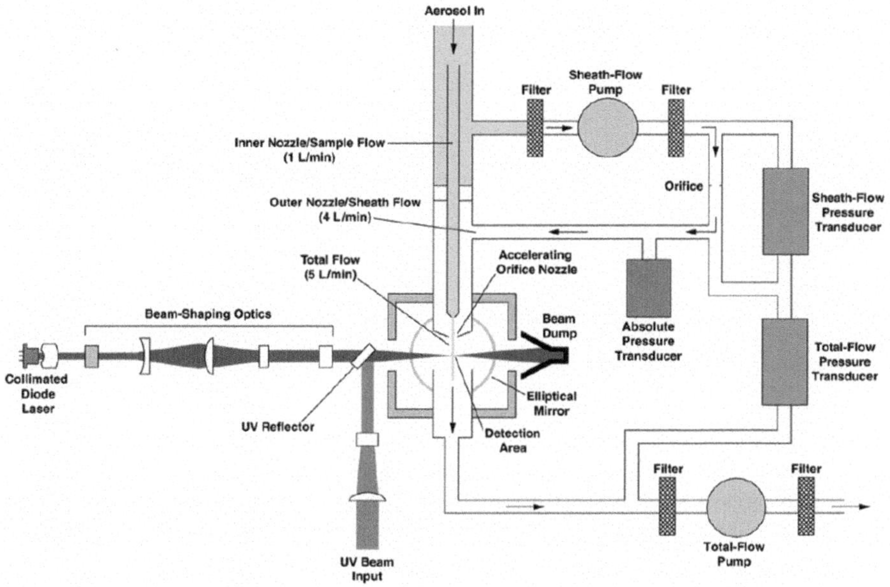

Fig. 2.10 Schematic representation of operating principle and aerosol flow through UVAPS

to the analysis. The following are some of the important concepts of QA/QC with reference to the analysis of vapour phase chemical composition in air.

Accuracy This is the comparison with a true value. This is established with the help of an external analytical standard, which are sold commercially with a certificate of analysis. This is used in the calibration of the instrument for the determination of concentrations. The analytical standards are typically available for an analyte for use in a specific instrument.

Precision This is also known as repeatability. The precision of an analytical instrument is checked by analysing the same sample several times, and the mean and standard deviation are verified.

Method Detection Limit (MDL) This is the lowest level of signal that the instrument will record of analyte that is clearly distinguishable from the noise of the system. In general, this is set at 3σ, where σ is the standard deviation of seven non-consecutive measurements of a blank or a standard. The MDL is also an indication of the sensitivity of the instrument. All analyte signals that are below the MDL are reported as BDL (or below detection limit) and not as ZERO.

Blank This is a sample that does not contain the analyte of interest. The purpose of a blank is to detect false positives. There are several types of blanks for various purposes. A solvent blank is just the solvent and is measured to obtain the contamination of the solvent with the analyte of interest. A method blank is the analysis of a sample that represents the entire sequence of sample collection but without the

Table 2.1 Common analytes in vapour phase and the methods for their analysis

Method ID	Species	Collection	Analysis
TO-1	VOCs	Tenax	GC/GC-MS
TO-2	VOCs	Molecular sieve adsorption	GC/GC-MS
TO-3	VOCs	Cryogenic preconcentration	GC
TO-4	PCBs pesticides	PUF	GC
TO-5	Aldehydes ketones	Impinger; reagents	HPLC
TO-6	Phosgene	Impinger; reagents	HPLC
TO-7	Nitroso dimethylamine (NDMA)	Thermosorb	GC
TO-8	Phenols/cresols	Impingers	HPLC

analyte. For example, in the use of a sorbent to collect a vapour phase sample, the method blank will be to use a sorbent, but not expose it to ambient air, but still go through the entire process of analysis. This will provide for any introduction of false positives. This usually occurs as a result of the contamination of the analytical workbench or the analytical instrument as a result of sample memory.

Surrogate Standards This is used to analyse the recovery of an analyte from different sampling or processing devices. For example, if there are adsorption losses of an analyte in a Tedlar bag, a surrogate standard can be used to find out what is the efficiency of the extraction. By definition, the surrogate is a representative of the analyte, but is not the analyte. For environmental applications, the suggested surrogate chemicals standards are radio-labelled chemicals. For example, the surrogate for naphthalene would be naphthalene with one C or hydrogen isotope. This is especially suitable for GC-MS analysis. A known amount of the surrogate is added to the sample at the beginning of the sampling process and goes all the way to the analysis, and the amount of surrogate recovered is applied to all the chemicals that are considered similar to the surrogate. If the analyte list is long, then multiple surrogates may be used. Table 2.1 shows a list of standard analytical methods for the measurement of common organic analytes in vapour phase in ambient air (http://www.epa.gov/ttn/amtic/airtox.html#compendium).

References

Agranovski V, Ristovski Z, Hargreaves M, Blackall PJ, Morawska L (2003a) Real-time measurement of bacterial aerosols with the UVAPS: performance evaluation. J Aerosol Sci 34(3):301–317

Agranovski V, Ristovski Z, Hargreaves M, Blackall PJ, Morawska L (2003b) Performance evaluation of the UVAPS: influence of physiological age of airborne bacteria and bacterial stress. J Aerosol Sci 34(12):1711–1727

Agranovski V, Ristovski ZD, Ayoko GA, Morawska L (2004) Performance evaluation of the UVAPS in measuring biological aerosols: fluorescence spectra from NAD (P) H coenzymes and riboflavin. Aerosol Sci Technol 38(4):354–364

Brosseau L, Welch V, Wells G, Tugwell P, Gam A, Harman K, Morin M (2000) Low level laser therapy for osteoarthritis and rheumatoid arthritis: a metaanalysis

Chowdhury Z, Campanella L, Gray C, Al Masud A, Marter-Kenyon J, Pennise D, Charron D, Zuzhang X (2013) Measurement and modeling of indoor air pollution in rural households with multiple stove interventions in Yunnan, China. Atmos Environ 67:161–169

Costa MAM, Carvalho JA, Neto TGS, Anselmo E, Lima BA, Kura LTU, Santos JC (2012) Real-time sampling of particulate matter smaller than 2.5 μm from Amazon forest biomass combustion. Atmos Environ 54, 480–489

CPCB (Central Pollution Control Board) (2014) http://cpcb.nic.in/air.php. Accessed Jan 2014

Giechaskiel B, Maricq M, Ntziachristos L, Dardiotis C, Wang X, Axmann H, Bergmann A, Schindler W (2014) Review of motor vehicle particulate emissions sampling and measurement: from smoke and filter mass to particle number. J Aerosol Sci 67:48–86

Gong W, Zhang M, Han G, Ma X, Zhu Z (2015) An investigation of aerosol scattering and absorption properties in Wuhan, Central China. Atmosphere 6:503–520

Hairston PP, Ho J, Quant FR (1997) Design of an instrument for real-time detection of bioaerosols using simultaneous measurement of particle aerodynamic size and intrinsic fluorescence. J Aerosol Sci 28(3):471–482

Hinds WC (1999) Aerosol technology: properties, behavior, and measurement of airborne particles, 2nd edn. Wiley, Hoboken

Huffman JA, Treutlein B, Pöschl U (2010) Fluorescent biological aerosol particle concentrations and size distributions measured with an ultraviolet aerodynamic particle sizer (UV-APS) in Central Europe. Atmos Chem Phys 10(7):3215–3233

IITM (Indian Institute of Tropical Meteorology) (2014) http://safar.tropmet.res.in. Accessed Jan 2014

Jiang R, Bell ML (2008) A comparison of particulate matter from biomass-burning rural and non-biomass-burning urban households in north-eastern China. Environ Health Perspect 116:907–914

Kanani HH, Klapa MI (2007) Data correction strategy for metabolomics analysis using gas chromatography–mass spectrometry. Metab Eng 9(1):39–51

Krecl P, Ström J, Johansson C (2007) Carbon content of atmospheric aerosols in a residential area during the wood combustion season in Sweden. Atmos Environ 41:6974–6985

Lack DA, Lovejoy ER, Baynard T, Pettersson A, Ravishankara AR (2006) Aerosol absorption measurement using photoacoustic spectroscopy: sensitivity, calibration, and uncertainty developments. Aerosol Sci Technol 40:697–708

Leskinen J, Tissari J, Uski O, Virén A, Torvela T, Kaivosoja T, Lamberg H, Nuutinen I, Kettunen T, Joutsensaari J et al (2014) Fine particle emissions in three different combustion conditions of a wood chip-fired appliance—particulate physico-chemical properties and induced cell death. Atmos Environ 86:129–139

Majumdar D, Mukherjeea AK, Sen S (2011) BTEX in ambient air of a Metropolitan City. J Environ Prot 2(01):11

Mellon D, King SJ, Kim J, Reid JP, Orr-Ewing AJ (2011) Measurements of extinction by aerosol particles in the near-infrared using continuous wave cavity ring-down spectroscopy. J Phys Chem 115:774–783

Nussbaumer T, Czasch C, Klippel N, Johansson L, Tullin C (2008) Particulate emissions from biomass combustion in IEA countries. In: Proceeding of the 16th European Biomass Conference and Exhibition, Zurich, 2–6 June 2008, p 40

Öztürk E (2015) Performance, emissions, combustion and injection characteristics of a diesel engine fuelled with canola oil–hazelnut soapstock biodiesel mixture. Fuel Process Technol 129, 183–191

Pettersson A, Lovejoy ER, Brock CA, Brown SS, Ravishankara AR (2004) Measurement of aerosol optical extinction at with pulsed cavity ring down spectroscopy. J Aerosol Sci 35:995–1011

Pöhlker C, Wiedemann KT, Sinha B, Shiraiwa M, Gunthe SS, Smith M, Elbert W (2012) Biogenic potassium salt particles as seeds for secondary organic aerosol in the Amazon. Science 337(6098):1075–1078

Prado GF, Zanetta DMT, Arbex MA, Braga AL, Pereira LAA, de Marchi MRR, de Melo Loureiro AP, Marcourakis T, Sugauara LE, Gattás GJF et al (2012) Burnt sugarcane harvesting: particulate matter exposure and the effects on lung function, oxidative stress, and urinary 1-hydroxypyrene. Sci Total Environ 437, 200–208

Santoro RJ, Shaddix CR (2002) Laser-induced incandescence. In: Applied combustion diagnostics, pp 252–286

Simões Amaral S, Carvalho J, Martins Costa M, Pinheiro C (2015) An overview of particulate matter measurement instruments. Atmosphere 6:1327–1345

Sivertsen B (2008) Monitoring air quality, objectives and design. Chem Ind Chem Eng Q 14. https://doi.org/10.2298/ciceq0803167s

Torvela T, Tissari J, Sippula O, Kaivosoja T, Leskinen J, Virén A, Lähde A, Jokiniemi J (2014) Effect of wood combustion conditions on the morphology of freshly emitted fine particles. Atmos Environ 87:65–76

Vincent JH (2007) Aerosol sampling: science, standards, instrumentation and applications. Wiley, Hoboken

Chapter 3
Air Quality Modelling

S. M. Shiva Nagendra, Uwe Schlink, and Mukesh Khare

3.1 Basic Concepts on Mathematical Model Formulation

Air pollution is a worldwide issue, especially in urban regions and demands for comprehensive multi-disciplinary scientific and technological efforts. The definition of efficient strategies to control and manage air quality cannot be achieved without a good and clear knowledge of the atmospheric processes involving emission, transport and transformation. Therefore, management decisions require a variety of modelling and simulation efforts, ranging from complex process models in three dimensional to simplified approaches that allow a quick overview of the quality of urban air.

There are two major chellanges associated with air quality modelling. The first challenge of modeling concerns the prediction of the air pollutant concentrations, which is used in operational forecasting as well as in studies of the effect of emission reduction, the comparative consideration of emission scenarios and other mitigation strategies. Such forecasts will be of different precision and statistical approaches are applied for the comparison and evaluation of the performance of different air quality models.

S. M. Shiva Nagendra (✉)
Department of Civil Engineering, Indian Institute of Technology Madras, Chennai, Tamil Nadu 600036, India
e-mail: snagendra@iitm.ac.in

U. Schlink
Department of Urban and Environmental Sociology, Helmholtz Centre for Environmental Research UFZ, 04318 Leipzig, Germany
e-mail: uwe.schlink@ufz.de

M. Khare
Department of Civil Engineering, The Indian Institute of Technology, Hauz Khas, New Delhi 110 016, India
e-mail: mukeshk@civil.iitd.ernet.in

S. M. Shiva Nagendra et al. (eds.), *Urban Air Quality Monitoring, Modelling and Human Exposure Assessment*, Springer Transactions in Civil and Environmental Engineering,
https://doi.org/10.1007/978-981-15-5511-4_3

A second challenge is the transfer of the air quality information into the context of management and control (link to Chap. 4). For the latter, not only atmospheric processes have to be considered but also the consequences of air pollution for the environment, including impact on agriculture, human health, urban, and indoor air quality. This challenge also includes the development of guidelines for an optimal design of a monitoring network, consisting of an optimum number of stations located at the most relevant sites.

The long-term urban air quality planning and episode forecasting depend upon the mathematical models which help in the prediction of air quality. Models are playing an important role for making of decision in the developed countries in the recent past. The current and future air quality as a result of emission sources could be assessed with the help of models for framing policy decisions. Further, a mathematical model may serve to gain insight into the relation between meteorological elements and source emissions. The dispersion model helps in forecasting the changes in air quality on the basis of different scenarios by taking the input of the current source emission and meteorological data. The past two decades have seen significant advancements in research and application of air quality modelling approaches. These modelling approaches include deterministic, statistical, hybrid of deterministic and statistical distribution, and artificial neural networks. In the present chapter, a brief review on air pollution meteorology, modelling approaches and their application in air quality management have been discussed. The limitations of mathematical modelling approaches are also presented.

3.2 Air Pollution Meteorology

Air pollution meteorology is the study of atmospheric processes, namely transport, dispersion, transformation, and removal that affect the fate of air pollutants. Therefore, knowledge of air pollution meteorology is used to manage and control the release of pollutants into the ambient air. The atmosphere surrounds the earth and rotates with the earth as it orbits the sun. The dry air consists of about 78% nitrogen, 21% oxygen, and 1% argon. Trace gases such as carbon dioxide, neon, and helium also exist. Although the water vapor content of the air is fairly small, it absorbs six times more radiation than any other atmospheric constituent and is therefore a very important component of the atmosphere (Lyons and Scott 1990).

3.2.1 Layers of the Atmosphere

The atmosphere is divided into four distinct layers: the troposphere, stratosphere, mesosphere, and thermosphere. The lowest layer is called the troposphere which accounts for about three quarters of the mass of the atmosphere and contains nearly all of the water associated with the atmosphere (vapor, clouds, and precipitation).

The troposphere, where air masses, fronts, and storms reside, is the most unsettled layer and provides earth its weather. The depth of the troposphere varies with latitude and season. The top of the troposphere (tropopause) is about 16.5 km (54,000 ft) on average over the equator and about 8.5 km (28,000 ft) over the poles. Seasonal changes affect the thickness of the troposphere causing it to be thicker in summer (when the air is warmer) than in winter. The depth of the troposphere changes constantly due to changes in atmospheric temperature.

Virtually, all air pollution are emitted within the troposphere. Air pollution transport is governed by the speed and direction of the winds. The rate of dispersion is influenced by the thermal structure of the atmosphere as well as by mechanical agitation of the air as it moves over the different surface features of the earth. Transformation of the emitted air pollutants is impacted by exposure to solar radiation and moisture as well as other constituents in the atmosphere. The removal of pollutants depends not only on the pollutants' characteristics but also on weather phenomena such as rain, snow, and fog. These interactive meteorological phenomena are studied as part of air pollution meteorology.

3.2.2 Importance of Air Pollution Meteorology

Since the atmosphere serves as the medium into which air pollutants are released, the transport and dispersion of these releases are influenced significantly by meteorological parameters. Understanding air pollution meteorology and its influence in pollutant dispersion is essential in air quality planning activities. Planners use this knowledge to help locate air pollution monitoring stations and to develop implementation plans to bring ambient air quality into compliance with standards. Meteorology is used in predicting the ambient impact of a new source of air pollution and to determine the effect on air quality from modifications to existing sources. When meteorological conditions develop that are not conducive to pollutant dispersion, governmental air pollution agencies must act fast to ensure that air pollutants do not build up to unacceptable levels in the air we breathe. When pollutant levels become excessively high, an air pollution episode results and emissions into the atmosphere must be curtailed. Donora, Pennsylvania, provides an extreme example of this situation. In 1948, Donora suffered a disastrous air pollution episode.

Air moves in an attempt to equalize imbalances in pressure that result from differential heating of the earth's surface. While moving from areas of high pressure to low pressure, wind is heavily influenced by the presence or absence of friction. Thus, surface winds behave differently than winds aloft due to frictional forces acting near the earth's surface. The rotation of the earth modifies atmospheric motion but does not cause it, since the atmosphere essentially rotates with the earth. The movement of air helps keep concentrations of pollutants that are released into the air from reaching dangerous levels.

Wind is the basic element in the general circulation of the atmosphere. Winds are always named by the direction from which they blow. When wind blows more frequently from one direction than from any other, the direction is termed the prevailing wind. Wind speed increases rapidly with height above the ground level, as frictional drag decreases. Eddies are variations from the main current of wind flow. Larger irregularities are caused by convection or vertical transport of heat. These and other forms of turbulence contribute to the movement of heat, moisture, and dust into the air aloft.

3.2.3 Topographical Influences

Topographical features not only influence the way the earth and its surrounding air heat up, but they also affect the way air flows. Terrain features, as you would expect, predominantly affect air flow relatively close to the earth's surface. Turbulence in the wind over flat terrain is limited to the amount of roughness of either natural or man-made features that are on the ground. These features induce a frictional effect on the wind speed and result in the well-known wind profile with height.

Mountain/valleys heat unevenly because of the sun's motion across the sky. In the morning, one side of a mountain or valley is lit and heated by the sun. The other side is still dark and cool. Air rises on the lighted side and descends on the dark side. At midday both sides are 'seen' by the sun and are heated. The late afternoon situation is similar to the morning. After dark, as the air cools due to radiational cooling, the air drains down into the valley from all higher slopes.

The land and water not only exhibit different roughness characteristics but different heating properties. The air flow and thus plume dispersion and transport can be very difficult to predict. The thermal properties of land and water are radically different. Land and objects on it will heat and cool quickly. However, water heats and cools slowly. As the sun shines down on the land/water interface, solar radiation will penetrate several feet through the water. On the other hand, solar radiation striking land will only heat the first few inches. Also, as the sun shines on the water surface, evaporation and some warming take place. The thin layer of water next to the air cools due to evaporation and mixes downward, overturning with the small surface layer that has warmed. This mixing of the water layer close to the surface keeps the water temperature relatively constant. On the other hand, land surfaces warm quickly causing the adjacent air to heat up, become less dense, and rise. The cooler air over the water is drawn inland and becomes the well-known sea breeze. At night, the air over the land cools rapidly due to radiational cooling, which causes the land temperature to fall faster than that of the adjacent water body. This creates a return flow called the land breeze. The wind speeds in a land breeze are light, whereas the wind speeds in sea breeze can be quite fast. Differential pressure over land and water causes sea breezes.

Urban areas have added roughness features and different thermal characteristics due to the presence of man-made elements. The thermal influence dominates the influence of the frictional components. Building materials such as brick and concrete absorb and hold heat more efficiently than soil and vegetation found in rural areas.

After the sun sets, the urban area continues to radiate heat from buildings, paved surfaces, etc. Air warmed by this urban complex rises to create a dome over the city. It is called the **heat island effect (UHI)**. The city emits heat all night. Just when the urban area begins to cool, the sun rises and begins to heat the urban complex again. Generally, city areas never revert to stable conditions because of the continual heating that occurs.

3.2.4 Vertical Motions and Atmospheric Stability

3.2.4.1 Vertical Motions

The high and low pressure systems, the lifting of air over terrain, and the different types of fronts and convection result in vertical motions. There are a number of basic principles related to vertical motion. These principles include instability, stability, and plume behavior.

The **lapse rate** is defined as the rate at which temperature of air changes with altitude. The actual lapse rate in the atmosphere is approximately −6 to −7 °C per km (in the troposphere) but different locations and times of day result in variation in the lapse rate. The decrease in temperature with altitude is called as a negative lapse rate and on the other hand, the increase in temperature with altitude is called as a positive lapse rate (Boubel et al. 1994).

Dry adiabatic lapse rate: The process in which heat or mass is not transferred across the air parcel boundaries is known as adiabatic process. The heating and cooling takes place due to compression and expansion of air parcel in an adiabatic process. The cooling of rising dry air parcel takes place at the dry adiabatic rate of 9.8 °C/1000 m which has a lapse rate of −9.8 °C/1000 m in the atmosphere. While the heating of sinking dry air parcel takes place at a dry adiabatic rate of 9.8 °C/1000 m in the atmosphere which has a lapse rate of 9.8 °C/1000 m. The presence of water in gaseous form in the air parcel makes it dry. The dry adiabatic lapse does not change with the temperature of ambient air as it is a fixed rate.

Wet adiabatic lapse rate: The cooling of rising air parcel which contains water vapor takes place at the dry adiabatic lapse rate up to the time of reaching to its condensation temperature, or dew point. The condensation of water vapor starts when the pressure of the water vapor equals the saturation vapor pressure of the air. The latent heat released in the parcel after condensation results in slowing of cooling rate of the parcel which is called as the **wet adiabatic lapse rate**. The wet adiabatic lapse rate is variable with temperature and pressure while the dry adiabatic lapse rate is constant.

Environmental lapse rate: The **environmental lapse rate** also known as **atmospheric lapse rate** is defined by the ambient air temperature. The complicated interplay of meteorological factors also plays an important role in the environmental lapse rate. The environmental lapse is the decrease in temperature with altitude. The rising

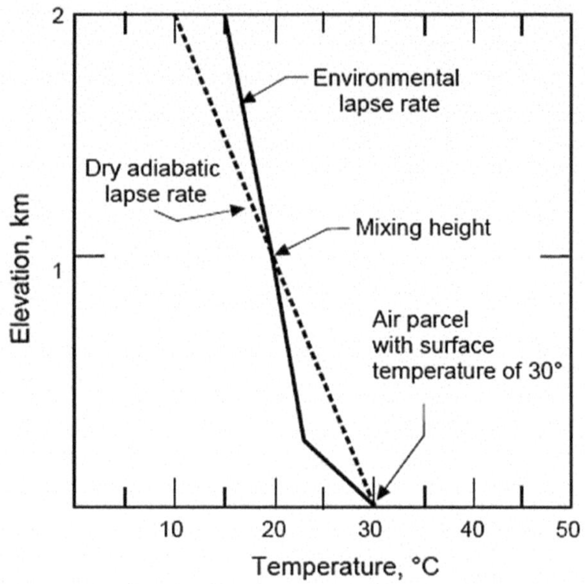

Fig. 3.1 Mixing height

or sinking of air parcels depends on the surrounding air temperature. The condition when temperature actually increases with altitude is referred to as a **temperature inversion**. The temperature inversion occurs at elevations of from 200 to 350 m.

Mixing height: The vertical depth of the atmosphere where mixing takes place is called the mixing layer. The top of the mixing layer is referred to as the mixing height. The mixing height determines the vertical extent of the dispersion process for air pollutants that are released below the mixing height. The mixing height at a given time may be estimated by use of the morning radiosonde ascent plotted on a thermodynamic chart. The surface temperature at the given time is plotted on the diagram. If the dry adiabat is drawn through this temperature, the height aboveground at the point where this dry adiabat intersects the morning sounding is the mixing height for that time (Fig. 3.1). Use of this sounding procedure provides an approximation because it assumes that there has been no significant advection since the time of the sounding. The mixing height is an important variable in air quality studies as it restricts vertical dispersion of pollutants measurements.

3.2.4.2 Atmospheric Stability

The degree of stability of the atmosphere is determined by the temperature difference between an air parcel and the air surrounding it. This difference can cause the parcel to move vertically (i.e., it may rise or fall). This movement is characterized by four basic conditions that describe the general stability of the atmosphere. In **stable** conditions, this vertical movement is discouraged, whereas in **unstable** conditions, the air parcel tends to move upward or downward and to continue that movement. When conditions neither encourage nor discourage air movement beyond the rate of adiabatic heating or cooling, they are considered **neutral**. When conditions are extremely stable, cooler air near the surface is trapped by a layer of warmer air above it. This condition, called an **inversion**, allows virtually no vertical air motion. These conditions are directly related to pollutant concentrations in the ambient air.

Inversions: The increasing air temperature with height results in inversion. There is usual occurence of inversion but limited to around lower layer only. The four major types of inversions are caused by different atmospheric interactions and can persist for different amounts of time are **radiation inversion, subsidence inversion, and frontal inversion and advection inversions**.

The **radiation inversion** is the most common form of surface inversion and occurs when the earth's surface cools rapidly. As the earth cools, so does the layer of air close to the surface. If this air cools to a temperature below that of the air above, it becomes very stable, and the layer of warmer air impedes any vertical motion. Radiation inversions usually occur in the late evening through the early morning under clear skies with calm winds, when the cooling effect is greatest.

The **subsidence inversion** is almost always associated with anticyclones (high pressure systems). Air in an anticyclone descends and flows outward in a clockwise rotation. As the air descends, the higher pressure at lower altitudes compresses and warms it at the dry adiabatic lapse rate. Often this warming occurs at a rate faster than the environmental lapse rate. The inversion layer thus formed is often elevated several hundred meters above the surface during the day. At night, because of the surface air cooling, the base of a subsidence inversion often descends, perhaps to the ground. In fact, the clear, cloudless day's characteristics of anticyclones encourage radiation inversions, so that there may be a surface inversion at night and an elevated inversion during the day. Subsidence inversions, unlike radiation inversions, last a relatively long time.

The **frontal inversion** is usually associated with both cold and warm fronts. At the leading edge of either front, the warm air overrides the cold, so that little vertical motion occurs in the cold air layer closest to the surface. The strength of the inversion depends on the temperature difference between the two air masses.

Advection inversions are associated with the horizontal flow of warm air. When warm air moves over a cold surface, conduction and convection cool the air closest to the surface causing a surface-based inversion. This inversion is most likely to occur in winter when warm air passes over snow cover or extremely cold land. Another type of advection inversion develops when warm air is forced over the top of a cooler air layer.

3.2.5 Stability and Plume Behavior

The degree of atmospheric stability and the resulting mixing height has a large effect on pollutant concentrations in the ambient air. The vertical air movement and horizontal air flow combination influence the behavior of plumes from point sources (stacks) (Arya 1999).

The **looping plume** occurs in highly unstable conditions and results from turbulence caused by the rapid overturning of air. While unstable conditions are generally favorable for pollutant dispersion, momentarily high ground-level concentrations can occur if the plume loops downward to the surface.

The **fanning plume** occurs in stable conditions. The inversion lapse rate discourages vertical motion without prohibiting horizontal motion, and the plume may extend downwind from the source for a long distance. Fanning plumes often occur when there is a radiation inversion in the early morning.

The **coning plume** is characteristic of neutral conditions or slightly stable conditions. There is occurence of these type of conditions on sunny days or cloudy days between the breakup of a radiation inversion and the development of unstable daytime conditions. When conditions are unstable above an inversion, the release of a plume above the inversion results in effective dispersion without noticeable effects on ground-level concentrations around the source. This condition is known as **lofting**.

If the plume is released just under an inversion layer, a serious air pollution situation could develop. As the ground warms in the morning, air below an inversion layer becomes unstable. When the instability reaches the level of the plume that is still trapped below the inversion layer, the pollutants can be rapidly transported down toward the ground. This is known as **fumigation**. There is presence of very high concentrations of ground-level pollutants when fumigation takes place. Fumigation can be prevented by developing adequately tall stacks in most of the cases.

The trapped plume occurs when the plume effluent is caught between two inversion layers. The diffusion of effluent is severely restricted to the unstable layer between the two stable regions. Trapping may also be associated with subsidence inversions lasting for several days, where almost all emissions are trapped below the inversion layer thus creating one of the worst pollution situations.

3.3 Air Quality Modelling Techniques

Several modelling approaches have been applied to elucidate the dispersion of pollutants in the urban environment. These include deterministic, stochastic, hybrid of deterministic and statistical distribution, and artificial neural-network-based modelling approaches.

3.3.1 Deterministic Modelling

The concentrations of air pollutants in urban areas can be predicted by applying a traditional approach which is known as deterministic modeling. There is a differential equation which represents air pollutant dispersion modeling and expresses the rate of change of pollutant concentration in terms of average wind speed and turbulent diffusion. The mass conservation principle is the basis for this mathematical formulation (Hickman and Colwill 1982). The air quality modelling is represented by the diffusion equation given below:

$$\frac{\partial C}{\partial t} = -\left(u\frac{\partial C}{\partial x} + v\frac{\partial C}{\partial y} + w\frac{\partial C}{\partial z}\right) + \frac{\partial}{\partial x}K_H\frac{\partial C}{\partial x} + \frac{\partial}{\partial y}K_H\frac{\partial C}{\partial y} + \frac{\partial}{\partial z}K_z\frac{\partial C}{\partial z} + Q + R. \quad (3.1)$$

where

C = pollutant concentration
t = time
x, y, z = position of the receptor relative to the source
u, v, w = wind speed coordinate in x, y and z direction
K_x, K_y, K_z = coefficients of turbulent diffusion in x, y and z direction
Q = source strength
R = sink (changes caused by chemical reaction).

3.3.2 *Statistical Modelling*

The meteorological and traffic parameters are used in the statistical methods for obtaining a statistical relationship from measured concentrations as compared to deterministic modeling. Some important methods which are used in statistical modeling are regression, multiple regression, and time series technique. The general form of a linear regression model considers the concentration of a pollutant as the dependent variable (predictant) and meteorological and traffic characteristics as independent variables (predictors). Pollution concentrations are highly influenced by meteorological data as shown by considerable evidence. Chock et al. (1975) pointed out the nonapplicability of regression technique in modelling air pollutants at Riverside, CA, USA. Further, their study indicated the applicability of linear stochastic models as alternative to linear regression modelling technique.

The stochastic models are empirical in nature. The dispersion of air pollutants on a local scale is described by the time series analysis technique which is known as Box--Jenkins' models. The abstraction of auto regression and moving average process is used by the Box--Jenkins approach which results in the autoregression moving average (ARMA) and autoregression-integrated moving average (ARIMA) and seasonal ARIMA models where dependent variable is lag-regressed. The current and past data related to one variable is used by a univariate time series model. The deportment of the desired series is described by the other variables which are emphatically used by a time series model and is known as a multiple time series model or multivariate time series model. There is a presence of dynamic relationship among these variables as expressed by the multivariate time series model which is known as a transfer function model. There is presence of a dependent variable as well as one or more explanatory variables which is the characteristics of both transfer function model and standard regression model.

3.3.3 Hybrid Modelling

The combination of both deterministic and statistical modelling techniques results in the hybrid approach as narrated in Jakeman et al. (1988) and Gokhale et al. (2003). The frequent occurrence of average pollutant concentration predicted by deterministic model is strengthened and facilitated by the hybrid approach. The analysis of parametric distribution of data of air pollutant to evaluate percentiles along with maximum values of the appropriate distribution is done by using the statistical technique. The causal association between meteorology, emissions, and average ground-level concentrations is made by deterministic models that are the basis of this approach. When the historical air pollution data has suitable distributional arrangement, the distribution of all occurence about mean is predicted by the statistical models. The hybrid approach first uses the pollutant concentration data produced by the deterministic model within its ambit of highest accuracy. Then, the specifications of expedient statistical model are predicted from these curtailed samples. The narration of air quality data for the whole extent of percentiles is provided by the statistical model when the data were parameterized by appropriate computation method.

3.3.4 ANN-Based Modelling

ANN-based modeling required familiar input data set without any postulations which has dominance over usual phenomenological or semi-empirical models. The quickly accessible inputs are used by the neural-network model which developed its own internal model and afterwards predicted the output. The adaptive nature is the most influential characteristics of neural networks, where ‘learning by example’ replaces ‘programming’ while fixing the problems. The computational models with such characteristics are very tempting in application domains where training data is easily accessible even with incomplete understanding of the problem to be solved. The input is accepted and produced by the neural network which is also like a ‘black box.’ The vehicular pollutant concentration is predicted by the multilayer perceptron (MLP) structure of the neural networks for line source modelling. The interconnected ‘neurons’ or ‘nodes’ are present in layer in the multilayer perceptron. The input layer is formed by arranging ‘nodes’ in a layer, the ‘hidden’ layers, and an output layer is also formed; the nodes present in each layer connected to all other nodes in neighboring layers. The distribution of input signals to the hidden layer is done by the input layer 'neurons' which act as a buffer. The input signals are added up by each 'neuron' in the hidden layer and processed with a simple nonlinear transfer or 'activation' function that passes the result to the output layer. The output signal in the output layer is also computed by 'neurons' in the similar manner. The output signals from each neuron in multilayer perceptron proliferate in forward direction; henceforth, multilayer perceptron is also known as 'feed-forward' neural network.

The multilayer perceptron can estimate any smooth, measurable function between input and output vectors by selecting a suitable set of connecting weights and transfer functions. Multilayer perceptron has the capability to acquire through training. A supervised back-propagation algorithm is usually selected in buildup of multilayer perceptron. Training requires a set of data, which consists of a series of input and associated output vectors. During training, the multilayer perceptron is repeatedly presented with training data and the weights in the network are adjusted until the desired input output mapping occurs. During training, the output from the multilayer perceptron is compared with the desired output. If the network output is not matched with desired output, an error signal is propagated back through the network. Training uses the magnitude of these error signals to adjust the weights and this process continues till the network output matches the desired output.

3.4 Source Apportionment Models

Receptor models represent a statistical evaluation of ambient measurements at different times and locations. The fundamental principle of receptor modelling is the assumption of mass conservation and a mass balance analysis to determine and apportion ambient pollutant concentrations to individual emission sources (Hopke 1991). For this purpose, a mass balance equation is written to account for all chemical species to be identified in the filter samples analyzed, as contributions from independent sources. Various techniques have been used for source apportionment of atmospheric PM (Watson et al. 2002). The selection of the appropriate technique depends on prior knowledge on the sources and source profiles. In general, there are two main approaches to receptor modeling (U.S.EPA 2004). If the sources are known and detailed information on source profiles is available, Chemical Mass Balance (CMB) models can be applied, whereas Positive Matrix Factorization (PMF) is best preferred where the sources are unknown and there is limited information on source profiles. Thus, in the source apportionments of PM_{10} and $PM_{2.5}$ mass concentrations in urban areas are done by both U.S.EPA-CMB-8.2 and U.S.EPA-PMF-3 models.

3.4.1 Source Identification Using CMB Model

The contribution of sources to ambient PM levels in an urban area can be estimated using U.S.EPA-CMB version 8.2 Model (Coulter 2004). The effective variance least-squares algorithm is used in the model to distribute the ambient data to chosen source profiles. Equation 3.2 expresses the concentration of species i at a receptor site k, C_{ik}, as below:

$$C_{ik} = \sum_{j=1}^{m} a_{ij} \cdot s_{jk} \quad i = 1, 2, \ldots I \tag{3.2}$$

where the relative concentration of chemical species is shown by a_{ij}, the emission of fine or coarse particle from sources by i, j, and fine/coarse particle mass contribution from source j, at receptor site k is represented by s_{jk}. The predicted concentrations C_{ik} are matched with measured concentrations at receptor site k resulted from the source contributions s_{jk} which is solved by the model. There will be the best result if the number of chemical species (i) is larger than or equal to the number of sources (m).

These are some postulates present in the CMB mode: (i) configurations of source emissions are steady at the time of ambient and source sampling; (ii) presence of inert chemical species; (iii) identification of all potential sources for contribution to the receptor and should have characterization of their emissions; (iv) the number of species should be larger than the number of sources or source categories; (v) linear independency of the source profiles; and (vi) random, and normal distribution of measurement uncertainties.

The CMB model is run with the following steps: (i) recognition and notation of the contributing sources and chemical species; (ii) computation of the part of each of the chemical species which is included in each source type (source profiles); (iii) computation of the uncertainty in both ambient concentrations and source profiles; and (iv) illustration of the CMB equations.

The inputs for the CMB model are the ambient concentration of the species and their uncertainty. The replacement of values should be done if they are below the detection limit. The uncertainty is calculated with the standard uncertainty values from the sampling with PM samplers, microbalance, pipette and repeatability of ICP–OES and IC for analyzed elements and ions, respectively. The other uncertainties related to calibration curves and calibration standard solutions may also be considerd.

The selection of indicator species is important in a CMB analysis as the preservation of species is one of the postulates of the CMB model. The various source types are represented by different earlier delineated species which help in the investigation of different case scenarios for selecting the indicator species (Hopke 1991; Chow et al. 1993; Lee et al. 1999; Kim and Henry 2000; Song et al. 2001; Abu-Allaban et al. 2002; Samara et al. 2003; Chow et al. 2004; Ward and Smith 2005; Lee et al. 2006; Gupta et al. 2007; Bullock et al. 2008; Callén et al. 2009; Yin et al. 2010). The trends and correlation among species are comprehended by conducting a preliminary data

Table 3.1 List of source profiles obtained from U.S.EPA Speciate database

For PM_{10}		For $PM_{2.5}$	
Source profile	Profile IDs	Source profile	Profile IDs
Diesel exhaust	3267	Diesel exhaust	3268
Gasoline exhaust	3,301,010	Gasoline exhaust	330102.5
Paved road dust	3327	Paved road dust	3328
Brake lining dust	3,400,410	Brake lining dust	340042.5
Brake pad dust	3,400,610	Brake pad dust	340062.5
Marine aerosols	4,310,110	Marine aerosols	431012.5
Cooking	3644	Cooking	4379

analysis as a part of the characterization and trend analysis that would impact CMB analysis. For example the representation of the PM contribution from diesel exhaust, gasoline exhaust, paved road dust, brake lining dust, brake pad dust, marine aerosols and cooking at the urban areas can be represented by selected source profiles. Table 3.1 lists the selected source profiles from the U.S.EPA Speciate 4.2 database (US EPA 2009). The accomplishment of CMB run is evaluated by five variables, namely the correlation coefficient (R^2), *t*-statistics, Chi-square (χ^2), percent mass and ratio of residual to uncertainty (*R*/*U* ratio). The ratio of the source contribution to the standard error is represented by *t*-statistics which should be >2.0. The fraction of the variance in the measured concentrations is illustrated by R^2 which is elucidated by the variance calculated in the concentrations of species and ranges between 0 and 1. Satisfactory results are represented by the correlation coefficient >0.6. The value of χ^2 should be <4 which is the weighed sum of squares of differences between estimated and measured species concentrations. The percent value of predicted or measured mass concentration should be between 80 and 120% which is also known as percent mass. The ratio of residual to uncertainty (R/U ratio) should be <2.

3.4.2 Source Identification Using PMF Model

The source contribution of ambient PM_{10} and $PM_{2.5}$ concentrations can also be identified using the U.S.EPA-PMF model version 3. PMF is a type of multivariate factor analysis tool that decomposes a matrix of speciated sample data into two matrices viz., factor contributions and factor profiles (Paatero 1997). The two matrices like factor contributions and factor profiles are formed from the matrix of speciated sample data with the help of PMF which is a type of a multivariate factor analysis tool. Henceforth, a speciated data set can be considered as a data matrix x of i by j dimensions, in which i = number of samples and j = measured chemical species. The computed concentration of chemical species is represented by row in the x matrix and each compound concentration is represented by the column of x matrix. The 'profile' for each source is indicated by the row of the computed factor matrix f and the amount of emission is specified by the corresponding column of g (Eq. 3.3).

$$x_{ij} = \sum_{k=1}^{p} g_{ik} \cdot f_{kj} + e_{ij} \quad (3.3)$$

where e_{ij} is the residual for each sample/species.

The sum of squares (loss function) Q with respect to the non-negative factor elements (Eq. 3.4) is minimized by applying the PMF model. Each of the sample data has the uncertainties which are represented by $u_{ij.}$ Henceforth, the smaller weights are usually assigned to the larger values of u_{ij} to accommodate more uncertain values of x_{ij} in the least squares fit.

$$Q = \sum_{i=1}^{n} \sum_{j=1}^{m} \left[\frac{x_{ij} - \sum_{k=1}^{p} g_{ik} \cdot f_{kj}}{u_{ij}} \right]^2 \quad (3.4)$$

3.4.2.1 Chemical Concentration and Uncertainty

The ambient concentrations and uncertainties corresponding to the concentration are required for the running of PMF. The PM_{10} and $PM_{2.5}$ data sets mass concentration and chemical make-up are to be restructured in MS Access where one sample is represented by each row and one species is represented by each column. The median value can be used for the missing data and four times the median value can be used for their missing uncertainties. Large standard deviations are allocated to the used median value. The observations having few missing concentrations need not to be rejected in such a way. The samples are precluded from the investigation if all the metals or ions and carbon species are absent. The half of the minimum value recorded replaces the values below detection limits (BDL) and the values which are 5/6 times the detection limit value replace the uncertainties (Polissar et al. 2001). The species having values with below detection limit more than 50% of samples and having higher missing values are not incorporated in the data set for PMF analysis.

3.4.2.2 Categorization of Species Through Signal-to-Noise (S/N) Ratio

The ambient mass concentration and uncertainty data are used for the calculation of *S/N* ratio for each species. Equation 3.5 is used for the calculation of the *S/N* ratio as per EPA PMF v3.0 (Norris et al. 2008).

$$\left(\frac{S}{N}\right)_j = \sqrt{\frac{\sum_{i=1}^{n}\left(x_{ij} - s_{ij}\right)^2}{\sum_{i=1}^{n} s_{ij}^2}} \tag{3.5}$$

The concentration for species j on day i is represented by x_{ij} and uncertainty is represented by S_{ij}. The S/N value represent the species characteristics where $S/N > 2$ represent the strong species, the S/N value of 0.2 represents weak species, and $S/N \leq 0.2$ represents bad species. The variability in the data will be absent and noise will be introduced in the data if there is presence of species with low S/N ratio and if there is presence of high percentage of data below detection. Henceforth, when PMF model run, there is exclusion of species (for PM_{10}-Co, Cr, Mn, Ni, F^-; for $PM_{2.5}$-Co, Ni, F^-) with very low S/N (≤ 0.2), while weak species ($0.2 \leq S/N \leq 2$) are considered for analysis. Some species with low S/N values are still useful as they act as tracer for identification of known local source while species with higher S/N values are classified as weak. Several runs are required to contemplate the species category from 'strong' to 'weak.'

3.4.2.3 Source Signature

The source signature is an important variable in receptor modelling and it has to be selected wisely. The sources are identified by the tracer elements and molecular markers which are referred as source signature. There is comparative study done between the source profiles from PMF analysis and the key indicator species narrated worldwide from the source apportionment studies for the identification of the source type (Lee et al. 1999; Chueinta et al. 2000; Ramadan et al. 2000; Song et al. 2001; Watson et al. 2001; Qin et al. 2002; Kim et al. 2003; Chow et al. 2004; Kim et al. 2004; Almeida et al. 2005; Yuan et al. 2006; Gildemeister et al. 2007; Bullock et al. 2008; Yatkin and Bayram 2008; Callén et al. 2009).

3.5 Statistical Distribution Models

Several statistical techniques are currently being used to identify the nature of trend in air quality data and for its analyses. The statistical techniques used in analyzing air quality data vary from simple regression and multiple regression models to time series techniques. Statistical models are used for various applications such as predicting the trend of air pollutants concentration emitted from vehicular, industrial, and residential sources into the atmosphere, assessment of air quality in the urban region, identifying the effects of air pollution concentration to the receptors, forming a framework to maintain the air quality management in the local and regional level and to check the

Table 3.2 Summary of the statistical distribution models to air pollutants concentration

S. No.	Type of distribution model	Time average	Pollutants	References
1.	Lognormal distribution	1 and 24 h	CO, NO, NO_2, NO_x, SO_2, and HC	Larsen (1971)
		1-h	SO_2	Cats and Holtslag (1980)
		24-h	SO_2	Berger et al. (1982)
		24-h	NO, NO_x and SO_2	Jakemen et al. (1986)
2.	Weibull distribution	8 and 24-h	CO and O_3	Jakemen et al. (1986)
		8-h	CO	Stefano Cernuschi et al. (1998)
3.	Gamma distribution	8 and 24-h	CO, NO_2 and O_3	Jakemen et al. (1986)
		24 h	SO_2	Taylor and Jakemen et al. (1985)
4.	Log logistics distribution	1-h	CO	Gokhale and Khare (2007)

violation of air pollutants concentrations at an air quality control regions (Longhurst et al. 2006).

In general, air pollutant concentrations are autocorrelated and exhibit systematic and random fluctuations due to changes in meteorological conditions and source emissions. In the past, the frequency distributions of air pollutant concentrations were carried out by several researchers. Larsen (1971) carried out a comprehensive analysis of the air quality data, namely CO, NO, NO_2, NO_x, SO_2, and hydrocarbons (HCs) monitored at eight cities during the years 1962–1968. The data analysis of the study concluded that lognormal distribution is accurate in fitting the air quality data. Knox and Pollack (1974), Gifford (1974), Kahn (1973), Mage (1974), Benarie (1974), and Ott and Mage (1976) were reported that lognormal distribution is accurate in fitting the observed ambient air quality data. Bencala and Seinfeld (1976) were investigated the effect of wind speed on frequency distribution using the lognormal, Weibull, gamma, and uniform distributions. The study concluded that the pollutant concentration follows lognormal distribution when the wind speed is also lognormally distributed.

Cats and Holtslag (1980) derived an empirical relation using the hourly SO_2 data sets observations from 1971 to 1974 in the Rijnmond industrial/urban area in the Netherlands. They concluded that frequency distribution of each subset follows lognormal distribution. Holland and Fitz-Simons (1982) developed a computer program (MAXFIT software) for fitting statistical distributions such as Johnson s, normal, three parameters lognormal, gamma, Weibull and four parameter betas to air pollution data using maximum likelihood estimation. Later, the software was evaluated with the hourly ozone data (ppm) collected from July to August in 1979 at the Chequamegon national forest in Wisconsin. The results indicated that the three-parameter lognormal distribution fitted accurately to the observed data set. Georgopoulos and Seinfeld (1982) presented a critical review of statistical distributions of air pollutant concentrations and concluded that the two-parameter lognormal distribution was widely used in urban air pollutant modelling. Berger et al. (1982) analyzed 24 h-average SO_2 and found that the cumulative frequency of the observed data fitted well with lognormal distribution. Mage and Ott (1984) analyzed the effectiveness of parameter estimation using the method of moments, the method of fractiles, maximum likelihood method, and empirical rollback approach. This study found that the maximum likelihood method is best suited for fitting lognormal distribution to air quality data.

Simpson et al. (1983) examined the cumulative frequency distribution of pollutants such as O_3, NO, and TSP concentrations using four different, concentrations recorded between January 1, 1977, and March 31, 1979, at 12 stations in Gent region of Belgium types of data set (i) continuous one-year data, (ii) continuous one week data in a month for the whole year, (iii) random sampling at any time, and (iv) random sampling only on weekdays between 9.00 a.m. and 5.00 p.m. The results showed that option (ii) yields best results indicating a cost-effective way of recording results. Simpson and Jakemen (1984) examined the air pollution data in order to understand the relationship between the parameters of the statistical distributions of air pollution data and wind speeds and the parameters used in the ATDL model. The collected data of TSP for one year from 1978 to 1979 were examined and suggested that the relationship implied by the ATDL model is relevant to the statistics of the data sets. The key parameter in the ATDL model can be determined by using the median values of air pollution levels and wind speed rather than the means. In another study, Simpson et al. (1984) proposed a model to estimate frequency distributions for SO_2 concentrations of an isolated point source. One-year air quality data collected in New South Wales, Australia for 1-h, 3-h, 8-h and 24-h averages of SO_2 were analyzed and found that the models performs were better than other statistical models. The results also indicated that the 24-h time average values were not able to reproduce shorter time averaged data.

Jakemen and Taylor (1985) developed a hybrid air quality model combining reliable features of a deterministic ATDL-based model with statistical distributional approaches to predict the distribution of concentrations of acid gas in an urban air shed for the data collected from 1972 to 1981 and identified as the gamma distribution was the best one. Taylor et al. (1985) used three statistical models to predict the upper percentiles of the distribution of air pollutant concentrations from restricted data sets recorded over yearly time intervals. They used first model for selection of data set using Kolmogorov--Smirnov test, second model for selection of distribution model for selected air quality data set, third model for model identification using best fit method. The results concluded that the first model is a very useful tool in air pollution monitoring. Taylor et al. (1986) explained the procedures by which a distributional model may be identified using the goodness of fit test method. The air pollution data of suspended particulates, NO, nitrogen dioxide (NO_2), nitrogen oxides (NO_x), ozone (O_3), sulfur dioxide (SO_2), and CO were recorded in Melbourne,

Australia, at 19 sites were examined. The results concluded that (a) lognormal distribution is appropriate for particulate data and the majority of the nitric oxide, oxides of nitrogen and sulfur dioxide data sets, (b) gamma distribution is best for both carbon monoxide, nitrogen dioxide, and ozone, and (c) Weibull distribution is appropriate for a significant number of CO and ozone data sets.

Jakemen et al. (1986) examined the parameter estimation problem using Monte Carlo simulation to compare methods for fitting statistical distributions to air quality data and investigates the method for estimating the parameters of the lognormal distribution, two methods for the exponential distribution, three methods for the gamma (ν) distribution, and four methods for the Weibull distribution. The results showed that the method of maximum likelihood provides the best estimators of the parameters of the lognormal, gamma (ν) distribution, and Weibull distribution. Jakemen et al. (1988) proposed hybrid modelling approach to predict the distribution of air pollutant concentrations (upper percentiles). This approach combines the features of both deterministic and statistical distribution models. The results showed that hybrid model has worked well producing estimates of pollutant concentration.

Jakeman et al. (1991) examined a hybrid approach that combines the features of statistical and deterministic model to predict the CO concentrations during winter conditions that can satisfy the requirements for providing air quality advice to the planning of urban areas. Cernuschi et al. (1995) assessed the effect of emissions reduction on air quality, by applying a relationship between the number of yearly exceedances and the mean annual concentration. Cernuschi et al. (1998) conducted an analysis for the CO concentration data collected from 22 monitoring stations for five years from 1990 to 1995 in the Milan urban area. A model was developed using simple rollback relationship in assessment of reduction in CO concentration. The weibull and two-parameter lognormal distributions were fitted to the 8 h average of CO concentration using method of moments. Based on the goodness of fit test analysis (Kolmogorov–Smirnoff test) concluded that Weibull distribution is accurate in fitting the air quality data.

Dixon et al. (2000, 2001) analyzed the relationship between NO_2 and NOx and used measurements of NO_x to estimate the emission controls required to meet the UK hourly standard for NO_2. Singh et al. (2001) introduced a generalized log logistic distribution as a general model fitting into the environmental pollutant concentration data which includes the distributions such as lognormal, Weibull, and gamma. Rumburg et al. (2001) analyzed the probability distribution of particulate matter for the collected air quality data from January 1995--December 1997 in Spokane, Washington. Their study concluded that the $PM_{2.5}$ and PM_8 concentration data were best fit by a three-parameter lognormal distribution and a generalized extreme value distribution. Hurairah et al. (2005) introduced a new extreme value distribution model to identify the problems of air pollution. They used the maximum likelihood method to estimate the parameters of the new extreme value theory for the CO concentration in Malaysia. The results showed that model can achieve a higher accuracy and significant impact of CO and the model can be used as an air quality management tool in by providing the decision makers to determine the required reduction of sources.

Jakeman et al. (2006) explained ten iterative steps to develop an environmental model for air quality management and evaluation of the models. These steps are used as learning process and they explained a need for wider awareness of good model-development practice including reporting of models to users and sceptical review of models by users. Mavroidis et al. (2007) described a statistical modelling approach for the pollutants such as CO collected from 1989 to 1999 and NO_2 collected from 1990 to 2003 in the Athens area for the assessment of the emissions reduction level required to meet the air quality standards. Table 3.2 presents summary of statistical distribution models applied to air pollutants concentration in the past.

3.6 Limitations of Air Quality Models

The deterministic modelling approach is the most rational and conventional approach for the prediction of air pollution concentrations, yet it is not free from limitations. These include (i) insufficient dispersion parameters, (ii) no treatment for plume rise for vehicular emissions, (iii) there is a tendency of over forecasting when the winds are parallel to line source, (iv) not applicable to very low wind speeds (calm conditions), (v) neglect of asymmetry due to the wind direction relative to traffic directions, and (vi) insufficient treatment of dispersion upwind of the road. The Gaussian model also requires a cumbersome numerical integration especially when the wind forms a small angle with the roadways. The Gaussian dispersion equation has a peculiarity at zero wind speeds. Henceforth, all Gaussian-based models do not act properly when wind speeds are less than 1 m/s.

There is presence of limitations in the statistical models which are like lack of physical perception and the long historical data sets is required. Statistical models are usually formulated to resolve the fundamental association between a set of input data and targets. The statistical modelling approach also includes regression modelling which generally underperforms when there is presence of nonlinear systems (Gardner and Dorling 1998). The response of levels of pollutant to the control of emission is not provided by the statistical model which is a limitation of this model. However, the objective of air quality management is fulfilled with the statistical distribution modelling where it helps in development of simple rollback formulae which determined preferable control of emission from sources.

The best practical method for time series analysis is the Box--Jenkins algorithm. The substantial knowledge is required in time series statistics for a time series model as pointed out by Kapoor and Terry (1983). The suitable time series model is identified by employing statistical graphs of the autocorrelation function (ACF) and partial autocorrelation function (PACF) by the analyst because disparity in pollution from vehicles are usually not simple autoregressive (AR) or moving average (MA) models. The individual understanding and familiarity of time series statistics decide the quality of culminated model in the identification stage. The same data used by different investigator delivering the conflicting results which indicate the site-specific nature of the statistical models. Moreover, there is no guaranty as to the reliability of the model once it is extrapolated beyond the range of the input data used to construct it.

The network architecture (i.e., number of hidden layers, number of nodes in hidden layers and their interconnection) is the main limitation in structuring the ANN-based line model. Currently, there is procedure of training of several network architecture in which the best network is chosen. The approximation of any smooth measurable function between inputs and outputs can be done by one hidden layer. The availability of training data, the amount of noise in the data, and the complexity of the input and output mapping are connected to optimal nodes present in the hidden layer which is problem dependent. The performance of multilayer neural network is good enough in case of interpolation, but poor in case of extrapolation. Apart from these, the data set for training, testing, and validation of neural-network-based model is not selected by thumb rule. The limitations of a source-receptor models are mainly due to the complexity in source compositions, micrometeorology, PM measurement, analysis and secondary chemical reactions.

References

Abu-Allaban M, Gertler AW, Lowenthal DH (2002) A preliminary apportionment of the sources of ambient PM10, PM2.5, and VOCs in Cairo. Atmos Environ 36(35):5549–5557

Almeida SM, Pio CA, Freitas MC, Reis MA, Trancoso MA (2005) Source apportionment of fine and coarse particulate matter in a sub-urban area at the Western European Coast. Atmos Environ 39(17):3127–3138

Arya SP (1999) Air pollution meteorology and dispersion. Oxford University Press, UK

Benarie M (1974) The use of the relationship between wind velocity and ambient pollutant concentration distributions for the estimation of average concentrations from gross meteorological data. In: Proceedings of the symposium on statistical aspects of air quality data, vol 5, Oct 1974

Bencala KE, Seinfeld JH (1976) On frequency distributions of air pollutant concentrations Atmos Environ 10(11):941–950 (1967)

Berger A, Melice JL, Demuth CL (1982) Statistical distributions of daily and high atmospheric SO_2-concentrations. Atmos Environ (1967) 16(12):2863–2877

Boubel RW, Fox DL, Turner DB, Stern AC (1994) Fundamentals of air pollution, 3rd edn. Academic Press, New York

Bullock KR, Duvall RM, Norris GA, McDow SR, Hays MD (2008) Evaluation of the CMB and PMF models using organic molecular markers in fine particulate matter collected during the Pittsburgh air quality study. Atmos Environ 42(29):6897–6904

Callén MS, De La Cruz MT, López JM, Navarro MV, Mastral AM (2009) Comparison of receptor models for source apportionment of the PM10 in Zaragoza (Spain). Chemosphere 76(8):1120–1129

Cats GJ, Holtslag AAM (1980) Prediction of air pollution frequency distribution—Part I. The lognormal model. Atmos Environ 14(2):255–258 (1967)

Cernuschi S, Giugliano M, Cemin A, Giovannimi (1995) Modal analysis of vehicle emission factors. Sci Total Environ 169:175–183

Cernuschi S, Giugliano M, Lonati G, Marzolo F (1998) Development and application of statistical models for CO concentration and duration events in the Milan urban area. Sci Total Environ 220(2–3):147–156

Cernuschi S, Giugliano M, Ozgen S, Consonni S (1998) Number concentration and chemical composition of ultrafine and nanoparticles from WTE (waste to energy) plants. Sci Total Environ 420:319–326

Chock DP, Terrell TR, Levitt SB (1975) Time-series analysis of Riverside, California air quality data. Atmos Environ (1967) 9(11):978–989

Chow JC, Watson JG, Kuhns H, Etyemezian V, Lowenthal DH, Crow D, ... & Green MC (2004) Source profiles for industrial, mobile, and area sources in the big bend regional aerosol visibility and observational study. Chemosphere 54(2):185–208

Chow JC, Watson JG, Pritchett LC, Pierson WR, Frazier CA, Purcell RG (1993) The DRI thermal/optical reflectance carbon analysis system: description, evaluation and applications in US air quality studies. Atmos Environ Part A General Top 27(8):1185–1201

Chueinta W, Hopke PK, Paatero P (2000) Investigation of sources of atmospheric aerosol at urban and suburban residential areas in Thailand by positive matrix factorization. Atmos Environ 34(20):3319–3329

Coulter CT (2004) EPA-CMB8.2 users manual. Report No. EPA-452/R-04e011. U.S. Environmental Protection Agency, Research Triangle Park, N.C.

Dixon J, Middleton DR, Derwent RG (2000) Using measurements of nitrogen oxides to estimate the emission controls required to meet the UK nitrogen dioxide standard. Environ Monit Assess 65(1–2):3–11

Dixon J, Middleton DR, Derwent RG (2001) Sensitivity of nitrogen dioxide concentrations to oxides of nitrogen controls in the United Kingdom. Atmos Environ 35(21):3715–3728

EPA US (2004) Air quality criteria for particulate matter. US Environmental Protection Agency, Research Triangle Park

Gardner MW, Dorling SR (1998) Artificial neural networks (the multilayer perceptron)—a review of applications in the atmospheric sciences. Atmos Environ 32(14–15):2627–2636

Georgopoulos PG, Seinfeld JH (1982) Statistical distributions of air pollutant concentrations. Environ Sci Technol 16(7):401A–416A

Gifford FA (1974) The form of the frequency distribution of air pollution concentrations. Environmental Research Laboratories, p 171

Gildemeister AE, Hopke PK, Kim E (2007) Sources of fine urban particulate matter in Detroit, MI. Chemosphere 69(7):1064–1074

Gokhale S, Khare M, Pavageau M (2003) Modelling distributions of air pollutant concentrations from vehicular exhausts in urban environment: a hybrid approach, PHYSMOD2003. In: International workshop on physical modelling of flow and dispersion phenomena, Prato, Italy, pp 3–5, Sept 2003

Gokhale S, Khare M (2007) Statistical behavior of carbon monoxide from vehicular exhausts in urban environments. Environ Model Softw 22(4):526–535

Gupta S, Khare M, Goyal R (2007) Sick building syndrome—a case study in a multistory centrally air-conditioned building in the Delhi City. Build Environ 42(8):2797–2809

Hickman AJ, Colwill DM (1982) The estimation of air pollution concentrations from road traffic (No. LR 1052 Monograph)

Holland DM, Fitz-Simons T (1982) Fitting statistical distributions to air quality data by the maximum likelihood method. Atmos Environ (1967):16(5):1071–1076

Hopke PK (ed) (1991) Receptor modeling for air quality management, vol 7. Elsevier

Hurairah A, Akma Ibrahim N, Bin Daud I, Haron K (2005) An application of a new extreme value distribution to air pollution data. Manag Environ Qual Int J 16(1):17–25

Jakeman AJ, Bai J, Miles GH (1991) Prediction of seasonal extremes of one-hour average urban CO concentrations. Atmos Environ Part B Urban Atmos 25(2):219–229

Jakeman AJ, Letcher RA, Norton JP (2006) Ten iterative steps in development and evaluation of environmental models. Environ Model Softw 21(5):602–614

Jakeman AJ, Simpson RW, Taylor JA (1988) Modeling distributions of air pollutant concentrations—III. The hybrid deterministic-statistical distribution approach. Atmos Environ (1967) 22(1):163–174

Jakeman AJ, Taylor JA, Simpson RW (1986) Modeling distributions of air pollutant concentrations—II. Estimation of one and two parameter statistical distributions. Atmos Environ (1967), 20(12):2435–2447

Kahn HD (1973) Note on the distribution of air pollutants. J Air Pollut Control Assoc 23(11):973

Kapoor SG, Terry WR (1983) A comparison of two automatic systems for building vector time-series models in air pollution research. Time-Series Anal Theory Pract 7:200–211

Kim BM, Henry RC (2000) Extension of self-modeling curve resolution to mixtures of more than three components: Part 3. Atmospheric aerosol data simulation studies. Chemom Intell Lab Syst 52(2):145–154

Kim E, Hopke PK, Edgerton ES (2004) Improving source identification of Atlanta aerosol using temperature resolved carbon fractions in positive matrix factorization. Atmos Environ 38(20):3349–3362

Kim E, Larson TV, Hopke PK, Slaughter C, Sheppard LE, Claiborn C (2003) Source identification of PM2.5 in an arid Northwest US City by positive matrix factorization. Atmos Res 66(4):291–305

Knox JB, Pollack RI (1974) An investigation of the frequency distribution of surface air pollutant concentrations. In: Proceedings of the symposium on Statistical aspects of air quality data. EPA-650/4-74-038, U.S. Environmental Protection Agency, Research Triangle Park, NC, U.S.A, pp. 9-1–9-17

Larsen RI (1971) A mathematical model for relating air quality measurements to air quality standards, vol 89. Environmental Protection Agency, Office of Air Programs

Lee E, Chan CK, Paatero P (1999) Application of positive matrix factorization in source apportionment of particulate pollutants in Hong Kong. Atmos Environ 33(19):3201–3212

Lee SL, Wong WHS, Lau YL (2006) Association between air pollution and asthma admission among children in Hong Kong. Clin Exp Allergy 36(9):1138–1146

Longhurst JWS, Beattie CI, Chatterton TJ, Hayes ET, Leksmono NS, Woodfield NK (2006) Local air quality management as a risk management process: Assessing, managing and remediating the risk of exceeding an air quality objective in Great Britain. Environ Int 32(8):934–947

Lyons TL, Scott WD (1990) Principles of air pollution meteorology. CRC Press

Mage DT (1974) On the lognormal distribution of air pollutants. In: Proceedings of the Fourth Meeting of the Expert Panel on Air Pollution Modeling, Paper No. 35, NATOJCCMS, Roskilde, Denmark, June 1974

Mage DT, Ott WR (1984) An evaluation of the methods of fractiles, moments and maximum likelihood for estimating parameters when sampling air quality data from a stationary lognormal distribution. Atmos Environ (1967) 18(1):163–171

Mavroidis I, Gavriil I, Chaloulakou A (2007) Statistical Modelling of CO and NO_2 Concentrations in the Athens area-evaluation of emission abatement policies (7 pp). Environ Sci Pollut Res Int 14(2):130–136

Norris G, Vedantham R, Wade K, Brown S, Prouty J, Foley C, Martin L (2008) EPA positive matrix factorization (PMF) v 3.0- fundamentals and user's guide. Prepared by U.S.EPA, Office of Research and Development, Washington, DC

Ott WR, Mage DT (1976) A general purpose univariate probability model for environmental data analysis. Comput Oper Res 3(2–3):209–216

Paatero P (1997) Least squares formulation of robust non-negative factor analysis. Chemometr Intell Lab Syst 37(1):23–35

Polissar AV, Hopke PK, Poirot RL (2001) Atmospheric aerosol over Vermont: chemical composition and sources. Environ Sci Technol 35(23):4604–4621

Qin Y, Oduyemi K, Chan LY (2002) Comparative testing of PMF and CFA models. Chemometr Intell Lab Syst 61(1–2):75–87

Ramadan Z, Song XH, Hopke PK (2000) Identification of sources of Phoenix aerosol by positive matrix factorization. J Air Waste Manag Assoc 50(8):1308–1320

Rumburg B, Alldredge R, Claiborn C (2001) Statistical distributions of particulate matter and the error associated with sampling frequency. Atmos Environ 35(16):2907–2920

Samara C, Kouimtzis T, Tsitouridou R, Kanias G, Simeonov V (2003) Chemical mass balance source apportionment of PM10 in an industrialized urban area of Northern Greece. Atmos Environ 37(1):41–54

Simpson RW, Jakeman AJ (1984) A model for estimating the effects of fluctuations in long term meteorology on observed maximum acid gas levels. Atmos Environ (1967) 18(8):1633–1640

Simpson RW, Butt J, Jakeman AJ (1984) An averaging time model of SO_2 frequency distributions from a single point source. Atmos Environ (1967) 18(6):1115–1123

Simpson RW, Daly NJ, Jakeman AJ (1983) The prediction of maximum air pollution concentrations for TSP and CO using Larsen's model and the ATDL model. Atmos Environ (1967) 17(12):2497–2503

Singh KP, Bartolucci AA, Bae S (2001) Mathematical modeling of environmental data. Math Comput Model 33(6–7):793–800

Song XH, Polissar AV, Hopke PK (2001) Sources of fine particle composition in the northeastern US. Atmos Environ 35(31):5277–5286

Taylor JA, Jakeman AJ (1985) Identification of a distributional model. Commun Stat B 14:497–508

Taylor JA, Jakeman AJ, Simpson RW (1986) Modeling distributions of air pollutant concentrations I. Identification of statistical models. Atmos Environ (1967) 20(9):1781–1789

Taylor JA, Simpson RW, Jakeman AJ (1985) A hybrid model for predicting the distribution of pollutants dispersed from line sources. Sci Total Environ 46(1–4):191–213

U.S. EPA (2009) Particulate matter—speciation data system, speciate 4.2 Office of Research and Development, Research Triangle Park, NC

Ward TJ, Smith GC (2005) The 2000/2001 Missoula Valley $PM_{2.5}$ chemical mass balance study, including the 2000 wildfire season—seasonal source apportionment. Atmos Environ 39(4):709–717

Watson JG, Chow JC, Houck JE (2001) $PM_{2.5}$ chemical source profiles for vehicle exhaust, vegetative burning, geological material, and coal burning in Northwestern Colorado during 1995. Chemosphere 43(8):1141–1151

Watson JG et al (2002) Receptor modeling application framework for particle source apportionment. Chemosphere 49(9):1093–1136

Yatkin S, Bayram A (2008) Source apportionment of PM_{10} and $PM_{2.5}$ using positive matrix factorization and chemical mass balance in Izmir, Turkey. Sci Total Environ 390(1):109–123

Yin J, Harrison RM, Chen Q, Rutter A, Schauer JJ (2010) Source apportionment of fine particles at urban background and rural sites in the UK atmosphere. Atmos Environ 44(6):841–851

Yuan ZB, Yu JZ, Lau AKH, Louie PKK, Fung JCH (2006) Application of positive matrix factorization in estimating aerosol secondary organic carbon in Hong Kong and its relationship with secondary sulfate. Atmos Chem Phys 6(1):25–34

Chapter 4
Air Quality Management and Control

S. M. Shiva Nagendra, Mukesh Khare, Uwe Schlink, and M. Diya

4.1 Principles of Air Quality Management

Air pollution is one of the key issues in developing countries as well as in developed countries. Urban air pollution tends to cause negative impact on human health, environment, economy, and social life. According to world health organization (WHO), 2 million deaths and different types respiratory issues per year are estimated because of air pollution all over the world. Urban air pollution is the after effect of fast population growth, economic growth, and rapid migration of rural populations into towns and cities. Industries, transportation, open burning, etc., are the major anthropogenic sources identified, out of which transportation is the source which most contributed to air pollution in many megacities worldwide.Seventy–eighty percentage of urban air pollution in developed countries from transportation sector is due to large number of old vehicles, poor maintenance, alteration of fuels, and poor enforcement systems. The criteria pollutants responsible for urban air pollution are particulate matter (PM_{10} and $PM_{2.5}$), oxides of nitrogen (NO_x), and oxides of sulfur (SO_x), carbon monoxide, and volatile organic matter (VOCs) (Gulia et al. 2015).

S. M. Shiva Nagendra (✉) · M. Diya
Department of Civil Engineering, Indian Institute of Technology Madras, Chennai, Tamil Nadu 600036, India
e-mail: snagendra@iitm.ac.in

M. Khare
Department of Civil Engineering, The Indian Institute of Technology, Hauz Khas, New Delhi 110016, India
e-mail: mukeshk@civil.iitd.ernet.in

U. Schlink
Department of Urban and Environmental Sociology, Helmholtz Centre for Environmental Research UFZ, 04318 Leipzig, Germany
e-mail: uwe.schlink@ufz.de

S. M. Shiva Nagendra et al. (eds.), *Urban Air Quality Monitoring, Modelling and Human Exposure Assessment*, Springer Transactions in Civil and Environmental Engineering,
https://doi.org/10.1007/978-981-15-5511-4_4

4.2 Air Quality Management Framework

There have been different management practices adopted from 1993 to present in many countries for effective management of their deteriorating air environment. The effective formulation of management frameworks primarily depends on scale of air pollution either at spatial scale or temporal scale. Urban Air Quality Management Plan (UAQMP) in developed countries and developing countries are not the same. In developing countries, it mainly depends on the ruling government, the existing laws, policies, standards, the public commitment, etc. The main components of UAQMP are the air quality objectives, air quality monitoring, emission inventory and source apportionment, modelling, control strategies and public participation (Gulia et al. 2018).

4.3 Air Quality Standards and Legislations

Most of the countries follow their own air quality standards. WHO air quality standard is the most stringent standard when compared to all the standards available. In developing countries such as the standards are always higher than the standards of developed countries such as USA, UK, EU countries etc. The present air quality information and the existing policies and its effectiveness of application are obtained from air quality monitoring. The monitoring network design is country specific, which depends on the common guidelines and standards followed by those countries. In developing countries, the primary objective of monitoring network is to ensure the compliance of the policies. In India, there are 1000 continuous ambient air quality monitoring stations employed across the country for the effective evaluation of the air quality similarly some other countries such as china-2000, Hong kong-14 (online), and South Africa-94 monitoring stations are being operated for this purpose.

The effective UAQM strategies are planned and implemented based on the pollutant contribution from different sources in the monitoring site. The net release of pollutants by each source, the rate of emission, proportion of each pollutants, the chemical characteristics, etc., are very important in planning and implementing appropriate control strategies at source, therefore, source apportionment is considered as one of the significant section of UAQMP. The emission inventory helps to estimate the total contributions of emissions to environment to from different sources. In mobile sources, the emissions are calculated by multiplying the emission factor and vehicular volume. The emission factors depend on the types of vehicles, model, year of make, type of engine, etc. There are different mathematical models such as MOVES, MOBILE-6, and NMIM used to estimate the emission from mobile sources. In India, the emission factors are developed by Indian Institute of Petroleum (IIP), Automotive Research Association of India (ARAI), and Central

Pollution Control board (CPCB) based on standard driving cycle for all types of vehicles. Emission inventory is a key step in UAQMP for developing control strategies for criteria pollutants.

The mathematical modelling plays vital role in efficient planning of UAQMP. Models provide the spatial and temporal variations of individual pollutants incorporating different environmental conditions. There are different mathematical models developed in different countries depending upon type of sources, geographical conditions, and other influencing parameters. AERMOD, CALPUFF, CALINE-4, etc., are most popular models recommended by USEPA. Past studies showed visible correlation between air pollution and hospital admissions. Exposure assessment studies showed that short-term and long-term exposure to polluted environment can cause various respiratory diseases and chronic respiratory diseases which may even lead to death. The major objective of UAQMP is to make the effect of air pollution on human health and environment minimal.

For a well-organized UAQMP, appropriate control strategies should be implemented and its effectiveness should be analyzed in terms of health impact and socio-economic impacts. The effectiveness can be analyzed by comparing before and after scenarios involving the implementation of specific strategies in a certain location. The health impacts and socio-economic behaviour of people affected by the UAQMP measures after their implementation are compared to the same group of people before strategies were in place. These comparisons are possible through the application of surveys. The various UAQMPs implemented in some developing countries and developed countries and its effectiveness are listed below.

4.4 Air Quality Management Practices in Developed Countries

Alternative fuel option in vehicles, improvement in fuel quality, fitting of catalytic converters in vehicles, inspection and maintenance program, use of low emission or zero emission vehicles, banned street parking, support for cycling, congestion charging, specific bus corridors, pedestrian corridors, low emission zones, banned on entry of old vehicle in air pollution control areas, banned on entry of heavy duty vehicle in air pollution control areas, banned on idling of vehicles at signals, split cycle offset optimization technique (SCOOT) system, intelligent transport system, banned on smoky vehicles, onboard diagnostic system in vehicle, subsidy on registration of environmental friendly vehicles, mass rapid system, encouraging carpools, maintenances of road infrastructure, stringent emission norms, and reduction in diesel vehicles.

4.5 Air Quality Management Practices in Developing Countries

Except some of the strategies which are not implemented completely or some strategies implemented partially, many of the strategies adopted in developing countries are as same as developed countries, which are listed below.

Alternative fuel option in vehicles, improvement in fuel quality, fitting of catalytic converters in vehicles, banned street parking, support for cycling, specific bus corridors, pedestrian corridors, banned on entry of old vehicle in air pollution control areas, banned on entry of heavy duty vehicle in air pollution control areas, banned on idling of vehicles at signals, banned on smoky vehicles, mass rapid system, maintenances of road infrastructure, and stringent emission norms.

4.5.1 Air Quality Management in India

In India, the air quality management practices started from the year 1972. The first legislation in this regard was, Air (Prevention and Control of pollution) 1981 which empowered the Central Pollution Control Board (CPCB) and State Pollution Control Boards (SPCBs) to control and prevent the air pollution. In the year 1982, the first National Ambient Air quality Standard (NAAQS) came into picture with standard fixed for three criteria pollutants SPM, SO_2 and NO_2 in residential areas, industrial areas, and sensitive areas. The NAAQS were revised eventually in 1994 and 2009; presently, it accounts for 12 pollutants. In the year 1984, National Ambient Air quality Monitoring Program (NAAMP) was started with seven stations and later increased to 591 stations across the country along with individual monitoring stations of SPCBs or other environmental-related organizations.

In 1990, more focus has given to the emissions from motor vehicle exhaust and hence proposed mass emission norms at manufacturing stage of new vehicles. From the Government of India has taken various steps such as adopting new policies, improving the engine quality by introducing Barat stage (BS) norms, improving fuel quality, restricting vehicles older than 10–15 years in cities, restricting entry of heavy duty vehicles on city roads during day time, declaration of low emission zones, etc., to curtail the exhaust emissions. GOI decided to switch to BS VI norm from BS IV directly by 2020 with an amendment in Central motor Vehicles Rule 1989.

Delhi is considered as one of the most polluted cities in the world. Due to high level of pollution, Delhi Government has taken many emergency measures like ban of registration of diesel vehicles, restriction of private cars in the city based on their registration number (odd even policy), etc.

4.5.2 Air Quality Management in China

China ranks 12th in the list of most polluted (air pollution) countries of the world. The year 2014 was a very crucial year in the history of china in terms of air pollution control and management. Ten leading technical institutes in china collectively found the Clean Air Alliance of China (CAAC) in the year 2013. The government of china introduced the Air Pollution Prevention and Control Action plan (2013–2017) which is also called "Ten measures of Air" also in the same year. According to this plan, the country has divided into different provinces/cities and well-built and strict actions have been adopted to reduce the emissions (Wang and Hao 2012). According to CAAC annual report, one year after launch of this plan, the overall air quality of the china was improved tremendously. The average reduction of $PM_{2.5}$ over the country was 12%. The AQM action plans designed during this period are given below.

1. China issued 15 standards on air pollution consisting sampling and monitoring standards for volatile organic chemicals/semi volatile organic chemicals for industries, exhaust gas monitoring standards for stationary pollution sources.
2. Introduced/upgraded air quality monitoring stations measuring six major pollutants (PM_{10}, $PM_{2.5}$, SO_2, NO_2, O_3 and CO) to a total number of 1436 across the country.
3. The continuous air quality data and air quality index values are made available in government website to create public awareness.
4. In seven critically polluted cities, the pollution discharge fee increased more than national standard fee.
5. The coal consumption is reduced up to 3% by replacing the coal with other sources of energy.
6. Along with the pollutants, PM_{10}, $PM_{2.5}$, SO_2, NO_2, O_3, and CO emission of Hg are also reduced considerably.
7. Banned yellow-labelled vehicles and outdated vehicles in the country. It is reported that out of total air pollutants emission, 50% contributions are from yellow-labelled vehicles. In 2014, the state council of People's Republic of China (PRC) assigned the mission of eviction of 6 million yellow-labelled vehicles from Chinese roads, and the task was completed with fulfilment ratio of about 120%.

4.5.3 Air Quality Management in UK

In UK and other EU countries, UAQMP has been introduced in 1980 with directive 80/779/EEC, which set limit and guide values for SO_2 and suspended particulates (SP). The European Commission (EC) oversight the implementation of EU legislation, including enforcement action against non-complied member state under the EC treaty. Longhurst et al. (1996) have introduced the concept of LAQMP (Fig. 4.1) to reduce the air pollution levels below permissible limits within specified air quality management area (AQMA). They have suggested that local government should be

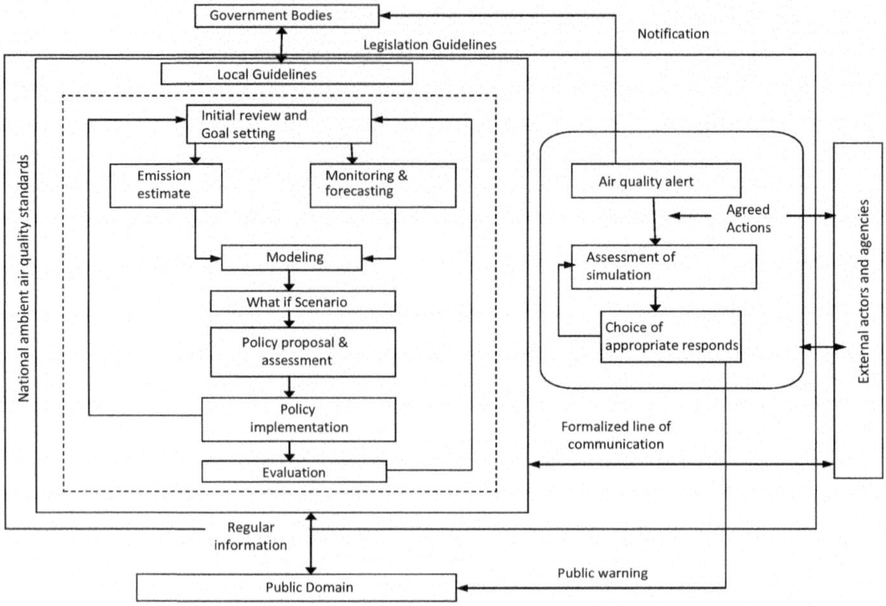

Fig. 4.1 A framework of LAQMP

actively involved in implementation of management plans. Further, local air quality management (LAQM) technical guidelines have been issued in 1998 to support local authorities in carrying out their duties under the Environment Act 1995 and the Environment (Northern Ireland) Order 2002. These guidelines later updated in year 2000, 2003 and 2009. Beattie et al. (2001) have suggested that local actions are required to reduce air pollution at urban *hot-spots*. Further, Woodfield et al. (2003) have found that identification of an AQMA is a very complex approach and does not uniform throughout the UK. It requires an advanced air quality modelling and monitoring tools. Later, Woodfield et al. (2006) have appraised the regional grouping between local authorities to manage urban air quality in London, the West Midlands, and former-Avon area of Southwest England. They have observed variations in the methods and tools used in developing LAQMP.

4.6 Challenges in Air Quality Management

The national ambient air quality standards (NAAQS) could be achieved by regulating the emission of pollutants through air quality management. The pollutant emissions estimates, ambient air pollution quantification, atmospheric dispersion model, and characterization of air pollutant concentrations are also available besides the air quality goals for the management of air quality. If the concept of risk assessment is also included in the approach of air quality management where alternative options, social costs and benefits are also valuated, then also the same above information is requisite.

The pollutant emitted from vehicles causes detrimental environmental and health effects which are growing concern from last five decades. The intense pursuit has been taken by public, administrators, and many lawmakers for improving the air quality. Several laws were also adopted for protection of the environment from vehicular emissions by government of India. Some of adopted laws are such as Air (Prevention and Control of Pollution) Act, 1981, Environment (Protection) Act, 1988, the Motor Vehicles Act, 1988, and Central Motor Vehicle Rules, 1989. The air pollution is also controlled by taking necessary preventive actions by implicated authorities as briefed by Supreme Court due to various Public Interest Litigations. However, the air quality is regularly deteriorated due to increasing urbanization, supply of poor quality fuel, deficit of proper public transport system, and traffic congestion at busy straight roads in urban centers.

Air quality management has two important outlooks—'decisive' and 'discreet.' The pollution is reduced at all extent of pollution problem from local to global scale in the former case. The long-term plan of 5, 10, or 15 years were made to accomplish the objectives for improving the air quality in case of decisive approach. In discreet approach, the inevitable hazards were prevented by making plan on a short notice for control of short-term pollution episodes. These short-term episodes were mostly of short duration of 34 h to 5 or 6 days.

The air quality can be improved by taking different approaches into considerations which can be used alone or with other approach for the national or regional requirements. There are four parameters through which air quality could be contemplated. These are geography and topography, meteorology and climatology, urban form, and emission source density and intensity. Emissions are the one of the important among all parameters which can be managed by human conciliation. There are various direct and indirect actions which affect the density of emission sources and also intensity of emissions. Thus, for achieving the required improvement in air quality, the density and intensity of emission sources should be addressed. Such a response will require the management of urban environment and the processes affecting its development. The air quality management could be done by taking into account the four important criteria which are emissions, impacts, cost-benefit analysis, and controls. These are established and expressed within a scientific evaluatory context. Thus, air quality management is involved with the emission and impact of air pollutants released from anthropogenic sources.

The air quality management in general may be defined as 'the application of a systematic approach to the control of air quality issues.' The air quality could be considered in a holistic way by incorporating the aspects of integration, cooperation, and communication as a system designed. Griffin expresses air quality management as comprising of five continuous steps: The definition, planning, control, implementation, and evaluation are five important steps for the air quality management as expressed by Griffin. This can be translated into air quality management terms as shown in Fig. 4.1. Figure 4.2 shows the important components involved in the air

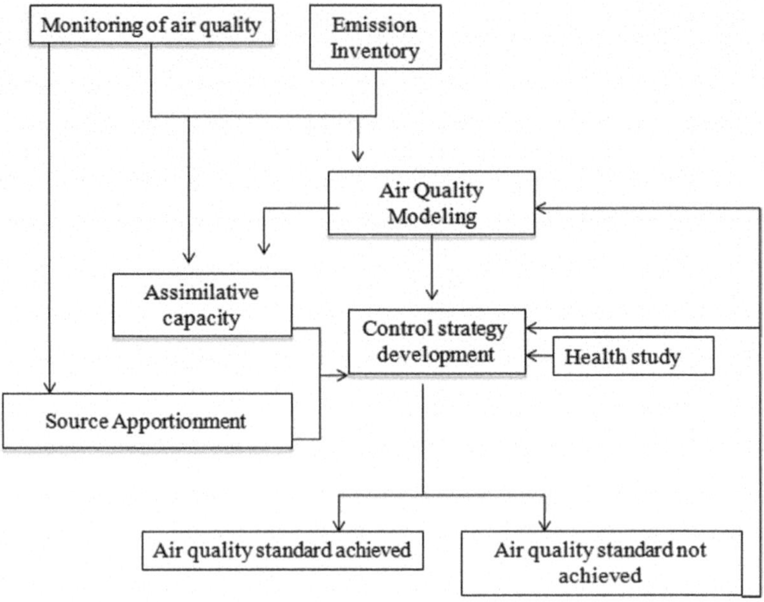

Fig. 4.2 Components of air quality management

quality management. The local air quality can be determined by establishing alliance between agencies implicated in the air quality management plan. The problems of air quality could be controlled by applying the systematic approach in the air quality management plan.

Clean air is an implicit requirement for health and well-being of every human being. Since 1960s, increase in industrialization resulted an alarming air quality-related problem, there after attracting the interest of scientific community to protect the air quality. The United Nations Conference on Environmental and Development (UNCED) held in 1992 in Brazil emphasized the need of air pollution control by adopting the Agenda prescribed in Framework Climate Convention. Further, United Nations Conference on Human Settlements underlined the sustainable development and sustainable human settlement for the protection of environment. Clean air is one of the key considerations to put forward in sustainable human settlement. To achieve the clean air quality, an effective air quality management is required which varies significantly with the type, magnitude, and intensity of the air pollution. Air quality management for an industrial sector can be achieved at source, pathway, and receptor levels. Previous studies reported the prevention of air pollution at source is the most effective way to control industrial emissions. Source control in many cases involves reduction in raw material losses thereby is considered as most cost-effective management method. Other source control methods include increasing the stack heights, change of fuels, installation of control devices, cleaner production, etc. (Goyal et al. 2006). Pollution control technology, conceptual framework, public support, governmental motivation, legislation, and regulatory infrastructure are some of the key elements required for air quality management (Yhdego 1995). Table 4.1 describes the UAQM definitions/concepts (Laxen et al. 1993; Longhurst et al. 1996; Fedra and Haurie 1999; Beattie et al. 2001; Karatzas 2002; Gokhale and Khare 2007; Vlachokostas et al. 2009, Sivertsen and Bartonova 2012, Gulia et al. 2015).

Table 4.1 The UAQM definition/concept

References	Air quality management definition/concept
Laxen et al. (1993)	AQMP is a program to ensure the control of emissions to protect public health and welfare
Longhurst et al. (1996)	Concept of the local air quality management plan (LAQMP) emphasizes the distribution of power at community level and strong communication and cooperation between *actors* of air quality management
Steinar et al. (1997)	AQMP is a system for design and implementation of monitoring, management and policies within an urban area
Fedra and Haurie (1999)	A system that can establish a robust and integrated environmental management life cycle for urban areas of interest
Beattie et al. (2001)	The UAQM emphasizes on local action to deal urban hot spots of air pollution (*episode*)
Karatzas (2002)	A software tools that combine air quality models with other software modules like geographical information systems, databases, expert systems and statistical analysis tools. It helps to interpret the complex interactions between various atmospheric, emission, land use and topographic parameters
Woodfield et al. (2003)	The UAQM identifies specific pollution hot spots based on advanced and complex air-quality dispersion modeling and monitoring techniques
Gokhale and Khare (2007)	Episodic urban air quality management plan (e-UAQMP)—it uses hybrid model to forecast extreme levels of pollutants during peak traffic hours
Vlachokostas et al. (2009)	AQMP is an integrated assessment methodological scheme for the evaluation of air pollution control measures that are put forward in order to reduce air pollution levels in urban areas. This approach brings together air quality modeling and mathematical programming techniques, provides a decision support system for the determination of optimal bundles of air pollution control options according to the particular features and needs of the areas examined
Sivertsen and Bartonova (2012)	The UAQM is a set of action that helps in attaining air quality goals in a specified geographical area. It requires actions by government, business, industry, NGO's and the population

References

Beattie CI, Longhurst JWS, Woodfield NK (2001) Air quality management: evolution of policy and practice in the UK as exemplified by the experience of English local government. Atmos Environ 35(8):1479–1490

Fedra K, Haurie A (1999) A decision support system for air quality management combining GIS and optimisation techniques. Int J Environ Pollut 12(2–3):125–146

Gokhale S, Khare M (2007) A theoretical framework for the episodic-urban air quality management plan (e-UAQMP). Atmos Environ 41(36):7887–7894

Goyal SK, Ghatge SV, Nema PSMT, Tamhane SM (2006) Understanding urban vehicular pollution problem vis-a-vis ambient air quality–case study of a megacity (Delhi, India). Environ Monit Assess 119(1–3):557–569

Gulia S, Nagendra SS, Barnes J, Khare M (2018) Urban local air quality management framework for non-attainment areas in Indian cities. Sci Total Environ 619:1308–1318

Gulia S, Nagendra SS, Khare M, Khanna I (2015) Urban air quality management-a review. Atmospheric Poll Res 6(2):286–304

Karatzas KD (2002) Theoretical investigation of urban air quality management systems performance towards simplified strategic environmental planning. Water Air Soil Pollut Focus 2(5–6):669–676

Laxen D (1993) An introduction to local air quality management. A supplement to Clean Air 23:12

Longhurst JWS, Lindley SJ, Watson AFR, Conlan DE (1996) The introduction of local air quality management in the United Kingdom: a review and theoretical framework. Atmos Environ 30(23):3975–3985

Sivertsen B, Bartonova A (2012) Air quality management planning (AQMP). Chem Ind Chem Eng Quart/CICEQ 18(4–2):667–674

Steinar L, Grqnskei KE, Hannegraaf MC, Jansen H, Kuik OJ, Oosterhuis FH, Olsthoorn XA, Jitendra JS, Tanvi N, Carter JB (1997) Urban air quality management strategy in Asia – a Guidebook. The World Bank Press. https://doi.org/10.1596/0-8213-4032-8

Vlachokostas Ch, Achillas Ch, Moussiopoulos N, Hourdakis E, Tsilingiridis G, Ntziachristos L, Sidiropoulos C (2009) Decision support system for the evaluation of urban air pollution control options: application for particulate pollution in Thessaloniki. Greece Sci Total Environ 407:5937–5938

Wang S, Hao J (2012) Air quality management in China: Issues, challenges, and options. J Environ Sci 24(1):2–13

Woodfield NK, Longhurst JWS, Beattie CI, Laxen DPH (2003) Regional variations in the implementation of the local air quality management process within Great Britain. J Environ Planning Manage 46(1):49–64

Woodfield NK, Longhurst JWS, Beattie CI, Chatterton T, Laxen DPH (2006) Regional collaborative urban air quality management: case studies across Great Britain. Environ Model Softw 21(4):595–599

Yhdego M (1995) Environmental pollution management for Tanzania: towards pollution prevention. J Clean Prod 3(3):143–151

Chapter 5
Indoor Air Quality

S. M. Shiva Nagendra and V. S. Chithra

5.1 Basics of Indoor Air Pollution

Indoor air quality (IAQ) comes under consideration in recent times as we are spenting large amount of time indoors in modern times. Over the last four decades, most of the air quality management programmes have been targeted only to regulate outdoor air pollution and the indoor air pollution (IAP) was typically under-reported and less regulated than its counterpart (Ott and Roberts 1998). Most people are aware that outdoor air pollution can impact their health, but indoor air pollution can also have significant, harmful effects. The U.S. Environmental Protection Agency (U.S. EPA) observed the exposure of human to air pollutants which indicate that indoor pollutant concentrations may be two to five times and sometimes 100 times higher than the outdoor levels. A comparative risk studies conducted by the EPA have ranked IAP among the top five environmental risks to public health (U.S. EPA 2013). The World Health Organization (WHO) has reported IAP as the 8th most important risk factor and responsible for 2.7% of the global burden of ill health measured as disability adjusted life years (DALYs). Every year, IAP is responsible for the death of 1.6 million people in the world. The indoor smoke is responsible for 3.7% of the overall disease burden in developing countries, making the IAP as the most lethal killer after malnutrition, unsafe sex and lack of safe water and sanitation (WHO 2002). The Global Burden Disease study 2010, published recently, found that household air pollution (HAP) from solid fuels accounted for 2.6–4.4 million deaths and 3.4–5.3%

S. M. Shiva Nagendra (✉)
Department of Civil Engineering, Indian Institute of Technology Madras, Chennai, Tamil Nadu 600036, India
e-mail: snagendra@iitm.ac.in

V. S. Chithra
Department of Civil Engineering, Mar Baselios Christian College of Engineering and Technology, Kuttikkanam, Kerala 685531, India
e-mail: drchithravs@mbcpeermade.com

S. M. Shiva Nagendra et al. (eds.), *Urban Air Quality Monitoring, Modelling and Human Exposure Assessment*, Springer Transactions in Civil and Environmental Engineering,
https://doi.org/10.1007/978-981-15-5511-4_5

of global DALYs (Lim 2013). In India, approximately 1.04 million premature deaths and 31.4 million DALYs to be attributable to HAP, and account for 6% of the total national burden of disease (IHME 2013).

5.2 Sources of Indoor Air Pollutants

The origins of poor IAQ issues are well known. In the early 1970s, public has recognized the importance of energy saving because of severe energy crisis, across the world. As a result, the insulation of building is started and more airtight building is built for saving the energy. The energy consumption is also done by reducing the fresh air flow in air conditioning systems (Jones 1999; Spengler and Chen 2000). Meanwhile, economic growth and urban development changed people's lifestyle and usage of fabricated materials and chemical products (e.g. furnishings, carpeting and decorating materials) has been increased in indoor work environment. Technological changes have made copiers, printers, computers, etc. becomes common in office buildings. The bacteria, fungi, dust and volatile organic compounds (VOCs) concentration increased indoor due to presence of low rates of ventilation and large amount of synthetic chemicals.

In developed countries, the major contributor of poor IAQ is low ventilation rate, emissions from building materials and furnishings, and imprudent maintenance and operation of building ventilation systems. In approximately 500 IAQ investigations operated under the control of the National Institute for Occupational Safety and Health (NIOSH) found that not enough ventilation and the release of contaminants from indoor and outdoor sources are the primary reasons for poor IAQ. Inadequate ventilation, contributes 53% of IAQ problems. Indoor and outdoor sources contribute 16% and 10% of IAQ problems, respectively. While, 13% are due to unknown sources, 5% are from microbial contamination and 3% are due to contamination from building fabric (NIOSH 1989). In the developing world, the types, sources, various indoor air pollutants concentrations and their exposure profiles are significantly different from the developed world. The higher biomass fuels are used for cooking in developing countries which result in higher concentrations of indoor air pollutants and more people exposure to air pollutants (Smith 2000; Colbeck et al. 2010; Lim 2013). Exposures to indoor air pollutants and their sources varied over a wide range in different regions of the world and they are linked to the socioeconomic developments.

5.3 Indoor Air Quality and Human Comfort

The nature of air that circulated all over the area where we work and live is referred to as IAQ. According to American Society of Heating, Refrigeration, and Air-Conditioning Engineers (ASHRAE), an acceptable IAQ is defined as "air in which there are no known contaminants at harmful concentrations as determined by cognizant authorities and with which a substantial majority (80% or more) of the people exposed do not express dissatisfaction" (ASHRAE 2001). This definition implies compliance is required for both objective criteria (pollutant concentration) to prevent illness and subjective to provide comfort. IAQ depends on various factors like indoor sources and sinks; infiltrations and ventilations; design, maintenance and operation of building ventilation systems; moisture and humidity; ambient air quality and meteorological parameters; and occupant perceptions and susceptibilities. In addition, there are many other psychological factors (physical, emotional and mental reactions) that affect perception of IAQ.

Exposure to indoor air pollutants can cause infections, lung cancer, and chronic lung diseases such as asthma. In addition, it can cause headaches, dry eyes, nasal congestion, nausea and fatigue. There is consistent proof that exposure to IAP increase the risk of pneumonia, acute lower respiratory infections (ALRI), chronic obstructive pulmonary disease (COPD), cataracts and tuberculosis. There is association between IAP and adverse pregnancy outcomes, in particular low birth weight, ischaemic heart disease and nasopharyngeal and laryngeal cancers as demonstrated by many studies (WHO 2004; ALA 2013). According to World Health Report 2002, about 35% of ALRI, 21% of COPD and 3% of cancers of the trachea, lung and bronchus are caused by exposure to IAP. The COPD and lung cancer are the main causes of death associated with IAP in China, whereas ALRI account for more than 80% of the deaths and DALY's lost in India (WHO 2002). Poor IAQ will not only affect health but also the productivity and morale of the occupants. The consequences include the loss of concentration and an increase in absenteeism which gives rise to direct and indirect financial losses associated with medical costs, material and equipment damages (Paevere and Brown 2008; Leblebici 2012).

5.4 Indoor Air Quality Modelling

The modelling techniques are developed for prediction of airflow and pollutant concentrations in the indoor environments as continuous measurements of indoor air pollutants is difficult. The single-sided and cross ventilation are two major types of natural ventilation where air is going through in and out of the building as a result of pressure differences across the openings, which is due to the combined action of wind and buoyancy driven forces. Natural ventilation is conceptually simple, its detailed design is very challenging due to the outdoor conditions (like as temperature, wind speed and direction); and building's and surrounding aerodynamic factors (like as building shape and form, building orientation, number and size of openings, location, surrounding topography, etc.). This complexity can make difficult to predict and design natural ventilation (Evola and Popov 2006; Visagavel and Srinivasan 2009). Two approaches are generally operated for observing the natural ventilation: multi zone network models and numerical modelling with CFD.

The type of microenvironment and its structure are the parameters which are taken into account for predicting indoor air pollution by using Indoor air quality models (IAQM). In the past, several IAQM are used which range from simple regression to complex computational fluid dynamics. The mass balance model is a simpler indoor air quality model (IAQM) used in single or multiple compartments and predict the indoor concentrations. The well-mixed indoor environment is assumed and average concentration is calculated in different zones of the building for the mass balance model (Hayes 1989, 1991). The spatial variation of air velocity and concentration of pollutant in a room is predicted by CFD models instead of the average concentration predicted by mass balance models.

5.5 Indoor Air Management

The indoor air pollutants include wide variety of compounds with varying concentrations and are emitted from several sources. Several instruments and techniques are available for measuring air pollutants in the ambient environment. But, the physical size, noise, airflow rates, power consumption, difficulty in installation etc. of these instruments restricts their applicability in the indoor environment (U.S. EPA 1990). Therefore, it is important to know the sampling and measurement techniques that can be used in indoor environments. Many governmental agencies and research organizations have developed indoor air monitoring protocols for industrial workplaces. However, the instrumentation requirements for non-industrial microenvironments (residential, commercial and institutional buildings) differ from those of ambient or industrial applications. At present, developed countries have set up guidelines for IAQ Assessment (U.S. EPA 1990; U.S. EPA and NIOSH 1991; Willeke and Macher 1999; HKEPD 2003; U.S. EPA 2003; ASTM 2012), which may not be suitable for developing nations, because of the difference in climatic and socioeconomic conditions and building design and construction practices.

References

ALA (2013) ALA State of the Air 2012 A.L. association. American Lung Association, Washington, DC

ASHRAE (2001) Ventilation for acceptable indoor air quality, standard 62. American Society for Heating, Refrigerating and Air Conditioning Engineers, Atlanta, GA

ASTM (2012) Standard guide for using indoor carbon dioxide concentrations to evaluate indoor air quality and ventilation. The American Society for Testing and Materials, West Conshohocken (D6245-12)

Colbeck I, Nasir ZA, Ali Z (2010) The state of indoor air quality in Pakistan—a review. Environ Sci Pollut Res 17(6):1187–1196

EPA (1990) Cancer risk from outdoor exposure to air toxics. Environmental Protection Agency, U.S

Evola G, Popov V (2006) Computational analysis of wind driven natural ventilation in buildings. Energy Build 38(5):491–501
Hayes SR (1989) Estimating the effect of being indoors on total personal exposure to outdoor air pollution. JAPCA 39(11):1453–1461
Hayes SR (1991) Use of an indoor air quality model (IAQM) to estimate indoor ozone levels. J Air Waste Manag Assoc 41(2):161–170
HKEPD (Hong Kong Environmental Protection Department) (2003) Guideline notes for the management of indoor air quality in offices and public places. Indoor Air Quality Management Group, Hong Kong
IHME (2013) The global burden of disease 2010: generating evidence and guiding policy. Institute for Health Metrics and Evaluation, Seattle, USA
Jones AP (1999) Indoor air quality and health. Atmos Environ 33(28):4535–4564
Leblebici D (2012) Impact of workplace quality on employee's productivity: case study of a bank in Turkey. J Bus Econ 1(1):38–49
Lim GB (2013) Public health: global burden of cardiovascular disease. Nat Rev Cardiol 10(2):59
Willeke K, Macher JM (1999) Bioaerosols: assessment and control. In: Cincinnati OH (ed) American conference of governmental industrial hygienists, vol 8
NIOSH (1989) Method 7400: Fibers. Niosh manual of analytical methods, 3rd edn. U.S. Department of Health, Education and Welfare, Cincinnati
NIOSH (1991) NIOSH manual of analytical methods. Centers for disease control and prevention, national institute for occupational safety and health. Division of Physical Sciences and Engineering, Cincinnati, OH
Ott WR, Roberts JW (1998) Everyday exposure to toxic pollutants. Sci Am 278(2):86–91
Paevere P, Brown S (2008) Indoor environment quality and occupant productivity in the CH_2 building: post-occupancy summary. March, CSIRO, Report No. USP2007/23
Smith KR (2000) National burden of disease in India from indoor air pollution. Proc Natl Acad Sci 97(24):13286–13293
Spengler JD, Chen Q (2000) Indoor air quality factors in designing a healthy building. Annu Rev Energy Env 25(1):567–600
U.S. EPA (2003) A standardized EPA protocol for characterizing indoor air quality in large office buildings. U.S. Environmental Protection Agency, Washington, DC
U.S. EPA (Environmental Protection Agency) (2013) Typical indoor air pollutants, IAQ reference guide, appendix E. Accessed in Oct 2013
Visagavel K, Srinivasan PSS (2009) Analysis of single side ventilated and cross ventilated rooms by varying the width of the window opening using CFD. Sol Energy 83(1):2–5
WHO (2002) World health report: reducing risks promoting healthy life. World Health Organization, Geneva
World Health Organization (WHO) (2004) Global burden of disease due to indoor air pollution. WHO, Geneva. http://www.who.int/indoorair/health_impacts/burden_global/en/

Chapter 6
Personal Exposure and Health Risk Assessment

S. M. Shiva Nagendra, Uwe Schlink, Andrea Müller, Jyothi S. Menon, and V. S. Chithra

6.1 Air Pollution and Chemical Toxicity

Air quality and associated effects in urban areas are major issues in both developed and developing countries. Recent assessments by the Central Pollution Control Board (CPCB) showed that PM levels in many of the Indian cities are exceeding the national ambient air quality standards (NAAQS) and WHO standards. During recent years, the importance of environmental factors for the development of diseases associated with toxic compounds has been increased evidently. Airborne particulate matter (PM), especially fine particles, and bound chemical compounds are potential mediators of health effects. For risk assessment the knowledge about the size,

S. M. Shiva Nagendra (✉)
Department of Civil Engineering, Indian Institute of Technology Madras, Chennai, Tamil Nadu 600036, India
e-mail: snagendra@iitm.ac.in

U. Schlink
Department of Urban and Environmental Sociology, Helmholtz Centre for Environmental Research UFZ, 04318 Leipzig, Germany
e-mail: uwe.schlink@ufz.de

A. Müller
Department Molecular Systems Biology, Helmholtz Centre for Environmental Research UFZ, 04318 Leipzig, Germany
e-mail: a.mueller@ufz.de

J. S. Menon
Public Health Foundation of India, Gurugram 122002, India
e-mail: jyothi.menon@phfi.org

V. S. Chithra
Department of Civil Engineering, Mar Baselios Christian College of Engineering and Technology, Kuttikkanam, Kerala 685531, India
e-mail: drchithravs@mbcpeermade.com

S. M. Shiva Nagendra et al. (eds.), *Urban Air Quality Monitoring, Modelling and Human Exposure Assessment*, Springer Transactions in Civil and Environmental Engineering,
https://doi.org/10.1007/978-981-15-5511-4_6

morphology, composition and toxicity of PM and bounded chemical compounds is imperative. Polycyclic aromatic hydrocarbons (PAHs) play an important role as air contaminant bounded on PM. Because of their high toxic and mutagenic potential, these compounds are used as indicators for risk evaluation. The concentration of PAHs associated with PM are analyzed using new advanced methods. For risk assessment, the biological impact on human cells lines are investigated. Different bioassays for cytotoxicity, oxidative stress and apoptosis are used to acquire more information about the underlying mechanism of particle-associated health effects. In a health risk assessment perspective, personal exposure is more representative as it gives the most accurate estimation of exposure at a personal level.

Emissions of PM-bound polyaromatic hydrocarbons (PAHs) from different combustion activities have emerged as one of the threats to the environment and public health in low middle-income countries. PAHs are a particular set of hydrocarbons that are genotoxic and carcinogenic to humans (Gupta et al. 1998; Venkataraman et al. 2002; Bhargava et al. 2004). PAHs can exist in both particulate and gaseous phase. However, they predominantly present as particulate matter, and particulate bound PAHs are considered to be most hazardous to human health. It is evident from the past studies that persistent exposure to higher concentrations of carcinogenic PAHs could impact the acuteness of pulmonary health problems specifically lung cancer. The major anthropogenic source of PAH includes petroleum refineries, incomplete combustion of petroleum products, biomass burning, combustion of organic products, petroleum gas, coal and wood, which are the major sources of energy production in India (Haritash and Kaushik 2009).

PAHs' concentrations were observed to be high in developing countries particularly in India (Ravindra et al. 2008; Sharma et al. 2008). Total PAH emissions from Asian countries were reported to be 290 Gg/year with 114 and 90 Gg/year contribution solely from China and India, respectively (Mohanraj et al. 2011a). PAHs with 2–7 aromatics rings are classified as critically hazardous (Sharma et al. 2007; Ravindra et al. 2008). India accounts 3.6% of high molecular PAH emissions which is significantly higher than the global average emissions (Mohanraj et al. 2011b). There are over 100 different aromatic hydrocarbons listed in critically hazardous group. However, the studies on particulate-bound PAH at industrial clusters in India are scanty. Further, there are studies that have been carried out to understand PAHs levels in rural India. However, most of the studies have been limited to the villages located in the Indo-Gangetic plain (IGP) of North India (Bhargava et al. 2004; Ansari et al. 2010) and very few studies in south India (Satsangi et al. 2014). Since air pollution varies significantly with climatology, geography, population and fuel usage, the disparity in terms of the quantum of the studies between south and north India provides an opportunity to make villages in south India as a subject for the current study. Furthermore, Municipal solid waste (MSW) open burning has been reported to be a significant source of PM-bound and gaseous PAHs in the urban and suburban areas. Most of the emitted pollutants from the open burning are persistent organic pollutants (POPs) with mutagenic and carcinogenic nature. Backyard (household) burning of waste and landfill fires were reported to be a prominent source of polychlorinated biphenyls (PCBs), PAHs, PCDDs and PCDFs (Hedman et al. 2005; Ruokojärvi et al. 1995).

The $PM_{2.5}$ and PAH concentrations in the ambient environment are largely affected by local climatic conditions, i.e., season, mixing height, wind speed and direction, rainfall, humidity, temperature and emissions sources (industry, traffic, petrol stations, open burning). The spatiotemporal distribution of air contaminants as well as their health risks can be estimated with respect to the urban meteorological conditions. For the latter, data gathered by monitoring are used in combination with meteorological measurements. Recent studies showed modifications of air pollution by humidity and temperature, which point out the relevance of meteorological information for risk evaluation.

6.2 Air Pollutants and Health Effects

The health effects due to air pollution can range from minor disturbances to death. The epidemiological studies have established a direct relationship between air pollution and health hazards. The studies have shown that the ambient air pollution is associated with impaired lung function and increased hospital admission (Guttikunda and Goel 2013; Oanh et al. 2012). The health impacts of air pollution depend on factors like the concentration of pollutants, nature of pollutants, exposure duration, state of receptor health and the age group. Some of the health impacts associated with major urban air pollutants are discussed below.

6.2.1 Health Impacts of Gaseous Pollutants (SO_2 and NO_x)

SO_2 can affect those people who are suffering from asthma and chronic lung diseases and can cause impaired breathing in sensitive individuals (Jiang et al. 2016). The SO_2 is hygroscopic and can also be converted to various forms. It can also react with other compounds in the atmosphere leading to the formation of toxic aerosols. SO_2 is also reported to induce asthma especially among children and elderly (Amancio and Nascimento 2014). Nitrogen oxides are immune toxic and increase the susceptibility to respiratory tract infection. It can reduce the immunity to lung infections, thus resulting in cold, wheezing, etc. The nitrogen dioxide has been found to be associated with increased mortality (Lipfert et al. 2006), hospitalizations due to respiratory and cardiovascular diseases (Hinwood et al. 2006) and asthma emergency room visits (Boutin-Forzano et al. 2004).

6.2.2 Health Impacts of O_3

O_3 is formed in the lower atmosphere (troposphere) by the action of sunlight on nitrogen dioxide (NO_2). The health effects associated with ozone are reduced lung function, permanent damage to lungs, aggravate lung diseases and asthma, damage cell lining of the lungs, etc. The ozone effect on asthma occurrence was mostly reported in early ages which shows its great oxidant capacity (Galan et al. 2003).

6.2.3 Health Impacts of PM

As already mentioned, the health impacts of PM are attributable to its size range. The health impacts of PM have been studied widely as it is one of the major pollutants in an urban area. Baldasano et al. (2003) presented an assessment of air quality for the principal cities in developed and developing countries which indicated that the major problem in Asia is that the particulate matter exceeds 300 $\mu g/m^3$ in many cities, while SO_2 maintaining a downward tendency throughout the world and NO_2 levels very close to WHO guideline. PM_{10} has been associated with emergency hospital admission for asthma, bronchitis, pneumonia, etc. Studies showed that for every 10 $\mu g/m^3$ increase of PM_{10}, mortality from all causes increases by 0.51% (Samet et al. 2000). The PM and related health effects are a widely discussed topic mostly in connection with the Asian cities. A study comprising 25 cities in Europe for the period 2004–2006 showed that the largest health burden was attributable to the chronic exposure to $PM_{2.5}$ (Pascal et al. 2013). The public health impacts of 25 European countries were assessed as a part of Aphekom project (Pascal et al. 2013). The results showed that the largest health burden was attributable to the impacts of chronic exposure to $PM_{2.5}$. Oanh et al. (2012) studied the urban PM pollution characteristics in the six major cities in Asia, which show high levels of $PM_{2.5}$ and PM_{10}, especially in dry season. A health-based assessment of particulate air pollution carried out in urban areas of Beijing during the year 2000–2004 evaluated the mortality and morbidity effects of PM_{10} pollution based on statistical data and the epidemiological exposure–response function (Zhang et al. 2007).

In another study, it was estimated that high PM concentration in Delhi city resulted in 7350–16,200 premature deaths and 6.0 million asthma attacks per year (Guttikunda and Goel 2013). In Chennai city, the average annual PM_{10} concentration for the year 2010 was found to be 73.1 $\pm$ 33.7 $\mu g/m^3$ and a premature death was estimated to be 3950 (Guttikunda and Jawahar 2012). According to the special report by Health Effects Institute, some of the highest levels of outdoor air pollution in the world are found in Asian cities and some of the cities in China and India have recorded world's highest PM_{10} levels. According to Pope and Dockery (2006) among all other pollutants in ambient air, PM affects more people than any other pollutant. The mortality and morbidity associated with air pollution are primarily due to the exposure to PM.

The assessment of the health effects from air pollution exposure at indoor environment is faced with a number of challenges: seasonal variability, pollutant interactions, heterogeneity, effects of daily variations in physical activity on air pollutant inhalation rates and the contribution of non-school-based exposures. Few studies have assessed relationships of various health outcomes among occupants with indoor environmental factors in microenvironments. Most studies considered respiratory health such as asthma, current asthma, wheezing, or allergies as assessed by standardized self-administered questionnaires. Fewer studies considered lung function, nasal patency or acoustic rhinometry. There are concerns that health problems caused by poor IAQ may impair performance and reduce attendance of students. Mendell and Heath (2005) reviewed the existing evidence for direct associations between indoor environmental quality and performance or attendance of school children. Persuasive evidence links higher indoor concentrations of NO_2 to reduced school attendance, and suggestive evidence links low ventilation rates to reduced performance. These evidences suggest that poor IAQ in schools is adversely affecting the performance and attendance of students, primarily through health effects from indoor pollutants. Most of the scientific literature providing evidence for the health impacts of IAP comes from studies conducted in developed country settings within North America and Western Europe, which was used for exposure and risk assessment. However, the differences between developed and developing countries in exposure concentrations, the nature of pollutants, baseline health and determinants of susceptibility add uncertainties while extrapolating exposure–response relationships across countries.

6.3 Personal Exposure to Air Pollutants

The long-term (annual) as well as short-term (daily and sub-daily) exposure to PM has been reported to cause significant health effects. The elevated PM concentration in urban centers mostly in the developing countries has raised concerns regarding the sub-daily (hours) exposure (EPA 2009). The street canyons, city centers, areas near busy roads, etc., are potential air pollution hotspots where short-term exposure to peak levels of air pollutants can result in adverse health effects (Rakowska et al. 2014; Zwack et al. 2011). Personal exposure refers to the actual pollutant level that an individual is exposed to as a part of their daily activity. Due to the ubiquitous nature of air pollutants particularly PM, as well as the spatiotemporal variability, the assessment of personal exposure is evolving quickly (Steinle et al. 2013). With the advancement of technology, it is possible to measure the pollutant concentrations while simultaneously tracking the individuals. The personal exposure assessment can be a better way to represent the short-term exposure, particularly at hotspots.

The direct and indirect approach can be used for measuring the personal exposure to the pollutants. The exposure level to an individual can be observed in direct approach with the help of personal sampler. The models are used for measuring exposure levels in the indirect approach. The direct method is a simple method where pollution monitoring is done at the personal level by carrying a monitoring device by the person or by analyzing the biological marker and provides the clear picture of the personal exposure levels of individual. The micro-environmental (ME) modeling is used for measuring the indirect personal exposure and required information related to chosen MEs like indoor area, i.e., bedroom or kitchen, in transit and outdoor area and also the account of people's activity. However, the narration of all MEs of individual or population is tough.

6.4 Health Risk Assessment of Polyaromatic Hydrocarbons

The earlier studies have described risk assessment calculations of exposure to PAHs (Hong et al. 2016). The relative potency factor (RPF) was also applied for risk calculation of PAH mixtures as given by US EPA. The BaP equivalent (BaPeq) concentration for each PAH compounds was estimated by multiplying its concentration by its RPF (Hong et al. 2016; Wang et al. 2012).

$$\sum \mathrm{BaPeq} = \sum_{1}^{n=1} (C_i * \mathrm{TEF}_i) \tag{6.1}$$

RPF values for nine PAH compounds (FLA, BaA, CHR, BbF, BkF, IcdP, DBaA and BghiP) were obtained from US EPA. Excess lifetime cancer risk (LCR) for PAH mixture was calculated by multiplying the sum of BaPeq concentration of each PAHS by inhalation unit risk (IUR) of exposure to BaP.

$$\mathrm{LCR} = C_{\mathrm{BaPeq}} * \mathrm{UR}_{\mathrm{BaP}} \tag{6.2}$$

where C_{BaPeq} is BaPeq concentration and $\mathrm{UR}_{\mathrm{BaP}}$ is inhalation unit risk of exposure to BaP. Two different $\mathrm{UR}_{\mathrm{BaP}}$ values of 1.1×10^{-6} per ng/m^3 (OEHHA 2005) and 8.7×10^{-5} per ng/m^3 (WHO 2002) can be used to calculate excess LCR.

6.5 Health Risk Assessment Among School Children

The naturally ventilated school building of Kendriya Vidyalaya School was chosen for the study which is present near a busy road (SP road) in Chennai city. The low-volume handy air samplers (Model APM 821, Envirotech Instruments Private Limited, India) were used for monitoring the indoor and outdoor suspended particulate matter (SPM) concentrations. The 24-hourly measurements were done in monsoon (October–November) of 2011, winter (January–February) and summer (April–May) seasons of 2012. The indoor sampling for SPM was carried out for 60 days (20 days each in three seasons), and outdoor sampling was done for 25 days (9 days in monsoon, 8 days in winter and 8 days in summer).

The PTFE filters having 25 mm diameter and 0.2 μm pore size were used for sampling of SPM samples from indoor and outdoor area. These filters were used for the determination of elements (Ag, Al, As, Ba, Be, Bi, Ca, Cd, Co, Cr, Cu, Fe, Ga, K, Li, Mg, Mn, Mo, Na, Ni, Pb, Rb, Se, Sr, Te, Ti, V and Zn) and ions (F^-, Cl^-, NO_2^-, NO_3^-, SO_4^{2-}, PO_4^{3-}, Na^+, NH_4^+, K^+, Mg^{2+} and Ca^{2+}). The quartz filters were used for the investigation of organic carbon (OC) and elemental carbon (EC) samples. The USEPA Compendium of Methods IO-3.1 (USEPA 1999a) and IO-3.5 (USEPA 1999b), respectively, were used for the extraction and analysis of elements. The ion chromatography (Dionex ICS-2500) was used for the determination of ions in the samples by using standard operating procedures provided by Central Pollution Control Board (CPCB), New Delhi. The thermal/optical analyzer which is based on IMPROVE thermal optical reflectance (TOR) method was used for the determination of OC and EC. The scientific methods for the analysis of different components adopted in this study were found elsewhere (Chithra and Nagendra 2013).

6.5.1 Health Risk Assessment

There are some elements which cause carcinogenic effects (Cr, Ni, Pb) and some elements that are non-carcinogenic but cause toxic effects (Al, Cr, Mn, Ni, V). The health risk assessment study was done only for exposure of particulate inside the classroom. The Eqs. (6.3) and (6.4), respectively, (USEPA 2009) were used for calculating the concentration of non-carcinogenic elements (EC_{nc}) and carcinogenic elements (EC_c) exposure.

$$EC_{nc} = \frac{CA * ET * EF * ED}{AT * ED} \tag{6.3}$$

$$EC_c = \frac{CA * ET * EF * ED}{AT * LT} \tag{6.4}$$

where CA = contaminant concentration in the indoor air (μg/m^3); ET = exposure time (6 h/24 h); EF = exposure frequency (200 days/year); ED = exposure duration (12 years) and AT = averaging time (365 days/year) and LT = lifetime (70 years).

For non-carcinogenic elements Hazard Quotient (HQ) was estimated as follows

$$HQ = \frac{EC_{nc}}{(Rfc * 1000\,\mu g/mg)} \tag{6.5}$$

where *Rfc* = inhalation toxicity value (mg/m^3).

The following equation can be used for the estimation of excess cancer risk (ECR) for a receptor which has been exposed via the inhalation pathway

$$ECR = IUR * EC_c \tag{6.6}$$

where IUR = Inhalation Unit Risk $(\mu g/m^3)^{-1}$.

There is USEPA database for Integrated Risk Information System (IRIS 2013) which can be used for obtaining the *Rfc* (Cr (VI), Mn) and IUR (Cr (VI), Ni) values of the metals. California Environmental Protection Agency (CALEPA 2013) database provided the *Rfc* and IUR values for Ni and Pb, respectively. Risk Assessment Information System (RAIS 2013) and Agency for Toxic Substances and Disease Registry (ATSDR 2013) are used for getting *Rfc* values for Al and V, respectively.

IAQ and health symptoms can be correlated by carrying out a survey in the schools by filling of questionnaire form from students. The questionnaire for the health-related studies was made by taking into account the already used questionnaire in many studies related to health (Meklin et al. 2002; Gupta et al. 2007). There are three parts in the questionnaire: The first is general information regarding age and gender. The second is about living conditions and classroom conditions (thermal, acoustic, visual, indoor air conditions, etc.). The third part is regarding effect on children as a result of the school environment. The questions in the questionnaire were related to school environment like indoor paint or any changes in the interior during the last 3 months. The acute respiratory symptoms were also accounted which help in evaluation of changes in human health due to environment. These are some symptoms which were comprehended like allergies, difficulty in concentrating, back pain, hearing problems, dizziness, skin irritation, heart burn, sinus congestion, sneezing, high stress levels, eye irritation, headache and fatigue.

6.5.2 Children's Health Risk

The carcinogenic metals such as Cr, Ni and Pb, while non-carcinogenic but having toxic effects includes Al, Cr, Mn, Ni and V. Cr and Ni come under both carcinogenic and non-carcinogenic category. Cr (VI) and Cr (III) are two forms of chromium in which it exists, where Cr (III) is toxic. The concentration ratio of Cr (VI) to Cr (III) was observed as about 1–6 in the air (USEPA 1998). Therefore, it lowers the risk of Cr (VI) to one-seventh of total Cr concentration (Hieu and Lee 2010; Massey et al. 2013). Table 6.1 shows the carcinogenic and non-carcinogenic risks linked with elements. Cr (3.77×10^{-6}) and Ni (0.8489) element represent the largest carcinogenic risk and hazard quotient, respectively, while Pb (9.919×10^{-9}) and Al (0.0673) represent lower carcinogenic risk and hazard quotient as shown from ECR values, respectively. The recommended value is 1.0, while the total HQ value of elements is 1.29 which was higher than the recommended value (1.0), although the individual HQ values were under the prescribed limit. It was observed that the total excess cancer risk (4.2×10^{-6}) is 4 times higher than the acceptable value of 1×10^{-6} which is recommended by USEPA.

It was observed from the table that Cr has higher excess cancer risk than the admissible value of 1×10^{-6} as commended by USEPA, while Pb and Ni both are the ECR values which were within the prescribed limit. The nickel refinery dust and

Table 6.1 Carcinogenic and non-carcinogenic risk associated with elements (Chithra and Nagendra 2018)

Elements	CA (μg/m^3)	Non-carcinogenic risk		Carcinogenic risk	
		EC (μg/m^3)	HQ	EC(μg/m^3)	ECR
Al	2.4555	0.3364	0.0673		
Cr (VI)	0.0143	0.0019	0.0196	0.0003	3.7704E−06
Mn	0.0693	0.0095	0.1899		
Ni	0.0868	0.0119	0.8489	0.0019	4.564E−07
Pb	0.0377	–		0.0008	9.9187E−09
V	0.1204	0.0165	0.1649		
Total			1.2907		4.2367E−06

nickel subsulfide which are human carcinogens could be included in group A materials as per USEPA. The evidence for considering lead as a human carcinogen (group B2) is not enough. Cr (VI) is showing the higher excess cancer risk as representing from higher ECR values although it has lower concentration in the PM. Cr (VI) is designated as group A—human carcinogen—as according to USEPA. The studies on effect of exposure of chromium exposure to children are unavailable. Chromium would have same health effect on children as on adults (ATSDR 2012). The main source of Cr is road dust in indoor as indicated by the CMB modeling. The catalytic converter based on Cr is the main source for the higher concentration of Cr in the road dust. The exposure of PM to the children in the classroom has detrimental effect on their health as suggested by risk assessment study.

The outdoor vehicular air pollution is influencing the indoor environment as communicated by 45% of respondents in the questionnaire survey at the school. It was reported from survey that sneezing, headache and allergies are associated with 50% of students. The health problems like headache and sneezing are observed mostly among students that may be building-related. Some other health problems include allergies which affected about 17% students, and some 14% students are complaining regarding difficulty in concentrating. The health of children is affected by the higher concentration of particles in the classrooms as observed by the study.

6.6 Exposure Assessment

The process of estimation or measurement of magnitude, frequency and duration of exposure to a representative is the exposure assessment (Zartarian et al. 2007). Exposure depends mainly on the magnitude (pollutant concentration) and the duration of exposure (time). Generally, the health outcome is directly proportional to the dose or concentration to which the subject is treated. But the duration of exposure also affects the outcomes. For example, the health risk will not be same if a person is exposed to the same pollutant concentration for a few minutes and for few hours.

As the duration of exposure increases, the health risk associated with the exposure also increases. Similarly, the frequency of exposure also has considerable effects. So, it is very important to take these factors into account while performing exposure assessment. The exposure measurements can be of two types, i.e., personal exposure measurement and population exposure measurement. The pollutant concentrations that an individual is exposed during normal daily activities are referred to as personal or individual exposures. It varies according to the locations that the individual moves around and the time spent there. Population exposure refers to the aggregated exposure for a community or a group of people.

6.6.1 Personal Exposure Assessment

Personal exposure refers to "measurement of a pollutant of concern performed by a monitor (or sampler) worn by a person while the sample is taken from a point near the breathing zone of the person" (Braniš et al. 2010). The personal exposure assessment determines the air pollutants concentration in the inhaled air as a function of time and space coordinates, and it also provides association between high air pollution concentrations and detrimental effect on human health. The individual exposure is estimated using portable personal monitors which are capable of direct measurements of pollutant concentration in a person's breathing zone (Steinle et al. 2013). Many large-scale exposure studies have been conducted in the USA (NHEXAS: National Human Exposure Assessment Survey) and Europe (EXPOLIS: Air Pollution Exposure Distribution of Adult Urban Populations in Europe) to assess the human exposure (Jantunen 1999; Pellizzari et al. 1999). A study on personal exposure of particulates in a transport environment shows that the fixed-site monitors to be a poor marker for PM_{10} concentrations recorded during various modes of transport (Gulliver and Briggs 2004). Various personal monitoring studies are conducted worldwide, but they are not usually linked to local air quality monitoring data (Lioy et al. 2009). In urban areas, the spatial distribution of air pollutants and movement of the people determine the personal exposure. The air pollutant concentration especially PM has high spatial and temporal variability in an urban area, and the personal exposure measurement gives a more reliable result for health risk assessment. It was also found that the variability in PM concentrations was mainly determined by traffic conditions, local sources and topographic characteristics, leading to large inter- and intra-urban variations, with severe implications for the characterization of population exposure (Kassomenos et al. 2014). Further, the variation of particle concentrations across locations within countries appeared to be considerably larger than the variation across countries (Hoek et al. 2002).

The higher concentration of PM was observed in some of the site-specific measurements carried out near a major road in Chennai city (Srimuruganandam and Nagendra 2011). The people and pedestrians are always exposed to higher concentration of pollutants at the busy road, traffic intersections and urban corridors. The personal exposure was observed in past by conducting study and understanding the effect of air pollutants. The type of microenvironments has different effects on the individuals as observed from different studies in different types of microenvironments (Yang et al. 2015) viz. near road, transport microenvironments (Li et al. 2017), street canyons (Zwack et al. 2011), indoor microenvironments, etc.

References

Amancio CT, Nascimento LF (2014) Environmental pollution and deaths due to stroke in a city with low levels of air pollution: ecological time series study. Sao Paulo Med J 132(6):353–358

Ansari FA, Khan AH, Patel DK, Siddiqui H, Sharma S, Ashquin M, Ahmad I (2010) Indoor exposure to respirable particulate matter and particulate-phase PAHs in rural homes in North India. Environ Monit Assess 170(1–4):491–497

ATSDR (2012) Public health statement chromium CAS # 7440-47-3. Division of Toxicology and Human Health Sciences

ATSDR (2013) http://www.atsdr.cdc.gov/substances/toxsubstance.asp?toxid=50. Accessed on 31 July 2013

Baldasano JM, Valera E, Jimenez P (2003) Air quality data from large cities. Sci Total Environ 307(1–3):141–165

Bhargava A, Khanna RN, Bhargava SK, Kumar S (2004) Exposure risk to carcinogenic PAHs in indoor-air during biomass combustion whilst cooking in rural India. Atmos Environ 38(28):4761–4767

Boutin-Forzano S, Adel N, Gratecos L, Jullian H, Garnier JM, Ramadour M, Lanteaume A, Hamon M, Lafay V, Charpin D (2004) Visits to the emergency room for asthma attacks and short-term variations in air pollution. Respiration 71(2):134–137

Braniš M, Vyškovská J, Malý M, Hovorka J (2010) Association of size-resolved number concentrations of particulate matter with cardiovascular and respiratory hospital admissions and mortality in Prague, Czech Republic. Inhal Toxicol 22(sup2):21–28

Chithra VS, Nagendra SMS (2013) Chemical and morphological characteristics of indoor and outdoor particulate matter in an urban environment. Atmos Environ 77:579–587

Chithra VS, Shiva Nagendra SM (2018) A review of scientific evidence on indoor air of school building. Asian Journal of Atmospheric Environment 12(2): 87–108

EPA (2009) Integrated science assessment for particulate matter. US Environmental Protection Agency, Washington, DC

Galan I, Tobias A, Banegas JR, Aranguez E (2003) Short-term effects of air pollution on daily asthma emergency room admissions. Eur Respir J 22(5):802–808

Gulliver J, Briggs DJ (2004) Personal exposure to particulate air pollution in transport microenvironments. Atmos Environ 38(1):1–8

Gupta S, Khare M, Goyal R (2007) Sick building syndrome—A case study in a multistory centrally air-conditioned building in the Delhi City. Build Environ 42(8):2797–2809

Gupta S, Saksena S, Shankar VR, Joshi V (1998) Emission factors and thermal efficiencies of cooking biofuels from five countries. Biomass Bioenerg 14:547–559

Guttikunda SK, Goel R (2013) Health impacts of particulate pollution in a megacity—Delhi, India. Environ Dev 6:8–20

Guttikunda SK, Jawahar P (2012) Application of SIM-air modeling tools to assess air quality in Indian cities. Atmos Environ 62:551–561

Haritash AK, Kaushik CP (2009) Biodegradation aspects of polycyclic aromatic hydrocarbons (PAHs): a review. J Hazard Mater 169(1–3):1–15

Hedman B, Näslund M, Nilsson C, Marklund S (2005) Emissions of polychlorinated dibenzodioxins and dibenzofurans and polychlorinated biphenyls from uncontrolled burning of garden and domestic waste (backyard burning). Environ Sci Technol 39(22):8790–8796

Hieu NT, Lee BK (2010) Characteristics of particulate matter and metals in the ambient air from a residential area in the largest industrial city in Korea. Atmos Res 98(2–4):526–537
Hinwood AL, Berko HN, Farrar D, Galbally IE, Weeks IA (2006) Volatile organic compounds in selected micro-environments. Chemosphere 63(3):421–429
Hoek G, Brunekreef B, Goldbohm S, Fischer P, van den Brandt PA (2002) Association between mortality and indicators of traffic-related air pollution in the Netherlands: a cohort study. Lancet 360(9341):1203–1209
Hong WJ, Jia H, Ma WL, Sinha RK, Moon HB, Nakata H, Minh NH, Chi KH, Li WL, Kannan K, Sverko E (2016) Distribution, fate, inhalation exposure and lung cancer risk of atmospheric polycyclic aromatic hydrocarbons in some Asian countries. Environ Sci Technol 50(13):7163–7174
Jantunen M (1999) Air pollution exposure distribution of adult urban populations in Europe (EXPOLIS). Final Report, Environment & Climate Research Programme (1994–1998)
Jiang XQ, Mei XD, Feng D (2016) Air pollution and chronic airway diseases: what should people know and do? J Thoracic Dis 8(1):E31
Kassomenos PA, Vardoulakis S, Chaloulakou A, Paschalidou AK, Grivas G, Borge R, Lumbreras J (2014) Study of PM_{10} and $PM_{2.5}$ levels in three European cities: analysis of intra and inter urban variations. Atmos Environ 87:153–163
Li Z, Che W, Frey HC, Lau AK, Lin C (2017) Characterization of $PM_{2.5}$ exposure concentration in transport microenvironments using portable monitors. Environ Pollut 228:433–442
Lioy PJ, Isukapalli SS, Trasande L, Thorpe L, Dellarco M, Weisel C, Georgopoulos PG, Yung C, Brown M, Landrigan PJ (2009) Using national and local extant data to characterize environmental exposures in the National Children's Study: Queens County, New York. Environ Health Perspect 117(10):1494–1504
Lipfert FW, Baty JD, Miller JP, Wyzga RE (2006) $PM_{2.5}$ constituents and related air quality variables as predictors of survival in a cohort of US military veterans. Inhal Toxicol 18(9):645–657
Massey DD, Kulshrestha A, Taneja A (2013) Particulate matter concentrations and their related metal toxicity in rural residential environment of semi-arid region of India. Atmos Environ 67:278–286
Meklin T, Husman T, Vepsäläinen A, Vahteristo M, Koivisto J, Halla-Aho J, Hyvärinen A, Moschandreas D, Nevalainen A (2002) Indoor air microbes and respiratory symptoms of children in moisture damaged and reference schools. Indoor Air 12(3):175–183
Mendell MJ, Heath GA (2005) Do indoor pollutants and thermal conditions in schools influence student performance? A critical review of the literature. Indoor Air 15(1):27–52. https://doi.org/10.1111/j.1600-0668.2004.00320.x
Mohanraj M, Muraleedharan C, Jayaraj S (2011a) A review on recent developments in new refrigerant mixtures for vapour compression-based refrigeration, air-conditioning and heat pump units. Int J Energy Res 35(8):647–669
Mohanraj R, Solaraj G, Dhanakumar S (2011b) Fine particulate phase PAHs in ambient atmosphere of Chennai metropolitan city, India. Environ Sci Pollut Res 18(5):764–771
Oanh NTK, Phuong MTT, Permadi DA (2012) Analysis of motorcycle fleet in Hanoi for estimation of air pollution emission and climate mitigation co-benefit of technology implementation. Atmos Environ 59:438–448
OEHHA (2005) Toxicity criteria database. California Environmental Agency, Sacramento
Pascal M, Corso M, Chanel O, Declercq C, Badaloni C, Cesaroni G, Henschel S, Meister K, Haluza D, Martin-Olmedo P, Medina S (2013) Assessing the public health impacts of urban air pollution in 25 European cities: results of the Aphekom project. Sci Total Environ 449:390–400
Pellizzari ED, Clayton CA, Rodes CE, Mason RE, Piper LL, Fort B, Pfeifer G, Lynam D (1999) Particulate matter and manganese exposures in Toronto, Canada. Atmos Environ 33(5):721–734
Pope CA III, Dockery DW (2006) Health effects of fine particulate air pollution: lines that connect. J Air Waste Manag Assoc 56(6):709–742
RAIS (2013) http://rais.ornl.gov/. Accessed on 31 July 2013

Rakowska A, Wong KC, Townsend T, Chan KL, Westerdahl D, Ng S, Močnik G, Drinovec L, Ning Z (2014) Impact of traffic volume and composition on the air quality and pedestrian exposure in urban street canyon. Atmos Environ 98:260–270

Ravindra K, Sokhi R, Van Grieken R (2008) Atmospheric polycyclic aromatic hydrocarbons: source attribution, emission factors and regulation. Atmos Environ 42(13):2895–2921

Ruokojärvi P, Ettala M, Rahkonen P, Tarhanen J, Ruuskanen J (1995) Polychlorinated dibenzo-p-dioxins and-furans (PCDDs and PCDFs) in municipal waste landfill fires. Chemosphere 30(9):1697–1708

Samet JM, Dominici F, Curriero FC, Coursac I, Zeger SL (2000) Fine particulate air pollution and mortality in 20 US cities, 1987–1994. N Engl J Med 343(24):1742–1749

Satsangi PG, Yadav S, Pipal AS, Kumbhar N (2014) Characteristics of trace metals in fine ($PM_{2.5}$) and inhalable (PM_{10}) particles and its health risk assessment along with in-silico approach in indoor environment of India. Atmos Environ 92:384–393

Sharma H, Jain VK, Khan ZH (2007) Characterization and source identification of polycyclic aromatic hydrocarbons (PAHs) in the urban environment of Delhi. Chemosphere 66(2):302–310

Sharma H, Jain VK, Khan ZH (2008) Atmospheric polycyclic aromatic hydrocarbons (PAHs) in the urban air of Delhi during 2003. Environ Monit Assess 147(1–3):43–55

Srimuruganandam B, Nagendra SS (2011) Chemical characterization of PM_{10} and $PM_{2.5}$ mass concentrations emitted by heterogeneous traffic. Sci Total Environ 409(17):3144–3157

Steinle S, Reis S, Sabel CE (2013) Quantifying human exposure to air pollution—Moving from static monitoring to spatio-temporally resolved personal exposure assessment. Sci Total Environ 443:184–193

USEPA (1998) Toxicological review of hexavalent chromium. In: Support of summary information on the integrated risk information system (IRIS). Washington, DC

USEPA (1999a) Compendium of methods for the determination of inorganic compounds in ambient air

USEPA (1999b) Compendium of methods for the determination of toxic organic compounds in ambient air. Second Edition, EPA 625/R-96/010b

USEPA (2009). Risk assessment guidance for superfund: vol I—Human health evaluation manual (Part F, supplemental guidance for inhalation risk assessment). Washington, DC

Venkataraman C, Negi G, Sardar SB, Rastogi R (2002) Size distributions of polycyclic aromatic hydrocarbons in aerosol emissions from biofuel combustion. J Aerosol Sci 33(3):503–518

Wang J, Chen S, Tian M, Zheng X, Gonzales L, Ohura T, Mai B, Simonich SLM (2012) Inhalation cancer risk associated with exposure to complex polycyclic aromatic hydrocarbon mixtures in an electronic waste and urban area in South China. Environ Sci Technol 46(17):9745–9752

WHO (2002) The world health report 2002: reducing risks, promoting healthy life. World Health Organization, Geneva, Switzerland

Yang F, Kaul D, Wong KC, Westerdahl D, Sun L, Ho KF, Tian L, Brimblecombe P, Ning Z (2015) Heterogeneity of passenger exposure to air pollutants in public transport microenvironments. Atmos Environ 109:42–51

Zartarian VG, Ott WR, Duan N (2007) Basic concepts and definitions of exposure and dose. Expos Anal 33–63

Zhang M, Song Y, Cai X (2007) A health-based assessment of particulate air pollution in urban areas of Beijing in 2000–2004. Sci Total Environ 376(1–3):100–108

Zwack LM, Paciorek CJ, Spengler JD, Levy JI (2011) Characterizing local traffic contributions to particulate air pollution in street canyons using mobile monitoring techniques. Atmos Environ 45(15):2507–2514

Chapter 7
Air Quality Measuring Sensors

S. M. Shiva Nagendra, Uwe Schlink, and Mukesh Khare

7.1 Air Quality Measuring Sensors

In many cities, the air quality information is provided to public in the form of air quality index (AQI). Various air quality indices were developed worldwide to provide air quality information to the public. The AQI is estimated from fixed stations. In most epidemiologic studies, the ambient concentrations measured at fixed monitoring sites are also used as surrogates for human exposure assessment. As these fixed stations were unable to capture the air quality beyond certain radius (up to 500 m to 1000 m), the air quality index cannot represent actual personal exposure to air pollutants in the urban environment. Therefore, existing air quality measuring techniques having the limitations like lack of spatial-temporal variations, target-specific monitoring, personal exposure, location-specific measurement, real-time alertness, feedback and control mechanism and many more, of course, they are more accurate. With the negligible amount of compromise in the accuracy makes the availability of sensors for the measurement of air quality.

Measurement of air quality becomes easy and comfortable with the availability of sensors for all the criteria pollutants (particulate matter, NO_2, SO_2, CO, etc.) (Vincent 2007). Table 7.1 provides some of the commercial low-cost air quality measuring

S. M. Shiva Nagendra (✉)
Department of Civil Engineering, Indian Institute of Technology Madras, Chennai, Tamil Nadu 600036, India
e-mail: snagendra@iitm.ac.in

U. Schlink
Department of Urban and Environmental Sociology, Helmholtz Centre for Environmental Research UFZ, 04318 Leipzig, Germany
e-mail: uwe.schlink@ufz.de

M. Khare
Department of Civil Engineering, The Indian Institute of Technology, New Delhi 110016, India
e-mail: mukeshk@civil.iitd.ernet.in

S. M. Shiva Nagendra et al. (eds.), *Urban Air Quality Monitoring, Modelling and Human Exposure Assessment*, Springer Transactions in Civil and Environmental Engineering,
https://doi.org/10.1007/978-981-15-5511-4_7

Table 7.1 Some of the low-cost air quality measuring sensors in the market

S. No.	Sensor	Measured	Range	Company	Technology
1	SDS011	$PM_{2.5}$	0.0–999.9 μg/m^3	Nova	Optical
2	OPC-N3	PM_1, $PM_{2.5}$, PM_{10}	0–2000 μg/m^3	Alphasense	Optical
3	HPMA115S0-XXX	$PM_{2.5}$, PM_{10}	0–1000 μg/m^3	Honeywell	Optical
4	Mics 2614	O_3	10 ppb–1 ppm	Sensortech	Semiconductor
5	3SP-O3-20	O_3	0–20 ppm	SPEC sensors	Electrochemical
6	A3OZ	O_3	0–10 ppm	CITY Technology	Electrochemical
7	Mics 2714	NO_2	50 ppb–5 ppm	Sensortech	Semiconductor
8	NO2 B43F	NO_2	0–20 ppm	Alphasense	Electrochemical
9	ULPSM-NO2 968-047	NO_2	0–20 ppm	SPEC sensors	Electrochemical
10	MQ-7	CO	20–2000 ppm	Hanwei Electronics	Semiconductor
11	CO-B4	CO	0–1000 ppm	Alphasense	Electrochemical
12	CO-A4	CO	0–500 ppm	Alphasense	Electrochemical
13	UL 2034	CO	1000 ppm	SPEC sensors	Electrochemical
14	SO2-B4	SO_2	0–100 ppm	Alphasense	Electrochemical
15	ULPSM-SO2 968-006	SO_2	0–20 ppm	SPEC sensors	Electrochemical
16	C03-0973-100	SO_2	0–20 ppm	Honeywell Rare SYS	Electrochemical

sensors. The future technological advancements will make sensor measurements more accurate. In this chapter, we focused on the air quality measurement with sensors and their limitation in real time. A sensor is a device which senses (responds) a physical quantity (stimulus) and provides a measurable output. With the advancement of technology, most of the sensors provide electrical output, hence, the manipulation and controlling of other devices made simple. In order to provide the electrical output, the sensing element of a sensor is associated with an electrical signal conditioning circuit called as transducer. Today, the sensor and transducer resemble same (Neubert 2003). Figure 7.1a illustrates RTD (PT1000, temperature sensor) operation upon the wheat stone bridge to provide electrical voltage to the corresponding temperature. Nonlinearity in the temperature voltage relationship as shown in Fig. 7.1b of RTD can be improved by other linearization signal conditioning methods.

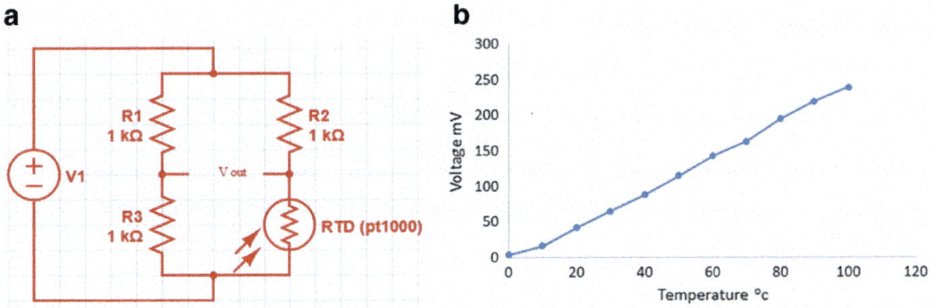

Fig. 7.1 Sensors with electrical circuit **a** RTD (Pt 1000) operation upon Wheatstone bridge, **b** temperature versus voltage graph

7.2 Types of Smart Air Quality Sensors

According to the external power requirement, the sensors are classified into two categories.

7.2.1 Active Sensors

Active sensors are those which do not require external source (voltage, current) to produce measurable output. Thermocouple and piezoelectric crystal are examples of active sensors. Thermocouple arrangement as shown in Fig. 7.2 for temperature measurement did not require any external supply. The EMF generated at the cold or reference junction is proportional to the temperature at hot junction.

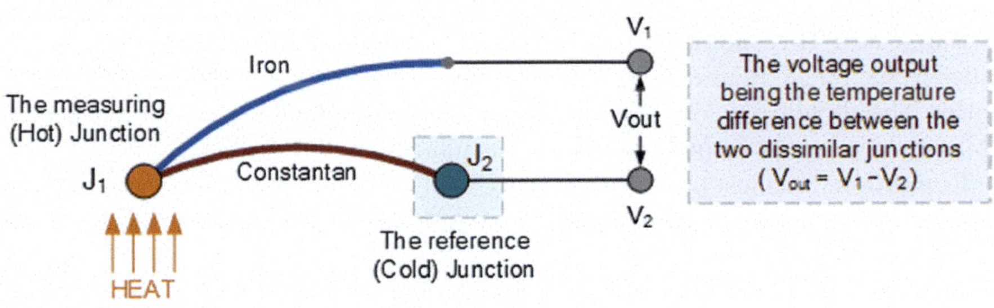

Fig. 7.2 Temperature measurements with thermocouple

7.2.2 *Passive Sensors*

Passive sensors are those which requires external source (voltage, current) for the transduction (input physical quantity to output measured) purpose. Thermistors, RTD and resistive strain gauges are the examples of passive sensors. RTD arrangement as shown in Fig. 7.1a requires external supply for the temperature measurement, hence, it is a passive sensor.

7.3 Specifications of Sensors

For the effective usage of sensors, the manufacturer provides the specifications in the form of a sheet called specification sheet. This sheet contains the detailed information about the sensor (Usher and Keating 1996). In general, all the specifications are brought into three categories. Apart from these sensors must be operated within the electrical specifications.

1. Static specifications
2. Dynamic specifications
3. Environmental specifications.

7.3.1 *Static Specifications*

Sensor characteristics when static input (input which is not changing continuously with respect to time) is applied to sensor are mentioned under static specifications. Accuracy, range, threshold, resolution, linearity, sensitivity, etc., fall into static specifications.

Range and Threshold: The minimum to maximum value measured by sensor is called as the range. Figure 7.3a shows the range specification of the SDS011 $PM_{2.5}$ sensor. Threshold is the lowest concentration of a pollutant that a sensor can detect. This may or may not be provided by the manufacturer.

Resolution: Smallest input change that can produce a measurable output without ambiguity is called as the resolution. Figure 7.3b shows the resolution specification of SDS011 $PM_{2.5}$ sensor.

Accuracy: The closeness of the measured value with the true value is called as accuracy. In general, the accuracy is specified in terms of full-scale error.

$$\text{Error} = \text{Measured value} - \text{True value} \tag{7.1}$$

$$\%\text{Error} = \frac{\text{Measured value} - \text{True value}}{\text{True value}} \times 100$$

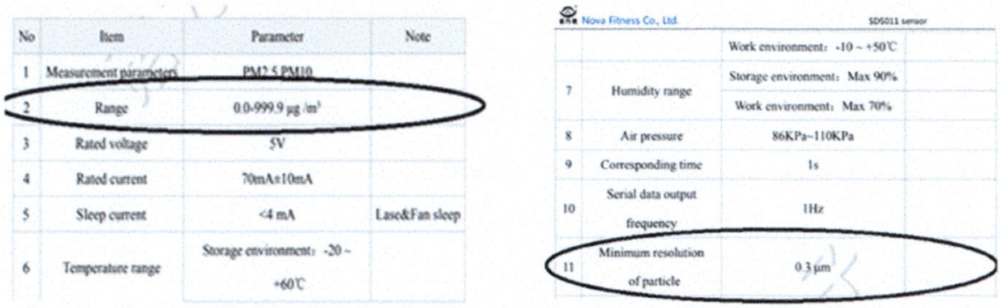

No	Item	Parameter	Note
1	Measurement parameters	PM2.5,PM10	
2	Range	0.0-999.9 μg /m³	
3	Rated voltage	5V	
4	Rated current	70mA±10mA	
5	Sleep current	<4 mA	Lase&Fan sleep
6	Temperature range	Storage environment：-20～+60℃	

a)

Nova Fitness Co., Ltd. SDS011 sensor

		Work environment：-10～+50℃	
7	Humidity range	Storage environment：Max 90%	
		Work environment：Max 70%	
8	Air pressure	86KPa~110KPa	
9	Corresponding time	1s	
10	Serial data output frequency	1Hz	
11	Minimum resolution of particle	0.3 μm	

b)

Fig. 7.3 Specification sheet of SDS011 $PM_{2.5}$ sensor **a** range specification, **b** resolution specification

$$\text{Full scale error} = \frac{\text{Measured value} - \text{true value}}{\text{Full scale value}} \times 100$$

$$\%\text{Error (reading)} = \%\text{Full scale error} \times \frac{\text{Full scale value}}{\text{Measured value (read)}}$$

Example 7.1 1000 $\mu g/m^3$ CO_2 gas sensor having scale error of ±0.5% is used to measure the 200 $\mu g/m^3$ target concentration, then find the accuracy lies between?

$$1000 \times \frac{0.5}{100} = 5\ \mu g/m^3$$

5 $\mu g/m^3$ is the error in the reading measured by sensor.

∴ The accurate value between (200 and 5) $\mu g/m^3$ and (200 + 5) $\mu g/m^3$.

Sensitivity: The change in output to the change input ratio is called sensitivity. The sensitivity is the slope of the sensor characteristic graph.

Example 7.2 Figure 7.4 shows the improved linearization signal conditioning output of an RTD as shown in Fig. 7.1a, b. Find out the sensitivity of RTD characteristics as shown in Fig. 7.4.

$$\begin{aligned}\text{Sensitivity} &= \frac{\text{change in output}}{\text{chnage in input}} \\ &= \frac{601.9 - 406.9}{60 - 40} = 9.75\end{aligned}$$

∴ Sensitivity = 9.75 mV/°C.

Example 7.3 An RTD having the specification like $Rt = 1000(1 + 0.00385 \pm \Delta T)$, here, 0.00385 will indicate the temperature sensitivity of RTD in Ω/°C.

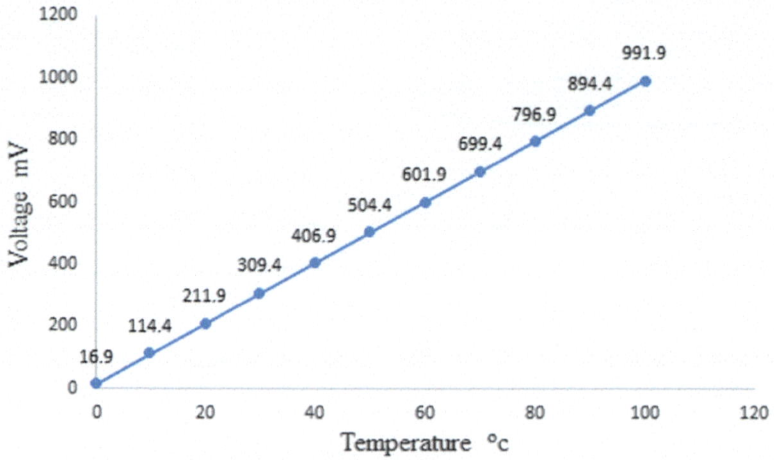

Fig. 7.4 Linearized characteristics of RTD

Precision: Precision refers to how well the sensor reproduces the measurement of a pollutant under identical circumstances. The sensor with precision and accurate is preferable.

7.3.2 Dynamic Specifications

Sensor characteristics when dynamic input (input which is changing continuously with respect to time) is applied to sensor are mentioned under dynamic specifications. Rise time, settling time, slew rate, etc.

Settling time: Time taken for the sensor output to reach and stay within the range of final value (steady state) is called settling time. In general, 2% or 4% is the considered range of the final value for settling time calculation.

Rise time: Time taken for the sensor output to rise from 10% of the steady-state value to 90% of the steady-state value is called rise time.

Slew rate: The maximum rate at which the output of sensor can change.

$$\text{slew rate} = \frac{\text{change in output}}{\text{change in time}} = \text{slope of the output.}$$

SPEC SENSORS

ULPSM-SO2 968-006

October 2016

ABSOLUTE MAXIMUM RATINGS

Parameter	Conditions	Min.	Rec.	Max.	Units
Supply Voltage		2.7	3	3.3	V
Storage Temperature	Vapor sealed @ 50% RH	5	20	30	°C
Storage Humidity	Non-condensing, Vapor sealed	20	50	80	% RH
Storage Pressure	Vapor sealed	0.8	1	1.2	atm.
Storage Time	Vapor sealed	-	12	-	Months
Operating Temperature	< 10 hours	-40	-	50	°C
Operating Humidity	< 10 hours, Non-condensing	0	-	100	% RH
Operating Temperature	Continuous	-20	25	40	°C
Operating Humidity	Continuous, Non-condensing	15	50	95	% RH
Operating Pressure	Continuous	0.8	1	1.2	Atm.

Fig. 7.5 SPEC sensors and SO_2 sensor data sheet for environmental specifications

7.3.3 Environmental Specifications

Environmental specification tells about the performance of the sensors in the harsh environments, because the ambient atmospheric conditions will affect the output of the sensor. All the air quality sensors need to be operated under the rated temperature, humidity and pressure. Some of the manufacturers will indicate both the operating as well as storage environmental specifications. Different environmental specifications of SO_2 sensor under rated conditions manufactured by SPEC sensors are shown in Fig. 7.5.

7.3.4 Electrical Specification

Operating current, voltage and power, etc., are come under electrical specifications. If the sensor is having multiple pins/ports, each pin will be specified with the rated operating conditions. Electrical specification of SO_2 sensor manufactured by SPEC sensors is shown in Fig. 7.6. Before integrating the sensor in the real-time monitoring

Parameter	Conditions	Min.	Typ.	Max.	Units
Supply Current	V+ = 3.0 V	5	10	15	μA
Power Consumption	V+ = 3.0 V	15	30	45	μW
Vref			V+/2 +0.1		V
Vgas Zero		(V+/2 + 0.1) – 0.005	(V+/2 + 0.1)	(V+/2 + 0.1) + 0.005	V
Vgas Span (M)	Room temperature	2.5	3.0	3.5	mV/ppm

Fig. 7.6 SPEC sensors and SO_2 sensor data sheet for electrical specifications

sensor board (PCB), the electrical compatibility with battery/line should be maintained. If source and sensor are not in the same range of electrical specifications, voltage/current regulators, etc., will be useful for electrical compatibility.

7.4 Development of Air Quality Sensors

Based on the physical, chemical and other important characters of air pollutants, the sensors will be developed (Maag et al. 2018). In general, particulate matter sensors are based on light scattering principles and gaseous sensors are utilizing semiconductor and electrochemical principles. Apart from these, other techniques like infrared radiation absorption, etc., are also popular.

7.4.1 Semiconductor

Metal oxide-based semiconductors such as ZnO and SnO_2 are highly sensible for both reducing gasses like H_2, H_2S, CO, ethanol, etc., and oxidizing gasses like O_2, NO_2. Whenever these gases exposed to the above-mentioned material-based sensors, the conductivity changes. Hence, the flow of current changes in the electrical circuit associated with sensor, by which we can measure target gas concentration. Most of the metal oxide semiconductor sensors are the passive sensors. Thermistor is the best example of this category.

7.4.2 Electrochemical

An electrochemical sensor consists of two electrodes, a working electrode and counter electrode in an electrolyte solution. When target gas is exposed to the electrochemical gas sensor the gaseous molecules either oxidized or reduced at working electrode, which causes charge generation or consumption. If we provide a closed loop, the current will flow. The amount of current flow is proportional to the targeted gas concentration.

7.4.3 Optical Principle (Light Scattering)

The pollution laden air is captured in a small chamber. When light passed into the chamber, i.e., either LED or laser, its gets scattered by the particulate matter. Depending upon the concentration and size of the particulates, the intensity of scattered light will change. In general, visible light (~600 nm) is used in the light scattering instruments to measure particulate matter.

7.5 Calibration of Sensors

Adjustment of sensor readings against the standard measurement for a target concentration is called as calibration. Calibration is important, because sensor performance can change over time. As the air quality sensors can be used for different air pollution applications, their performance requirement will also be different depending upon the application (Artiola et al. 2004).

Currently, there is no Indian regulatory (Central Pollution Control Board, CPCB) guidelines and EPA (United States Environmental Protection Agency) also does not provide any guidelines for performance requirement of the indicative measurements using sensors for supplementary monitoring, as in USA, there is no defined role for supplementary sensor measurements in regulatory monitoring. However, European Union (EU) directive on "Ambient air quality and cleaner air for Europe" provides for the use of "indicative measurement" (Equivalence 2010). These measurements can be used to supplement "fixed" (or, "regulatory") measurements to provide information on the spatial variability of pollutant concentrations. These supplementary measurements have less stringent requirements for data quality. The EU requirements specify a maximum uncertainty, and do not address precision and bias separately. Table 7.2 gives the EU performance requirements for the fixed and indicative measurements. Table 7.3 gives the suggested performance goals for each tier (Tier I–V) in comparison with regulatory monitoring requirements.

The studies shown that the calibration can be done in two ways

1. Pre-deployment calibration
2. Post-deployment calibration.

7.5.1 Pre-deployment Calibration

Calibrating sensor before deployment in the real-time environment is called as pre-deployment calibration. Generally, this calibration will be done under laboratory

Table 7.2 EU performance requirements for the fixed and indicative measurements

Type of measurement	Maximum uncertainty allowable in pollutant measurement			
	SO_2, NO_2, CO (%)	Benzene (%)	PM and lead (%)	Ozone (%)
Regulatory (fixed)	15	25	25	15
Supplemental (indicative)	25	30	50	30

Table 7.3 EU suggested performance goals for different application areas

Tier	Application area	Pollutants	Precision and bias error (%)	Data completeness (%)
I	Education and information	All	<50	≥50
II	Hotspot identification and characterization	All	<30	≥75
III	Supplemental monitoring	Criteria pollutants, air toxics (incl. VOCs)	<20	≥80
IV	Personal exposure	All	<30	≥80
V	Regulatory monitoring	O_3, CO, SO_2, NO_2, $PM_{2.5}$, PM_{10}	<7 <10 <15 <10	≥75

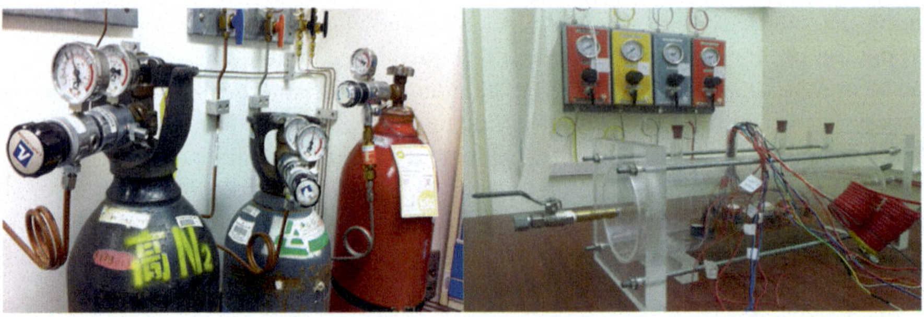

Fig. 7.7 Pre-deployment calibration setup in air quality research laboratory, IITM

conditions for the gain and offset adjustment by drawing the calibration curve, which shows the relationship between input gas concentration and output values (Williams et al. 2017). Offset indicates the sensor response value when target gas is absent. Figure 7.7 shows the pre-deployment calibration setup for NO_2, CO, O_3 gasses in air quality research laboratory, IIT Madras.

7.5.2 Post-deployment Calibration

Calibrating sensors after deployment in the field is called as post-calibration. The frequent post-calibration will reduce the errors in the measurement (Maag et al. 2018). The duration for the post-calibration depends upon the technology used to manufacture sensors (in general, semiconductor sensors exhibit more drift with time), the real-time weather conditions, concentration at which we are measuring, etc.

The studies have been shown different post-deployment calibration techniques like liner regression, multivariate linear regression, artificial neural networks, machine learning, etc.

7.6 Data Acquisition

Data acquisition system for air pollution measurement starts with the identification of right sensor (measuring range, accuracy range, sensitivity, etc.) for the pollutant. Once the sensors identified, they should have to calibrate under the laboratory conditions (temperature, humidity, pressure, etc.) by exposing sensors to the known gas concentrations. Compare the calibrated sensors output with the standard measuring devices output. Check for the errors and correct them either by recalibrating or any controlled mechanism. Then, all the sensor used for pollutants measurement is placed on a board (in general, PCB) called sensor board, which is either battery powered, or line powered along with the necessary signal conditioning circuit for every sensor. At present, the manufacturer is providing inbuilt signal condition for most of the sensors. Inbuilt signal condition is beneficial to reduce errors. The output from the sensors is integrated into the microcontroller or processor (personal computer) through the digital interface (serial (RS232) or parallel (IEEE 1284)) for further processing and storage. If any sensor is giving analog output, then analog-to-digital converters (ADC) must use to convert the obtained response in digital format. By using the transceiver, the data can be transferred to the central server. Different wireless communication techniques used for data transfer in the wireless sensors network listed in Table 7.4. Finally, the obtained information is used by the end user to understand, control and policy making. Figure 7.8 shows the data acquisition system for single sensor node. It is possible to create a sensor network by integrating large number sensor nodes to a single server to monitor air pollution at large area with spatial and temporal variations. The flowchart of sensor data acquisition is shown in Fig. 7.9.

Table 7.4 Wireless communication techniques used in wireless sensor networks

Technology	WIFI	Bluetooth	Zigbee
Standard	IEEE 802.11b	IEEE 802.15.1	IEEE 802.15.4
Range	100 m	(10–100) m	10–100 m
Data rate	(10–100) Mbps	125 kbps–3 Mbps	(20–250) kbps
Operating frequency range	(2.4/5) GHz	2.4 GHz	2.4 GHz

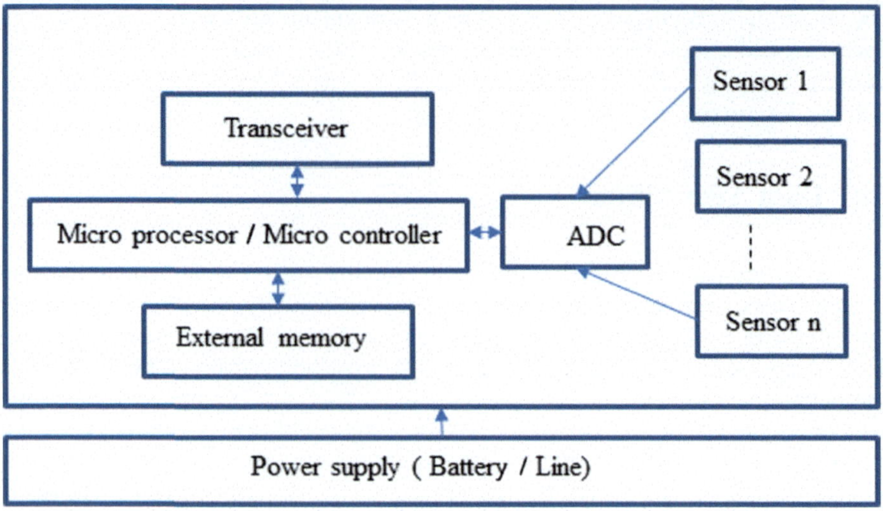

Fig. 7.8 Sensor node preparations

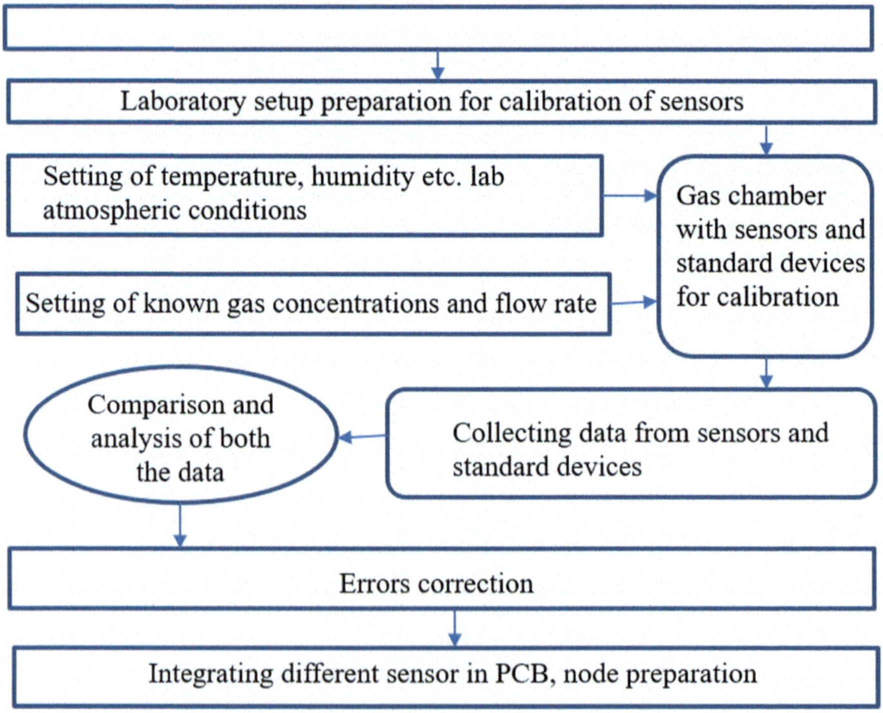

Fig. 7.9 Flowchart and data acquisition system preparation

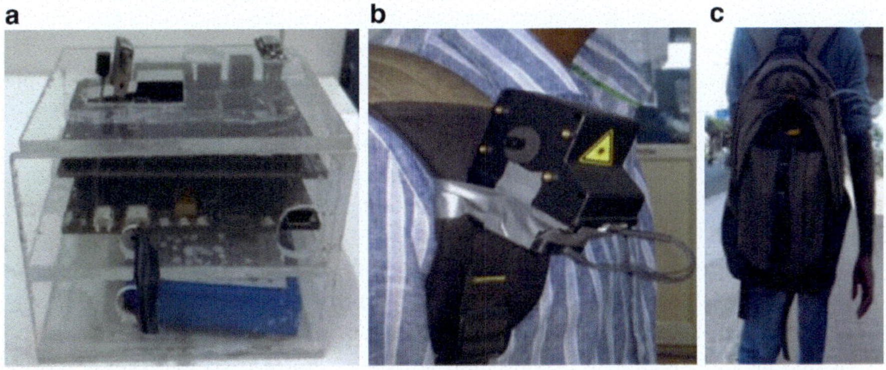

Fig. 7.10 Personal monitoring **a** personal monitoring sensor device, **b** attachment of device to a person, **c** pedestrian exposure monitoring

7.7 Smart Sensors for Mobile and Personal Monitoring

Personal exposure monitoring needs to take into consideration the individual exposure to air pollution (Ueberham and Schlink 2018). Personal monitoring can provide detailed insight into a person's individual short-term exposure in a specified area (SM et al. 2019). The recent studies showed the ways of personal exposure monitoring. The integrated sensor device with memory along with GPS device is carried out by the individual to get the personal exposure at different locations in different scenarios like inside the bus, pedestrian exposure with and without traffic scenarios. The miniaturization of the personal monitoring device will help in energy saving, hence, battery life, weight reduction, hence, carrying fatigue reduction, etc. Figure 7.10 shows the way of personal exposure monitoring.

7.8 Air Quality Maps

Air quality maps help us to visualize the air quality levels and spatial variation of pollutant concentrations in representative areas. With the advent of low-cost sensors measuring various air pollutants, there is significant potential for carrying out high-resolution mapping of air quality in the environment (Schneider.et al. 2017). Real-time maps are prepared by using real-time air quality observations of pollutant concentrations from ground stations. Using historical data, spatial and temporal variations of pollutant concentrations can also be depicted through air quality maps. This can be done with the help of interpolation techniques and GIS. Monitoring represents air pollution level for a particular point. Thus, spatial interpolation techniques are used to create a surface grid or contour map (Fig. 7.11). Interpolation processes estimate concentrations in the study area using known concentrations at fewer points. The other methods are regression techniques and dynamic modelling.

Some of the interpolation techniques used are the inverse distance weighted method and the kriging interpolation method. Inverse distance weighted (IDW) interpolation determines cell values using a linearly weighted combination of a set of sample points. The weight is a function of inverse distance. The surface being interpolated should be that of a locationally dependent variable. This method assumes that the variable being mapped decreases in influence with distance from its sampled location (Wang.et al. 2014). Kriging is an advanced geostatistical procedure that generates an estimated surface from a scattered set of points with *z*-values. Kriging assumes that the distance or direction between sample points reflects a spatial correlation that can be used to explain variation in the surface. Kriging is most appropriate when there is a spatially correlated distance or directional bias in the data.

Air quality maps can also be prepared with the help of dispersion models. But these models only provide estimates for known emission sources (Briggs et al. 1997). Even in urban areas, locally derived pollutants may make up only a small proportion of the total concentrations recorded at any site. It makes field validation difficult since what is monitored is not the same as what is modelled. Therefore, the modelled concentrations may represent only part of total exposures.

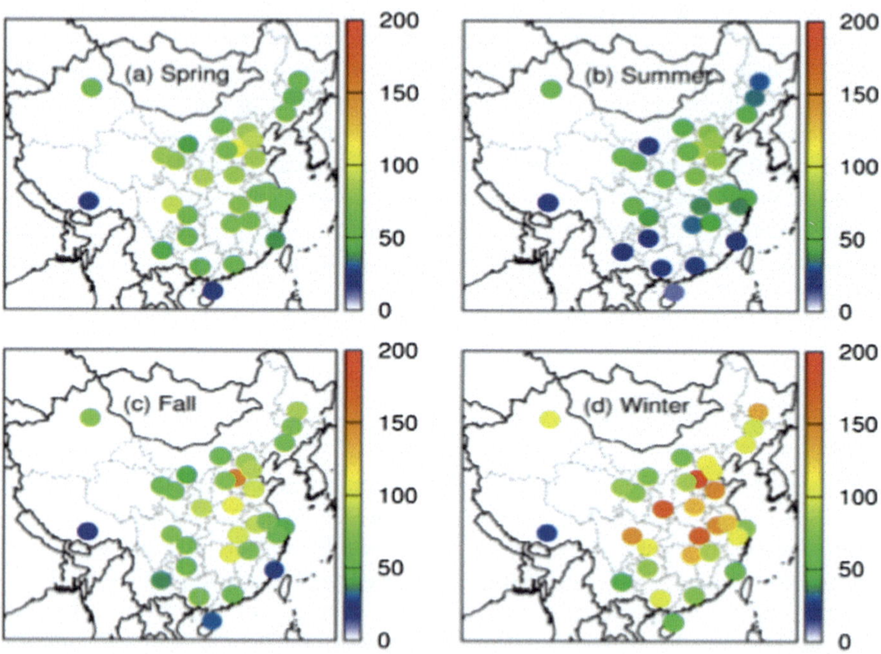

Fig. 7.11 Air quality maps showing temporal and spatial variation of pollutants generated using interpolation techniques. *Source* Wang et al. 2014

7.9 Limitations of Smart Air Quality Sensors

Sensor drift: Sensor drift refers to a gradual change in a sensor's response characteristics over time. Instruments drift may lead to wrongly conclude that concentrations have increased or decreased over time. Drift can be positive or negative, and it may occur due to a variety of reasons. One way to reduce drift is to calibrate the sensor frequently so that the instrument only drifts a small amount between each recalibration. The frequency of calibration needed will depend on how much drift occurs.

Accuracy limit: It is a known fact that the air quality sensors are not as accurate as standard instruments. But all the applications do not require cent percentage accuracy. Regular calibration will make sensors error-free and accurate.

Cross-sensitivity: Sensor response to other gases (which are not target gases) is called cross-sensitivity. The cross-sensitivity shows either positive (amplified value) or negative response (attenuated or reduced value) on the sensor performance. For example, CO sensor has positive response to the H_2 gas and SO_2 sensor has negative response to the NO_2 gas.

Apart from the above-mentioned issues, there exist software issues like data incompatibility, data interference, heterogeneous platforms, etc., and hardware issues like power failures, devices incompatibilities, network failures (for data transfer), etc.

References

Artiola J, Pepper IL, Brusseau ML (2004) Environmental monitoring and characterization. Elsevier

Briggs DJ, Collins S, Elliott P, Fischer P, Kingham S, Lebret E, Van Der Veen A (1997) Mapping urban air pollution using GIS: a regression-based approach. Int J Geogr Inf Sci 11(7):699–718

Equivalence DO (2010) Guide to the demonstration of equivalence of ambient air monitoring methods

Maag B, Zhou Z, Thiele L (2018) A survey on sensor calibration in air pollution monitoring deployments. IEEE Internet Things J 5(6):4857–4870

Neubert HK (2003) Instrument transducers-an introduction to their performance and design, 2nd edn. Oxford University Press, Oxford

Schneider P, Castell N, Vogt M, Dauge FR, Lahoz WA, Bartonova A (2017) Mapping urban air quality in near real-time using observations from low-cost sensors and model information. Environ Int 106:234–247

SM SN, Yasa PR, Narayana MV, Khadirnaikar S, Rani P (2019) Mobile monitoring of air pollution using low cost sensors to visualize spatio-temporal variation of pollutants at urban hotspots. Sustain Cities Soc 44:520–535

Ueberham M, Schlink U (2018) Wearable sensors for multifactorial personal exposure measurements – A ranking study. Environ Int 121:130–138

Usher MJ, Keating DA (1996) Sensors and transducers: characteristics, applications, instrumentation, interfacing. Macmillan International Higher Education

Vincent JH (2007) Aerosol sampling: science, standards, instrumentation and applications. Wiley

Wang Y, Ying Q, Hu J, Zhang H (2014) Spatial and temporal variations of six criteria air pollutants in 31 provincial capital cities in China during 2013–2014. Environ Int 73:413–422

Williams R, Conner T, Clements A, Foltescu V, Nthusi V, Jabbour J, Srivastava M (2017) Performance evaluation of the united nations environment programme air quality monitoring unit. United States Environmental Protection Agency, Washington, DC, USA

Part II
Case Studies

Chapter 8
Short-Term Variation of Particulate Matter and Black Carbon During Deepawali Festival in an Urban Area

Abhilash T. Nair, S. Devaanandan, and S. M. Shiva Nagendra

8.1 Introduction

The urban atmosphere usually comprises of high concentrations of particulate matter (PM) and black carbon (BC) from sources like biomass burning, vehicular emissions and industrial processes. The epidemiological studies related to air pollution have proved that high concentration of PM can significantly affect the respiratory system and pose several health impacts to humans (Jayanthi and Krishnamoorthy 2006). Similarly, PM impacts on ecological systems and climate and the associated risks are also well studied (Feng et al. 2016). During the festival, excessive burning of crackers increases the PM and BC concentrations in the urban environment. The fireworks generate smoke which usually contains black dust comprising of charcoal, potassium nitrate and sulphur, formed from the emission of trace gases and BC particulates (Sharma et al. 2016; Nishanth et al. 2012). Further, due to the influence of meteorological conditions, these emissions affect the visibility in the urban area up to several hours (Yerramsetti et al. 2013).

A. T. Nair (✉) · S. Devaanandan · S. M. Shiva Nagendra
Department of Civil Engineering, Indian Institute of Technology Madras, Chennai, Tamil Nadu 600036, India
e-mail: nairabhilasht@gmail.com

S. Devaanandan
e-mail: dev.27anand@gmail.com

S. M. Shiva Nagendra
e-mail: snagendra@iitm.ac.in

A. T. Nair
Department of Applied Sciences and Humanities, National Institute of Foundry and Forge Technology, Ranchi 834003, India

S. Devaanandan
Puducherry Pollution Control Committee (PPCC), Puducherry, India

S. M. Shiva Nagendra et al. (eds.), *Urban Air Quality Monitoring, Modelling and Human Exposure Assessment*, Springer Transactions in Civil and Environmental Engineering,
https://doi.org/10.1007/978-981-15-5511-4_8

The firework display during Deepawali festival is an increasing concern as they contribute significantly to air pollution in India (Mandal and Prakash 2012). The fireworks generally contain several heavy metals and toxic compounds like copper, cadmium, lead, magnesium, sodium, zinc, nitrate and nitrite which can have a detrimental effect on human health (Mandal and Prakash 2012). The smoke emitted by the fireworks are responsible for eye irritation, coughing, sneezing, asthma, wheezing, etc. (Mandal and Prakash 2012; CPCB 2014). The previous studies by Central Pollution Control Board (CPCB), India, have reported PM concentration higher than 1000 $\mu g/m^3$ across different parts of the country during Deepawali celebrations (CPCB 2014).

Several studies across India have confirmed that use of fireworks have degraded the air quality. In Delhi, the PM_{10} concentration increased up to 317–616 $\mu g/m^3$ due to use of fireworks during festivals (Sarkar et al. 2010). Similarly, the average level of PM_{10} and $PM_{2.5}$ throughout Deepawali in Nagpur was reported to be 930 and 271 $\mu g/m^3$, respectively, exceeding the National Ambient Air Quality (NAAQ) Standards of 100 and 60, $\mu g/m^3$ respectively (Rao et al. 2012). In 2005, the 24-h average PM_{10} level in Lucknow was determined to be 753 $\mu g/m^3$ (Barman et al. 2008). Hyderabad city witnessed an increase in concentrations of barium by 1091 times, potassium by 25, aluminium by 18 and strontium by 15 times due to firework combustion during Deepawali in 2002 (Kulshrestha et al. 2004). Similarly, the concentrations of several heavy metals like barium, potassium, strontium, magnesium, sodium, sulphur, aluminium, chloride, manganese, calcium and elemental carbon (EC) in Delhi increased by 264, 18, 15, 5.8, 5, 4, 3.2, 3, 2.7, 1.6 and 4.3 times, respectively, on Deepawali in comparison with background values (Sarkar et al. 2010). Air quality monitoring carried out in Delhi in 2010, showed that samples collected during Deepawali are comprised of sphere-shaped aerosols containing 56.95% metals, 1.89% BC and 1.11% aromatic organic carbon (AOC) (Agrawal et al. 2011). These studies prove that use of fireworks during festival are a big pollutant releasing events, releasing high quantities of metals and carbon into the atmosphere which affect the atmospheric chemistry for a long time (Agrawal et al. 2011). The present study aims to measure the variation in concentration of PM and BC during the 2016 Deepawali festival in Chennai city.

8.2 Materials and Methods

8.2.1 Study Area

Chennai is the fourth-largest city in India with 4.68 million population (Krishnamurthy and Desouza 2015). Foreign embassies, multinational companies, educational organizations, automobile manufacturing and cultural societies, etc., have contributed tremendously to the growth of the city. The spatial area of Chennai has expanded from 68 to 426 km^2 since 1901 (Krishnamurthy and Desouza 2015).

Table 8.1 Meteorological data of Chennai city during the study period

Date	Temperature (°C)	Humidity (%)	Wind speed (m/h)	Rainfall (mm)
27–Oct	29	58	1.4	0
28–Oct	29	57	0.8	0
29–Oct	29	69	2.2	0
30–Oct	27	84	1.7	3.05
31–Oct	29	77	1.7	0.25
01–Nov	28	82	5	0
02–Nov	27	89	2	0
03–Nov	27	86	5	2.03

Air quality monitoring was carried out during the 2016 Deepawali festival at two locations in Chennai. The characteristics of monitoring locations are residential-cum-commercial area. The detailed description of the locations and monitoring protocols is given below. Both the monitoring locations were approximately 8 km away from each other. Various meteorological factors like wind speed, humidity, temperature and precipitation data of Chennai city collected from wunderground.com during monitoring period are shown in Table 8.1.

8.2.1.1 Monitoring Location 1

Air quality monitoring station was located near to Velachery Gate of IIT Madras campus (12°59′16.07″N, 80°13′23.46″E). The location experiences vehicular traffic due to the vehicles moving in and out of the campus. Similarly, the traffic roads near the monitoring station are one of the busy roads with high traffic density (Fig. 8.1). Samples of PM_{10} and $PM_{2.5}$ were collected together during 27/10/2016–03/11/2016 to account for pre-Deepawali, during Deepawali and post-Deepawali days.

The mass concentrations of PM_{10} and $PM_{2.5}$ were determined using high volume sampler, APM 460 NL and APN 550 (Envirotech Instruments Pvt. Lt. India), respectively. Teflon filters with pore size 0.2 μm supported by polypropylene ring (Whatman International Limited, USA) were used to collect $PM_{2.5}$ samples, while glass microfiber filters (Whatman International Limited, USA) were used to collect PM_{10} samples (Whatman International Limited, USA). The weights of Teflon filters were measured prior and after exposure to ambient air using an analytical microbalance (Sartorius, ME5-F) having ±1 μg sensitivity. The gravimetric errors while handling the filters were estimated by using field and laboratory blanks. The mass concentrations of PM were calculated gravimetrically, from the difference in weight of filter paper divided by air flow rate.

Fig. 8.1 Ariel view of monitoring locations 1 and 2 in Chennai during Deepawali festival. *Source* Google earth

8.2.1.2 Monitoring Location 2

Ambient air quality monitoring was also conducted on the terrace of Kesari Primary Telugu School (13° 2′23.26″N, 80° 14′44.51″E) on Deepawali day (29/10/2016). The school is located on Theyagaraya Road (Fig. 8.1) situated near to Theyagaraya Nagar (T Nagar) which is considered to be one of the most prominent shopping locations in Chennai. The monitoring was carried out from 29/10/2016, 6:00 am to 30/10/2016, 6:00 am. The site is located approximately 150 m away from the roads with a high density of vehicles of all types, i.e. heavy, medium and light.

The concentration of PM and its size distribution at monitoring location 2 were obtained using Portable Aerosol Spectrometer Model 1.108 (GRIMM Aerosol Technik Gmb & Co, Germany). The instrument uses the principle of scattering of light sourced by a semiconductor laser by particles to measure the PM concentrations.

Aethalometer (Model AE42) (Magee Sci., Inc., USA,) was used for continuous real-time monitoring of BC concentration. In aethalometer, the particulates are continuously sampled on the filter, and the light absorption at 880 nm is measured.

At 880 nm, light absorption by other aerosol particles is considerably low, and hence, the absorption was considered to be of BC alone.

8.3 Results and Discussion

8.3.1 Variation in PM Concentrations

To analyse the variation in PM concentrations released during Deepawali festival, samples were collected for 7 days through the Deepawali day. The deviations in PM levels measured during the monitoring period in location 1 are presented in Fig. 8.2. The PM_{10} and $PM_{2.5}$ concentration varied from 86–119 $\mu g/m^3$ to 28–90 $\mu g/m^3$. However, the results indicated that between 28 and 30 October 2016, the level both PM_{10} and $PM_{2.5}$ exceeded the NAAQS of 100 $\mu g/m^3$ and 60 $\mu g/m^3$, respectively. This increased concentration of PM could be due to the additional effect of heavy traffic density and use of fireworks during the festival period. Usually, during festivals, public travel within the city for buying new clothes and household materials and visit close friends and relatives to exchange gifts, leading to increased vehicular density on the roads.

The $PM_{2.5}/PM_{10}$ ratio varied from 0.7 to 0.86 between 28 and 30 October 2016 indicating high concentration of fine particles throughout the monitoring period. This can be attributed to higher traffic flow in the adjacent road and excessive use of fireworks. After 30 October 2016, the concentrations of PM_{10} and $PM_{2.5}$ were less than the prescribed NAAQ standards. Hence, it can be concluded that the fireworks and

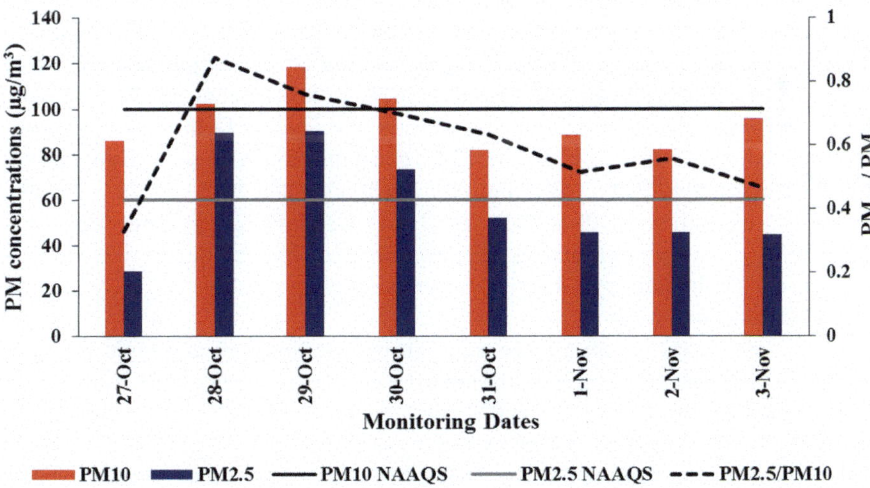

Fig. 8.2 Daily average PM_{10} and $PM_{2.5}$ concentrations at monitoring location 1 during Deepawali festival

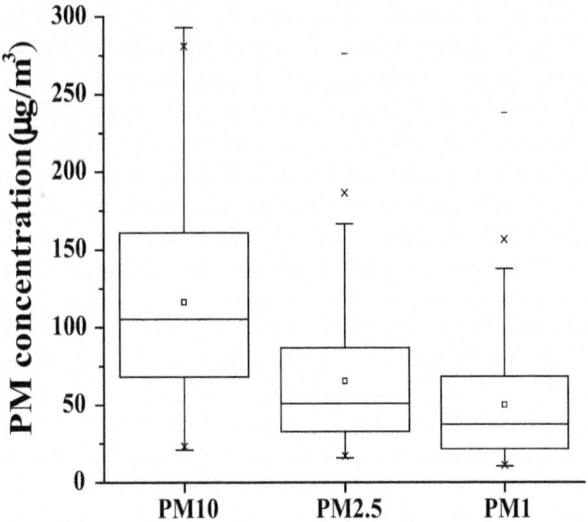

Fig. 8.3 Box plot indicating the concentration of PM_{10}, $PM_{2.5}$ and PM_1 at monitoring location 2 during Deepawali festival (28th October, 2016)

increased traffic density during Diwali festival had increased in PM mass concentration during Deepawali days. PM_{10} is emanated from crushing, or grinding operations and dust stirred up by vehicles on roads, while $PM_{2.5}$ is released into the atmosphere from various combustion activities including biomass burning in residential properties, exhaust emissions from vehicles, stack emissions power plants, forest fires and various other industrial processes.

The 24-h average concentrations of PM_{10}, $PM_{2.5}$ and PM_1 observed during Deepawali festival at monitoring station 2 are presented in Fig. 8.3. The mean PM_{10}, $PM_{2.5}$ and PM_1 levels during Deepawali celebrations were 116, 65, and 49 $\mu g/m^3$, respectively. The PM concentration reported in Chennai was less in comparison with the other cities of India. The PM_{10} concentration in Nagpur reached up to 500 $\mu g/m^3$ during Deepawali celebration in 2008 (Khaparde et al. 2012). Similarly, in 2009 Deepawali celebration, the 24-h average PM level in Delhi was 851 $\mu g/m^3$ (Agrawal et al. 2011). The 24-h average PM_{10} concentration in Chennai during Deepawali festival reported by CPCB in the year 2013, 2014 and 2015 was 215, 193 and 106 $\mu g/m^3$, respectively (CPCB 2014, 2013; CPCB, MoEF 2015). This decreasing trend in PM_{10} concentration during Deepawali festival may be due to reduced firework activity by the public as a result of intense awareness about its adverse effect on environment and health.

The use of firecracker on a large scale during the morning and night hours of Deepawali festival resulted in the emission of various toxic gases and PM into the air. Hence, considerable variation in the PM concentration was observed throughout the day (Fig. 8.4). The variation in concentrations of PM_{10}, $PM_{2.5}$ and PM_1 throughout the day showed a similar trend. The PM_{10}, $PM_{2.5}$ and PM_1 concentration peaked at morning (7:00–8:00 am) and night hours witnessed (18:00–21:00) as the people expressed their festive joy by bursting firecrackers during this time. The maximum

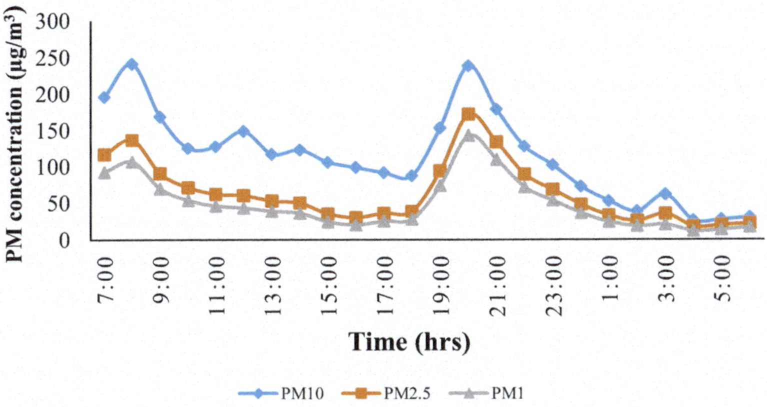

Fig. 8.4 Diurnal variations of particulate matter concentration at monitoring location 2 during Deepawali festival (28th October, 2016)

hourly average ambient concentration of PM_{10}, $PM_{2.5}$ and PM_1 was 238, 171 and 142 $\mu g/m^3$, respectively during the night hours. However, the city experienced rainfall of 3 mm (Table 8.1) during the night which helped to washout the pollutants from the atmosphere to some extent and reduce the PM_{10} concentration to below 50 $\mu g/m^3$.

It is evident from diurnal variation of PM concentration (Fig. 8.4) that during night hours (19:00–23:00), the city experienced very high PM concentration and that significant portion of PM is comprised of $PM_{2.5}$ and PM_1 which might be contributed by burning fireworks. Figure 8.5 shows that the substantial portion of $PM_{2.5}$ during the concentration is constituted of particle size between 0.265 and 0.575 µm. The previous studies in Pune also observed that the concentration of PM prevailing during Deepawali festival period was mostly of size 0.2 µm and from 3 to 4 µm (Devara et al. 2015). Scanning electron microscope (SEM) analysis of samples collected in Delhi city has shown that the aerosols released during Deepawali have a spherical shape with particle diameter of approximately 9.5 µm with settling velocity of 5.7 $\times 10^{-3}$ m/s (Agrawal et al. 2011).

8.3.2 *Variation in Concentration of BC*

BC particles absorb solar emission to a large extent. Therefore, BC is considered to be the second most significant contributor to global warming (Raju et al. 2014). The variations in BC concentrations with different time averages at monitoring locations 1 and 2 are presented in Figs. 8.6 and 8.7, respectively. The hourly average BC concentration increased up to 8800 ng/m^3. The higher levels of BC particles observed during Deepawali festival days were attributed to fireworks and increased vehicular movement, leaving the by-products of incomplete combustion within the

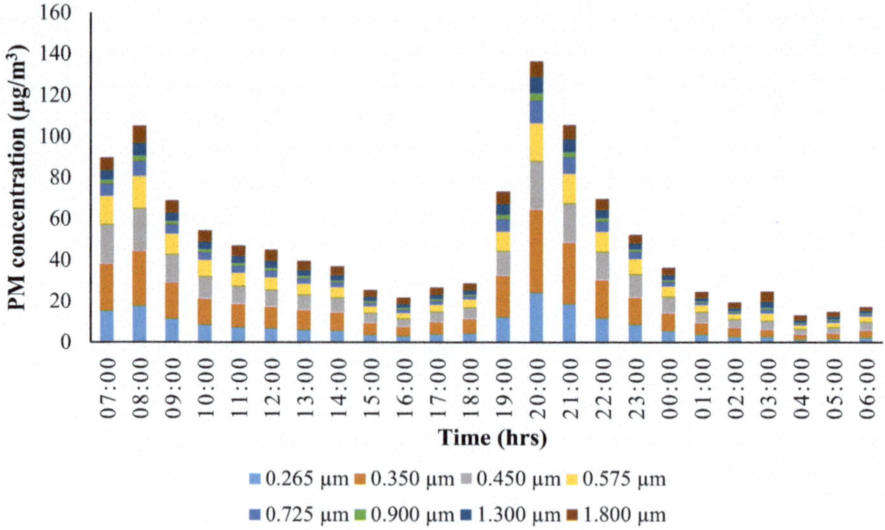

Fig. 8.5 Particle size distribution of $PM_{2.5}$ observed at monitoring location 2 during Deepawali festival (28th October, 2016)

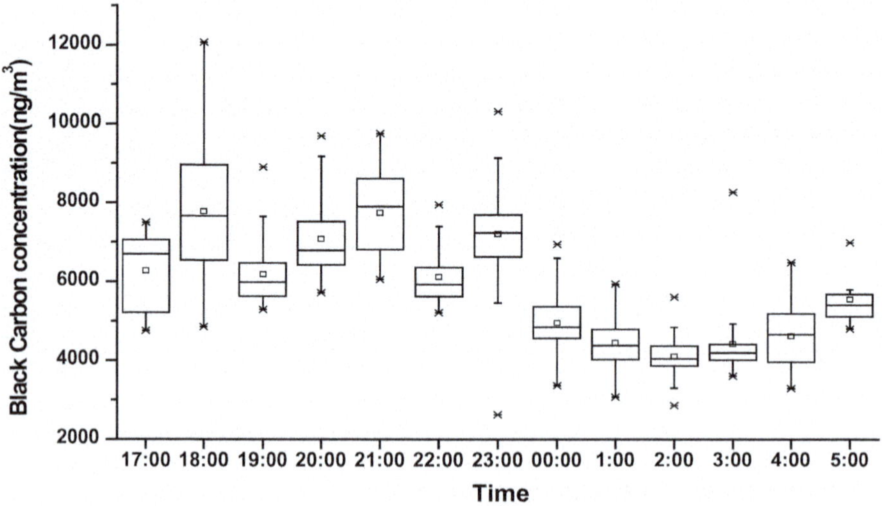

Fig. 8.6 Hourly average BC concentration at monitoring location 1

lower troposphere. The previous studies have reported that fireworks increased BC concentration in Hyderabad during the festival by 2–3 times (Yerramsetti et al. 2013). Daily average BC concentration increased by 5% during Deepawali festival due to fireworks (Raju et al. 2014).

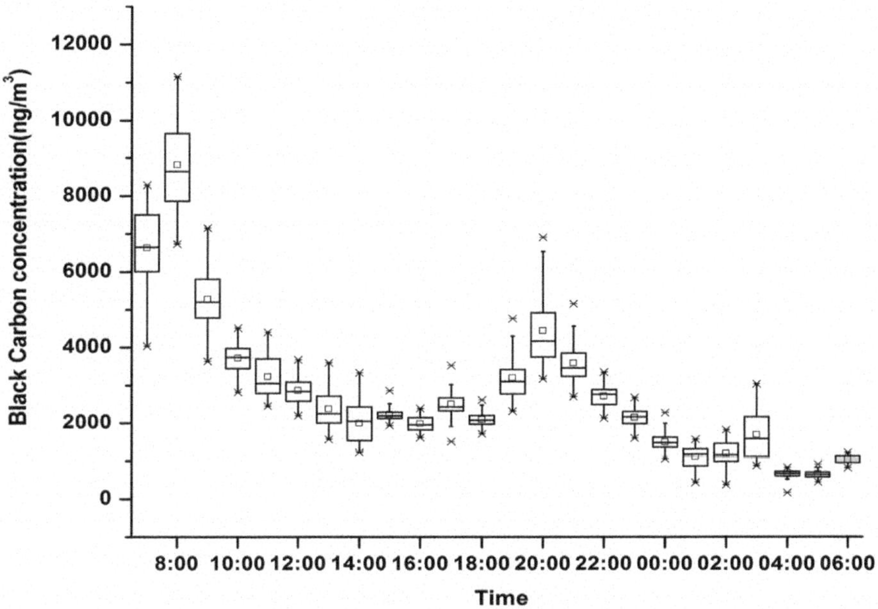

Fig. 8.7 Hourly average BC concentration at monitoring location 2

BC particles are generally small in size (<l μm) and hence can be easily inhaled and get collected in the respiratory systems (Raju et al. 2014). BC exposure has been reported as potential triggers of cardiorespiratory events (WHO 2012). It may act as carrier for transporting various toxic chemicals to the lungs (Raju et al. 2014; WHO 2012). The BC concentration started to rise from 16:00 h and attained maximum levels around 19:00 and 21:00 h at the monitoring locations 1 and 2, respectively. After 21:00 h, the BC concentration showed a steady dip in concentration throughout the night which may be due to reduced firecracker activity. The decreasing BC concentration was further supplemented by the scavenging effect of rains during the night. Also, there was a noticeable concentration dip in the hourly mean BC concentrations in the afternoon hours 12:00 and 13:00 h during Deepawali day. During night, due to radiation inversion, the BC concentration near the surface is found increasing because of lower mixing depth. Similarly, after sunset, a shallow boundary layer is formed near the surface which again raises the surface BC concentration. Whereas, during the afternoon, higher boundary layer and increased ventilation coefficient provided higher mixing depth which reduced accumulation of BC concentration near the surface (Bapna et al. 2013).

The correlation between BC concentration and PM concentration with their equations and coefficient of correlation (R^2) are presented in Fig. 8.8. The findings propose that BC mass concentration is strongly correlated to PM mass concentration with the R^2 value of 0.88, 0.78 and 0.75 for PM_{10}, $PM_{2.5}$ and PM_1, respectively. The results indicated that BC present in the atmosphere was also contributing to the

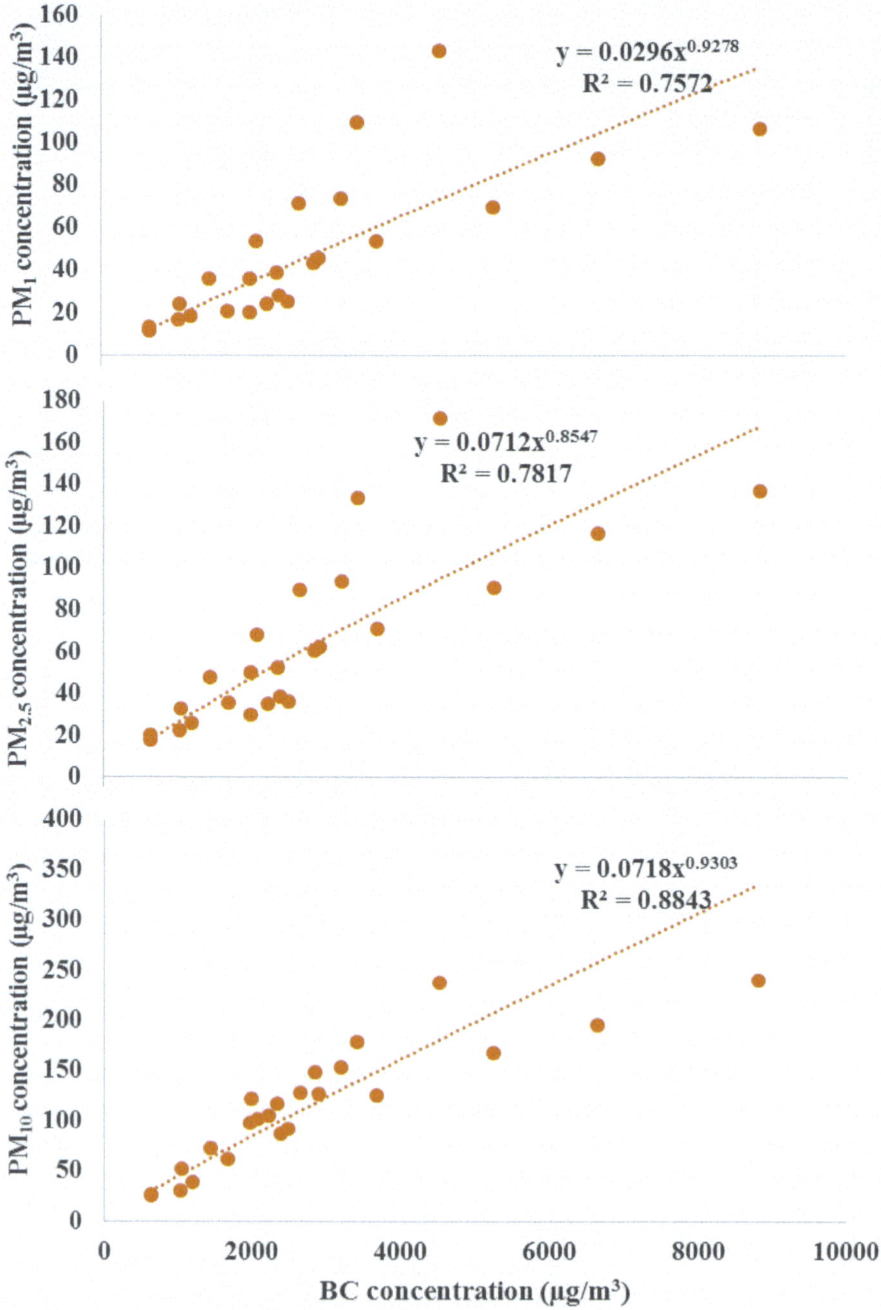

Fig. 8.8 Correlation plot between BC and PM concentration at location 2

PM concentration. Hence, high BC mass emissions correspond to high PM mass emissions.

8.4 Conclusion

The present study showed that the firework activity during the Deepawali festival had significantly increased the concentrations of both PM and BC over Chennai city. Daily mean levels of PM_{10}, $PM_{2.5}$ and PM_1 during Deepawali days were 116, 65 and 49 $\mu g/m^3$, respectively. The observed concentration levels of PM_{10} and $PM_{2.5}$ exceeded the NAAQS given by Central Pollution Control Board during the Deepawali celebrations due to the use of a large quantity of firecrackers that day and high vehicular density on the roads. The PM and BC concentrations were subsequently decreased after Deepawali festival. The detailed analysis showed that the PM concentration was very high during nighttime of Deepawali festival which is of grave concern. However, favourable meteorological conditions helped to lower the PM concentration subsequently. The interim exposure to the PM and BC above the NAAQ standards can increase the risk of severe health effects. This scientific study will help in raising consciousness among the public toward the ill effects of excessive fireworks on the ambient air quality and public health.

Acknowledgements First author would like to thank the Science and Engineering Research Board (SERB), India, for financial support through File no. PDF/2016/003683.

References

Agrawal A, Upadhyay VK, Sachdeva K (2011) Study of aerosol behavior on the basis of morphological characteristics during festival events in India. Atmos Environ 45:3640–3644

Bapna M, Sunder Raman R, Ramachandran S, Rajesh TA (2013) Airborne black carbon concentrations over an urban region in western India-temporal variability, effects of meteorology, and source regions. Environ Sci Pollut Res 20:1617–1631

Barman SC, Singh R, Negi MPS, Bhargava SK (2008) Ambient air quality of Lucknow City (India) during use of fireworks on Diwali Festival. Environ Monit Assess 137:495–504

CPCB (2013) Ambient air quality and noise levels during Deepawali 2013. Central Pollution Control Board, New Delhi

CPCB (2014) Deepawali festival 2014 : ambient air quality and noise levels. Central Pollution Control Board, New Delhi

CPCB, MoEF (2015) Final report on air quality status in the National Capital Region

Devara PCS, Vijayakumar K, Safai PD, Raju MP, Rao PSP (2015) Celebration-induced air quality over a tropical urban station, Pune, India. Atmos Pollut Res 6:511–520

Feng J, Yu H, Su X, Liu S, Li Y, Pan Y et al (2016) Chemical composition and source apportionment of PM2.5 during Chinese Spring Festival at Xinxiang, a heavily polluted city in North China: fireworks and health risks. Atmos Res 182:176–188

Jayanthi V, Krishnamoorthy R (2006) Key airborne pollutants—Impact on human health in Manali. Chennai Curr Sci 90:405–413

Khaparde VV, Pipalatkar PP, Pustode T, Rao CVC, Gajghate DG (2012) Influence of burning of fireworks on particle size distribution of PM10 and associated barium at Nagpur. Environ Monit Assess 184:903–911

Krishnamurthy R, Desouza KC (2015) Chennai, India. Cities 42:118–129

Kulshrestha UC, Nageswara Rao T, Azhaguvel S, Kulshrestha MJ (2004) Emissions and accumulation of metals in the atmosphere due to crackers and sparkles during Diwali festival in India. Atmos Environ 38:4421–4425

Mandal P, Prakash M (2012) Impact of Diwali celebrations on urban air and noise quality in Delhi City, India. Environ Monit Assess 184:209–215

Nishanth T, Praseed KM, Rathnakaran K, Satheesh Kumar MK, Ravi Krishna R, Valsaraj KT (2012) Atmospheric pollution in a semi-urban, coastal region in India following festival seasons. Atmos Environ 47:295–306

Raju MP, Safai PD, Rao PSP, Tiwari S, Devara PCS (2014) Impact of anthropogenic activity and cyclonic storm on black carbon during winter at a tropical urban city, Pune. Nat Hazards 71:881–894

Rao PS, Gajghate DG, Gavane AG, Suryawanshi P, Chauhan C, Mishra S et al (2012) Air quality status during Diwali festival of India: a case study. Bull Environ Contam Toxicol 89:376–379

Sarkar S, Khillare PS, Jyethi DS, Hasan A, Parween M (2010) Chemical speciation of respirable suspended particulate matter during a major firework festival in India. J Hazard Mater 184:321–330

Sharma SK, Sharma A, Saxena M, Choudhary N, Masiwal R, Mandal TK et al (2016) Chemical characterization and source apportionment of aerosol at an urban area of Central Delhi, India. Atmos Pollut Res 7:110–121

WHO (2012) Health effects of black carbon. Copenhagen Ø, Denmark

Yerramsetti VS, Sharma AR, Gauravarapu Navlur N, Rapolu V, Dhulipala NSKC, Sinha PR (2013) The impact assessment of Diwali fireworks emissions on the air quality of a tropical urban site, Hyderabad, India, during three consecutive years. Environ Monit Assess 185:7309–7325

Chapter 9
Surface O_3 and Its Precursors (NOx, CO, BTEX) at a Semi-arid Site in Indo-Gangetic Plain: Characterization and Variability

Nidhi Verma, Anita Lakhani, and K. Maharaj Kumari

9.1 Introduction

The term "Air pollution" can be defined as contamination of air by the gaseous and particulate matter which affects life on the earth. The air pollutants are generally categorized into gaseous and particulate pollutants. Pollutants emitted from natural sources like trees, volcanoes and dust storms do not make a significant contribution unlike emissions from anthropogenic sources which not only contribute primary emissions but also other harmful and non-degradable pollutants. The warming of the atmosphere has been observed since the preindustrial era due to the continuous rise in the levels of greenhouse gases (Pachauri and Meyer 2014). Among greenhouse gases, O_3 is the most important and paradoxical trace gas as in stratosphere, it absorbs harmful UV B radiations (250–300 nm) which can be absorbed by nucleic acids and can alter genetic sequences; hence the stratospheric ozone is good as it protects life on the earth. Stratospheric ozone defends us and life on earth cannot be imagined without it. However, in the troposphere O_3 is bad as it is a strong oxidant and it causes damage to crop and human health. Tropospheric ozone is a climate forcer; it is a greenhouse gas and traps outgoing long-wavelength radiation at 9.6 μm and contributes to radiative forcing after CO_2 and CH_4 (IPCC 2013). The effect of radiative forcing is more pronounced at higher altitudes in the troposphere (Gauss et al. 2003). Tropospheric ozone interacts with long-wave and short-wave solar radiation and is responsible for +0.40 W m^{-2} of mean global radiative forcing (IPCC 2013). Tropospheric ozone is secondary in origin which implies that it is not directly emitted from primary emission sources; two processes that predominantly contribute to tropospheric ozone levels are transferred from the stratosphere and

N. Verma · A. Lakhani · K. Maharaj Kumari (✉)
Department of Chemistry, Faculty of Science, Dayalbagh Educational Institute, Dayalbagh, Agra 282005, India
e-mail: maharajkumari.k@rediffmail.com

S. M. Shiva Nagendra et al. (eds.), *Urban Air Quality Monitoring, Modelling and Human Exposure Assessment*, Springer Transactions in Civil and Environmental Engineering,
https://doi.org/10.1007/978-981-15-5511-4_9

photochemical formation from its precursors (Lelieveld and Dentener 2000; Crutzen 1973). The photochemical generation of ozone is initiated by the attack of hydroxyl radicals on VOCs and CO to form peroxy radicals (RO_2) which react rapidly with nitric oxide (NO) to form nitrogen dioxide (NO_2). The generated NO_2 photolyzes to give oxygen atom which combines oxygen molecules to form ozone.

In the troposphere, nitrogen oxides are also released from biogenic and anthropogenic sources (Lee et al. 1997), the predominant source of NOx ($NO_2 + NO$) in the atmosphere is fossil fuel combustion and other sources are emissions from biomass-burning, lightning and soil (Seinfeld and Pandis 2006). Similar to NOx, CO is also emitted from natural (photochemical generation from biogenic hydrocarbons) and anthropogenic sources (incomplete combustion of fossil fuels released from vehicular and industrial sources) (Granier et al. 2000; Holloway et al. 2000). VOCs include a vast range of organic compounds and can be divided into different categories depending upon their source of emission and their composition (Patokoski et al. 2015). Aromatic hydrocarbons like Benzene, Toluene, Ethyl-Benzene and Xylenes represent an important fraction of VOCs emitted from anthropogenic sources. A group of these aromatic hydrocarbons is collectively known as BTEX. BTEX may contribute up to 60% of total non-methane hydrocarbons (Lee et al. 2002). Among BTEX, Xylenes are predominantly released from printing industries, textile industries, polishing and leather industries (Parra et al. 2009; Ling et al. 2011). However, Benzene is predominantly from vehicular sources rather than solvent evaporation (Yuan et al. 2013). Depending upon the duration of exposure BTEX can have different harmful health effects (Ueno et al. 2001; Han and Naeher 2006).

The study site is located in the North-central part of India and lies in Indo-Gangetic Plain (IGP) which is a highly polluted and populated region of India. IGP extends from the Indus river system to Ganges Delta. According to INTEX-B inventory given by Zhang et al. (2009), the emission fluxes of CO, NOx and NMHCs over IGP were 200–700, 5–20, 30–50 mol $km^{-2}\ h^{-1}$, respectively, however, emission fluxes of CO, NOx and NMHCs over rest of India were 50–200, 3–10, 5–30 mol $km^{-2}\ h^{-1}$, respectively. The emission fluxes over IGP are much higher than rest of the India. Being a secondary pollutant, ozone levels are significantly influenced by geography and meteorological conditions of a particular site. The chemical reactions affecting ozone formation also varies with site location. Earlier studies have not reported simultaneous measurements of ozone, NOx, CO and BTEX at the present site. Therefore, to understand ozone chemistry, the present study was aimed to simultaneously measure ozone and its precursors. In addition, source apportionment of BTEX using inter-species ratio and their ozone formation potential was also calculated. BTEX profile, their source apportionment, OFP and reactivity on the basis of propylene-equivalent concentration has not yet been reported at the site by earlier studies.

9.2 Methodology

9.2.1 Site Description

Measurements of surface O_3 and its precursors (NO, NO_2, CO and BTEX) were carried out at a semi-urban site, Dayalbagh, Agra (27° 10′N, 78° 05′E) during Jan–Dec, 2015 (Fig. 9.1). The study site is mostly surrounded by agricultural fields. The industrial area of the city lies about 6 km south of the site. According to the wind direction, the site is predominantly upwind in all the seasons except in monsoon season. On 2/3rd of its periphery, Agra is surrounded by Thar Desert of Rajasthan which makes it semi-arid region.

According to meteorological conditions, the site has four distinct seasons; summer (March to June), monsoon (July to September), post-monsoon (October and November) and winter (December to February). The average temperature was the highest in summer while lowest in winter season. The average humidity was highest in monsoon followed by winter, post-monsoon and summer. Surface ozone is a photochemically produced secondary air pollutant; therefore, its formation should be favourable under high temperatures and solar radiation conditions. For primary pollutants like BTEX, NOx and CO, stagnant weather conditions should be favourable for their accumulation, however, variation in emission sources and atmospheric dynamics also play significant role tow. Map of sampling site (shown by a star) at Dayalbagh, Agra, India.

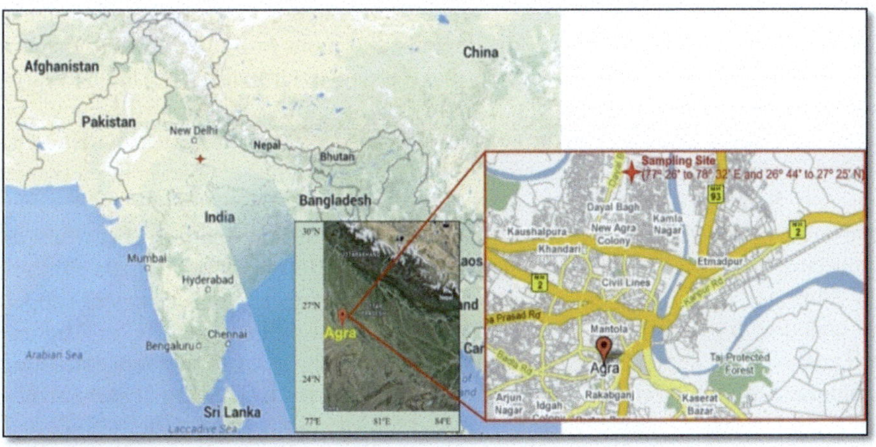

Fig. 9.1 Map of sampling site (shown by star) at Dayalbagh, Agra, India

9.2.2 O_3, NOx and CO Measurement

The levels of surface O_3, NOx and CO were measured through continuously operating O_3 (Thermo Fischer Model 49*i*), NOx (Thermo Fischer Model 42*i*) and CO (Teledyne T300) analyzers, respectively. The functioning of 49*i* is based on the absorption of UV radiations by ozone molecules at a wavelength of 254 nm. The analyzer determines ozone concentration using Lambert-Beer's law. NOx analyzer (42*i*) works on the principle of chemiluminescence. In this analyzer excited NO_2 molecules emit a characteristic luminescence and concentration of NO molecules is directly proportional to intensity of IR beam. The intensity of IR radiation is linearly proportional to the NO concentration. The NOx analyzer is specific for NO measurements and it converts NO_2 to NO through a molybdenum convertor which works at 325 °C. In CO analyzer, the beam of radiation passes through a bandpass filter which allows only light of 4.7 μm wavelength to pass. Then the light falls over a solid-state photo-detector which converts light signals into voltage signals. To maintain the accuracy of analyzers, they were regularly calibrated through a dynamic gas calibrator (Teledyne Model T700).

9.2.3 VOCs Measurement

The sampling of BTEX was carried out using activated charcoal tubes (Anasorb CSC, Coconut charcoal 226-01) procured from SKC Pvt Ltd, USA. Charcoal is a good adsorbing material for BTEX (Harper 2000). As it is electrically non-polar in character therefore, it adsorbs organic vapours and gases in preference to moisture. The advantage of using Anasorb CSC is that it has an extensive surface area, as large as 1200 m^2 per gram of material which greatly enhances its adsorption efficiency. The dimensions of the tubes used in the study are 70 mm length, 6 mm outer diameter and 1 mm thickness. These tubes contain 150 mg (100 + 50 mg) of adsorbent (activated charcoal). Samples were collected through an active sampling technique by drawing air using a pump (Model 224-PCXR8 SKC, USA) through an adsorbent tube at a constant flow rate (100 ml min^{-1}) for one hour. Before sampling, the sampler was calibrated using rotameter. After sampling, both the ends of tubes were capped and sealed with parafilm M and then kept in sealed plastic bags at <4 °C in a freezer until analysis. For analysis, desorption of tubes was done using the solvent extraction technique. Extraction process was initiated by transferring the content of sorbent tube into a 2 ml brown coloured glass vial then 1 ml of carbon disulphide (CS_2) (99% pure) was added and ultrasonicated for 15 min by placing the vial into ice-cooled beaker.

Meteorological parameters were recorded using Automatic Weather Station (WM271). PBL height data was taken from ERA-interim reanalysis. Principal component analysis (PCA) was applied using a statistical tool SPSS 16.0 (Statistical Packages for Social Sciences).

9.3 Results and Discussion

9.3.1 *Diurnal Variation of O_3 and Its Precursors (NO_2 and CO)*

Figure 9.2a shows the diurnal pattern of ozone, NO and NO_2. The average diurnal pattern of ozone was characterized by a minimum value of 17.9 ± 9.7 ppb during 7:00 h, reached the highest value of 53.7 ± 24.9 ppb during ~15:00 h. The nighttime low levels of ozone were due to loss through NO (Nishanth et al. 2014). NO and NO_2 showed bimodal patterns; morning and evening peak. The morning peak of NO (6.1 ppb) was observed at around 8:00 h and that of NO_2 (11.6 ppb) appeared 1 h after NO peak. The evening peaks of NO (7.7 ppb) and NO_2 (13 ppb) were observed around 22:00–24:00 h. Similar to NO_2, CO is also a precursor of surface O_3, therefore, it showed an almost reverse diurnal variation pattern to O_3. The diurnal variation of CO was characterized by bimodal peaks and low concentrations during the daytime (15:00–17:00 h). The peak levels of CO observed during the morning (09:00–10:00 h) and night hours (22:00–23:00 h). These peaks correspond to peak traffic emissions hours and boundary layer processes. At the study site, the nighttime peak (586 ± 471 ppb) was dominant as compared to the morning peak (523 ± 410 ppb). This disparity in the morning and nighttime peak may be attributed to the change in the planetary boundary layer (PBL) height. During morning hours, although traffic emission is high the dilution of air takes place due to rising PBL height whereas during nighttime, traffic emissions are almost similar but PBL height is low which results in the accumulation of pollutants near-surface (Chandra et al. 2015). The daily variation of PBL height is shown in Fig. 9.2b which clearly indicates high PBL height (171.7 m) during 09:00 h and comparatively low PBL height (54.9 m) during 21:00 h.

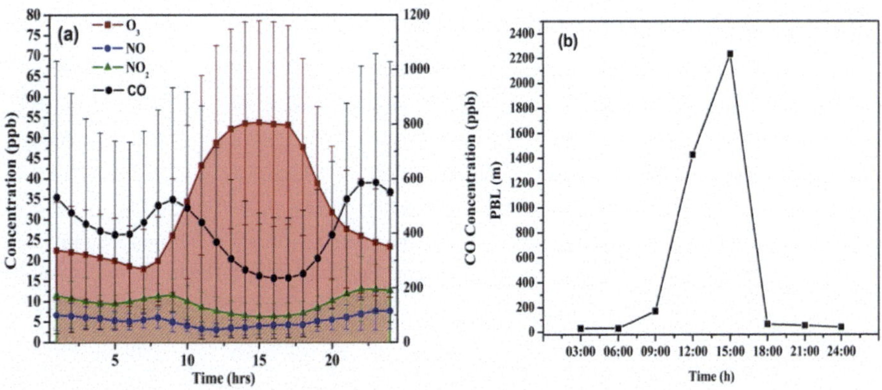

Fig. 9.2 **a** Diurnal variation of O_3, NO, NO_2 and CO during the study period **b** daily variation of PBL height

9.3.2 *Seasonal Variation of O_3 and Its Precursors (NO_2 and CO)*

Ozone showed distinct seasonal variation with the highest levels in summer season (46.6 ± 23.5 ppb) followed by post-monsoon (37.6 ± 24 ppb), monsoon (28.0 ± 15.1 ppb) and winter (25.1 ± 19.1 ppb) season. Low levels in winter season were attributed to a decrease in solar intensity and duration of solar radiation due to frequent foggy conditions (Ojha et al. 2012). In addition, in the winter season rate of dry deposition is fast in low and stable PBL. The maximum levels in summer season were attributed to high temperature, intense solar radiation and non-cloudy conditions which favour photochemical generation of ozone. Similar seasonal variation was also observed by Ojha et al. (2012) at a semi-urban site, Pantnagar.

Levels of both NO and NO_2 were high in winter season because PBL height was low which results in minimal convective activities and turbulent mixing which causes stagnation of primary pollutants near the surface. The highest morning peaks of NO and NO_2 were observed in February (11.5 ppb) and November (16 ppb) months, respectively. The highest evening peaks of both NO and NO_2 were observed in December (NO = 19.9 ppb, NO_2 = 21.4 ppb). The seasonal cycle of CO is affected by the strength of emission source, chemistry and transport. The highest levels of CO were observed in wither season (770 ± 466 ppb) followed by post-monsoon (466 ± 301 ppb), summer (232 ± 223 ppb) and monsoon (153 ± 122 ppb) season. The highest levels of CO in winter season may be due to (a) increase in local emission sources like coal or wood-burning to combat cold (in absence of any other domestic heating source) (b) stagnant weather conditions (c) and weak photochemical removal. In the winter season, foggy and hazy conditions are very frequent over IGP (Indo-Gangetic Plain) region (Gautam et al. 2007) and low wind speed causes trapping of pollutants near the surface. Table 9.1 shows a monthly variation of ozone, NOx and

Table 9.1 Monthly variation of O_3, NOx and CO

Month	O_3 (ppb)	NOx (ppb)	CO (ppb)
Jan	18.9 ± 14.8	15.6 ± 9.8	687.2 ± 543.5
Feb	32.2 ± 21.6	11.0 ± 4.2	463.0 ± 323.0
Mar	33.1 ± 19.2	13.3 ± 7.2	533.0 ± 353.0
Apr	39.3 ± 22.5	13.9 ± 5.5	472.8 ± 414.9
May	42.6 ± 22.8	14.6 ± 7.1	273.2 ± 335.9
Jun	46.9 ± 18.7	14.9 ± 8.2	158.6 ± 74.7
Jul	33.5 ± 15.3	8.6 ± 3.3	130.3 ± 90.5
Aug	21.0 ± 12.4	10.0 ± 5.2	161.2 ± 150.2
Sep	29.7 ± 16.2	9.3 ± 3.6	280.9 ± 200.8
Oct	40.1 ± 22.1	12.9 ± 8.3	642.2 ± 272.3
Nov	34.1 ± 25.5	11.6 ± 8.9	1174.4 ± 375.2
Dec	16.7 ± 10.5	11.8 ± 11.2	1268.6 ± 631.2

Table 9.2 Mean, median and range of BTEX ($\mu g\ m^{-3}$) at Dayalbagh, Agra

VOC	Mean ($\mu g\ m^{-3}$)	Median ($\mu g\ m^{-3}$)	Range ($\mu g\ m^{-3}$)
Benzene	39.7 ± 31.3	30.9	2.4–107.2
Toluene	45.5 ± 25.4	52.6	1.2–97.4
Ethyl-Benzene	5.5 ± 4.6	4.1	0.1–18.7
m, *p*-Xylene	5.1 ± 4.1	4.2	0.8–15.0
o-Xylene	4.3 ± 3.1	3.6	0.5–11.1
BTEX	64.3 ± 46.6	56.2	11.3–181.3

CO during the study period. In the monsoon season, frequent rainfall results in a sharp decline in pollutant levels due to wet scavenging. Frequent cloud cover also causes a decrease in solar intensity which may result in a decrease in ozone levels. In addition, clean winds from Arabian Sea and the Indian Ocean were also frequent during monsoon season.

9.3.3 BTEX: Ozone Precursors

BTEX are potent hazardous volatile organic compounds in the atmosphere and they also contribute to surface ozone formation. The mean, median and range of BTEX species at the study site are shown in Table 9.2. Among BTEX, Toluene (45.5 ± 25.4 $\mu g\ m^{-3}$) showed the maximum average concentration. A similar trend was observed by Bauri et al. (2016) in Dehradun; Mohan and Ethirajan (2012) in Chennai; Majumdar et al. (2011) in Kolkata and Kerbachi et al. (2006) in Algiers. The average concentration of Toluene was followed by Benzene (39.7 ± 31.3 $\mu g\ m^{-3}$), Ethyl-Benzene (5.5 ± 4.6 $\mu g\ m^{-3}$), *m*, *p*-Xylene (5.1 ± 4.1 $\mu g\ m^{-3}$) and *o*-Xylene (4.3 ± 3.1 $\mu g\ m^{-3}$). The sum of BTEX ranged from 11.3 to 181.3 $\mu g\ m^{-3}$ with an average concentration of 64.3 ± 46.6 $\mu g\ m^{-3}$.

9.3.4 Seasonal Variation of BTEX

Seasonal variation of BTEX is governed by a number of factors viz. strength of emission sources, meteorological parameters and variation of $OH^{\cdot}$ radicals. Figure 9.3 shows that maximum levels of TEX species were observed in winter season while minimum levels were observed in the monsoon season. However, maximum levels of Benzene were observed in post-monsoon while minimum in monsoon season. The difference in winter and post-monsoon season was not statistically significant ($p < 0.542$). The high levels in the winter season may be due to an increase in emissions from the combustion of fossil fuel. In addition, calm conditions and low PBL height (due to temperature inversion) reduces the rate of dispersion and dilution

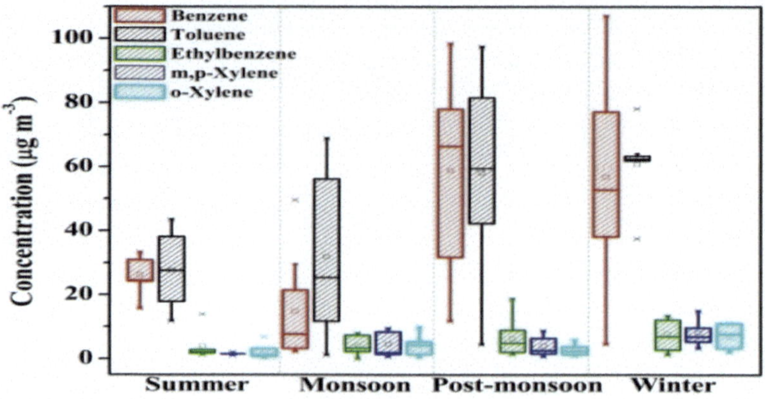

Fig. 9.3 Seasonal variation of benzene, toluene, ethyl-benzene, *p*, *m*-Xylene and *o*-Xylene

of the pollutants. In the summer season, high temperatures, high wind speed and unstable weather conditions result in a decrease in BTEX levels. In addition, high solar radiation favours reaction of OH˙ with hydrocarbons to generate ozone.

High temperature in summer season may contribute BTEX through evaporation loss (fugitive emission) from liquid petroleum gas (LPG), propellants, secondary formation in aged air-masses and biogenic emissions (Na et al. 2003). In monsoon season rain showers and washout is the most probable reason for the low levels of BTEX.

9.3.5 Interspecies Ratios

Interspecies ratios are used to identify the sources of BTEX as they have different emission sources and different photochemical age. The "Photochemical age" is defined as lifetime of VOC against reaction with OH˙ radical (Nelson and Quigley 1967). The low ratio of less reactive to more reactive VOC is an indicator of small photochemical activities and ageing of air-masses. In order to determine contribution of vehicular sources, Benzene is used as a reference VOC because (i) Benzene has low ozone formation potential (Carter 2000) and it is photochemically less active as compared to other aromatic hydrocarbons and (ii) mobile sources are considered as a dominant source of Benzene. Xylenes are usually emitted from paint and solvent usage industries (Barletta et al. 2005). Generally, B/T ratio close to 1.0 is an indicator of traffic emissions and it tends to decrease near the emission sources (Gelencsér et al. 1997). In 43 cities of China, Barletta et al. (2005) identified B/T ratio ranged from 0.2 to 2.2. For some cities, the ratio was 0.6 indicating vehicular emission is the major source of VOCs while for some cities it was higher than 1.0; Benzene emission was much higher than Toluene indicating dominance of biofuel and coal burning. In the US and UK, B/T ratio ranged from 0.2 to 0.7 (Derwent et al. 2000; Baker et al.

2008) indicating the dominance of vehicular emission. The B/T ratio at the study site was 0.9 which indicates that the study site was not directly influenced by traffic emissions. As shown in Table 9.3, at the traffic site of Dehradun, B/T ratio was 0.26 and at commercial and campus sites where traffic density was low, the ratio was 0.31 and 0.5, respectively (Bauri et al. 2016). At the study site, the ratio was higher than Shizouka, Japan (Ohura et al. 2006); Kaohsiung, Taiwan (Liu et al. 2008); Yokohama, Japan (Tiwari et al. 2010); Foshan, China (Tan et al. 2012); Algiers (Kerchich and Kerbachi 2012); Poland (Marć et al. 2014) and Gorakhpur, India (Masih et al. 2016). The ratio of Ethyl-Benzene and Xylenes are used to determine photochemical reactivity of the atmosphere and photochemical age of air-masses because Xylenes are more reactive as compared to Ethyl-benzene (Table 2.5). Low B/X and E/X ratio indicates ageing of airmass and low photochemical activities as Xylenes are photochemically more reactive (nearly three times) than Ethyl-benzene (Miller et al. 2012). During the study period, B/E, B/X, T/E and T/X ratios were high which suggests that air-masses were photochemically aged in nature. E/X ratio was also greater than one.

Table 9.3 Comparison of interspecies ratios with other sites around the world

Site	B/T	B/E	B/X	T/E	T/X	E/X	References
Shizuoka, Japan Industrial (summer) Industrial (winter)	0.13 0.17		0.62 0.87			0.86 1.06	Ohura et al. (2006)
Kaohsiung, Taiwan Urban (rush hours) Industrial (rush hours)	0.11 0.07		0.91 1.08			1.67 2.63	Liu et al. (2008)
Yokohama, Japan Industrial Urban	0.22 0.21		0.37 0.53			0.8 0.88	Tiwari et al. (2010)
Foshan, China (urban)	0.43–0.48						Tan et al. (2012)
Algiers Roadside Urban background Semirural	0.42 0.45 0.63					0.23[a] 0.42[a] ND	Kerchich and Kerbachi (2012)
Poland Gdynia Gdansk	0.6 0.35		0.76[b] 0.58[b]			0.29[a] 0.30[a]	Marć et al. (2014)
Dehradun Traffic Commercial Campus	0.26 0.31 0.5	1.89 2.63 2.2	0.96[b] 1.37[b] 1.75[b]				Bauri et al. (2016)
Gorakhpur	0.6	4.1	5.6	7.3	9.9	1.4	Masih et al. (2016)
Dayalbagh, Agra	0.9	8.5	9.4	12.9	10.4	1.1	Present study

[a]EthlyBenzene/*m-p*, Xylene
[b]Benzene/*m,p*-Xylene

9.3.6 Ozone Formation Potential

VOCs are precursors of surface ozone and the formation of ozone depends on amount of specific VOC emitted and its chemical reactivity. Based on this, Carter (1994) proposed maximum incremental reactivity (MIR) scale to assess ozone formation potential (OFP). OFP depends on the volatile organic compound added to the system as well as on the atmospheric conditions in which reactions are taking place (Carter et al. 1995). It has always been a major research objective to find out ozone formation potential of different species in order to devise more effective environmental regulations and control technologies. Various methods have been developed to determine the potential of ozone generation of individual VOCs (e.g. Carbon Mass Approach and Reactive Organic Gas Approach) (Rubin 2001; Saunders et al. 2003) but these methods have several flaws. MIR scale is the most simplistic and associated with minimal flaws. The MIR coefficients were taken from the literature (Carter 1994) and OFP can be calculated using the formula given below.

$$\mathrm{OFP}_i = \mathrm{MIR\ coefficient}_i x \mathrm{Concentration}_i$$

Figure 9.4 shows OFP of BTEX species in different seasons. In all the seasons, Toluene showed maximum OFP while Benzene and Ethyl-Benzene showed minimum OFP. Although, Xylenes were lowest in levels they have OFP higher than Benzene and Ethyl-Benzene. Therefore, BTEX have a significant role in tropospheric ozone generation and earlier studies have also reported similar results (Alghamdi et al.

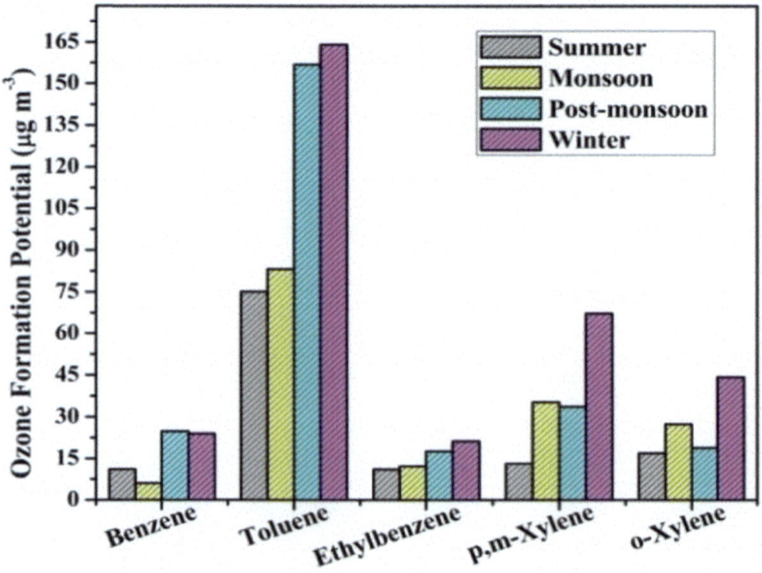

Fig. 9.4 Ozone formation potential in different seasons

Table 9.4 Ozone formation potential and propylene-equivalent for benzene, toluene, ethyl-benzene, *p*, *m*-Xylene and *o*-Xylene

Hydrocarbon	Concentration (ppbC)	Concentration (μg m^{-3})	10^{-12} × k_{OH}	MIR	Prop-equiv (ppbC)	OFP (μg m^{-3})
Benzene	74.7	39.7	1.23	0.42	3.5	16.7
Toluene	84.6	45.5	5.96	2.7	19.2	123
Ethyl-benzene	10.1	5.5	7.1	2.7	2.7	15.1
m, *p*-Xylene	9.4	5.1	19.0	7.4	6.8	41.8
o-Xylene	7.9	4.3	13.7	6.5	4.1	28.0

2014; Kumar et al. 2017). Bauri et al. (2016) also reported similar contributions of BTEX towards OFP at the urban site of Dehradun in winter season, while in summer season, Xylenes showed the highest OFP and Benzene showed the lowest.

The reactivity of hydrocarbons in terms of ozone formation can also be calculated through propylene-equivalent (prop-equiv) concentration (So and Wang 2004). The prop-equiv (*j*) is calculated using a formula given below:

$$\text{Prop} - \text{Equiv}(j) = \text{Conc}(i) * \frac{k_{OH(j)}}{k_{OH(prop)}}$$

where concentration of hydrocarbon is in ppbc, $k_{OH(j)}$ = rate constant for reaction of hydrocarbon (*j*) with OH˙ radicals and $k_{OH(propene)}$ = rate constant for reaction of Propene with OH˙ radicals. k_{OH} for propene is 26.32 × 10^{-12} cm^{3}molecle^{-1}s^{-1}.

"Prop-Equiv (*j*) is a measure of the concentration of hydrocarbon *j* on an OH-reactivity based scale normalized to the reactivity of propene". Table 9.4 shows maximum prop-equivalent for Toluene followed by Xylene, Benzene and minimum for Ethyl-Benzene.

9.3.7 Principal Component Analysis (PCA)

PCA was used to identify possible sources of BTEX and relationship of BTEX with ozone, CO and NOx. Earlier studies have also used PCA as a statistical tool for source apportionment of VOCs (Guo et al. 2004, 2007). This statistical technique produces mutually orthogonal new variables by maximizing the correlation between the original variables. These new variables are produced by the linear combinations of original variables and known as principal components (PCs). Although PCs are equals in equal number of original variables mostly first few PCs can explain most of the variation in data set. This can be achieved using varimax rotation which confirms that one variable has a maximum correlation with one PC and almost no association with other PC. The first PC always explains the largest variance. For PCA

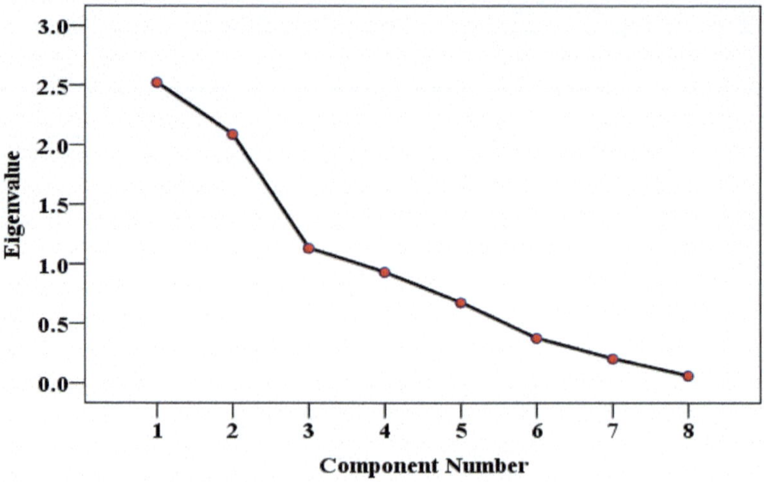

Fig. 9.5 Scree plot showing Eigenvalue associated with each principal component

simultaneously collected data of Benzene, Toluene, Ethyl-benzene, *p*, *m*-Xylene, *o*-Xylene, O_3, NOx and CO in μg m^{-3} units were arranged. After applying varimax rotation three PCs with Eigenvalue greater than one was obtained (Fig. 9.5) and these components explain maximum variance (Table 9.5). PC_1, PC_2 and PC_3 explain 29.5, 26.6 and 15.7% variance, respectively.

Table 9.6 shows the rotated component loadings of three PCs associated with all variables. Although all the variables are included in PCA and the variables with high loadings (greater than 0.5) are denoted in bold. In first PC, Benzene, NOx and CO have significant positive loadings. Second PC is heavily loaded on Ethyl-benzene, *p*, *m*-Xylene and *o*-Xylene while third PC is heavily loaded on O_3. The significant loadings associated with different PCs define different emission sources of variables.

Benzene, NOx and CO is placed in one component which suggests their similar emission source, i.e. combustion. Ethyl-benzene, *p*, *m*-Xylene and *o*-Xylene are placed in second component which is an indicator of emissions from solvent usage. Ozone is placed in third component which is photochemical in origin therefore this component is an indicator of photochemical generation. No other variable is placed in this component means other variables were predominantly primary in origin. Figure 9.6 shows different components in rotated space.

9.4 Conclusions

The photochemical generation of surface ozone takes place through nonlinear reactions in the presence of NOx and VOCs. O_3 formation can be categorized into NOx and VOC sensitive regime. In NOx sensitive regime, ozone formation increases with

Table 9.5 Total variance explained

PC	Initial Eigenvalues			Extraction sums of squared loadings			Rotation sums of squared loadings		
	Total	% of variance	Cumulative (%)	Total	% of variance	Cumulative (%)	Total	% of variance	Cumulative (%)
1	2.5	31.5	31.5	2.5	31.5	31.5	2.4	29.5	29.5
2	2.1	26.1	57.7	2.1	26.1	57.7	2.1	26.6	56.1
3	1.1	14.2	71.8	1.1	14.2	71.8	1.3	15.7	71.8
4	0.9	11.7	83.5						
5	0.7	8.4	91.9						
6	0.4	4.7	96.7						
7	0.2	2.6	99.2						
8	0.1	0.8	100.0						

Table 9.6 Rotated component loadings

	PC_1	PC_2	PC_3
Benzene	**0.746**	0.116	0.262
Toluene	0.147	0.440	0.250
Ethyl-benzene	0.051	**0.837**	−0.171
p, m-Xylene	−0.142	**0.936**	0.001
o-Xylene	0.252	**0.582**	0.437
NOx	**0.932**	−0.013	−0.203
O_3	−0.213	−0.011	**0.909**
CO	**0.886**	0.056	−0.195

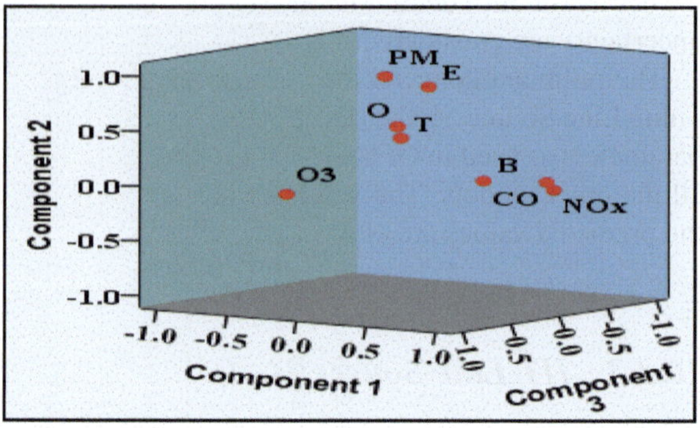

Fig. 9.6 Component plot in rotated space

an increase in NOx concentration while in VOC sensitive regime, ozone formation increases with an increase in VOC concentration. The sensitivity of a site shows diurnal and seasonal variation, therefore, simultaneous measurement of O_3, NOx, CO and VOCs is essential. Very few studies in India and no study at the present site reported simultaneous measurements of O_3, NOx, CO and VOCs. Ozone diurnal variation is characterized by peak concentration during daytime while its precursors show minimum levels during the daytime. Ozone showed maximum levels in June (46.9 ppb) while minimum in December (16.7 ppb). High levels of ozone were observed in summer months because of intense solar radiation and high temperature which favour photochemical generation of surface ozone. Benzene, Toluene and Xylene (BTEX) also contribute to surface ozone formation. The average concentration of Benzene, Toluene, Ethyl-benzene, *m,p*-Xylene and *o*-Xylene were 39.7, 45.5, 5.5, 5.1 and 4.3 μgm^{-3}, respectively. Interspecies ratio was used to identify the influence of emission sources at the site and it was observed that the study site was not directly influenced by traffic emissions and air-masses were aged in nature. Based on the MIR scale, Toluene (123) showed the highest ozone formation potential followed by *m, p*-Xylene (41.8), *o*-Xylene (28) Benzene (16.7) and Ethyl-benzene (15.1). Principal Component Analysis was performed to carry out source apportionment of ozone precursors and to determine their relationship with surface ozone.

Acknowledgements The authors are thankful to the Director, Dayalbagh Educational Institute, Agra and the Head, Department of Chemistry for necessary help. The authors gratefully acknowledge the financial support for this work, which is provided by ISRO GBP under AT-CTM.

References

Alghamdi MA, Khoder M, Abdelmaksoud AS, Harrison RM, Hussein T, Lihavainen H, Al-Jeelani H, Goknil MH, Shabbaj II, Almehmadi FM, Hyvärinen AP (2014) Seasonal and diurnal variations of BTEX and their potential for ozone formation in the urban background atmosphere of the coastal city Jeddah, Saudi Arabia. Air Qual Atmos Health 7(4):467–480

Baker AK, Beyersdorf AJ, Doezema LA, Katzenstein A, Meinardi S, Simpson IJ, Blake DR, Rowland FS (2008) Measurements of nonmethane hydrocarbons in 28 United States cities. Atmos Environ 42(1):170–182

Barletta B, Meinardi S, Rowland FS, Chan CY, Wang X, Zou S, Chan LY, Blake DR (2005) Volatile organic compounds in 43 Chinese cities. Atmos Environ 39(32):5979–5990

Bauri N, Bauri P, Kumar K, Jain VK (2016) Evaluation of seasonal variations in abundance of BTXE hydrocarbons and their ozone forming potential in ambient urban atmosphere of Dehradun (India). Air Qual Atmos Health 9(1):95–106

Carter WP (1994) Development of ozone reactivity scales for volatile organic compounds. Air Waste 44(7):881–899

Carter WP (2000) Documentation of the SAPRC-99 chemical mechanism for VOC reactivity assessment. Contract 92(329):95–308

Carter WP, Pierce JA, Luo D, Malkina IL (1995) Environmental chamber study of maximum incremental reactivities of volatile organic compounds. Atmos Environ 29(18):2499–2511

Chandra N, Lal S, Venkataramani S, Patra PK, Sheel V (2015) Temporal variations in CO_2 and CO at Ahmedabad in western India. Atmos Chem Phys Discuss 15:32185–32238

Crutzen PJ (1973) Photochemical reactions initiated by and influencing ozone in the unpolluted troposphere. Tellus 26:47–57

Derwent RG, Davies TJ, Delaney M, Dollard GJ, Field RA, Dumitrean P, Nason PD, Jones BMR, Pepler SA (2000) Analysis and interpretation of the continuous hourly monitoring data for 26 C2–C8 hydrocarbons at 12 United Kingdom sites during 1996. Atmos Environ 34(2):297–312

Gauss M, Myhre G, Pitari G, Prather MJ, Isaksen IS, Berntsen TK, Brasseur GP, Dentener FJ, Derwent RG, Hauglustaine DA, Horowitz LW (2003) Radiative forcing in the 21st century due to ozone changes in the troposphere and the lower stratosphere. J Geophys Res Atmos 108(D9)

Gautam R, Hsu NC, Kafatos M, Tsay SC (2007) Influences of winter haze on fog/low cloud cover over the Indo-Gangetic plains. J Geophys Res 1112:D05207. https://doi.org/10.1029/2005JD007036

Gelencsér A, Siszler K, Hlavay J (1997) Toluene–benzene concentration ratio as a tool for characterizing the distance from vehicular emission sources. Environ Sci Technol 31:2869–2872

Granier C, Petron G, Mueller JF, Brasseur G (2000) The impact of natural and anthropogenic hydrocarbons on the tropospheric budget of carbon monoxide. Atmos Environ 34:5255–5270

Guo H, Wang T, Simpson IJ, Blake DR, Yu XM, Kwok YH, Li YS (2004) Source contributions to ambient VOCs and CO at a rural site in eastern China. Atmos Environ 38(27):4551–4560

Guo H, So KL, Simpson IJ, Barletta B, Meinardi S, Blake DR (2007) C1–C8 volatile organic compounds in the atmosphere of Hong Kong: Overview of atmospheric processing and source apportionment. Atmos Environ 41(7):1456–1472

Han X, Naeher LP (2006) A review of traffic-related air pollution exposure assessment studies in the developing world. Environ Int 32(1):106–120

Harper M (2000) Sorbent trapping of volatile organic compounds from air. J Chromatogr A 885(1):129–151

Holloway T, Levy H II, Kasibhatla P (2000) Global distribution of carbon monoxide. J Geophys Res 105:12123–12147

IPCC: Climate Change (2013) The physical science basis. Cambridge University Press, Cambridge, 1552p

Kerbachi R, Boughedaoui M, Bounoua L, Keddam M (2006) Ambient air pollution by aromatic hydrocarbons in Algiers. Atmos Environ 40(21):3995–4003

Kerchich Y, Kerbachi R (2012) Measurement of BTEX (benzene, toluene, ethybenzene, and xylene) levels at urban and semirural areas of Algiers City using passive air samplers. J Air Waste Manage Asso 62(12):1370–1379

Kumar A, Singh D, Anandam K, Kumar K, Jain VK (2017) Dynamic interaction of trace gases (VOCs, ozone, and NOx) in the rural atmosphere of sub-tropical India. Air Qual Atmos Health 1–12

Lee DS, Köhler I, Grobler E, Rohrer F, Sausen R, Gallardo-Klenner L, Olivier JGJ, Dentener FJ, Bouwman AF (1997) Estimations of global no, emissions and their uncertainties. Atmos Environ 31(12):1735–1749

Lee SC, Chiu MY, Ho KF, Zou SC, Wang X (2002) Volatile organic compounds (VOCs) in urban atmosphere of Hong Kong. Chemosphere 48(3):375–382

Lelieveld J, Dentener FJ (2000) What controls tropospheric ozone? J Geophys Res Atmos 105:3531–3551. https://doi.org/10.1029/1999jd901011

Ling ZH, Guo H, Cheng HR, Yu YF (2011) Sources of ambient volatile organic compounds and their contributions to photochemical ozone formation at a site in the Pearl River Delta, southern China. Environ Pollut 159(10):2310–2319

Liu PW, Yao YC, Tsai JH, Hsu YC, Chang LP, Chang KH (2008) Source impacts by volatile organic compounds in an industrial city of southern Taiwan. Sci Total Environ 398(1–3):154–163

Majumdar D, Mukherjeea AK, Sen S (2011) BTEX in ambient air of a Metropolitan City. J Environ Prot 2(01):11

Marć M, Namieśnik J, Zabiegała B (2014) BTEX concentration levels in urban air in the area of the Tri-City agglomeration (Gdansk, Gdynia, Sopot), Poland. Air Qual Atmos Health 7(4):489–504

Masih A, Lall AS, Taneja A, Singhvi R (2016) Inhalation exposure and related health risks of BTEX in ambient air at different microenvironments of a terai zone in north India. Atmos Environ 147:55–66

Miller L, Xu X, Grgicak-Mannion A, Brook J, Wheeler A (2012) Multi-season, multi-year concentrations and correlations amongst the BTEX group of VOCs in an urbanized industrial city. Atmos Environ 61:305–315

Mohan S, Ethirajan R (2012) Assessment of hazardous volatile organic compounds (VOC) in a residential area abutting a large petrochemical complex. J Trop For Environ 2(1)

Na K, Kim YP, Moon KC (2003) Diurnal characteristics of volatile organic compounds in the Seoul atmosphere. Atmos Environ 37(6):733–742

Nelson PF, Quigley SM (1967) The m, p-xylenes: ethylbenzene ratio. A technique for estimating hydrocarbon age in ambient atmospheres. Atmos Environ 17(3):659–662

Nishanth T, Praseed KM, Kumar MS, Valsaraj KT (2014) Influence of ozone precursors and PM_{10} on the variation of surface O_3 over Kannur, India. Atmos Res 138:112–124

Ohura T, Amagai T, Senga Y, Fusaya M (2006) Organic air pollutants inside and outside residences in Shimizu, Japan: levels, sources and risks. Sci Total Environ 366(2):485–499

Ojha N, Naja M, Singh KP, Sarangi T, Kumar R, Lal S, Lawrence MG, Butler TM, Chandola HC (2012) Variabilities in ozone at a semi-urban site in the Indo-Gangetic Plain region: association with the meteorology and regional processes. J Geophys Res Atmos 117(D20)

Pachauri RK, Meyer LA (eds) (2014) IPCC, 2014: climate change: synthesis report. Contribution of working group I, II and III to the fifth assessment report of the intergovernmental panel on climate change. IPCC, Geneva, Switzerland, p 151

Parra MA, Elustondo D, Bermejo R, Santamaria JM (2009) Ambient air levels of volatile organic compounds (VOC) and nitrogen dioxide (NO_2) in a medium size city in Northern Spain. Sci Total Environ 407(3):999–1009

Patokoski J, Ruuskanen TM, Kajos MK, Taipale R, Rantala P, Aalto J, Ryyppö T, Nieminen T, Hakola H, Rinne J (2015) Sources of long-lived atmospheric VOCs at the rural boreal forest site, SMEAR II. Atmos Chem Phys 15(23):13413–13432

Rubin ES (2001) Introduction to engineering and the environment, 1st edn. McGraw-Hill

Saunders SM, Jenkin ME, Derwent RG, Pilling MJ (2003) Protocol for the development of the master chemical mechanism, MCM v3 (part A): tropospheric degradation of non-aromatic volatile organic compounds. Atmos Chem Phys 161–180

Seinfeld JH, Pandis SN (2006) Atmospheric chemistry and physics. Wiley, New York

So KL, Wang T (2004) C3–C12 non-methane hydrocarbons in subtropical Hong Kong: spatial–temporal variations, source–receptor relationships and photochemical reactivity. Sci Total Environ 328(1):161–174

Tan JH, Guo SJ, Ma YL, Yang FM, He KB, Yu YC, Wang JW, Shi ZB, Chen GC (2012) Non-methane hydrocarbons and their ozone formation potentials in Foshan, China. Aerosol Air Qual. Res. 12(3):387–398

Tiwari V, Hanai Y, Masunaga S (2010) Ambient levels of volatile organic compounds in the vicinity of petrochemical industrial area of Yokohama, Japan. Air Qual Atmos Health 3(2):65–75

Ueno Y, Horiuchi T, Morimoto T, Niwa O (2001) Microfluidic device for airborne BTEX detection. Analyt Chem 73(19):4688–4693

Yuan B, Hu WW, Shao M, Wang M, Chen WT, Lu SH, Zeng LM, Hu M (2013) VOC emissions, evolutions and contributions to SOA formation at a receptor site in eastern China. Atmos Chem Phys 13(17):8815–8832

Zhang Q, Streets DG, Carmichael GR, He KB, Huo H, Kannari A, Klimont Z, Park IS, Reddy S, Fu JS, Chen D (2009) Asian emissions in 2006 for the NASA INTEX-B mission. Atmos Chem Phys 9(14):5131–5153

Chapter 10
Predicting Particulate Air Pollution Using Line Source Models

Selvakumar Madhavan and S. Geetha

10.1 Introduction

Air quality is worsening day by day due to various reasons like increased motor vehicle use, industrial growth, impact of population growth, and various activities resulting in the release of gases beyond their prescribed limits. During transportation by various motor vehicles, the energy conversion has resulted in the emission of air pollutants into the atmosphere and together with industrial emissions, it accounts for the bulk of the air pollutant emission. This has resulted in deterioration of health to the urban folk causing illness and also degrading its surrounding environment. Air quality is worsening day by day and is an alarming issue in almost all countries today. When meteorological and emission data are available for a region, its air quality levels can be modeled based on different scenarios, such as the formation of photochemical smog, the impact of population growth, or increased motor vehicle use.

The economic activities taking place around the globe has resulted in the use and conversion of energy in which transportation occupies a major role. As a result, it is accompanied by emission of air pollutants into the atmosphere thus degrading the environment and causing illness to the urban folk. The compound that is emitted into the atmosphere from an industry in the form of flue gas as it comes from the exhaust or chimney and the unburnt fuel in the form of smoke from vehicles is known as emission. These are also known as primary pollutants. As these pollutants are emitted, they are dispersed in the atmosphere through various chemical reactions, depending on the weather conditions. These chemical reactions result in stable compounds

S. Madhavan (✉) · S. Geetha
Rajalakshmi Engineering College, Chennai 600062, Tamil Nadu, India
e-mail: selvakumar.m@rajalakshmi.edu.in

S. Geetha
e-mail: geetha.s@rajalakshmi.edu.in

S. M. Shiva Nagendra et al. (eds.), *Urban Air Quality Monitoring, Modelling and Human Exposure Assessment*, Springer Transactions in Civil and Environmental Engineering,
https://doi.org/10.1007/978-981-15-5511-4_10

that can be detected and analyzed using equipment like High Volume sampler and other devices. These are the pollutants that we breathe or which are deposited on the ground.

The compounds formed as a result of chemical reactions in the atmosphere are generally of anthropogenic origin and result from combustion processes of fossil fuels. Also, natural pollutants like gases and dust from volcanoes, forest fires, or dust entrained by storms also exist. But the quantity of pollutants from natural sources is much less than that of the anthropogenic origin pollutants. Anthropogenic pollutants are the major source of pollution in urban areas. All pollutants exhibit a bimodal distribution pattern that is, decrease in concentration from morning towards midday with a subsequent build-up towards evening. Among them, the concentration of oxides of nitrogen and carbon monoxide is moderately higher during midday hours (Prasad et al 2003).

The emission of any toxic element in the surrounding atmosphere leads to serious harmful effects at the place of origin and also at distant places. Hence, it is necessary to make the pollutants get dispersed and diluted as soon as possible to minimize its ill effects (Dubey 1992). Based on the cause of pollution we can classify the sources as fixed or mobile sources, point sources, line sources and diffuse or area sources. Health impacts of air pollution depend on population growth, industrialization and increased vehicular traffic. Combustion of fossil fuels and their products is responsible for anthropogenic air pollution and this problem is particularly acute in urban areas (Shanker and Rao 2000). City dwellers are exposed regularly to pollution. Exposed people may be at risk and may face serious chronic health hazard problems since it has been well proven by researchers that there is a direct correlation between environmental pollution and mortality (Schwartz et al. 2001).

The pollutant concentration due to vehicular emission has been modeled by various line source models (Nagendra and Khare 2002). The CO concentration in Delhi due to traffic emission was predicted by Singh et al. (1990), using the GFLSM. Computational Fluid Dynamics (CFD) was also used by researchers (Xie et al. 2009) to investigate the traffic-related pollutants. Field investigations on vehicular pollutant concentrations based on the traffic volume were done by Qin and Kot (1993). Semi-empirical model which involved a box model in which the dispersion component was based only on wind speed and direction, for predicting CO concentration due to traffic flow was used by Kim et al (2003). Berkowicz et al. (1996) used the Danish Operational Street Pollution Model (OSPM) to model the pollutants concentration on different street configurations considering the influence of meteorological parameters on the dispersion of the pollutants. Apart from only monitoring the pollutant concentration, there are many works in identifying the health effects caused by air pollutants. Gaussian plume model was used by Karim (1999) to monitor and estimate air pollutants like CO, NOx, lead, particulate matter, and black smoke at street intersections of Dhaka. Line source models were approximated by Akul and Horst based on the wind direction using Horst–Venkatram approximation and compared with Luhar–Esplin (LE) approximation. This paper aims at finding a suitable line source model for modeling the vehicular pollution at Coimbatore city and validating the model by comparing it with the experimental observations.

Fig. 10.1 Locations of monitoring stations

10.2 Study Area Characteristics

Coimbatore is the second-largest city in Tamilnadu. The city has six major arterial roads and three National Highways. Most of the textile industries are situated in Coimbatore. There are about 40,000 small, medium and large-scale industries in the city. Due to industrialization and urbanization Coimbatore's air quality is worsening. The ambient air quality of Coimbatore has deteriorated with an increase in the number of vehicles and industrial pollution. The selected monitoring stations are shown in Fig. 10.1; (Table 10.1).

10.3 Methodology

The study involved 25 monitoring stations and the sampling duration was 12 h. One station was monitored each day. Periodic sampling was taken in each station in a cyclic manner. High volume sampler (Model APM 460 NL) was used in this study for collecting the air samples. The collected air samples were analyzed in the laboratory. The sampling time was from 8 am to 8 pm covering peak hours and moderate traffic flow hours. Method of estimation of suspended and respirable particulate matter was done by Gravimetric Method as per BIS: 5182 (part IV) 1999. Data on temperature, relative humidity, wind speed, and wind direction were collected for finding the concentration of the pollutants. The meteorological data was collected from the Surface and Groundwater Department at Taramani, Chennai. The behavior of weather patterns in a localized area (meteorology) is important in any air pollution study. When air pollutants are distributed into the atmosphere, they

Table 10.1 Geographical location details of monitoring stations

Location	Latitude (°N)	Longitude (°E)
ITI (Near Thudiyalur)	11.0551	76.9450
Near Agriculture University	11.0152	76.9326
Racecourse area	10.9991	76.9773
Singanallur	10.9987	77.0320
Near PSG College	11.0248	77.0028
Near TVS Nagar (Thadagam Road)	11.043	76.9239
Near Saibaba colony	11.0236	76.9426
Ramanathapuram	10.9975	76.9916
Selvapuram	10.9919	76.9369
Near Gandhipuram	11.0210	76.9663
Avinashi Road (Near SITRA)	11.0374	77.0372
Govt. Hospital	10.9963	76.9701
Kavundampalayam	11.0451	76.9475
Saravanampatti	11.0791	77.0017
Ganapathy	11.0494	77.0094
TataBad	11.0225	76.9606
Ukkadam	10.9902	76.9629
RS Puram	11.0104	76.9499
Sundarapuram	10.9616	76.9729
Ramnagar	11.0124	76.9629
Peelamedu	11.0303	76.9998
Udayampalayam	11.0193	76.9945
Ramanuja Nagar	11.0203	77.0259
Vadavalli	11.0247	76.8980
Near GCT college	11.0236	76.9426

The city is 432 m above MSL

are influenced by the movements and characteristics of the air mass into which they are emitted. The pollutants cannot disperse if the air is calm resulting in subsequent build-up of pollutant concentration near the pollution source. But if a turbulent wind blows, these pollutants will disperse rapidly resulting in lower concentrations in the atmosphere. Using the collected meteorological data and observed pollutant data, the following models were used to model the pollutants considering the traffic survey that was conducted along with the observed data.

10.3.1 General Finite Line Source Model (GFLSM)

Luhar and Patil in (1986) developed a simple general finite line source model overcoming the constraints of infinite line source.

This model was developed using a coordinate transformation between the wind coordinate system (x_1, y_1, z_1) and the line source coordinate system (x, y, z). The origin for both the coordinate system is the midpoint of the line source and z-axis is assumed to be the same for both the coordinate systems. Figure 10.2 shows the receptor position R with reference to both the coordinate systems. The parameters x, y, and z of the line source coordinate system can be calculated from road receptor geometry. The dispersion parameter as given in Eq. (10.1) is based on the wind-road orientation angle and the source distance.

Pollutant concentrations (μg/m^3) are calculated as given in the following equation:

$$C = \frac{Q}{\sqrt{2\pi}\sigma_Z u \ \sin\theta}\left[\exp\left\{-\frac{1}{2}\left(\frac{z-h_0}{\sigma_z}\right)^2\right\}+\exp\left\{-\frac{1}{2}\left(\frac{z+h_0}{\sigma_z}\right)^2\right\}\right]$$
$$x\left[\mathrm{erf}\left(\frac{\sin\theta(L/2-y)-x\ \cos\theta}{\sqrt{2}\sigma_y}\right)+erf\left(\frac{\sin\theta(L/2+y)+x\ \cos\theta}{\sqrt{2}\sigma_y}\right)\right] \tag{10.1}$$

where

Q	is emission rate per unit length, g/m-s;
u	the mean ambient wind speed, m/s;
σ_y and σ_z	are horizontal, vertical dispersion parameter, respectively, m;
x	the distance from the receptor to the line source, m;

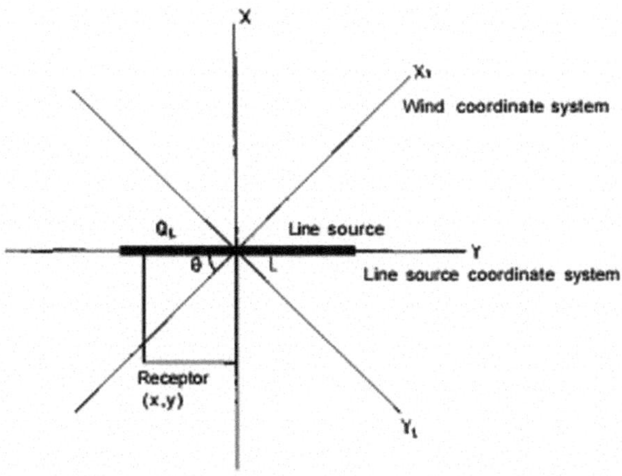

Fig. 10.2 Relationship between wind and line source coordinate system

Fig. 10.3 **a** Validation of GFLSM for respirable particulate matter. **b** Validation of GFLSM for total suspended particulate matter

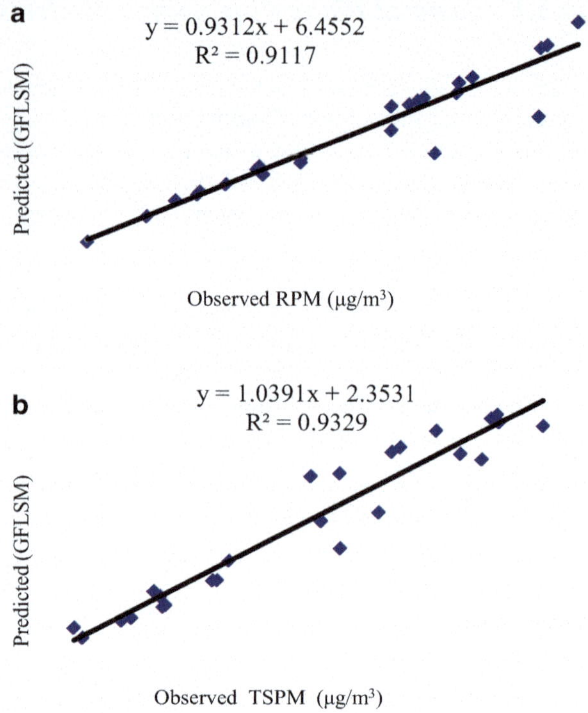

y	receptor distance from the roadway centre line along with the line source, m;
z	height of the receptor relative to the ground, 1.8 m;
h_0	is line source height (m) taken as 0.5 m
H	the plume centre height relative to the ground, m;
L	length of the source, m;
θ	angle between the ambient wind and the road; and
erf ()	is the error function.

The traffic data and observed wind speed data were used to calculate the particulate matter in the General Finite Line Source Model (GFLSM). The calculated data were compared with the observed data and validated using scatter plots as shown in Fig. 10.3a, b for RPM and TSPM, respectively. The results show a good correlation.

10.3.2 *Delhi Finite Line Source Model (DFLSM)*

Khare and Sharma (1999) developed the Delhi Finite Line Source Model (DFLSM) which was derived from the general finite line source model (GFLSM) by removing the error function that was used in that model.

$$C = \frac{Q}{2\sqrt{2\sigma_Z u_e}}\left[\exp\left\{-\frac{1}{2}\left(\frac{z-h_o}{\sigma_Z}\right)^2\right\} + \exp\left\{-\frac{1}{2}\left(\frac{z+h_o}{\sigma_Z}\right)^2\right\}\right] \quad (10.2)$$

where

Q	is emission rate per unit length, g/ms;
u	the mean ambient wind speed, m/s;
σ_y and σ_z	are horizontal, vertical dispersion parameter, m;
x	the distance from the receptor to the line source, m;
y	receptor distance from the roadway centre line along the line source, m;
z	height of the receptor relative to the ground, 1.8 m;
h_0	is line source height (m) taken as 0.5 m.

As per the GFLSM the receptor is located at 90° to the road segment considered and the length of the line source is three times that of the distance between receptor and road (Gokhale and Khare 2004). The hybrid GFLSM model combines the best features of deterministic and statistical distribution models as variability and uncertainty are considered.

The pollutant concentrations for particulate matter were calculated using Delhi Finite Line Source Model (DFLSM) considering the emission factors and dispersion parameters as used in GFLSM. This model showed better performance compared to all the other models. The validated scatter plots are shown in Fig. 10.4a, b. Hence the predicted values are closer to the observed values.

10.3.3 IIT Line Source Model (IITLSM)

A line source model, based on Gaussian formulation, has been developed in IIT Delhi, i.e., the IIT line source model (IITLS), by Goyal and Rama Krishna (1999) in which improvements have been made in the calculation of source strength at major intersections in the city by consideration of vehicular speed as well as acceleration or deceleration and idling emissions. The IITLS model is a Gaussian type computer-based line source model, developed to predict concentrations of gaseous pollutants CO, SO_2, NO_2, and particulates on various types of roads, speeds of different vehicles and different modes of driving at traffic intersections. Source inventory of IITLS model has been designed according to traffic data, available in urban cities of India. The present model does not include chemical reactions. The model formulation for finite and infinite line sources are developed from Gaussian plume equation, which are given by

$$C_p = \frac{Q}{\pi\ u\sigma_y\sigma_z}\exp\left[-\frac{y^2}{2\sigma_y^2}\right]\exp\left[-\frac{(z-h)^2}{\sigma_z^2}\right]\exp\left[-\frac{(z-h)^2}{\sigma_z^2}\right] \quad (10.3)$$

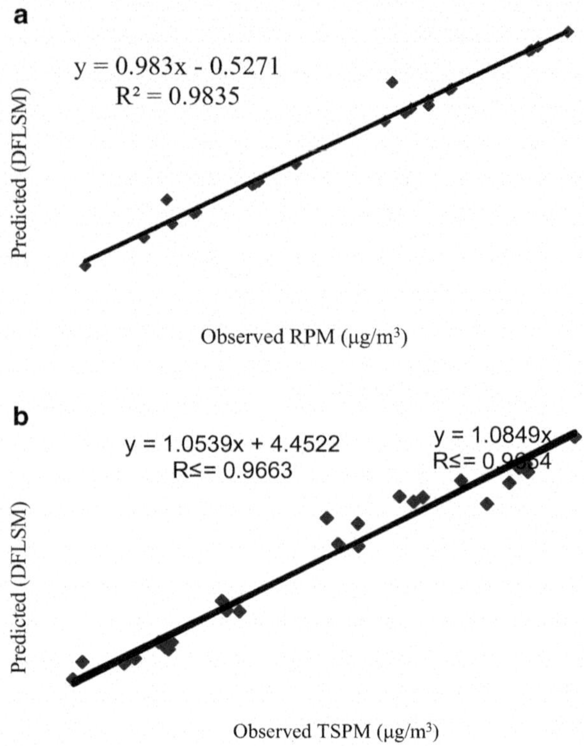

Fig. 10.4 **a** Validation of DFLSM for respirable particulate matter. **b** Validation of DFLSM for total suspended particulate matter

where

Q	is the source strength (g/s),
u	wind speed (m/s), y, crosswind distance (m),
z	vertical distance (m),
h	source height (m),
σ_y and σ_z	horizontal and vertical dispersion parameters (m).

The line source is assumed as a line of infinitesimal point sources. The ground level concentrations (glc's) of gaseous pollutants can be obtained by integrating Eq. (10.3) along the length of the line source, i.e., length of the road.

The IITLS model uses separate equations for calculating pollutant concentrations under crosswind and parallel wind conditions. In the crosswind case, when the wind is normal to the road, the ground level concentration of pollutants is given by

$$C_{X\,\text{Wind}} = \sum_{i=1}^{\text{NEL}} C_p \frac{2q}{(2\pi)^{1/2}\,\sigma_z\,u} \exp\left[-\frac{h^2}{2\sigma_z^2}\right] \tag{10.4}$$

where C_{xwind} is the ground level concentration (g/m^3) of pollutants in crosswind, q the source strength (g/ms) and the other parameters hold the same meaning as mentioned above. For parallel wind, the glc's are determined from:

$$C_{\|\mathrm{Wind}} = \sum_{i=1}^{\mathrm{NEL}} C_{\mathrm{p}} \tag{10.5}$$

where C|wind is the glc's of pollutant when wind is parallel to the road (g/m^3), NEL is the number of line elements along the length of roadway.

For other wind angles, i.e. 15° </< 60°, the glc's (C_{obl}) are determined by 'weighted average' of the 'pure' crosswind and parallel wind terms by

$$C_{\mathrm{obl}} = \sin^2\varphi C_{x\,\mathrm{wind}} + \cos^2\varphi C_{\mathrm{II\,wind}} \tag{10.6}$$

where C_{obl} is the glc's of pollutant for wind 15° </< 60° (g/m^3).

Dispersion parameters:

The horizontal and vertical dispersion parameters are expressed as below:

$$\sigma_y^2 = \sigma_{yo}^2 + \sigma_{yb}^2 \quad \text{and} \tag{10.7}$$

$$\sigma_z^2 = \sigma_{zo}^2 + \sigma_{zb}^2 \tag{10.8}$$

As per the study made by Kono and Ito (1990), the dispersion of pollutants in the initial stage for a distance of 500 m from the source is given as σ_{yo} and σ_{zo}, σ_{yb} and σ_{zb} are the Briggs dispersion parameters for distance beyond 500 m from the source. Twice σ_{zo} is taken as σ_{yo} and σ_{zo} which is the initial vertical dispersion is taken as 0.2 $Z_{i,}$ where Z_i is the vertical height (m).

The pollutants were calculated based on the IITLS model and compared with the observed data. The validated scatter plots are shown in Fig. 10.5a, b. R_2 values for the pollutants are lower than that obtained from GFLSM and DFLSM models. The predicted values were greater than the observed values.

10.3.4 Caline4 Model

California Line Source Model Version4 (Caline4) is the most recent version of the Caline models series, developed by the California Department of Transportation (Benson 1992). Caline4 is a Gaussian model, capable of predicting concentrations of CO (and other non-reactive gases), NO_2 (includes a sub-model for NO_2 formation) and aerosols. The model divides the road into a number of elements, each representing a part of it and the concentration at the receptor is calculated by summing the contributions from the up-wind elements. Effectively, the model represents the road as a series of finite line sources, positioned normal to the wind direction, and

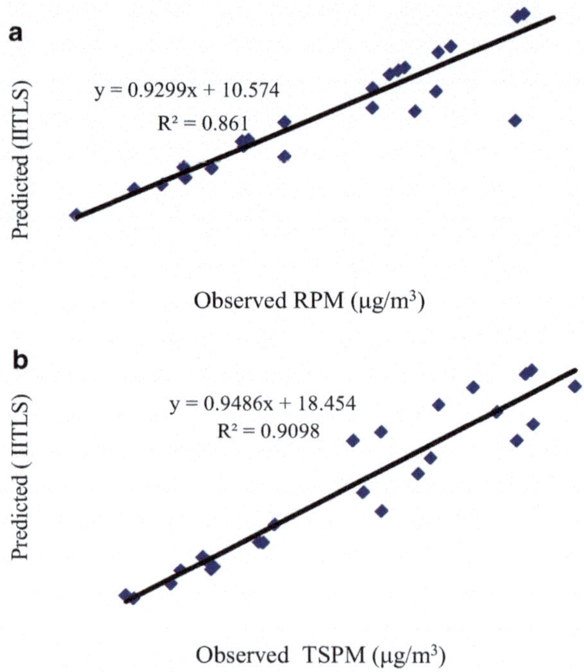

Fig. 10.5 **a** Validation of IITLSM for respirable particulate matter. **b** Validation of IITLSM for total suspended particulate matter

centred at the element midpoint. Incremental downwind concentrations are calculated according to the crosswind Gaussian formulation for a line source of finite length. Incremental downwind concentrations are calculated according to the crosswind Gaussian formulation for a line source of finite length,

$$C(x, y, 0; H) = \frac{Q}{\pi \sigma_z u} \int_{y_1-y}^{y_2-y} \exp\left(\frac{-y^2}{2\sigma_y^2}\right) \mathrm{d}y \tag{10.9}$$

where Q is linear source strength;

u is wind speed;

σ_y and σ_z are horizontal and vertical Gaussian dispersion parameters;

and y_1, y_2 are finite line source endpoints y-coordinates ($y_1 < y_2$).

For the computation of vertical Gaussian dispersion parameter, Caline 4 takes into account the vehicle induced and thermal turbulence, along with the ambient turbulence. The latter has almost no influence on dispersion in the immediate vicinity of the road and dominates as the distance from the road increases (Benson 1992). Horizontal Gaussian dispersion parameter is estimated directly from the wind direction standard deviation, σ_h, using a method developed by Draxler (1979). For depressed sections, greater values for initial vertical dispersion are used, and higher mixing zone concentrations are predicted, compared to equivalent at-grade and elevated sites. The channeling and eddying effects effectively decrease the rate of pollutant transport out of the depressed- section mixing zone. This means there is a higher residence time of the pollutant within or close to the depressed section (up to a distance of three times the section depth). As a result, concentrations within or close to the depressed section will be relatively high and will be relatively low outside of the depressed section, in the area in which the higher vertical dispersion dominates (Benson 1992).

Using the available data CALINE4 software is used to get the pollutant concentrations. The data are compared with the observed data and validated using a scatter plot as shown in Fig. 10.6a, b. The predicted data using CALINE 4 model was found to be higher than the observed data.

The pollutant dispersion is represented using GIS platform with the help of ARCView software which provides the raster and vector data. The respective map which contains the data of roads was digitized to a scale of 1:25,000. The road network consists of local city roads and highway road segments for which a road code was assigned to each road segment. The traffic volume in each road segment was estimated based on the traffic counts. The emission of pollutant E_i in kg was calculated using the equation $E_i = L \times C_{i,}$ where L is the vehicle kilometer travelled which is the product of the traffic volume and length of the road segment and the emission coefficient of pollutant i, is C_i (kg/km).

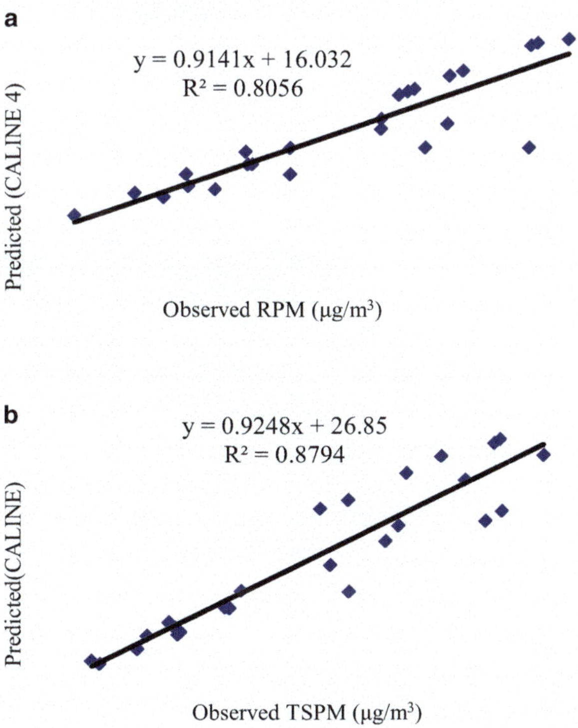

Fig. 10.6 **a** Validation of CALINE4 for respirable particulate matter. **b** Validation of CALINE4 for total suspended particulate matter

The dispersion model used here is CALINE4, a simple line source. The air pollution concentrations collected from the monitoring stations were analyzed and its spatial distribution is shown in the form of map (Fig. 10.7a, b). The levels of the concentrations are differentiated using various shades. It is found that Gandhipuram has a maximum concentration among all the places. Then the places like Kavundampalayam, Saravanampatti, Ganapathy, TataBad, Ukkadam, and Ramnagar have recorded higher concentrations. Next high concentrations were recorded near Thudialur, SITRA, Government Hospital, RS Puram, Sundarapuram, and near PSG

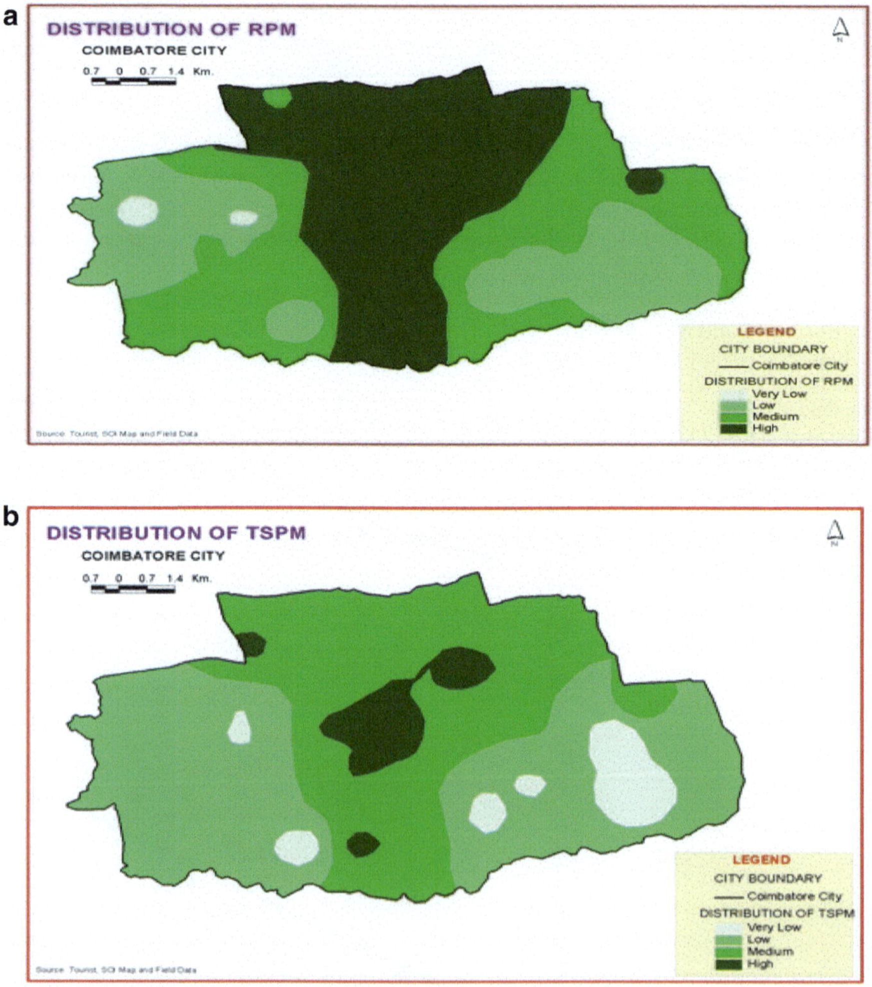

Fig. 10.7 **a** GIS Map showing the distribution of Particulate matter. **b** GIS map showing the distribution of particulate matter

College. The places near GCT, Agri college, Vadavalli, Ramanujar Nagar, Udayampallayam, Singanallur, Racecourse area, Near TVS Nagar (Thadagam Road), Near Saibaba colony, Ramanathapuram and Selvapuram have recorded low concentration.

In order to characterize the quality of air, a number has to be assigned for each level of pollutant concentration indicating their level of health concern. Such a number according to EPA is called as Air Quality Index (AQI). AQI is used by several governments and pollution agencies to characterize the quality of air at any given location. AQI is divided into various ranges and each range is assigned a color code and description starting from good condition up to hazardous condition. Each AQI range has got detailed advisory recommendations based on the type of public health agency. As the pollutant concentration increases in ambient atmosphere, the AQI value also increases which indicates severe adverse health effects. An air quality monitor or model is required to compute the concentration of air pollutants. In order to convert an air pollutant concentration to AQI values a function is required. This function varies for different pollutants and is different in different countries. Regarding the various environmental standards defined in the world, in this work the following AQI standard is used. The index for a pollutant is calculated using the mathematical expression (EPA 1999).

$$I_P = \frac{\mathrm{IH}_i - \mathrm{IL}_o}{\mathrm{BP}_{\mathrm{Hi}} - \mathrm{BP}_{\mathrm{Lo}}} \times [C_P - \mathrm{BP}_{\mathrm{Lo}}] + \mathrm{IL}_o \tag{10.10}$$

I_p Pollution index
C_P the rounded concentration of pollutant P
$\mathrm{BP}_{\mathrm{Hi}}$ the breakpoint that is greater than or equal to C_P
$\mathrm{BP}_{\mathrm{Lo}}$ the breakpoint that is less than or equal to C_P
IH_i the AQI value corresponding to $\mathrm{BP}_{\mathrm{Hi}}$
IL_o the AQI value corresponding to $\mathrm{BP}_{\mathrm{Lo}}$

Air quality index (AQI) for the recorded concentrations are calculated and depicted on the map (Fig. 10.8a, b). It is found that AQI falls into two categories, one is a moderate level and the next is Unhealthy for Sensitive Groups. The interior part which comprises Gandhipuram, RS Puram, the commercial area of Ganapathy, residential areas like Saibaba colony and in places near Ukkadam bus stand and Singanallur the AQI value shows that it is unhealthy for a sensitive group of people. In other areas, it is in a moderate level (Table 10.2).

10.4 Conclusions

The results from this study show that the Gaussian based line source models are reliable models for predicting the concentration of the pollutants caused by traffic emission. Also, the results of these theoretical models compared with monitored data has a good correlation. Four line source models were used in this work and

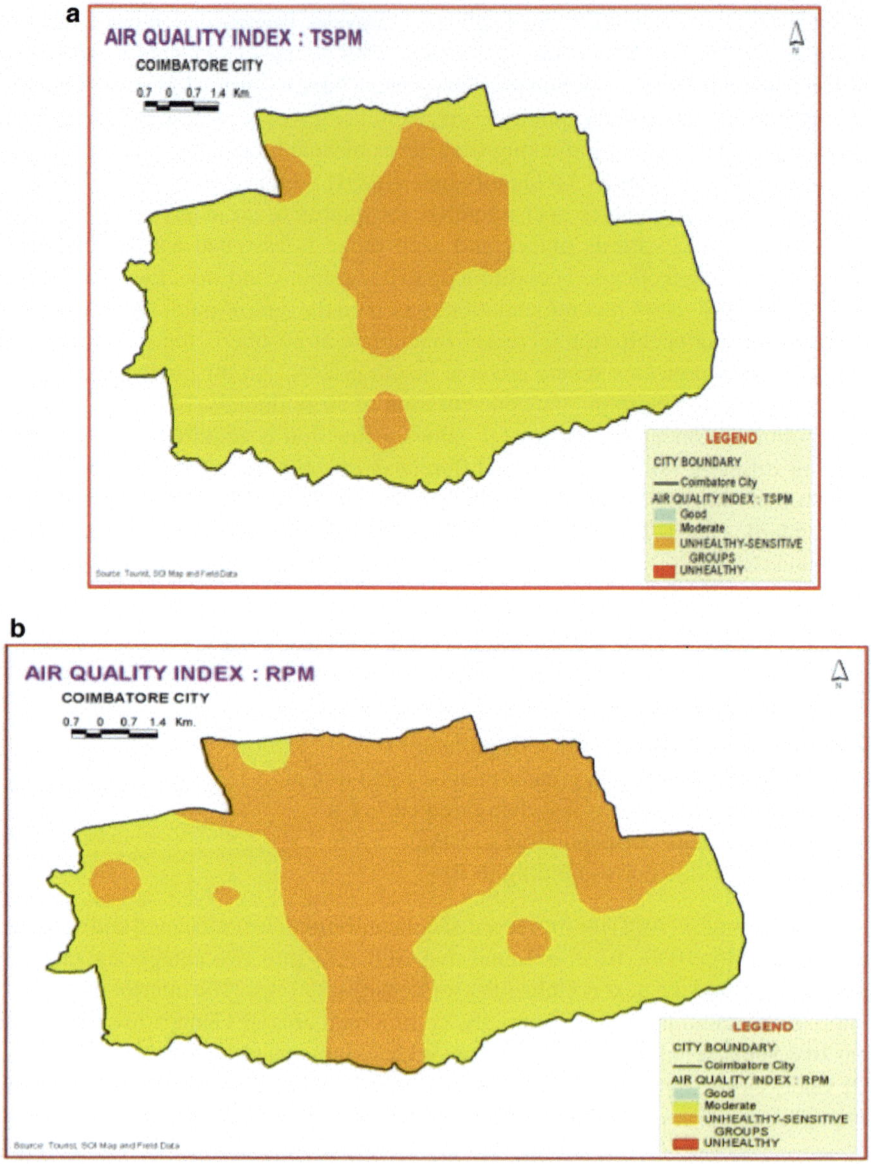

Fig. 10.8 **a** GIS map showing the air quality index of total suspended particulate matter. **b** GIS map showing the air quality index of respirable particulate matter

Table 10.2 Breakpoints for the air quality index (EPA 1999)

O_3 (ppm)	O_3 (ppm)	PM10	PM2.5	CO (ppm)	SO_2 (ppm)	NO_2 (ppm)	AQI	
8-hour	1-h[a]	(μg/m^3)	(μg/m^3)					
0.000–0.059	–	0–54	0.0–15.4	0.0–4.4	0.000–0.034	([b])	0–50	Good
0.060–0.075	–	55–154	15.5–40.4	4.5–9.4	0.035–0.144	([b])	51–100	Moderate
0.076–0.095	0.125–0.164	155–254	40.5–65.4	9.5–12.4	0.145 –0.224	([b])	101–150	Unhealthy for sensitive groups
0.096–0.115	0.165–0.204	255–354	65.5–150.4	12.5–15.4	0.225–0.304	([b])	151–200	Unhealthy
0.116–0.374	0.205–0.404	355–424	150.5–250.4	15.5–30.4	0.305–0.604	0.65–1.24	201–300	Very unhealthy
(0.155–0.404)[c]								
([d])	0.405–0.504	425–504	250.5–350.4	30.5–40.4	0.605–0.804	1.25–1.64	301–400	Hazardous
([d])	0.505–0.604	505–604	350.5–500.4	40.5–50.4	0.805–1.004	1.65–2.04	401–500	Hazardous

[a]Areas that are required to report the AQI based on 8-h ozone values. In these cases the index for both the 8-h and the 1-h ozone values may be calculated and the maximum AQI reported

[b]NO_2 has no short-term NAAQS and can generate an AQI only above a value of 200

[c]The numbers in parentheses are associated with 1-h values to be used in this overlapping category only

[d]8-h O_3 values do not define higher AQI values ($\geq$301). AQI values of 301 or higher are calculated with 1-h O_3 concentrations

among the models considered, Delhi Finite Line Source Model (DFLSM) predicted the pollutant concentration very close to the monitored data. Also, the representation of the pollutant concentration and AQI using GIS maps gives an overview of the quality of air prevailing in that area. Hence it can be concluded that a study on the traffic volume and meteorological data will be sufficient to predict the pollutant concentration in air due to vehicular emission using these Gaussian based line source models. However, GFLSM and DFLSM are considered as deterministic models and cannot be used at places where extreme concentrations may prevail. They can be used to access the quality of air in a particular area and for regulatory purposes.

References

Benson P (1992) A review of the development and application of the CALINE3 And 4 models. Atmos Environ 26B(3):379–390

Berkowicz R, Palmgren Finn, Hertel Ole, Vignati Elisabetta (1996) Using measurements of air pollution in streets for evaluation of urban air quality—meterological analysis and model calculations. Sci Total Environ 189:259–265

Draxler RR (1979) An improved Gaussian model for long term average air concentration estimates. Atmos Environ 14:597–601

Dubey Ashutosh (1992) Impact of trace element air pollution on human health. Indian J Environ Prot 12(7):512–514

EPA (1999) Air quality index reporting; final rule, federal register, part 3, CFR part 58

Gokhale S, Khare M (2004) A review of deterministic, stochastic and hybrid vehicular exhaust emission models. Int J Trans Manage 2:59–74

Goyal P, Rama Krishna TVBPS (1999) A line source model for Delhi. Transp Res Part D 4:241–249

Karim MM (1999) Traffic Pollution Inventories and Modeling in Metropolitan Dhaka, Bangladesh. Trans Res Part D Trans Environ 4:291–312

Khare M, Sharma P (1999) Performance evaluation of general finite length source model for delhi traffic conditions. Trans Res Part D Trans Environ 4:65–70

Kim N, Dirks Murray D, Johns John E, Hay Andrew P, Sturman (2003) A semi-empirical model for predicting the effect of changes in traffic flow patterns on carbon monoxide concentrations. Atmos Environ 37:2719–2724

Kono H, Ito S (1990) A micro scale dispersion model for motor vehicle exhaust gas in urban areas—OMG volume source model. Atmos Environ 24(2):243–251

Luhar A, Patil S (1986) A general finite line source model for vehicular pollution prediction. Atmos Environ 23:555–562

Nagendra SMS, Khare Mukesh (2002) Line source emission modelling. Atmos Environ 36(13):2083–2098

Prasad Rajendra, Aggrawal Rakesh, Singal SP (2003) Estimation of vehicular pollution at Satna. Indian J Air Pollut Cont 3(1):37–44

Qin Y, Kot SC (1993) Dispersion of vehicular emission in Street Canyons, Guangzhou City, South China (P.R.C.). Atmos Environ Part B Urb Atmos 27(3):283–291

Schwartz J, Ballester F, Saez M, Perez Heyos S, Bellido J, Cambra K, Arribas F, Canada A, Perez-Boillos M, Sunyer J (2001) The concentration response relation between air pollution and daily deaths. Environ Health Perspect 109(10):1001–1006

Shanker PR, Rao GR (2000) Impacts of air quality on human health: a case of Mumbai City, India, IUSSP. In: Regional conference on southeast population in a changing Asian contest, June 10–13

Singh MP, Goyal P, Basu S, Agarwal P, Nigam S, Kumari M, Panwar TS (1990) Predicted and measured concentrations of traffic carbon monoxide over Delhi. Atmos Environ 24(4):801–810

Xie X, Wang Jia-song, Huang Zhen (2009) Traffic emission transportation in street canyons. J Hydrodyn Ser B 21(1):108–117

Chapter 11
Prediction of Air Pollution by the Contribution of Road Traffic—Signal Processing and Higher-Order Statistics (HOS) Spectra Approach

S. Sangeetha, P. Venkatakrishnan, and G. Srikanth

11.1 Introduction

In the present scenario, there are two main air pollutants threatening the human life. One is CO_2 which reduces oxygen carrying capacity of blood in our body and the other one is SO_2 that can cause respiratory diseases. For the day-to-day survival of human in this world, the above-mentioned two systematic operations are inevitable. The other air pollutants like the inhaul of excessive CO_2 leads to reduce the ion content of blood becoming less than 7.35, and inhaul of NO_2 can reduce the immunity of lung infections. The frequent inhaul of these toxic gases causes health-related problems such as wheezing, coughing, colds, flu and bronchitis. The air pollutant VOC will also create problems such as irritated skin, eyes, nose and throat directly to human. The raised pollution level of PM can cause hectic lung irritation, sudden heart attack and blood clots. Due to the proliferation of total population and the vehicle motor utilization for the transportation in the urban and suburban areas, there is an imperative need for the research which deals with the changes of air pollution level (Zhong et al. 2012). Currently, throughout the world, the major scare to the society is global climate change because of the increasing trends in the temperature. Recently, the attention of the research leads toward the

S. Sangeetha
Department of EEE, CMR College of Engineering & Technology, Hyderabad, India
e-mail: sangeee12681@gmail.com

P. Venkatakrishnan (✉) · G. Srikanth
Department of ECE, CMR Technical Campus, Hyderabad, India
e-mail: pvkmephd@gmail.com

G. Srikanth
e-mail: hodececmrtes@gmail.com

S. M. Shiva Nagendra et al. (eds.), *Urban Air Quality Monitoring, Modelling and Human Exposure Assessment*, Springer Transactions in Civil and Environmental Engineering,
https://doi.org/10.1007/978-981-15-5511-4_11

greater challenges to greenhouse gas emissions. Imperative steps have been taken to keep the greenhouse gas emissions (CO_2, CO, SO_2, NO_x, VOC and PM) at a reasonable level (Chang et al. 2013). Nowadays, perfect air quality modeling is an essential tool for the learning of particular environmental-related changes. Hence, reliable prediction is helpful to describe the salient characteristics of pollution level changes and to forecast the possible pollution level index in plight situations. The complex phenomenology in the prediction of pollution level index involves several aspects such as emission particles, transport chemical—transformation processes and particles' dilution in air phenomena, and they are connected with the atmospheric layers either static or dynamic (Viney and Aiyyer 2011). The above observations have motivated to analyze and interpret the contribution of road traffic vehicle emission into air quality research. In this context, two parameters such as traffic capacity and traffic congestion have been consistently envisaged with great attention in road traffic-related researches (Morris et al. 2012). The strong evidence is that the roadside traffic is also a much more interesting research in terms of traffic-related emissions such as energy (fuel consumption), greenhouse gases and pollutants (Abo-Qudais and Qdais 2005).

In the past three decades, many research efforts have been deployed from the bottom-up approach to develop accurate vehicle emission models like EMission FACtors (EMFAC), Comprehensive Modal Emission Model (CMEM), Motor Vehicle Emission Simulator (MOVES), MOBILE and COPERT (Abo-Qudais and Qdais 2005; Rakha et al. 2003; Sharma and Khare 2001). In the literature, many researchers have analyzed the detailed information about the vehicle technology, driving pattern and large-scale survey of international vehicle emission model (Liu et al. 2007). Nowadays, the estimation of vehicle emission and its percentage of contamination in air have emerged as an active research field, due to the study of emission characteristics in government institutions, and these institutions are still effectively utilizing two techniques of air pollution model for initial studies. The first one is Environmental Protection Agency (EPA) and the other one is California Air Resources Board (CARB). The present government has been utilizing both the emission models, EPA and CARB (EPA 2009; CARB 2011), and they can be used to determine the impacts of future emission activity. These models will help to determine the emission characteristics and its linearity for specific scenarios. There are two critical problems persisting in the study of roadway traffic congestions, (1) vehicle density and (2) monitor the roadway network. To mitigate the above problems, static and dynamic simulation software packages have been used for vehicle traffic study. The emission software simulation packages including INTEGRATION (Rakha and Ahn 2004), Transportation Analysis SIMulation System (TRANSIMS) (Zietsman and Rilett 2001) and Visual Traffic Simulation (VISSIM) (10 User Manual 2005) are often used to study statistically about the effects on the emission policies (Li et al. 2012). The above-mentioned simulation-based emission models have not effectively estimated the perfect emission value in real time because of the complexities in the determination of vehicle arrival rate (Poisson distribution) and the speed of the same vehicle (Gaussian distribution). Based on the above information, still no tool is available to predict and calculate real-time transportation-related emissions. Because of the proliferation of vehicles, the air pollution is a growing problem in this world (Viney and Aiyyer 2011). As a result, it is not possible for the stakeholders to handle or plan future decisions for plight situation about the air quality prediction modeling.

In transportation, the vehicle density and its irregular structure are often carrying more information about the singularity pattern. In recent years, the singularity pattern analysis and its local variation are emerging fields in multiple area especially from agriculture to information technology sectors (Arneodo et al. 1995; Asada and Brady 1986; Mallat and Zhong 1992; Mallat 1992; Zhang and Zheng 1997). Hence, this singularity parameter is measured by Lipschitz exponent (LE), and it is a real number to manifest the local variations in traffic vehicle density data (Mallat and Hwang 1992). Using the mathematical procedure of LE, the sharpness of an edge variation pattern of vehicle traffic density can be described. The wavelet transform has been used to measure the local and global variations using LE value. The theorem and the relationship between the wavelet transform modulus maxima (WTMM) and the LE theory are also explained in detail (Mallat and Hwang 1992). The local maxima of the wavelet transform and its modulus are represented as WTMM. The values of WTMM across the scale are represented in the form of local regularity variations. The result presents that the LE determination based on WTMM provides promising performance even in the intricate problem which persists in the sampling nature of the reconstruction process. The organization of this chapter is as follows: The fundamental concepts are depicted in Sect. 2, methodology is presented in Sect. 3, the corresponding result analysis with the help of higher-order statistical test is provided in Sect. 4 and Sect. 5 gives the conclusion and further extension of the work with respect to the prediction of air quality model.

11.2 Fundamental Concepts

The fundamental theory of the wavelet transforms, orthogonal condition and the determination of LE are mathematically defined in this section. The detailed procedure for basic construction of continuous wavelet transform (CWT) is excerpted in (Grossmanm and Morlet 1984). Let the complex-valued wavelet function is represented by $\psi(t)$, and its Fourier transform of complex valued wavelet $\widehat{\psi}(\omega)$ satisfies

$$\int_{0}^{+\infty} \frac{\left|\widehat{\psi}(\omega)\right|^2}{\omega} \mathrm{d}\omega = \int_{-\infty}^{0} \frac{\left|\widehat{\psi}(\omega)\right|^2}{\omega} \mathrm{d}\omega = C_\psi < +\infty \tag{11.1}$$

This condition implies that

$$\int_{-\infty}^{+\infty} \psi(t)\mathrm{d}t = 0 \tag{11.2}$$

The CWT of a time-domain function *f(t)* ∈ *L*²*(R)* with respect to the complex valued wavelet $\psi(t)$ is represented as

$$Wf(u,s) \int_{-\infty}^{+\infty} f(t)\frac{1}{\sqrt{s}}\psi^*\left(\frac{t-u}{s}\right)\mathrm{d}t \tag{11.3}$$

where ψ^* denotes the complex conjugate of ψ (real and imaginary oscillatory vanishing moments)

A complex-valued wavelet $\psi(t)$ is specified as a n number of vanishing moments, and it satisfies the following condition if and only if for all positive integers and they are represented by the number of vanishing moments like $k < n$,

$$\int_{-\infty}^{+\infty} t^k \psi(t)\mathrm{d}t = 0 \tag{11.4}$$

In example, the wavelet with *n*th-order derivation of the Gaussian function is defined in terms of

$$\psi_n(x) = -\frac{\mathrm{d}^n}{\mathrm{d}x^n}\mathrm{e}^{\frac{-x^2}{2}} \tag{11.5}$$

The number of vanishing moments is very important, when the analysis of singularity uses wavelet transform. The reason is that it provides strength and upper bound measurement for the singularity characterization in road vehicle density. This work has been excerpted from the derivation of WTMM of signals contrived by Mallat and Hwang.

The definition of singular exponent is as follows: If a real-valued function $f(x)$ is said to be Lipschitz α and the range varying $0 \leq \alpha \leq 1$, then

$$|f(x) - f(x_0)| \leq A|x - x_0|^\alpha \tag{11.6}$$

The real-valued time-domain function *f(x)* is uniformly Lipschitz α (n times continuously derivative) for any $x_0 \in (a, b)$ and $x \in (a, b)$; then, there are four cases possible to evolve as follows:

case (i) The $f(x)$ is a form of singular in x_0; then, the representation x_0 in the value of Lipschitz is not equal to 1.

case (ii) The function $f(x)$ has a Lipschitz α and its value is greater than zero; then, for $\alpha > 0$, the function $f(x)$ is continuous in x_0.

case (iii) The function $f(x)$ is discontinuity in x_0 and bounding in a neighborhood of x_0; then, the Lipschitz value is 0 in x_0.

case (iv) The function $f(x)$ is continuously differentiable as well as the Lipschitz value is equal to unity and that point is not called as a singular.

11.2.1 Theorem

The wavelet function is well defined for the real-valued tempered distribution function $f(x)$ over (a, b), and let $x_0 \in (a, b)$. The condition of modulus maxima of $Wf(s, x)$ belongs to a cone defined by initial conditions that are represented as follows:

$$|x - x_0| \leq Cs \tag{11.7}$$

The above condition is satisfied; then, at all points, $x_1 \in (a, b)$, $x_1 \neq x_0$, $f(x)$ is uniformly Lipschitz n in a neighborhood of x_1. The LE value, which is less than the number of vanishing moments, is treated as a non-integer. When the function $f(x)$ is Lipchitz α at x_0, the above Eq. (7) can be defined as

$$|Wf(s, x)| \leq As^{\alpha} \tag{11.8}$$

Using simple derivations of linear constant coefficient difference equation and by substituting S_i and S_{i+1} into Eq. (9), the ratio of two logarithmic-scale parameters is expressed as LE with the following form

$$\alpha = \frac{\log_2\left|\frac{Wf\left(s_{(i+1)},x\right)}{Wf(s_i,x)}\right|}{\log_2\left|\frac{s_{(i+1)}}{s_i}\right|} \tag{11.9}$$

The degree of failure can be reflected through the value LE. For example, when the LE value is small, the derivations of the curve are much stronger at that time.

11.3 Methodology

The systematic procedure has been needed to predict the air pollution index level in terms of hectic road traffic contribution, and the consistent flow of the proposed methodology is shown in Fig. 11.1. Here, the sample network has been artificially simulated with six links using MATLAB (Ando et al. 1999). Table 11.1 shows the one-hour period of OD movements, and it represents the number of vehicles per link. There are three links, namely *A*, *B* and *C*, which are the corresponding random numbers of vehicles arrived, and their speed is generated using arrival rate λ and the

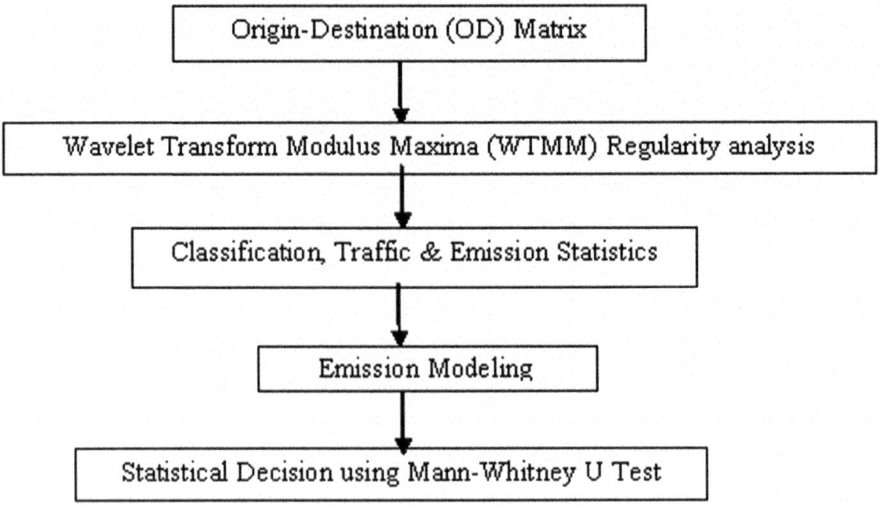

Fig. 11.1 Air pollution analysis methodology block diagram

Table 11.1 Origin–destination matrix used in the simulation (number of vehicles per hour) in Madurai city as on 01.05.2016 (8.45 am–9.45 am)

Origin–destination	Periyar junction (*A*)	Goripalayam junction (*B*)	Kalavasal junction (*C*)
Periyar junction (*A*)	0	410	379
Goripalayam junction (*B*)	631	0	298
Kalavasal junction (*C*)	562	530	0

sample mean μ, respectively. The singularity (discontinuity) behavior of a traffic data can be measured by the most popular technique known as LE and its characteristics distribution. It is a real number, and it is capable to characterize traffic data sample that is either the local regularity or smoothness is present or not.

The procedure for the magnitude of decay of wavelet modulus maxima across the scales has been taken into account for the local regularity measurement at the singularity regions. At each singularity point, the LE value rapidly changes with respect to vehicle density, and it provides local regularity measurement of the road vehicle traffic data. At this point, it is very difficult to choose the number of vanishing moments to analyze a particular class of signals and to select the suitable wavelet type. The constraint expressed clearly shows to estimate the LE up to a maximum value of n, and the concerned wavelet family must have at least n vanishing moments. For the purpose of analyzing the interesting irregular behavior of the signal and for minimizing the amount of computations, maximum number of vanishing moments is needed to be kept in the wavelet. This means that it is necessary to choose a wavelet with minimum number of vanishing moments as possible. But, it must be

Table 11.2 List of air quality parameters and emission measurement of the proposed study

S. No.	Data	Instrument	Company
1	CO_2 (Carbon dioxide) CO (Carbon monoxide)	Gas Alert Micro (5) PID (multigas analyzer)	BW Technology Honeywell, Canada
2	SO_2 (sulfur dioxide) NO_2 (nitrogen oxide) VOC (volatile organic compounds) PM (particulate matter)	Ambient find dust sampler	Polltech instruments Pvt. Ltd. Mumbai MH, India

ensured that few vanishing moments are capable to detect higher-order LEs. In road traffic vehicle data processing, it is needed to analyze the singularities and peaks that have LE smaller than one. Due to this reason, only one vanishing moment has been demonstrated with excellent understanding of Haar wavelet.

The road network example in Madurai city was monitored during April 2016 to July 2016 and recorded the data of total number of vehicles for the necessary study. In the present study, three particular highly traffic road stations such as Periyar, Goripalayam and Kalavasal have been considered. The linear air pollution models are considered based on the determination of vehicles arrival rate λ, and it is proposed by Poisson distribution, and the vehicle speed μ is proposed by normal distribution. The alternate hypothesis represents the level of pollutant which is linearly related to the emission rate, and the null hypothesis represents the concentration level of pollutant that is not related to the emission rate. The primary study of the work is whether the concentration level is a function of the total number of vehicles with heavy and less road traffic situation. It clearly reveals that the concentration of air pollutant level is linearly related to the vehicle's emission. The level of concentration is measured with the equipment, and the analyzed data recordings using higher-order statistics are provided in Table 11.2.

The main problem that exists in basic air pollution models (Gaussian, box, narrow plume, trajectory, gradient transport and complex computational fluid dynamic) is the input data that are time-varying parameters, and they are averaged over by either temporal or spatial distribution (Ando et al. 2000). Consequently, the identification of causes of air pollution is described without clear procedure of temporal and spatial distributions of road traffic analysis. Hence, it will create difficulty to understand, identify and discriminate the point and nonpoint of pollution sources. The above-mentioned pivotal problems are impetus to do research in the area of air pollution prediction modeling using advanced data processing techniques.

The attainment of the expected goal is the prediction of air pollution model by the enumeration measure of measures of effectiveness (MOE) technique, and it provides a quality basis analysis through the prediction of efficiency in air pollution estimation. The average delay, average link density and link flow have been taken into account to assess each OD matrix. The before-and-after (BAA) study is commonly used in the field visit to analyze the effect of air pollution changes in the peak and nonpeak hours. In the present study, among the BAA approaches, the naïve BAA study has been utilized for the emission recordings (Transportation Safety Council 2009). It determines a smooth value for the expected air pollutions and eliminates the randomness element of other information. The nonparametric test such as Mann–Whitney test has been used to (unpaired t-test) represent the study of null and alternate hypothesis that are significant or not. This test is based on two samples from the same population or alternatively; one sample population tends to lower than the observations in the other population. This test is particularly useful in air pollution prediction model for peculiar situation. The main advantage of this type of nonparametric test does assume that the two-sample populations are similar in shape (probability distribution) (Rosie 2004).

11.4 Results

The data such as the number of vehicles from *A* to *B* (blue color) and *C* to *B* (red color) or signal are represented in Fig. 11.2, and they are simulated using MATLAB with help of Poisson and Gaussian distribution parameters in the OD matrix. The local or global regularity variations are estimated using LE values, and they are shown in Fig. 11.3. When the traffic density is varying much faster, the LE values

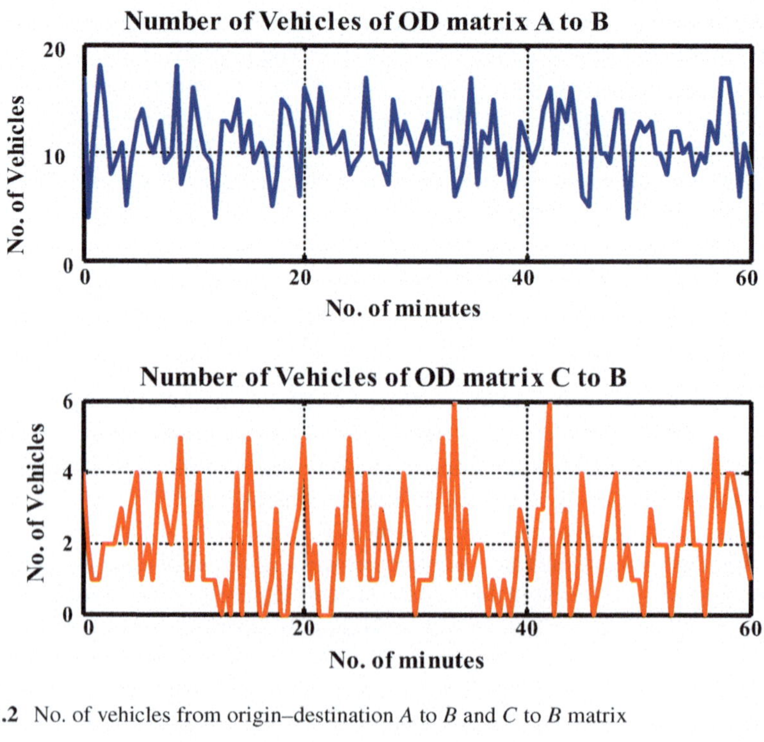

Fig. 11.2 No. of vehicles from origin–destination *A* to *B* and *C* to *B* matrix

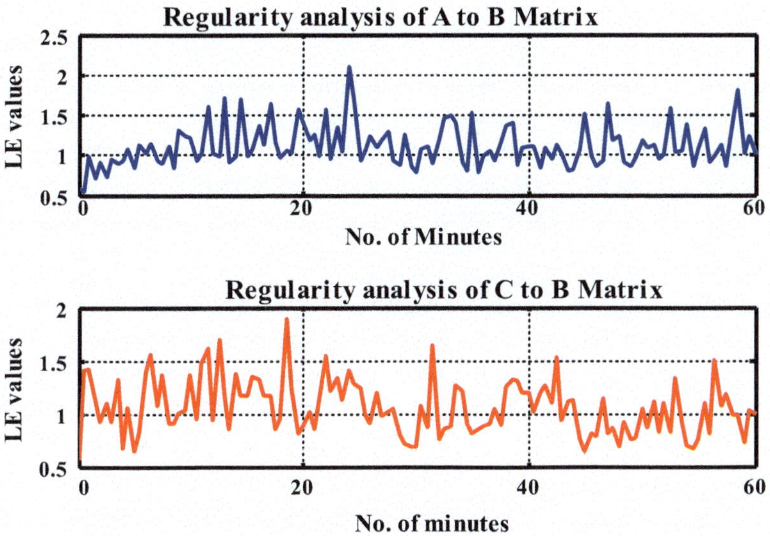

Fig. 11.3 Lipschitz exponent values for *A* to *B* and *C* to *B* matrix

become smaller and it is represented in the 50th time-bounded location that has an LE value of less than 0.750. The infer point related to OD matrix is rapidly changing regularly in *A* to *B*. In the rest of the time-bounded segment, the traffic density has a slow variation with LE values that represent almost equal to unity. The distributions of LE values for heavy and less traffic are shown in Fig. 11.4. From this figure,

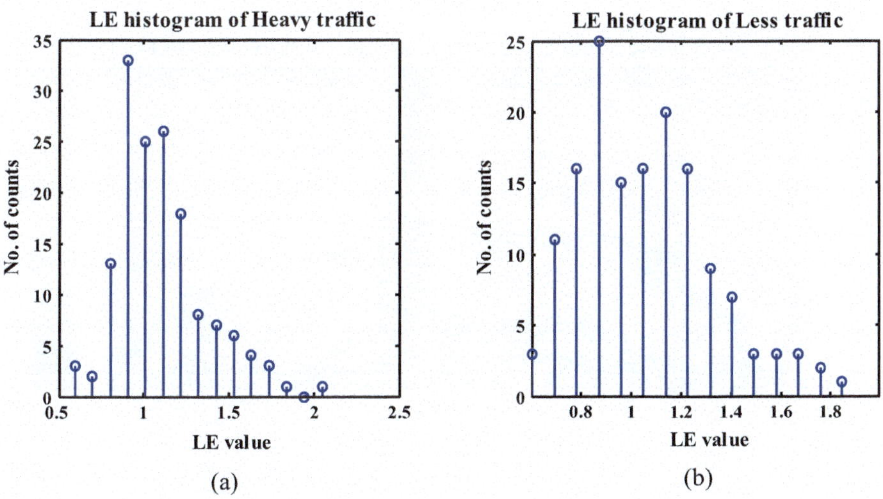

Fig. 11.4 LE value distribution for **a** heavy traffic and **b** less traffic

it is evident that some peculiar cases are not possible to detect singularity points because the LE values are always not influenced by any of high-frequency traffic components (road accidents). Table 11.2 manifests the list of pollution parameters and the corresponding measurement device name handled in the proposed study.

The spatial–temporal distribution of averaged CO_2 emission estimation sample values in the regions of *A* to *B* and *C* to *B* matrix is shown in Fig. 11.5. The *p*-values most importantly specify the research findings with desirable and favorable decision about the linear relationship between the level of pollution and the percentage of emission level, and it also shows the p-value of air pollutants. The null hypothesis (H_0) is pollution index that is always linearly related to the traffic density index with *p*-value, and it is <0.05 which is accepted. The information excerpted in Fig. 11.6 is the interpolation point of traffic intensity versus CO_2 emissions in percentage during peak hours. The spatial distributions of CO_2 emissions in road segments are

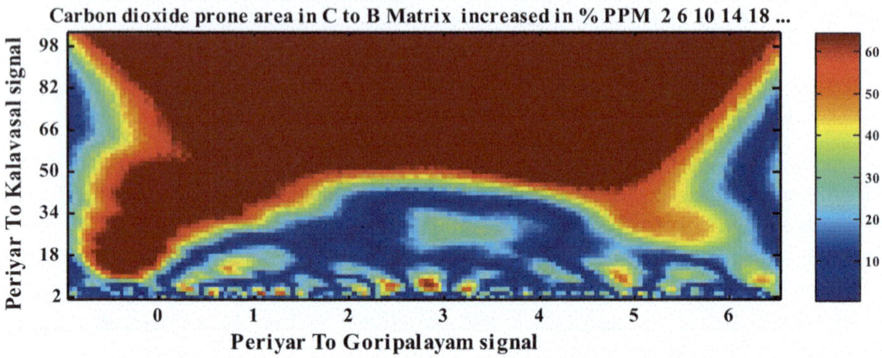

a)

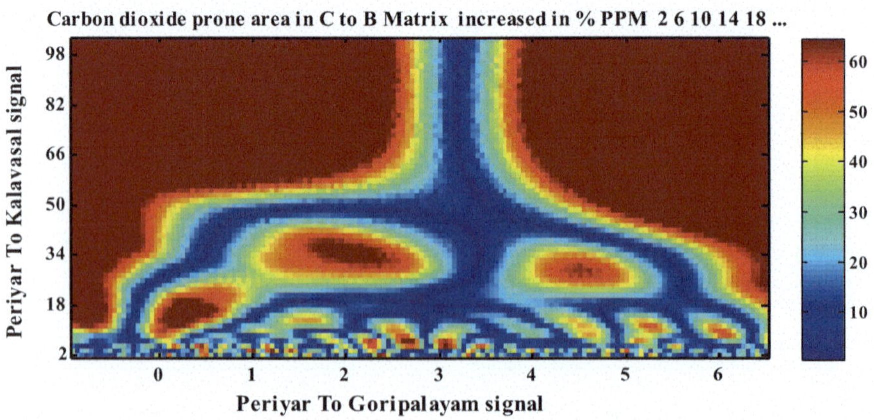

b)

Fig. 11.5 Geographical position of the area of average CO_2 in particular surrounding **a** *A* to *B* matrix and **b** *C* to *B* matrix in origin–destination road network

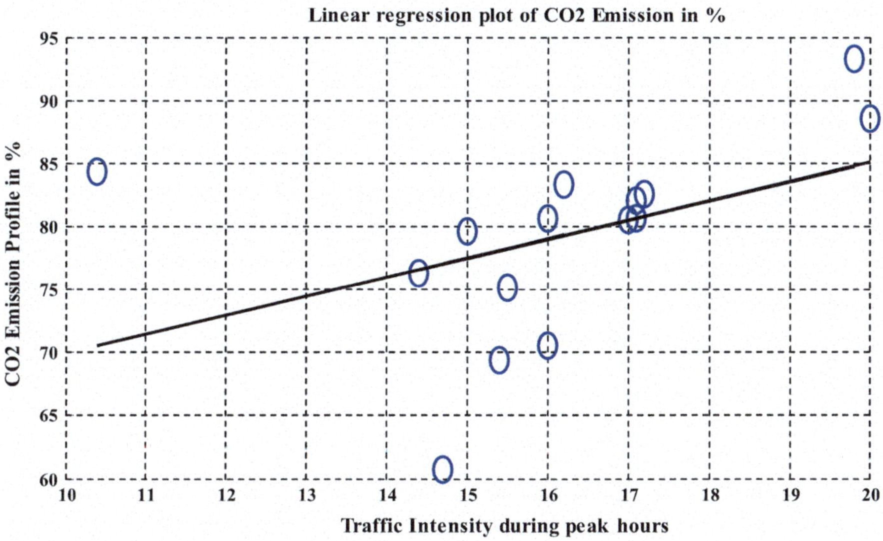

Fig. 11.6 CO_2 emission under various traffic intensity time intervals

Table 11.3 *p*-value for the statistical Mann–Whitney U test

Measure	Subject		
	A	*B*	*C*
CO_2	0.1578	0.126	0.310
CO	0.1735	0.2152	0.245
SO_2	0.7437	0.1765	0.1276
NO_2	0.0076	0.0031	0.0042
PM	0.9968	0.160	0.6035

normalized by their road lengths, due to the case study of different road lengths, and their corresponding CO_2 emissions have been investigated (Table 11.3).

11.4.1 Program Code

```
clc;clear all;close all;n=0:1:149;n1 = 0:.5:60;
    s=RandStream('mt19937ar');
    RandStream.setDefaultStream(s);
    % OD12 = poissrnd(500,1,150)';
    % OD13 = poissrnd(200,1,150)';
    % V = poissrnd(6.8,25,150);
    OD12 = poissrnd(10.8,1,150)';
```

```
OD13 = poissrnd(1.8,1,150)';
% OD12 = V (1,:);
% OD13 = V(10,:);
figure(1);subplot(211);
plot(n1(1:121),OD12(1:121),'b','LineWidth',2);
grid; xlabel('No. of minutes'); ylabel('No. of Vehicles');
title('Number of Vehicles of OD matrix A to B');
subplot(212);
plot(n1(1:121),OD13(1:121),'r','LineWidth',2);
grid; xlabel('No. of minutes'); ylabel('No. of Vehicles');
title('Number of Vehicles of OD matrix C to B');
sca = 2:4:256;
c =cwt(OD12,sca,'db5');
ac = abs(c);
lgsca = log2(sca);
x1 = min(lgsca);x2 = max(lgsca);
for i=1:150
dprc= ac(:,i);
lgdrc =log2(dprc);
% %slope
y1=min(lgdrc); y2=max(lgdrc);
LE1(i)=(y2-y1)/(x2-x1);
end
d =cwt(OD13,sca,'db3');
ac1 = abs(d);
for i=1:150
dprc= ac1(:,i);
lgdrc =log2(dprc);
% %slope
y1=min(lgdrc); y2=max(lgdrc);
LE2(i)=(y2-y1)/(x2-x1);
end
figure(2); subplot(211);
plot(n1,LE1(1:121),'b','LineWidth',2);
xlabel('No. of Minutes');ylabel('LE values');
title('Regularity analysis of A to B Matrix');
grid; subplot(212);
plot(n1,LE2(1:121),'r','LineWidth',2);
xlabel('No. of minutes');ylabel('LE values');
title('Regularity analysis of C to B Matrix');
grid; figure(3);
[n1,xx1]=hist(LE1,15);
stem(xx1,n1);
title('LE histogram of Heavy traffic');
xlabel('LE value');ylabel('No. of counts');
```

```
figure(4);
[n2,xx2]=hist(LE2,15);
stem(xx2,n2);
title('LE histogram of Less traffic');
xlabel('LE value');ylabel('No. of counts');
```

11.5 Conclusion

In this chapter, the findings related to air pollution in the field of road traffic using wavelet transform and higher-order statistical techniques are described clearly. The sink of traffic simulation with road traffic network link is usually very difficult, and it is impossible to predict certain features by empirical manner. Another problem is choice of wavelet which is a hindrance to the prediction performance of perfect road traffic regulation analysis. Due to the complexity in the nature of road traffic density study, these findings confidently provide contribution to enhance the diagnostic information about the road traffic analysis through wavelet transform. Further, it is possible to obtain the complete discrimination characteristics of road vehicle traffic density, and its corresponding linear relationship between the percentage of possible vehicle emission and the peak and nonpeak hour densities has also been studied. In this aspect, the cross-validation technique is very much useful for empirical data analysis for the validation of research findings related to air pollution prediction model. The above work can further be developed for the prediction of air quality in road traffic air–water samples by temporal and spatial distributions using foldscope instrument. The future study mainly focuses on the prediction model with the help of *Microsoft Azure* machine learning tool for real-time applications.

Acknowledgements On behalf of CMRTC Chairman and the Director, the corresponding authors would like to extend special thanks to Indo-US and Government of India, ministry of Science and Technology, Department of Biotechnology (DBT), New Delhi, for having supported with the grant *No.* ***BT/IN/Indo-US/ Foldscope /39/2015/dated on 20.03.2018/***for further research work using foldscope.

Behalf of ECE Department and CMRTC, we would like to thank the Chairman and the Director for their wonderful support, laboratory utilization and valuable suggestions that helped to improve the findings of this research.

Behalf of CMRTC, we express our sincere thanks to the Advanced Environmental Lab, Telangana Pollution Control Board and Dr.S.Rajamohan, MD of Enviro Care India Pvt. Ltd, Madurai, for their constant support, laboratory utilization and valuable suggestions which have enhanced the findings of this research.

References

Abo-Qudais S, Qdais H (2005) Performance evaluation of vehicles emissions prediction models. Clean Technol Environ Policy 7(4):279–284

Ando B, Cammarata G, Fichers A, Graziani S, Pitrone N (1999) A procedure for the optimization of air quality monitoring networks. IEEE Trans Syst Man Cybern C 29:157–163
Ando B, Salvatore B, Graziani S, Pitrone N (2000) Models for air quality management and assessment. IEEE Trans Syst Man Cybern Part C Appl Rev 30(3):358–363
Arneodo A, Bacry E, Muzy JF (1995) The thermodynamics of fractals revisited with wavelets. Phys A 213:232–275
Asada H, Brady M (1986) The curvature primal sketch. IEEE Trans Pattern Anal Mach Intell 8(1):2–14
CARB (2011) Emfac2011 technical documentation. California Environmental protection Agency Air Resources Board, Technical Report
Chang X, Chen BY, Li Q, Cui X, Tang L, Liu C (2013) Estimating real-time traffic carbon dioxide emissions based on intelligent transportation system technologies. IEEE Trans Intell Syst 14(1):469–479
EPA (2009) Motor vehicle emission simulator (moves) 2010: user guide, Office of Transportation and Air quality: United States Environmental Protection Agency, Technical Report
Grossmanm A, Morlet J (1984) Decomposition of hardy functions into square integrable wavelets of constant shape. SIAM J Math Anal 15(4):723–736
Li ZC, Lam WHK, Wong SC, Sumalee A (2012) Environmentally sustainable toll design for congested road networks with uncertain demand. Int J Sustain Transp 6(3), pp 127–155
Liu H, He KB, Wang QD, Huo H, Lents J, Davis N, Nikkila N, Chen CH, Osses M, He CY (2007) Comparison of vehicle activity between Beijing and Shanghai. J Air Waste Manage Assoc 57(10):1172–1177
Mallat S (1992) A wavelet tour of signal processing. Academic, San Diego, CA
Mallat S, Hwang WL (1992) Singularity detection and processing with wavelets. IEEE Trans Inf Theory 38(2):617–643
Mallat S, Zhong S (1992) Characterization of signals from multiscale edges. IEEE Trans Pattern Anal Mach Intell 14(7):710–732
Morris BT, Tran C, Scora G, Trivedi M, Barth MJ (20121) Real time video based traffic measurement and visualization system for energy/emissions. IEEE Trans Intell Syst 13(4):1667–1678
Rakha H, Ahn K (2004) Integration modeling framework for estimating mobile source emissions. J Transp Eng 130(2):183–193
Rakha H, Ahn K, Trani A (2003) Comparison of MOBILE5a, MOBILE6, VT-MICRO, and CMEM models for estimating hot-stabilized light-duty gasoline vehicle emissions. Can J Civ Eng 30(6):1010–1021
Rosie S (2004) Statistics, the Mann Whitney U Test. Mathematics Learning Support Centre (2004)
Sharma P, Khare M (2001) Modeling of vehicular exhausts—a review. Transp Res D Transp Environ 6(3):179–198
Transportation Safety Council (2009) Before and after study—technical brief. Institute of Transportation Engineer, Washington, pp 1–13
Viney P, Aiyyer A (2011) U.S–India collaboration on air quality and climate research and education. For publication in EM, Air and Waste Management Association, pp 1–15
VISSIM 4.10 User Manual (2005) PTV planning Transport Verkehr AG, Karlsruhe, Germany
Zhang J, Zheng CX (1997) Extracting evoked potentials with the singularity detection technique. IEEE Trans Biomed Eng 16(5):155–161
Zhong RX, Sumalee A, Maruyamab T (2012) Dynamic marginal cost, access control, and pollution charge: a comparison of bottleneck and whole link models. J Adv Transp 46(3):191–221
Zietsman J, Rilett LR (2001) Analysis of aggregation bias in vehicular emission estimation using TRANSIMS output. Transp Res Rec 1750:56–63

Chapter 12
Performance Evaluation of UK ADMS-Urban Model and AERMOD Model to Predict the PM_{10} Concentration for Different Scenarios at Urban Roads in Chennai, India and Newcastle City, UK

Prince Vijay, S. M. Shiva Nagendra, Sunil Gulia, Mukesh Khare, Margaret Bell, and Anil Namdeo

12.1 Introduction

The size of pollutants is linked to their potential of causing effects on human health. Particulate matter, which is derived from traffic and its resulting vehicle emissions, is one among six criteria pollutants and is of more importance due to its capability to penetrate into deep lungs and get deposited, thereby affecting the performance of lungs badly. The term particulate matter refers to the collection of both solid and liquid particles of different sizes. Particulate matter of size less than 10 μm in diameter is inhalable and can accumulate in the respiratory system. Particulate matter of size less than 2.5 μm in diameter, which are generally referred to as fine particles, can penetrate deep into the lungs and are supposed to cause largest health risks. Second reason is due to its synergistic effect with other pollutants present in the air.

P. Vijay (✉)
Jawaharlal College of Engineering and Technology, Palakkad, Kerala, India
e-mail: prince.vijayiit2011@gmail.com

S. M. S. Nagendra
Department of Civil Engineering, Indian Institute of Technology Madras, Chennai, Tamil Nadu 600036, India
e-mail: snagendra@iitm.ac.in

S. Gulia · M. Khare
Indian Institute of Technology, Hauz Khas, New Delhi, India

M. Bell · A. Namdeo
School of Civil Engineering and Geosciences, Newcastle University, Newcastle upon Tyne, UK

S. M. Shiva Nagendra et al. (eds.), *Urban Air Quality Monitoring, Modelling and Human Exposure Assessment*, Springer Transactions in Civil and Environmental Engineering,
https://doi.org/10.1007/978-981-15-5511-4_12

Air pollution related to particulate matter is now becoming a serious problem in developing as well as developed countries, factors being the vehicles and the resuspension caused by the vehicular movement. From the studies, it has been found that vehicles in major metropolitan cities of India accounts for 70% of CO, 30–40% of NO_x and 30% of PM of the total load of these cities, of which two-third is contributed by two wheelers alone (Sharma et al. 2005). Source apportionment studies of Chennai (Clean Air Asia: Air quality profile 2010 edition) showed that from the residential monitoring stations, levels of particulate matter in Chennai lies in the range of 51–70 $\mu g/m^3$. According to DoT, about 80% of the road emissions in UK are generated from particulate matter due to road traffic even though there is no factors like resuspension in this country.

12.2 Methodology

12.2.1 Domain Definition

The study region in Chennai is Sardar Patel road (SP road) which stretches from Adayar to Guindy. It is flanked by Indian Institute of Technology Madras and Cancer Institute on one side and Central Leather Research Institute on other side. The terrain of the area is flat terrain about 9.0 m above mean sea level. Study area was located at out gate of IIT Madras. The site is surrounded by a number of educational and research institutes. At this site, there is no local stationary emission source; all the emissions are from the traffic data. The instrument was kept at a height of 1.2 m above the ground level and about 7 m away from the centre line of SP road.

The study region in Newcastle is Newcastle City Centre which is the busiest intersection of Newcastle upon Tyne. The Newcastle International Airport is in close proximity to the city centre, about 6 km from the city. Here also all the emissions are from vehicles and there is no stationary sources. About 10,000 car parking spaces are available within the city centre to the Nexus Metro covering the region. The site selected here was 20 m away from city centre. The monitoring station is operated under Automatic Urban and Rural Network (AURN), the main air quality compliance network for DEFRA. The monitored data for both cities are used for the validation of the model.

The study area for both the cities is shown in Fig. 12.1a, b.

12.2.2 Traffic Characteristics

The traffic flow study on Sardar Patel road is based on visual study. The counting procedure involved video recording of the diurnal traffic followed by manual counting. It is very difficult to distinguish the different category of each vehicle

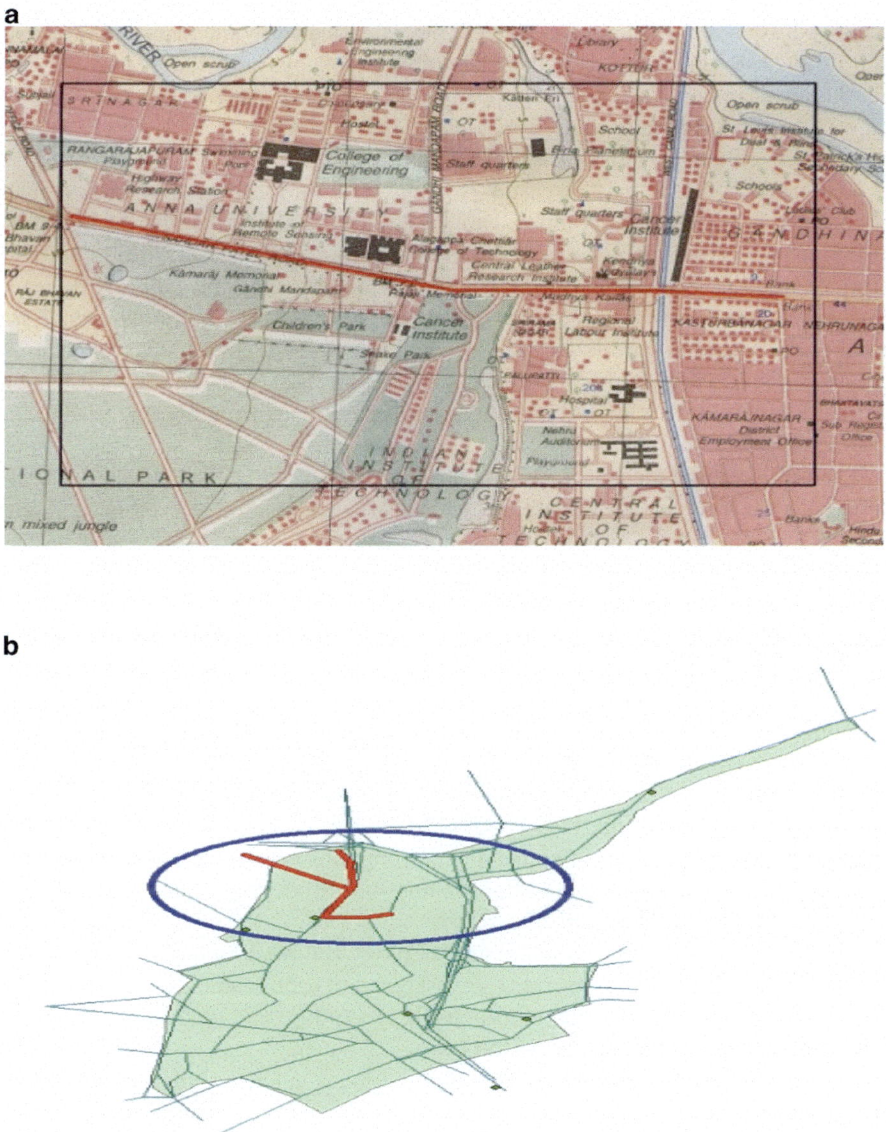

Fig. 12.1 **a** Study area: Chennai. **b** Study area: Newcastle City Centre, UK

while counting the traffic volume so there is a need to identify the exact composition of vehicle to predict correct emission factor for each category of vehicles. The vehicles are categorized as two wheelers, three wheelers, cars, trucks and buses but it does not represent the exact classification of each vehicle's category. For this reason, the data has been obtained from Chennai Pollution Control Board with different category of vehicles and different emission factors for each category. The peak flow occurs between 8 am–10 am and 5 pm–7 pm as indicated by the peaks as shown in Fig. 12.2a. The vehicle number at weekends and weekdays are 126,099 and 128,622 per day, respectively. The fleet composition is dominated by two-wheelers (about 51%). The second-largest share is held by cars and three-wheelers (34% and 8%, respectively). Bus, lorry and van form a small percentage.

The traffic pattern in Newcastle is somewhat uniform over weekdays and minimum in weekends as compared to weekdays, as shown in Fig. 12.2b. Most of the time we can even see a uniform flow of traffic flow. The maximum number of vehicles in a day is 25,537 which is one-tenth of the traffic flow of SP road. Newcastle diurnal traffic flow shows that morning peak flow occurs between 7 am and 11 am and evening peak occurs between 3 pm and 8 pm. The fleet composition of Newcastle is dominated by petrol car which comprises 60%. The second-largest share is by diesel car (30%). Above data has been obtained from SCOOT profile for the central cordon region of Newcastle City.

The traffic data of the year 2009 was used for both the cities and the analysis was done with the meteorological data of the critical winter period (December 2008 to February 2009).

12.2.3 Emissions

The emissions for the year, calculated using the number of vehicles and its emission factors given by Automatic Research Association of India (ARAI) were used in model calculations. Based on the emission factor for each category of vehicle and the percentage distribution given by CPCB between each category of vehicle, the emission loads are obtained in grams/day or kilograms per day. Emission values for Newcastle were calculated using Simple Emissions Modelling Framework (SEMF) developed by TORG, Newcastle University.

The emission estimate is done by methodology followed by Righi et al. (2009). In this methodology, emissions rate (g/s) assigned to each road based on volume of traffic (vehicles/hour), vehicular emissions factors (g/km) and road length (km) was found and applied to the formula given below.

Emission in g/day = Number of Vehicles × Emission Factor (g/km) × Length of the Road (km).

12.2.4 Meteorological Data

The main input meteorological parameters for ADMS are wind speed and wind direction plus of the following parameters cloud cover, heat flux, temperature, reciprocal of Monin–Obukhnov length, upper air data. The parameter solar radiation will be used if NO_x chemistry has to be used. (CERC 2006a, b). Sequential hourly meteorological data were obtained from Laga Systems. Data characterizing upper air

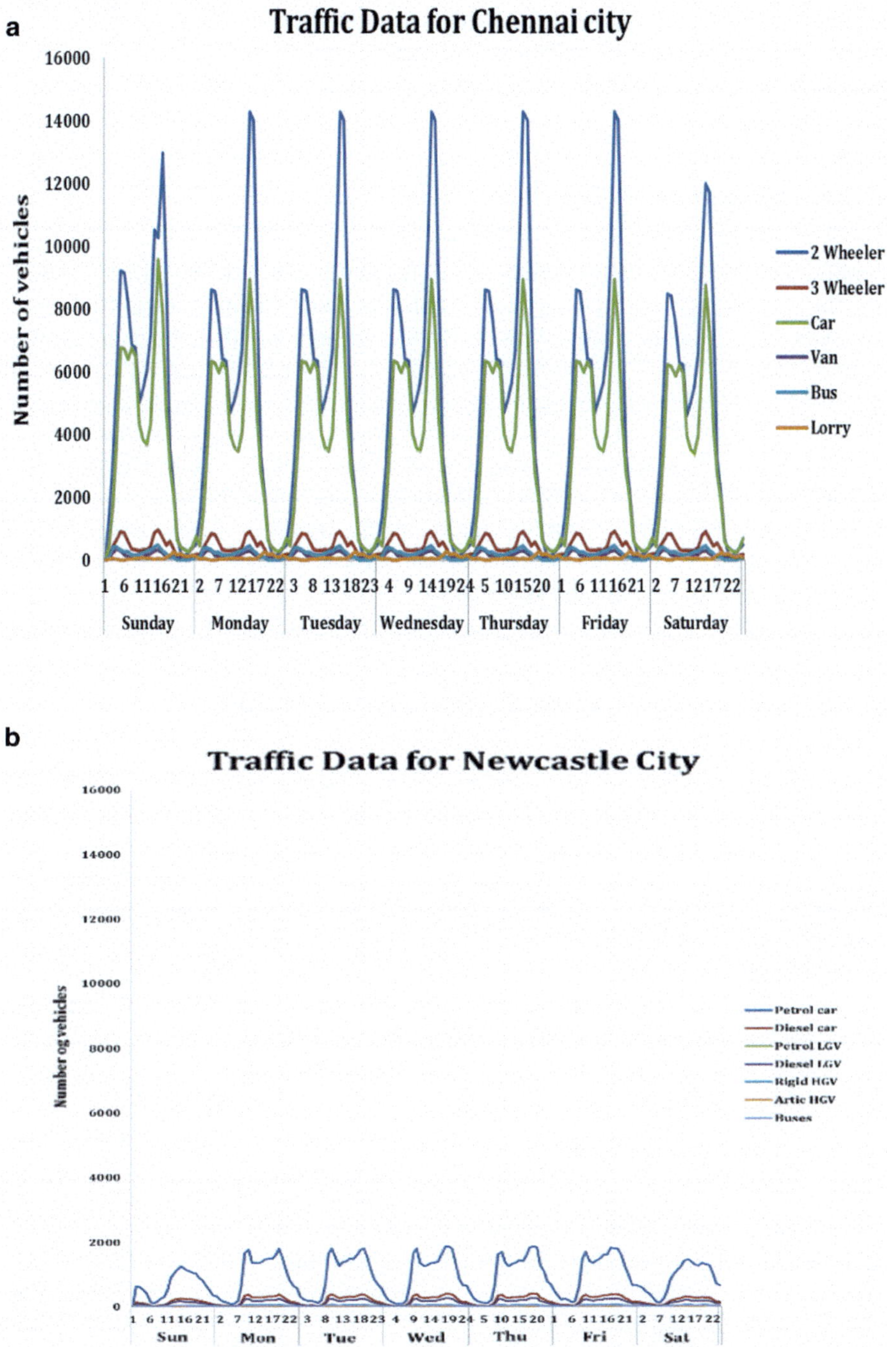

Fig. 12.2 **a** Daily average vehicle count for Sardar Patel Road. **b** Daily average vehicle count for Newcastle City

conditions were acquired from the same data set. The windrose diagram of the study period for Chennai and Newcastle City has been shown in Figs. 12.3a, b and 12.4a, b, respectively.

12.2.5 Monitored PM_{10}

Monitored data for the study period was collected on daily hourly concentrations from the monitoring stations set up at IIT Madras out gate, Chennai and Civic Centre, Newcastle City. The observed values of PM for Chennai was done by using Grimm monitor (Model No: 1.108) and for Newcastle, the observed values were collected from the data recorded by gravimetric monitor set up at Civic Centre. The same location was used as receptor point for monitoring station in the model so that comparisons between estimated and observed concentrations could be made. The study was done during critical winter period as pollutant concentrations are higher during this period. The outputs of the pilot model, predicted hourly sequential measurements of air pollutant data and dispersion derived from vehicular emissions, were compared against observed data within the study area.

12.2.6 Model Used: ADMS-Urban and AERMOD

The model used here is ADMS-Urban which is highly advanced and complex, allowing the user to define an array of parameters and outputs depending on requirements. Developed by Cambridge Environmental Research Consultants (CERC) Ltd ADMS-Urban is a state-of-the-art advanced atmospheric dispersion modelling software package. It adopts a Gaussian dispersion distribution in stable and neutral conditions and a non-Gaussian distribution in the vertical plane for convective conditions. ADMS-Urban can calculate concentrations of pollutants emitted both continuously from point, line, volume and area sources. The model has been extensively validated by over 70 UK local authorities and it is used to predict the dispersion of pollutants derived from on road traffic (Carruthers et al. 1999; Owen et al. 2000; Colvile et al. 2002; McHugh et al. 2004).

AERMOD is a steady-state plume model developed by American Meteorological Society (AMS) and US Environmental Protection Agency (EPA). This model also adopts a Gaussian approach to dispersion modelling. The basic input requirements can be broadly classified into meteorological data, emission data, geometrical data and pollutant background data. AERMOD is applicable to rural and urban areas, flat and complex terrain, surface and elevated releases and pollutant emissions from point, line, volume and area sources.

Both the models mentioned have been used to predict the PM_{10} concentrations in Newcastle City, UK and Sardar Patel Road, Chennai, India.

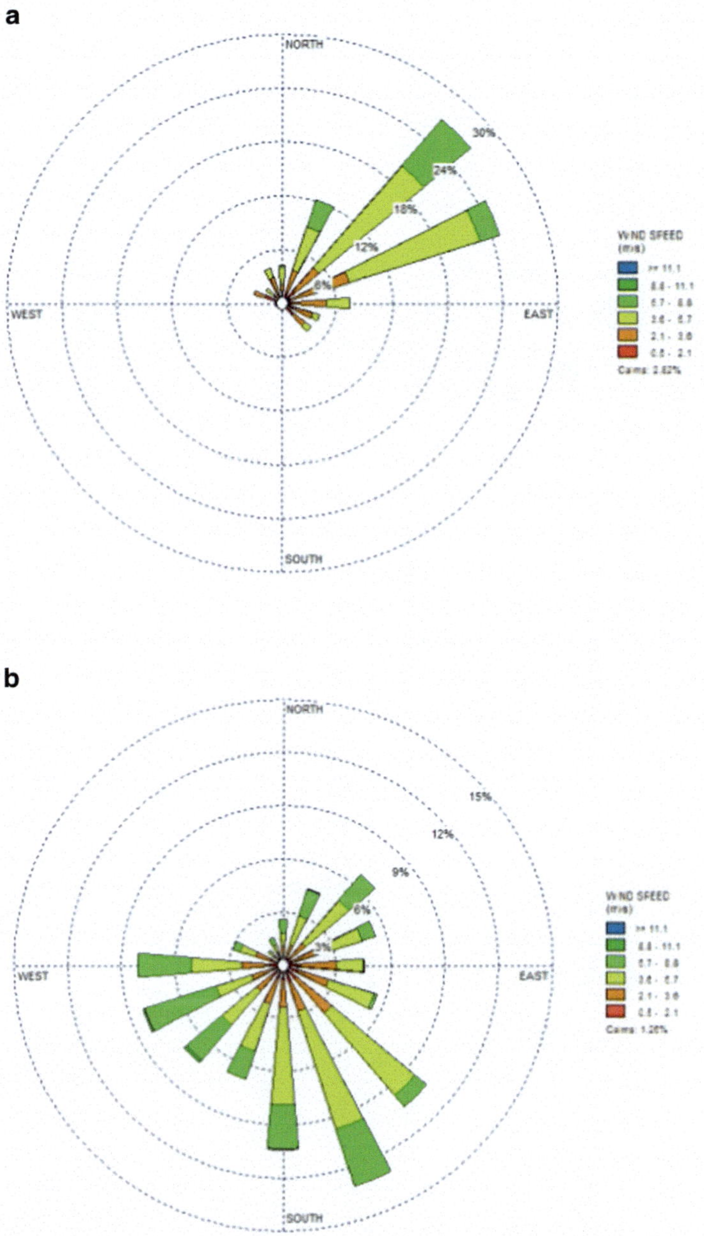

Fig. 12.3 **a** Windrose for Chennai for December 2008. **b** Windrose for Chennai for January–February 2009

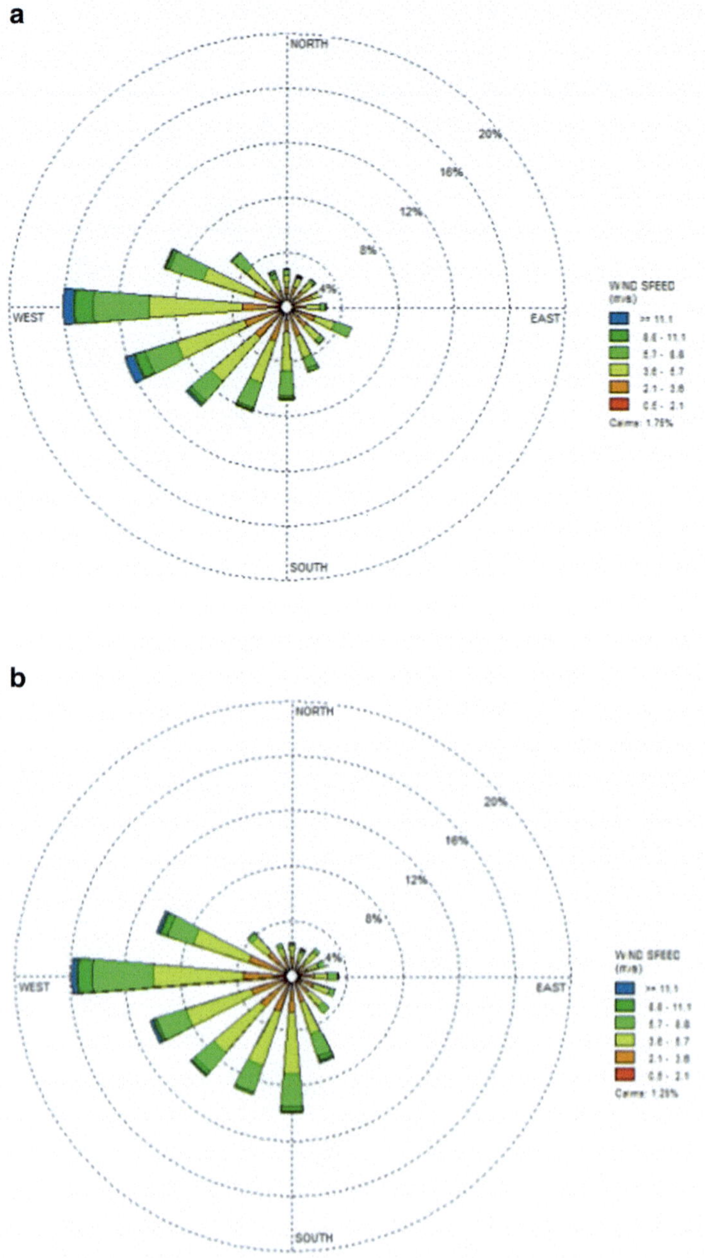

Fig. 12.4 **a** Windrose for Newcastle for December 2008. **b** Windrose for Newcastle for January–February 2009

Table 12.1 Emission factors for different category of vehicles for both the cities

Vehicle type	Chennai city
Petrol-cars	0.032
Diesel-cars	0.020
2W-2S	0.011
2W-4S	0.013
3W-P	0.004
3W-D	0.085
Trucks	1.443
Bus	0.755
Vehicle type	Newcastle City
Petrol-cars	0.023
Diesel-cars	0.047
Petrol LGV	0.034
Diesel LGV	0.072
Rigid HGV	0.149
Artic HGV	0.228
Bus	0.182

12.3 Results and Discussion

The meteorological data was analysed for three months (December 2008, January and February of 2009). The wind rose for the three months has been shown in Fig. 12.3a, b for Chennai and Fig.12.4a, b for Newcastle City. There is a consistency in wind speed and wind direction for the Newcastle City as compared to Chennai city. The results of the analysis indicated an average wind speed of 4.82 m/s and wind direction 230°. For Chennai, the wind speed averaged 3.66 m/s and wind direction 115°. The predominant wind speed, wind direction and the calm conditions (wind speed <1 m/s) (Table 12.2) are summarized in table below:

Table 12.2 Predominant wind speed, wind direction and the calm conditions (wind speed <1 m/s)

Month	Predominant wind direction	Calm period (%)	0.5–2.1 (%)	2.1–3.6 (%)	3.6–5.7 (%)	5.7–8.8 (%)	8.8–11.1 (%)	>11.1 (%)
December 2008	West	1.8	9.1	24.6	38.1	19.9	0.6	2.1
January–February 2009	West	1.3	9.1	25.2	39.8	19.6	0.5	1.3
December 2008	North-East	1.4	12.6	24.4	41.7	19.4	0.6	0.0
January–February 2009	South-East	1.3	10.9	21.1	42.5	23.9	0.5	0.0

12.3.1 Variation of PM_{10} Model Results with Monitored and Meteorological Data

The weekly modelled ground level concentrations for Sardar Patel road Chennai and Newcastle City Centre road were calculated for line sources for a particular week. The variation of modelled and monitored concentration with meteorological parameters like wind speed, wind direction, temperature and relative humidity is shown in Fig. 12.5. It is seen from the figure that meteorological parameters are more or less positively correlated with pollutant concentrations. The concentrations are more when the wind is towards the receptor location. The concentrations are high when the humidity is almost constant. For high wind speed, the concentration is lower and vice versa. This is due to the reason that pollutants get diluted by dispersion.

The figure shows the weekly average of PM_{10} over the two study regions and the relation between the observed and predicted values.

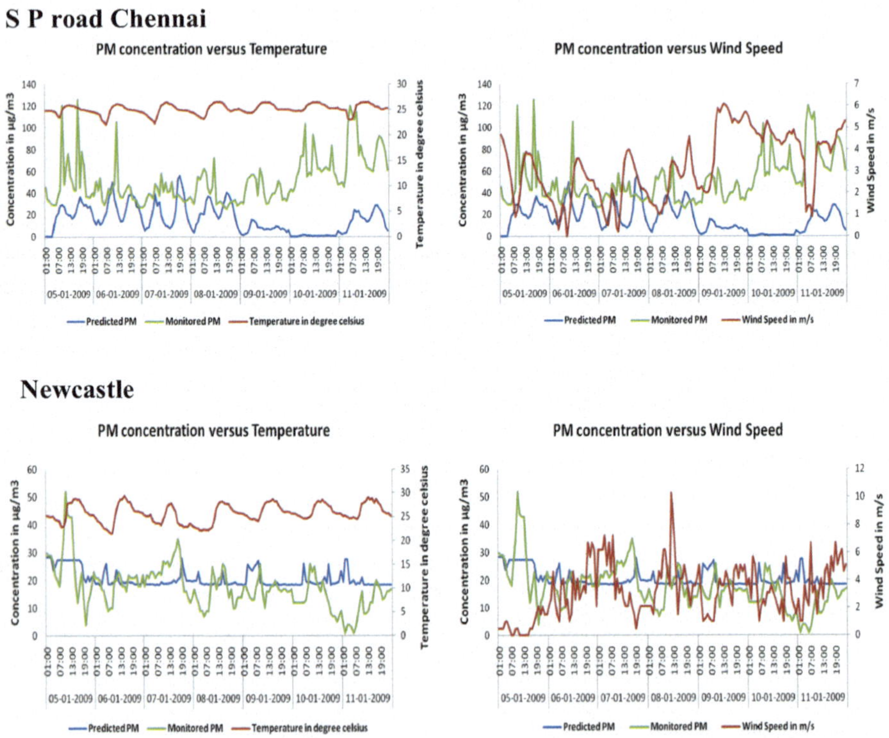

Fig. 12.5 Average of the predicted values and observed values for PM for both the cities

12.3.2 *Statistical Analysis for the Model Performance of PM_{10}*

Monitored data from the monitoring stations were used for evaluating the model performance. The statistical descriptors index of agreement (IA), fractional bias (FB), normalized mean square error (NMSE), geometric mean bias (MG) and geometric mean variance (VG) were used to evaluate the model performances.

Index of agreement (IA): Index of agreement indicates how much the predicted value departs from observed values. It has a theoretical value between 0–1. 1 indicates perfect agreement.

Normalized Root Mean Square Error (NRMSE): Normalized root mean square error is an estimator of the overall deviations between the observed and predicted values (Kumar et al. 2006). Smaller values of NRMSE indicate better performance and are not biased towards model that over predicts or under predict.

Fractional Bias (FB): The fractional bias or FB represents the relative difference between observed and modelled values in a bounded range (Cooper 1999). It is a symmetrical measure and a dimensionless number. It compares different concentration levels and FB value lies between −2 and +2, −2 indicates extreme over prediction and +2 under prediction.

Geometric Mean Bias (MG) and Geometric mean variance (VG): Geometric mean bias (MG) and geometric mean variance (VG) are measures of dispersion which find application when values in a set follow a log normal distribution. A perfect model will have both MG and VG equal to 1.0.

The criteria set by Kumar et al. (1993) were used for understanding the model performance. According to him, the model will be deemed acceptable if NMSE $\leq$ 0.5; $-0.5 \leq$ FB ≤ 0.5; $0.75 \leq$ MG ≤ 1.25; $1 \leq$ VG ≤ 1.25. The model performance is deemed perfect by the statistical analysis according to Righi et al. (2009) that the predicted values match the observed values, then fractional bias (FB), normalized mean square error (NMSE) will be zero and index of agreement (IA), geometric mean bias (MG) and geometric mean variance (VG) will be one. Table 12.3 shows the statistical analysis for the model performance of $PM_{10.}$

12.4 Conclusions

The performance of two Gaussian-based air quality models, namely ADMS and AERMOD models have been evaluated for Chennai city and Newcastle City. ADMS model performed fairly better for PM_{10} predictions for both the cities when compared with AERMOD which is indicated by the degree of agreement values. The IA values for ADMS are found to be 0.39 for Chennai city and 0.48 for Newcastle City.

Table 12.3 Statistical analysis for the model performance of PM_{10}

Pollutant PM_{10}	ADMS	AERMOD	Measured concentration
For Newcastle City			
IA	0.48	0.44	
MG	1.07	1.14	
VG	1.14	0.98	
FB	−1.19	−1.27	
NRMSE	0.09	0.11	
Mean ($\mu g/m^3$)	4.44	3.72	16.38
Maximum ($\mu g/m^3$)	28.10	24.74	59.00
Minimum ($\mu g/m^3$)	18.30	16.30	1.00
Pollutant PM_{10}	ADMS	AERMOD	Measured concentration
For Chennai city			
IA	0.39	0.37	
MG	1.12	1.19	
VG	1.04	1.01	
FB	−1.03	−1.15	
NRMSE	0.17	0.20	
Mean ($\mu g/m^3$)	41.33	20.68	130.39
Maximum ($\mu g/m^3$)	120.11	86.34	343.16
Minimum ($\mu g/m^3$)	25.61	25.61	25.61

Acknowledgements The research work was a part of UKIERI funded research project titled "Evaluation of Quantitative Dispersion Models for Urban Air Quality Assessment". We wish to thank UKIERI New Delhi. We also acknowledge CERC, UK who provided the ADMS for academic and research purpose; and Laga Systems Hyderabad for providing the meteorological data.

References

Carruthers, DJ, Singles, RJ, Nixon, SG, Ellis, KL, Pendrey, M, Harwood, J (1999). Modelling air quality in central London. Cambridge Environmental Research Consultants Ltd. Report FM327 to the London Borough of Camden, the Central London Cluster Group of local authorities and the UK Department of the Environment, Transport and the Regions

CERC (2006a) http://www.cerc.co.uk

CERC (2006b) Cambridge environmental research consultants, ADMS—Urban user guide, http://www.cerc.co.uk/environmental-software/ADMSUrban-model.html

Colvile, RN, Woodfield, NK, Carruthers, DJ, Fisher, BEA, Rickard, A, Neville, S, Hughes, A (2002). Uncertainty in dispersion modelling and urban air quality mapping, J Environmental Science & Policy 5:207–220

Kumar A, Dixit S, Varadarajan C, Vijayan A, Masuraha A (2006) Evaluation of the AERMOD dispersion model as a function of atmospheric stability for an urban area. Environ Progr 25(2)

Kumar, A., Luo, J, Bennet, FG (1993). Statistical evaluation of lower flammability distance, process safety progress. Wiley InterScience, vol 12, p 1–11

McHugh, TE, Connor, JA., Ahmad F (2004). An empirical analysis of the groundwater-to-indoor air exposure pathway: The role of background concentrations in indoor air. Environmental Forensics, 5, p 33–44

Owen, B, Edmunds, HA., Carruthers, DJ, Singles, RJ (2000). Prediction of total oxides of nitrogen and nitrogen dioxide concentrations in a large urban area using a new generation urban scale dispersion model with integral chemistry model. Atmospheric Environment, 14:397–406

Righi, S, Lucialli, P, Pollini, E (2009). Statistical and diagnostic evaluation of the ADMS-Urban model compared with an urban air quality. J Atmospheric Environment 43: 3850–3857

Sharma, N, Chaudhry KK, Chalapati Rao, CV (2005). Vehicular pollution modeling in India. J Institution of Engineers (India) 85:46–63

Chapter 13
Modeling of Atmospheric Mercury Deposition in India

Krish Vijayaraghavan, Shari Libicki, Ross Beardsley, and Sunil Ojha

13.1 Introduction

Mercury (Hg) is a hazardous pollutant that could affect human health and ecosystems. Human exposure to elemental mercury vapours is a health concern at very high concentrations because of toxic inhalation (Clarkson et al. 2003). In addition, inorganic Hg emitted to the atmosphere will ultimately deposit to the surface of the earth where it can be transformed to the more harmful form, methylmercury, that can bioaccumulate in fish and other food chains (Schroeder and Munthe 1998; Seigneur et al. 2001; Pirrone and Mahaffey 2005). While developing fetuses are particularly vulnerable because of its neurotoxicity, there are also concerns due to health effects on sensitive humans and wildlife (Driscoll et al. 2013).

Atmospheric inorganic Hg exists in three forms: (1) elemental mercury vapour, Hg^0, also referred to as gaseous elemental mercury (GEM), (2) gaseous divalent mercury, Hg^{II}, also known as gaseous oxidized mercury (GOM), and (3) particulate mercury, Hg_p, also referred to as particle bound mercury (PBM). Hg_p can arise from GOM becoming adsorbed to atmospheric particulate matter after it is emitted in vapour form or from divalent Hg being emitted into the atmosphere as particulate matter directly in the flue gas. These three forms of Hg vary in their physical and chemical properties and, therefore, have different deposition rates and atmospheric lifetimes. Because Hg^0 has low reactivity and solubility (e.g., Lindberg et al. 2007), it has a lifetime of several months to a year and may be transported across continents. In contrast, Hg^{II} and Hg_p have lifetimes ranging from hours to days because they have

K. Vijayaraghavan (✉) · S. Libicki · R. Beardsley
Ramboll, Novato and San Francisco, CA 94945, USA
e-mail: kvijayaraghavan@ramboll.com

S. Ojha
Ramboll India, Gurgaon 122002, India
e-mail: suojh@ramboll.com

S. M. Shiva Nagendra et al. (eds.), *Urban Air Quality Monitoring, Modelling and Human Exposure Assessment*, Springer Transactions in Civil and Environmental Engineering,
https://doi.org/10.1007/978-981-15-5511-4_13

high wet and dry deposition rates near their sources. The different forms of Hg also inter-convert between each other through gas- and aqueous-phase chemical reactions (e.g., Schroeder and Munthe 1998; Lin et al. 2006; Seigneur et al. 2006) which affects their lifetimes. Due to the long-range transport of Hg, recovery of ecosystems influenced by atmospheric deposition would be influenced both by reductions in local emissions as well as changes in global Hg emissions (e.g., Vijayaraghavan et al. 2014).

Emission inventories have been compiled in numerous studies for large sources in India including coal-based thermal power plants, waste incineration, ferrous and non-ferrous metal production, cement production and the chlor-alkali industry (Mukherjee et al. 2009; Streets et al. 2009; Pirrone et al. 2010; Sloss 2012; AMAP/UNEP 2013; Chakraborty et al. 2013; Rai et al. 2013). Examining the extent of Hg deposition in India allows us to understand the potential contributions of Hg air emissions in the country as well as upwind sources to deposition in the region.

In this paper, an overview of the atmospheric Hg model and application for this study is first presented. The emissions' inventory utilized is discussed, followed by a description of other model inputs and the model configuration. Results are presented for Hg wet and dry deposition across India in general and in specific areas. Major sources of uncertainties in the modeling study are discussed.

13.2 Modeling Methodology

Due to the potential for Hg to undergo long-range transport, it is essential to use a modeling approach that takes into account the global cycling of Hg. In this study, we apply the Goddard Earth Observing System Chemistry (GEOS-Chem) model (www.geos-chem.org), a global three-dimensional (3D) chemistry transport model that uses meteorology from the Goddard Earth Observing System (GEOS) of the National Aeronautics and Space Administration (NASA) Global Modeling and Assimilation Office (GMAO). GEOS-Chem was first developed by the Atmospheric Chemistry Modeling Group (ACMG) at Harvard University over fifteen years ago (Bey et al. 2001) and adapted for Hg cycling by Selin et al. (2007) and Strode et al. (2007). It dynamically couples a 3D atmosphere (Selin et al. 2007), a 2D terrestrial reservoir (Selin et al. 2008) and a 2D ocean module (Soerensen et al. 201). The model has since been extensively evaluated for air concentrations and/or wet deposition in several other global Hg deposition studies (e.g., Holmes et al. 2010; Corbitt et al. 2011; Amos et al. 2012; Chen et al. 2014; Song et al. 2015; Zhang et al. 2016).

We applied version 10-01 of GEOS-Chem using a modeling grid over the entire world with a horizontal grid resolution of 2 by 2.5° (latitude and longitude, respectively). The model simulates the emissions, dispersion, conversion, and wet and dry deposition of Hg^0, Hg^{II} and Hg_p in the atmosphere. An annual simulation was conducted for 2011 using year 2010 for model initialization (spin-up). Assimilated vertical and surface meteorological data were obtained from the NASA GMAO GEOS-5 data for 2011 and used for the modeling. Mercury deposition fluxes over

India, and parts of neighbouring countries were extracted from this GEOS-Chem global simulation output.

The model includes the gas-phase oxidation of Hg^0 to Hg^{II} by bromine and the aqueous-phase photoreduction of Hg^{II} to Hg^0 (Holmes et al. 2010). The dry deposition algorithm is from Wesely (1989) and is based on a resistances-in-series method. Hg^0 evasion includes volatilization from soil and rapid recycling (re-emission) of newly deposited Hg (Selin et al. 2008; Wang et al. 2016). The former is estimated as a function of soil Hg content and solar radiation. The latter is modeled by recycling a fraction of deposited Hg^{II} to the atmosphere as Hg^0 immediately after deposition. Wet deposition follows scavenging of Hg^{II} and Hg_p which is based on Liu et al. (2001). Below-cloud scavenging of Hg_p by snow is also included (Holmes et al. 2010; Amos et al. 2012; Zhang et al. 2012). The wet deposition of Hg^0 is negligible because it is sparingly soluble in water.

Worldwide Hg emissions from the following source categories are included in the modeling: (1) Anthropogenic emissions of Hg^0, Hg^{II} and Hg_p, (2) Hg^0 emissions from biomass burning, (3) Hg^0 emissions from land (including re-emissions of previously deposited Hg) and (4) Hg^0 emissions from oceans (including re-emissions).

13.2.1 Mercury Emissions

The anthropogenic emissions used in the modeling is based on the 2010 global Hg inventory from the Arctic Monitoring and Assessment Programme (AMAP)/United Nations Environment Programme (UNEP) 2013 global mercury assessment (AMAP/UNEP 2013, 2015).

The total estimated worldwide anthropogenic Hg emission in 2010 is 1960 Mg $year^{-1}$ (AMAP/UNEP 2013). Fossil fuel (mainly coal) combustion for power and heating is the largest single category of emissions from anthropogenic sources. Other important source categories are artisanal and small-scale gold production (mainly in Asia and South America), cement production and metal production (mainly in Asia) and waste incineration. Figure 13.1 (adapted from data in AMAP/UNEP 2013) shows the contributions of different source regions to global

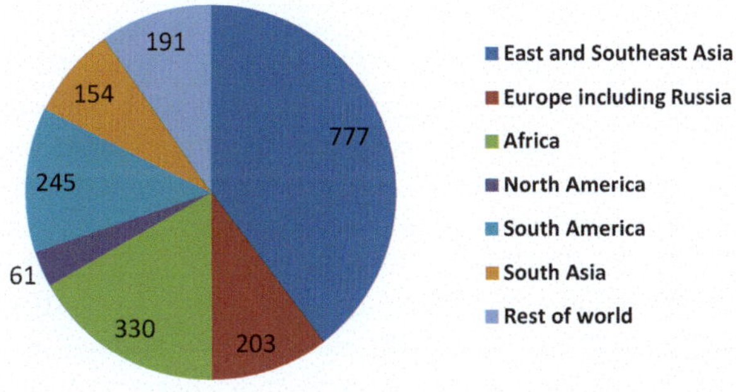

Fig. 13.1 2010 global anthropogenic mercury emissions inventory used in modeling (Mg $year^{-1}$). *Source* AMAP/UNEP (2013)

anthropogenic Hg emissions in 2010. East and Southeast Asia together constitute the largest contributor at 40% of the total (with the majority from China at 575 Mg year^{-1}), followed by Africa, South America and Europe. The total estimated anthropogenic Hg emission from South Asia (consisting of Afghanistan, Bangladesh, Bhutan, India, Maldives, Nepal, Pakistan and Sri Lanka) is 154 Mg year^{-1}, representing 8% of the worldwide anthropogenic inventory. Coal combustion and cement production are estimated to contribute 59% and 11%, respectively, to the South Asian total.

Large uncertainties exist, in general, in modeled global mass balances of Hg and, in particular, in air emission inventories due to variability in input parameters and process rates (e.g., Qureshi et al. 2011). The values reported above are best estimate values from the AMAP/UNEP inventory, with a range of 1010–4070 Mg year^{-1} for the global Hg emission total.

Table 13.1 shows estimated Hg emissions from India in 2010 from the AMAP/UNEP (2013) inventory as applied in the current study. The total estimated anthropogenic Hg emission from India is 145 Mg year^{-1} with the uncertainty ranging from 75 to 330 Mg year^{-1}. Emissions from coal combustion dominate the inventory with coal burning in power plants, industrial uses and domestic/residential use constituting 62% (or 90 Mg year^{-1}) of the total Hg emissions. Waste disposal and cement production constitute approximately 9% each, while copper and zinc production each represent approximately 7.5%.

Total coal consumption at 111 coal-burning thermal power plants in India during the 2010–2011 period was an estimated 500 million Mg year^{-1} (Guttikunda and Jawahar 2014). Rai et al. (2013) estimated uncontrolled Hg emissions from combustion of coal in India in 2010–11 to be 160 Mg year^{-1} based on total coal consumption of a similar amount (590 million Mg year^{-1}) and an average Hg content in coal of 0.272 ppm. The estimated emissions rate would be lower when considering particulate control devices in place. Historic estimates of Hg emissions in India are generally higher. For example, Mukherjee et al. (2009) estimated that 2004 Hg emissions in India were 222–310 Mg year^{-1}. Sloss (2012) summarized the Indian anthropogenic Hg emissions inventory from the prior 2008 AMAP/UNEP assessment that used a 2005-year datum. Emissions of Hg in India were estimated to be 161 Mg year^{-1} in 2005. However, a direct comparison between the 2005 and 2010 AMAP inventories is not possible because of changes in the methodology between the two assessments (AMAP/UNEP 2013).

Chakraborty et al. (2013) quantified anthropogenic Hg flows in India; air emissions were estimated to be 235 Mg year^{-1} in 2010. This value is at the higher end of the AMAP/UNEP range and higher than the value (145 Mg year^{-1}) in the current study. Differences are likely due to difference in methodology and activity data between the two studies. However, the fraction of the total due to coal combustion is comparable between the two inventories, at approximately 60%.

Hg emissions from biomass burning are based on version 3 of the Global Fire Emission Database (van der Werf et al. 2010) for carbon monoxide (CO) and a Hg:CO ratio of 100 nmol mol^{-1} (Holmes et al. 2010; Song et al. 2015). This results in a global total biomass burning emission of 210 Mg year^{-1}. The

Table 13.1 Estimated 2010 anthropogenic Hg emissions from India used in modeling (kg year^{-1})

Source category	Hg emissions (kg year^{-1})
Bituminous coal combustion in power plants	41,387.5
Hard coal (bituminous/anthracite) combustion in industrial use	22,677.2
Coal combustion for domestic/residential use, transport, etc.	16,033.7
Waste and other losses due to breakage and disposal in landfill, etc.	13,691.8
Cement production	13,420.8
Non-ferrous metal production: copper	10,939.6
Non-ferrous metal production: Zn	10,809.8
Lignite coal combustion in power plants	8056.3
Production of iron and steel	1934.3
Brown coal/lignite combustion in industrial use	1289.8
Artisanal and small-scale gold mining	1125.0
Caustic soda production	940.0
Oil refining	634.3
Human cremation (dental amalgams)	607.7
Non-ferrous metal production: Aluminium	464.0
Non-ferrous metal production: Lead	168.6
Non-ferrous metal production: gold large-scale	154.0
Heavy fuel oil combustion in industrial use	136.2
Light fuel oil combustion for domestic/residential use, transport	71.4
Heavy fuel oil combustion for domestic/residential use, transport	64.1
Waste incineration	42.5
Light fuel oil combustion ind. use	30.6
Heavy fuel oil combustion in power plants	28.9
Light fuel oil combustion in power plants	16.5
Natural gas combustion in power plants	6.4
Natural gas combustion in industrial use	1.5
Natural gas combustion for domestic/resident. Use, transport	0.5
Total	144,733

Source AMAP/UNEP (2013)

other non-anthropogenic Hg emission categories (Song et al. 2015) include land geogenic emissions (90 Mg year^{-1}), soil emissions parameterized as a function of solar radiation (1680 Mg year^{-1}), re-emissions from soil, vegetation and snow-pack (520 Mg year^{-1}), and net ocean emissions (3000 Mg year^{-1}). Thus, the total modeled worldwide mercury emission from biomass burning and natural sources is approximately 5500 Mg year^{-1}.

Figure 13.2 shows the spatial distribution of modeled anthropogenic Hg emissions in India and surrounding regions. Anthropogenic emissions are highest in the region

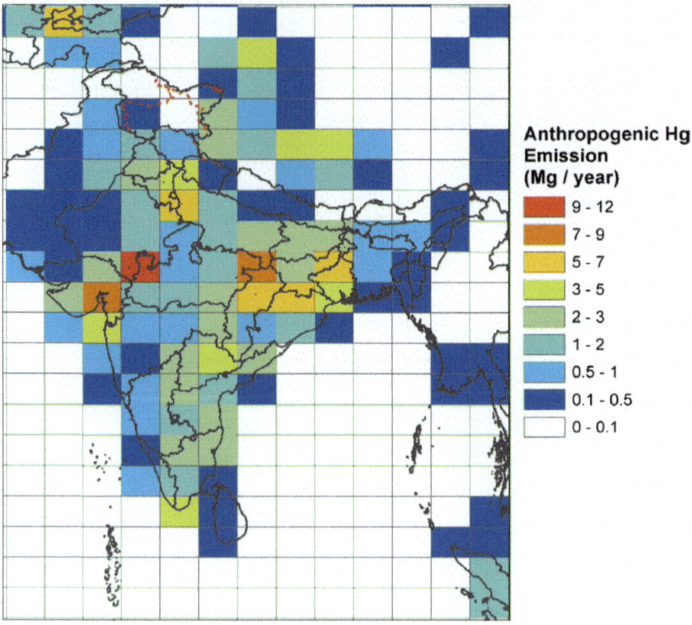

Fig. 13.2 Anthropogenic mercury emissions in the modeling grid in India and surrounding regions

near the southeastern part of the state of Rajasthan and the westernmost part of the state of Madhya Pradesh (MP) due to a combination of coal-based thermal power plants and non-ferrous metal (zinc and copper) production in the region. Hg emissions are also very high in the area encompassing eastern MP and southeastern part of the state of Uttar Pradesh (UP) and central/northern Chhattisgarh; this region has some of the highest density of coal-fired thermal power plants in the country due to proximity to coal mines.

The natural emissions and re-emissions of Hg are mainly from evasion from land (soil/vegetation emissions and re-emissions) with a total across India of 75 Mg year^{-1} (Fig. 13.3). Thus, when combined with the anthropogenic inventory of 145 Mg year^{-1}, the total annual Hg emission from India is estimated to be 220 Mg year^{-1} (Fig. 13.4).

13.3 Results and Conclusions

Figure 13.5 depicts the worldwide annual total (i.e., sum of wet and dry) deposition flux of Hg. Deposition shown is for total Hg, i.e., the sum of Hg^0, Hg^{II} and Hg_p. The annual Hg deposition flux ranges from 0 to 125 μg m^{-2} year^{-1} across the world, with high deposition (values exceeding 50 μg m^{-2} year^{-1}) over polluted regions including East and South Asia, western Europe, parts of Africa and South

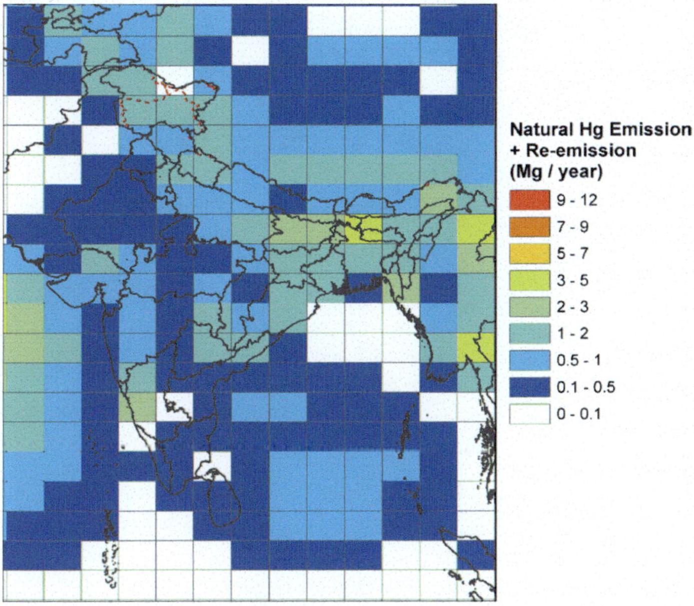

Fig. 13.3 Annual natural mercury emissions plus re-emissions in the modeling grid in India and surrounding regions

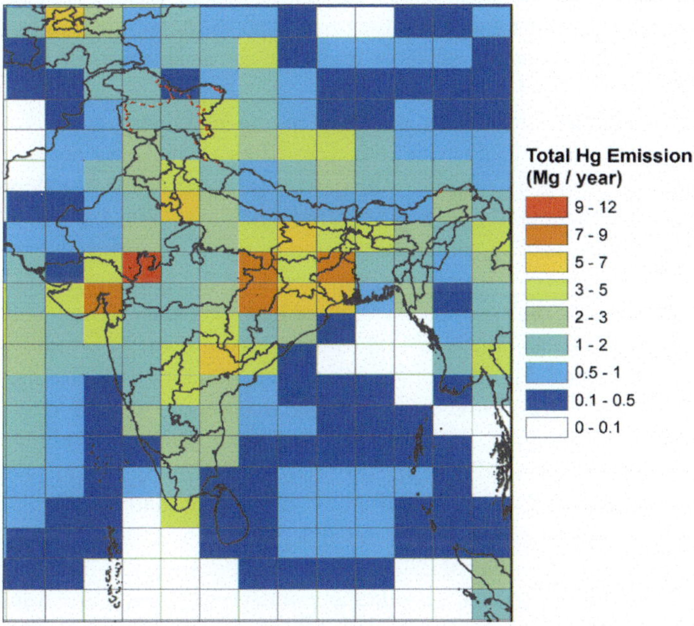

Fig. 13.4 Annual total (anthropogenic + natural + re-emitted) mercury emissions in the modeling grid in India and surrounding regions

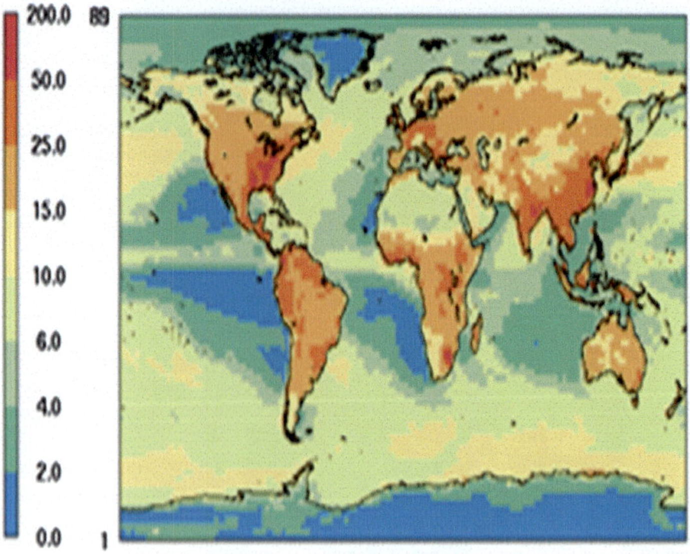

Fig. 13.5 Annual total deposition flux of total mercury (μg m^{-2} $year^{-1}$)

America, and the eastern United States. In addition to deposition in industrial regions due to local anthropogenic sources, the global deposition pattern also reflects the oxidation of Hg^0 and subsequent deposition as well as the air-surface exchange of Hg^0 associated with vegetated surfaces and high precipitation over some remote areas (AMAP/UNEP 2013).

The annual Hg deposition flux over the Indian mainland ranges from 10 to 57 μg m^{-2} $year^{-1}$ (Fig. 13.6) with over four-fifths of the land area with deposition exceeding 25 μg m^{-2} $year^{-1}$. In particular, the highest Hg deposition is seen in northern Chhattisgarh, eastern MP, Jharkhand and West Bengal, all regions with high Hg emissions. The annual deposition in those regions ranges from 50 to 57 μg m^{-2} $year^{-1}$. These estimates are comparable to the peak deposition fluxes of 49 to 61 μg m^{-2} $year^{-1}$ over south Asia reported by Giang et al. (2015) who modeled Hg deposition with GEOS-Chem v. 9-02 using a 2006 global Hg inventory.

The average annual Hg deposition flux over India in the current study is 27.6 μg m^{-2} $year^{-1}$, resulting in a total Hg deposition to land over India of approximately 82 Mg $year^{-1}$. This estimate is approximately 44% higher than the 57 Mg $year^{-1}$ total estimated by Qureshi (2016) using a Hg cycling box model. Deposition over India is due to both emissions within the country and long-range Hg transport from upwind sources. The modeled deposition over India is less than 40% of the total estimated emissions of 220 Mg $year^{-1}$, suggesting that a large fraction of Indian Hg emissions is exported outside the country via atmospheric transport. No separate modeling was performed here to quantify upwind source contributions.

The modeled wet deposition (Fig. 13.7) is due to a combination of precipitation and Hg emissions. The annual wet deposition flux over the Indian mainland

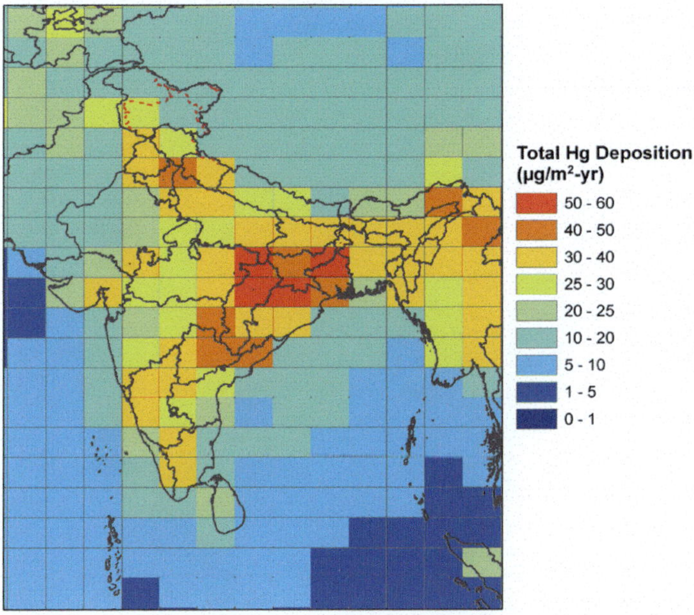

Fig. 13.6 Annual total deposition flux of total mercury ($\mu g\ m^{-2}\ year^{-1}$) in India and surrounding regions

typically ranges from 5 to 30 $\mu gm^{-2}\ year^{-1}$ with the highest wet deposition flux of 30 $\mu gm^{-2}\ year^{-1}$ occurring in the northeastern part of the states of Assam and Arunachal Pradesh, a region with extremely high rainfall ranging approximately from 500 to 1000 cm $year^{-1}$. Other regions that experience a concurrence of relatively high emissions and precipitation also show high wet deposition, including the states of Uttarakhand, Jharkhand, eastern MP and northern Chhattisgarh. Annual wet deposition in southeast Rajasthan is low despite high emissions due to scant rainfall in the arid climate. In general, wet deposition varies by season (not shown here) and is influenced by temporal variability in rainfall.

Huang et al. (2013) measured Hg wet deposition at Lhasa in Tibet. The total annual observed wet deposition over the year 2010, 8.2 $\mu g\ m^{-2}\ year^{-1}$, was influenced primarily by local industrial sources and human activities. The modeled wet deposition from our study is 5.4 $\mu g\ m^{-2}\ year^{-1}$. The under-estimate is likely due, in part, to the coarse grid resolution (2° latitude by 2.5° longitude) utilized in the GEOS-Chem model. Use of a finer grid spacing results in better resolution of local source emissions which would result in better spatial resolution of Hg deposition near such sources (e.g., Zhang et al. 2012). The modeled wet deposition over large parts of India (20–30 $\mu g\ m^{-2}\ year^{-1}$) is higher than the range of wet deposition (2.6–19.8 $\mu g\ m^{-2}\ year^{-1}$) measured in North America in 2015 at the Mercury Deposition Network (MDN) stations (NADP 2015). The difference would be larger with a finer

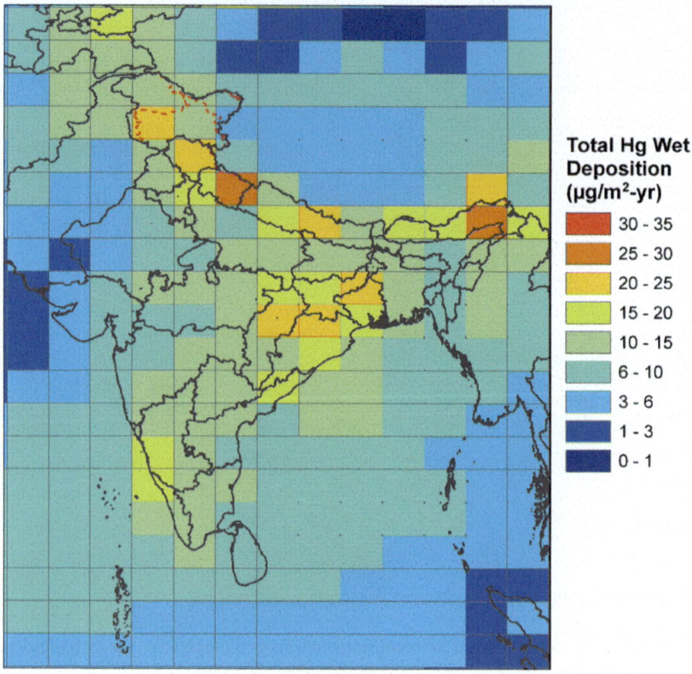

Fig. 13.7 Annual wet deposition flux of total mercury ($\mu g\ m^{-2}\ year^{-1}$) in India and surrounding regions

modeling grid and higher modeled wet deposition. The peak modeled Hg wet deposition worldwide is 30 $\mu g\ m^{-2}\ year^{-1}$ (not shown here). Thus, peak modeled wet deposition in India is comparable to the highest deposition modeled in the world and reflects the confluence of relatively high emissions and high precipitation in India.

The spatial distribution of dry Hg deposition in India (Fig. 13.8) closely reflects the emissions footprint discussed previously (Fig. 13.4).

The maximum annual dry deposition (35 $\mu g\ m^{-2}\ year^{-1}$) is predicted in the area spanning northern Chattisgarh, northern Odisha and southern Jharkhand with a 50% contribution from Hg^{II}, 47% from Hg^{II}, and 3% from Hg_P. The peak modeled dry deposition in India is considerably less than the worldwide maximum of 100 $\mu g\ m^{-2}\ year^{-1}$. Dry deposition over most of India is also less than the dry deposition modeled in China (typically 25–75 $\mu g\ m^{-2}\ year^{-1}$ in this study) reflecting the over fourfold higher Hg emissions in the latter.

Across India on average and at the area of peak total deposition (northern Chhattisgarh), dry deposition is higher by approximately 20% over wet deposition (Table 13.2), with gaseous Hg^{II} dominating wet deposition while dry deposition is dominated by both Hg^{II} and Hg^0.

There are several sources of uncertainty in these results. A relatively large grid resolution was utilized in this study. Use of a finer grid would better capture the

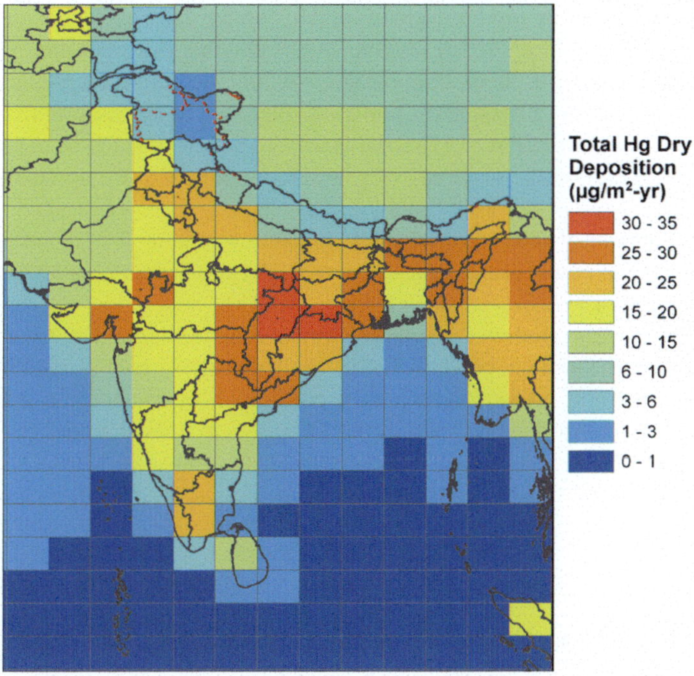

Fig. 13.8 Annual dry deposition flux of total mercury (μg m^{-2} $year^{-1}$) in India and surrounding regions

Table 13.2 Speciated annual mercury deposition fluxes in India (μg m^{-2} $year^{-1}$)

	Average across India	At location of maximum deposition
Total Hg	27.6	56.3
Dry Hg	15.2	35.0
Wet Hg	12.4	21.3
Hg^0 dry	7.9	16.5
Hg^{II} dry	7.0	17.7
Hg_p dry	0.3	0.8
Hg^{II} wet	11.0	18.3
Hg_p wet	1.4	3.0

large spatial variability of deposition near sources (Seigneur et al. 2003; Zhang et al. 2012). Future work should apply a regional chemistry transport model with finer grid spacing using boundary conditions from Hg air concentrations simulated by the global modeling system (e.g., Seigneur et al. 2004; Lin et al. 2006; Vijayaraghavan et al. 2007, 2008; Bullock et al. 2008; Zhu et al. 2015). More monitors for mercury wet deposition and air concentrations (e.g., Kumari et al. 2015; Pirrone et al. 2016) are required in India for model evaluation. This study does not consider litterfall Hg deposition (Wang et al. 2016) nor the small amounts of direct methylmercury deposition from the atmosphere, both of which would increase total predicted deposition amounts. Due to atmospheric transport over very long distances in the free troposphere, the contribution of Hg from sources upwind of India to deposition in India would depend on the accuracy of characterization of those upwind emission sources.

The Government of India promulgated emission standards for Hg emissions for coal-based thermal power plants in December 2015 (http://pib.nic.in/newsite/PrintRelease.aspx?relid=133726). Emission control measures adopted to meet these standards would lower the predicted Hg deposition rates across India. However, some of the potential reductions in deposition due to these measures may be offset by growth in coal-based thermal power plants and in production of cement and non-ferrous metals and other anthropogenic sources of mercury across the country.

References

Amos HM, Jacob DJ, Holmes CD, Fisher JA, Wang Q, Yantosca RM, Corbitt ES, Galarneau E, Rutter AP, Gustin MS, Steffen A, Schauer JJ, Graydon JA, Louis VLS, Talbot RW, Edgerton ES, Zhang Y, Sunderland EM (2012) Gas-particle partitioning of atmospheric Hg(II) and its effect on global mercury deposition. Atmos Chem Phys 12:591–603

AMAP/UNEP (Arctic Monitoring and Assessment Programme/ United Nations Environment Programme) (2013) Technical background report for the global mercury assessment. Arctic monitoring and assessment programme. Oslo, Norway/UNEP Chemicals Branch, Geneva, Switzerland. vi + 263 pp

AMAP/UNEP (2015) Global mercury modelling: update of modelling results in the global mercury assessment 2013. Arctic monitoring and assessment programme. Oslo, Norway/UNEP Chemicals Branch, Geneva, Switzerland. iv + 32 pp

Bey I, Jacob DJ, Yantosca RM, Logan JA, Field BD, Fiore AM, Li Q, Liu HY, Mickley LJ, Schultz MG (2001) Global modeling of tropospheric chemistry with assimilated meteorology: Model description and evaluation. J Geophys Res 106:23073–23095

Bullock OR, Atkinson D, Braverman T, Civerolo K, Dastoor A, Davignon D, Ku J, Lohman K, Myers T, Park R, Seigneur C, Selin N, Sistla G, Vijayaraghavan K (2008) The North American Mercury model intercomparison study (NAMMIS): study description and model-to-model comparisons. J Geophys Res 113:D17310. https://doi.org/10.1029/2008JD009803

Chakraborty LB, Qureshi A, Vadenbo C, Hellweg S (2013) Anthropogenic mercury flows in India and impacts of emission controls. Environ Sci Tech 47(15):8105–8113

Chen L, Wang HH, Liu JF, Tong YD, Ou LB, Zhang W, Hu D, Chen C, Wang XJ (2014) Intercontinental transport and deposition patterns of atmospheric mercury from anthropogenic emissions. Atmos Chem Phys 14:10163–10176. https://doi.org/10.5194/acp-14-10163-2014

Clarkson TW, Magos L, Myers GJ (2003) The toxicology of mercury—current exposures and clinical manifestations. New Engl J Med 349:1731–1737. https://doi.org/10.1056/NEJMra022471

Corbitt ES, Jacob DJ, Holmes CD, Streets DG, Sunderland EM (2011) Global source–receptor relationships for mercury deposition under present-day and 2050 emissions scenarios. Environ Sci Technol 45:10477–10484

Driscoll CT, Mason RP, Chan HM, Jacob DJ, Pirrone N (2013) Mercury as a global pollutant: Sources, pathways, and effects. Environ Sci Technol 47(4967–4983):2013. https://doi.org/10.1021/es305071v

Giang A, Stokes LC, Streets DG, Corbitt ES, Selin NE (2015) Impacts of the Minamata convention on mercury emissions and global deposition from coal-fired power generation in Asia. Environ Sci Technol 49(9):5326–5335

Guttikunda S, Jawahar P (2014) Atmospheric emissions and pollution from the coal-fired thermal power plants in India. Atmos Environ 92:449–460. https://doi.org/10.1016/j.atmosenv.2014.04.057

Holmes CD, Jacob DJ, Corbitt ES, Mao J, Yang X, Talbot R, Slemr F (2010) Global atmospheric model for mercury including oxidation by bromine atoms. Atmos Chem Phys 10:12037–12057

Huang J, Kang SC, Wang SX, Wang L, Zhang QG, Guo JM, Wang K, Zhang GS, Tripathee L (2013) Wet deposition of mercury at Lhasa, the capital city of Tibet. Sci Total Environ 447:123–132. https://doi.org/10.1016/j.scitotenv.2013.01.003

Kumari A, Kumar B, Manzoor S, Kulshrestha U (2015) Status of atmospheric mercury research in South Asia: a review. Aerosol Air Qual Res 15:1092–1109

Lindberg S, Bullock R, Ebinghaus R, Engstrom D, Feng X, Fitzgerald W, Pirrone N, Prestbo E, Seigneur C (2007) A synthesis of progress and uncertainties in attributing the sources of mercury in deposition. Ambio 36:19–32

Lin C-J, Pongprueksa P, Lindberg SE, Pehkonen SO, Byun D, Jang C (2006) Scientific uncertainties in atmospheric mercury models I: model science evaluation. Atmos Environ 40:2911–2928

Liu H, Jacob DJ, Bey I, Yantosca RM (2001) Constraints from 210Pb and 7Be on wet deposition and transport in a global three-dimensional chemical tracer model driven by assimilated meteorological fields. J Geophys Res 106:12109–12128

Mukherjee AB, Bhattacharya P, Sarkar A, Zevenhove R (2009) Mercury emissions from industrial sources in India and its effects in the environment (Chapter 4). In: Pirrone N, Mason R (eds) Mercury fate and transport in the global atmosphere. Report of the UNEP global partnership on atmospheric mercury transport and fate research, pp 81–112

National Atmospheric Deposition Program (NADP) (2015) Mercury deposition network (MDN). Champaign. Available from: http://nadp.isws.illinois.edu/mdn/

Pirrone N, Mahaffey KR (2005) Dynamics of mercury pollution on regional and global scales. Atmospheric processes and human exposures around the world. Springer, Berlin

Pirrone N, Cinnirella S, Feng X, Finkelman RB, Friedli HR, Leaner J, Mason R, Mukherjee AB, Stracher GB, Streets DG, Telmer K, (2010) Global mercury emissions to the atmosphere from anthropogenic and natural sources. Atmos Chem Phys 10:5951–30 5964. https://doi.org/10.5194/acp-10-5951-2010

Pirrone N, Sprovieri F, Ebinghaus R (2016) Global mercury observation system—atmosphere (GMOS-A). Atmos Chem Phys Special Issue 2016

Qureshi A, MacLeod M, Hungerbühler K (2011) Quantifying uncertainties in the global mass balance of mercury. Global Biogeochem Cycles 25:GB4012

Qureshi A (2016) Simple box modeling of mercury cycling in the Indian environment. Possible impacts of control scenarios and need for data (preliminary). Presented at the 2016 atmospheric mercury monitoring workshop, Bangkok, Thailand. Available at http://rsm2.atm.ncu.edu.tw/apmmn/PDF/2016/Presentation/21_Simple_box_modeling_of_mercury_cycling_in_the_Indian.pdf. July 27, 2016

Rai VK, Raman NS, Choudhary SK (2013) Mercury emissions control from coal fired thermal power plants in India: critical review & suggested policy measures. Int J Eng Res Technol (IJERT) 2(11)

Schroeder WH, Munthe J (1998) Atmospheric mercury—an overview. Atmos Environ 29:809–822

Seigneur C, Karamchandani P, Lohman K, Vijayaraghavan K, Shia R-L (2001) Multiscale modeling of the atmospheric fate and transport of mercury. J Geophys Res 106:27795–27809

Seigneur C, Karamchandani P, Vijayaraghavan K, Shia R-L, Levin L (2003) On the effect of spatial resolution on atmospheric mercury modeling. Sci Total Environ 304:73–81

Seigneur C, Vijayaraghavan K, Lohman K, Karamchandani P, Scott C (2004) Global source attribution for mercury deposition in the United States. Environ Sci Technol 38:555–569
Seigneur C, Vijayaraghavan K, Lohman K (2006) Atmospheric mercury chemistry: Sensitivity of global model simulations to chemical reactions. J Geophys Res 111:D22306
Selin NE, Jacob DJ, Park RJ, Yantosca RM, Strode S, Jaeglé L, Jaffe D (2007) Chemical cycling and deposition of atmospheric mercury: global constraints from observations. J Geophys Res 112:D02308
Selin NE, Jacob DJ, Yantosca RM, Strode S, Jaeglé L, Sunderland EM (2008) Global 3-D land-ocean-atmosphere model for mercury: Present-day versus preindustrial cycles and anthropogenic enrichment factors for deposition. Global Biogeochem Cycles 22:GB2011
Sloss L (2012) Mercury emissions from India and South East Asia. ISBN 978-92-9029-528-0. IEA Clean Coal Centre
Soerensen AL, Sunderland EM, Holmes CD, Jacob DJ, Yantosca RM, Skov H, Christensen JH, Strode SA, Mason RP (2010) An improved global model for air-sea exchange of mercury: high concentrations over the North Atlantic. Environ Sci Technol 44:8574–8580
Song S, Selin NE, Soerensen AL, Angot H, Artz R, Brooks S, Brunke E-G, Conley G, Dommergue A, Ebinghaus R, Holsen TM, Jaffe DA, Kang S, Kelley P, Luke WT, Magand O, Marumoto K, Pfaffhuber KA, Ren X, Sheu G-R, Slemr F, Warneke T, Weigelt A, Weiss-Penzias P, Wip DC, Zhang Q (2015) Top-down constraints on atmospheric mercury emissions and implications for global biogeochemical cycling. Atmos Chem Phys 15:7103–7125
Streets DG, Zhang Q, Wu Y (2009) Projections of global mercury emissions in 2050. Environ Sci Technol 43:2983–2988
Strode SA, Jaeglé L, Selin NE, Jacob DJ, Park RJ, Yantosca RM, Mason RP, Slemr F (2007) Air-sea exchange in the global mercury cycle. Global Biogeochem Cycles 21:GB1017
van der Werf GR, Randerson JT, Giglio L, Collatz GJ, Mu M, Kasibhatla PS, Morton DC, DeFries RS, Jin Y, van Leeuwen TT (2010) Global fire emissions and the contribution of deforestation, savanna, forest, agricultural, and peat fires (1997–2009). Atmos Chem Phys 10:11707–11735. https://doi.org/10.5194/acp-10-11707-2010
Vijayaraghavan K, Seigneur C, Karamchandani P, Chen S-Y (2007) Development and application of a multi-pollutant model for atmospheric mercury deposition. J App Meteorol Climatol 46:1341–1353
Vijayaraghavan K, Karamchandani P, Seigneur C, Balmori R, Chen S-Y (2008) Plume-in-grid modeling of atmospheric mercury. J Geophys Res 113:D24305. https://doi.org/10.1029/2008JD010580
Vijayaraghavan K, Levin L, Parker L, Yarwood G, Streets D (2014) Response of fish tissue mercury in a freshwater lake to local, regional, and global changes in mercury emissions. Environ Toxicol Chem 33:1238–1247
Wang X, Bao Z, Lin C-J, Yuan W, Feng X (2016) Assessment of global mercury deposition through litterfall. Environ Sci Technol. https://doi.org/10.1021/acs.est.5b06351
Wesely ML (1989) Parameterization of surface resistances to gaseous dry deposition in regional-scale numerical models. Atmos Environ 23:1293–1304
Zhang Y, Jaeglé L, van Donkelaar A, Martin RV, Holmes CD, Amos HM, Wang Q, Talbot R, Artz R, Brooks S, Luke W, Holsen TM, Felton D, Miller EK, Perry KD, Schmeltz D, Steffen A, Tordon R, Weiss-Penzias P, Zsolway R (2012) Nested-grid simulation of mercury over North America. Atmos Chem Phys 12:6095–6111. https://doi.org/10.5194/acp-12-6095-2012
Zhang Y, Jacob DJ, Horowitz HM, Chen L, Amos HM, Krabbenhoft DP, Slemr F, St. Louis VL, Sunderland EM (2016) Observed decrease in atmospheric mercury explained by global decline in anthropogenic emissions. Proc Natl Acad Sci (PNAS) 113(3)
Zhu J, Wang T, Bieser J, Matthias V (2015) Source attribution and process analysis for atmospheric mercury in eastern China simulated by CMAQ-Hg. Atmos Chem Phys 15:8767–8779. https://doi.org/10.5194/acp-15-8767-2015

Chapter 14
Risk-Based Optimal Ranking of Air Quality Monitoring Stations in a Fuzzy Environment: A Case Study

Jyoti Yadav and Ashok Deshpande

14.1 Introduction

The present status of poor air quality in cities, in several countries across the globe, has been a talking point and calls for initiating and prioritizing air pollution mitigation measures. CAQI and newly developed fuzzy logic-based ZD formalism have been successfully applied in areas such as water quality classification ambient air quality assessment, academic performance evaluation of students, and health monitoring of process control instruments. Because of deteriorating air quality in several cities in the world, we believe that there is a need to initiate an action plan on the reduction in air pollution risk. In CAQI, numeric values of criteria air pollutants are considered and air quality status described in linguistic terms. The human brain does not describe air or water quality directly in linguistic terms. While in ZD method similar constraints are considered and the air quality status is straightway described linguistically with a numeric certainty measure associated with each phonological description. However, there are some other constraints such as wind velocity, population density, temperature, humidity and alike which might affect air quality status, but these are not considered. These constraints could also be fuzzy. For example, there could be several AQMS in a city with poor air quality but with varying numeric certainty measure. In such a scenario, an optimal ranking of AQMS which involves risk, for the identification of pollution mitigation measures assumes importance. This

J. Yadav (✉)
Department of Computer Science, Savitribai Phule Pune University, Pune, India
e-mail: jyo.yadav@yahoo.co.in

A. Deshpande
Berkeley Initiative Soft Computing (BISC)—Special Interest, Group (SIG)—Environmental Management Systems (EMS), University of California, Berkeley, CA, USA
e-mail: ashok_deshpande@hotmail.com

College of Engineering, Pune, India

S. M. Shiva Nagendra et al. (eds.), *Urban Air Quality Monitoring, Modelling and Human Exposure Assessment*, Springer Transactions in Civil and Environmental Engineering,
https://doi.org/10.1007/978-981-15-5511-4_14

is fundamentally a multifaceted decision-making problem in an imprecise environment. It is surprising that there are no references in the literature on the application of BZ method in environmental management systems. In this paper, we demonstrate the application of a combination of ZD and Bellman-Zadeh (BZ) method in final risk-based optimal ranking of AQMS which is basically a complex multi-constraint decision-making problem. Furthermore, classification of twelve cities in Maharashtra State, India into different air quality clusters to demonstrate the importance of the concept of supervised learning followed by unsupervised learning using reference level concept is presented.

14.2 Techniques Used

Following is a brief description of the methodology used:

14.2.1 Conventional Air Quality Index (CAQI)

The CAQI was formulated by the US Environmental Protection Agency (US-EPA) which is used by the state and local organizations. The CAQI describes Air quality index in terms of numbers to classify air quality at monitoring stations. A very large section of the population is likely to experience adverse health effects due to bad air quality. The AQI is computed with the help of parametric data of air pollutants available from various monitoring stations. The computation of AQI using pollutant concentration differs according to the country. Once the indices are computed they are divided into ranges, wherein each range is assigned a particular colour code which further classifies the quality of air. Air pollutant with maximum concentration is considered as the "accountable pollutant".

14.2.2 Zadeh-Deshpande (ZD) Formalism

It is our belief that fuzzy logic is a precise framework that can be used in decision-making since it is modelled as per the human thinking process and deals with perception-based modelling. We discuss the novel method named ZD formalism to straightway describe the quality of air linguistically with linguistic degree of certainty associated with each description briefly in the following sections. The attributes for the defined purpose including number of sampling stations, time and frequency of observations etc., are decisive and are unvaryingly based on the experts' knowledgebase. As a first step, pollution parametric data from various monitoring stations was collected. For probability distribution fitting, mean and variance values are also used in decision analysis. If the collected pollutant data from monitoring stations are insufficient for carrying out statistical analysis, then to ensure data reliability, a defined statistical technique called bootstrapping is employed. The proposed formalism exhibits the application of Fuzzy Inference System (FIS) via degree of match which is a level 1 complexity in computing with words for classifying air quality directly in linguistic term with a linguistic degree of certainty associated with each classification. ZD formalism is analogous to that of vague thinking of humans which actually does not work with numbers. The reliability of air quality described using the ZD formalism is ensured with the concept of linguistic degree of certainty.

14.2.3 Convex Normalized Fuzzy Number (CNFN)

Gaussian distribution is used to arrive at the probability density function using Eq. (14.1) in order to simulate uncertainty in pollutant data with mean as μ and σ as the standard deviation for any random variable x.

$$f(x) = \frac{1}{\sigma\sqrt{2\pi}} e^{\frac{-1}{2}\left(\frac{x-\mu}{\sigma}\right)^2} \tag{14.1}$$

The function $\mu_A(x)$ is defined as follows:

$$\mu_A(x) = p(x)/p(x_i) \tag{14.2}$$

where x_i is some point on the parametric domain x, with the highest value of $p(x_i)$.

The random variable x is converted to CNFN A using Eq. (14.2) with membership grade $\mu_A(x)$. The fuzzy sets are formed by modelling the perceptions of the domain experts' as follows:

- Referencing all parametric values using linguistic terms.
- Classify the pollutant data and associate a linguistic term to each class with the help of experts' reflecting uncertainty in their thinking.
- Assignment of membership grade is done as follows:

1. If the set of values are categorized using the same linguistic term by all experts' the membership grade assigned is $\mu = 1.0$.
2. If none of the experts' assign that term, it is given membership grade as $\mu = 0.0$.
3. All the points between membership grades 0.0 and 1.0 are connected by a continuous monotonic function which assumes that the agreement index amongst the experts' increases as the pollutant data gradually approaches the core of fuzzy number for a particular linguistic term (Deshpande and Raje 2003; Mckone and Deshpande 2005).

Figure 14.1 depicts the CNFN for PM_{10} pollutant reflecting possibility distribution. The graph also shows the probability distribution for PM_{10} concentration.

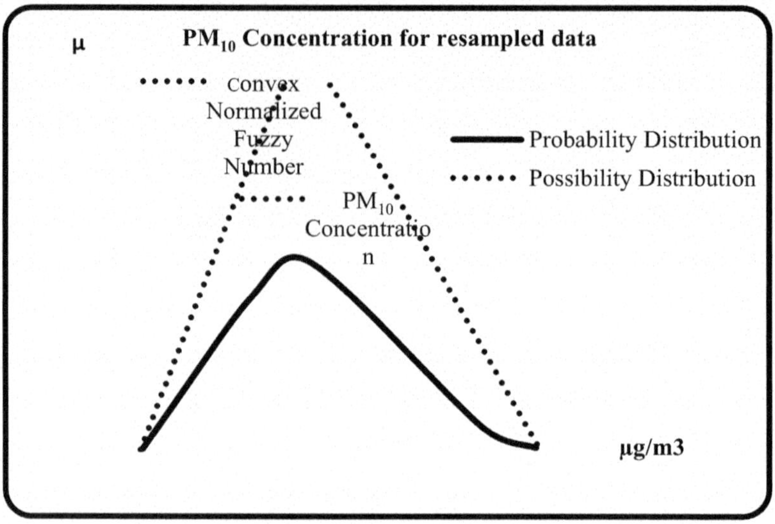

Fig. 14.1 Convex normalized fuzzy number A

Probability distribution is converted to possibility distribution using CNFN with membership grade function $\mu_A(x)$, in order to compare probability distribution with fuzzy sets given by the domain experts'.

14.2.4 *Degree of Match (DM)*

A matching between the expert's thinking and the antecedent part of the rule is achieved with the help of DM, thus describing air quality fuzzily with certain reliability value expressed as a number. The fuzzy numbers (A) and (A') characterizing parametric data and the linguistic terms, respectively are superimposed to get DM which is defined (Fig. 14.2) using Eq. (14.3).

$$\mathrm{DM}_{ff}(A, A') = \frac{\int \mu_{A\cap A'}(x)\mathrm{d}x}{\int \mu_A(x)\mathrm{d}x}, \quad x \in X \tag{14.3}$$

where X is the universe of discourse and $\mu_{A\cap A'}(x)$ specifies the membership grade for the intersection of A and $A\prime$. In the case of discrete possibility distributions of A and $A\prime$ DM can be defined by using Eq. (14.4).

$$\mathrm{DM}_{ff}(A, A') = \frac{\sum \mu_{A\cap A'}(x)\mathrm{d}x}{\sum \mu_A(x)\mathrm{d}x}, \quad x \in X \tag{14.4}$$

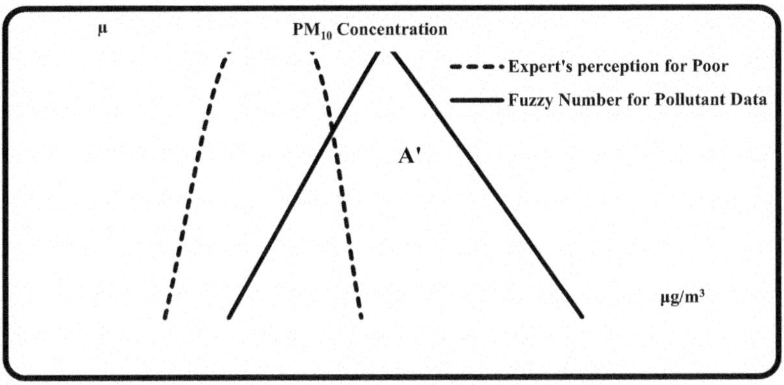

Fig. 14.2 Degree of match for PM_{10} parametric data for fuzzy set *poor*

14.2.5 Type I Fuzzy Inference System

Fuzzy rules are created for describing air quality as *Good, Moderate, Unhealthy* etc., in order to combine the set of attributes. Each rule is composed of an 'if' part (antecedent) and a 'then' part (consequence). The antecedent part is made up of proposition coupled with conjunction and disjunction operators, subsequently giving rise to some result or consequence. These consequences expressed linguistically are fuzzy keeping in view the noisy parametric data. The reasoning process or the knowledge representation is carried out by creating the fuzzy if-then rule-based system. The reliability value for each description is given by DM of each classification rule. For example, DM = 0.8 signifies the greater possibility of the air quality getting classified in that particular class. The best acceptance policy is the one for which the degree of assertion is highest.

14.2.6 Soft Clustering

Soft Clustering is an unsupervised learning methodology of partitioning data elements into collections (or clusters) depending on their locality and connectivity within an m-dimensional feature space. In the clustering process, it should be more similar to the sets of members in the cluster, but the similarities should be less to members outside the cluster. Fuzzy clustering thus partitions data elements into overlapping fuzzy partitions. The data element thus belongs to a cluster with a membership grade between 0 and 1. Unsupervised FCM algorithm is a technique based on the objective function.

FCM ignores the importance of features in the clustering process. This affects its accurateness and exactness. This problem can be overcome by appropriately assigning weights to features according to their clustering importance. The weighting factor, usually denoted by m controls the fuzziness of FCM clustering results and the experimental results show that m value interval is [2, 2.4]. On the basis of the distance between the cluster centre and the data element, FCM assigns membership grade to each data element. Membership grade is high if the data element is near the cluster centre. The sum of all the membership grades of each data element should be one.

14.2.7 Bellman-Zadeh (BZ) Formalism

Bellman and Zadeh (1970) applied the fuzzy set theory to many decision-making problems. They represented each objective as a fuzzy subset over a set of alternatives U. The membership grade of ith objective A denoted by $\mu_{Ai}(u)$, specifies the degree to which u satisfies the criterion specified by this objective. The objectives and then incorporated in a decision function δ. Furthermore, the overall objective function δ of a set of objectives $A = \{A_i/i = 1 \ldots p\}$ can be represented as:

$$\delta = A_l \text{AND } A_2 \text{AND} \ldots \text{AND } A_p$$

With a lack of information, the AND operator can be mathematically expressed using Min operator. Hence, the decision function $\delta(u)$ can be expressed as follows:

$$\delta(u) = \text{Min}\big[A_1(u), A_2(u) \ldots A_p(u)\big] \quad \text{for each } u \in U \tag{14.5}$$

The final optimal solution represented by $u*$ is given by the following equation:

$$\delta\left(u^*\right) = \text{Max}\, D(u) \quad \text{for each} u \in U \tag{14.6}$$

If the objectives have different prominence, then the decision function is specified as:

$$\delta(u) = \text{Min}_{i=1..p}[a_i A_i(u)] \tag{14.7}$$

with a_i representing the weighting factor, focusing on the importance of objective A_i such that:

$$\sum_{i=1}^{n} a_i = 1 \tag{14.8}$$

It is pertinent to mention that BZ method has been used, in this paper in a different setting.

Figure 14.3 presents an overall framework for the combined ZD and BZ Approach. The air quality classification obtained for each city using ZD formalism is described

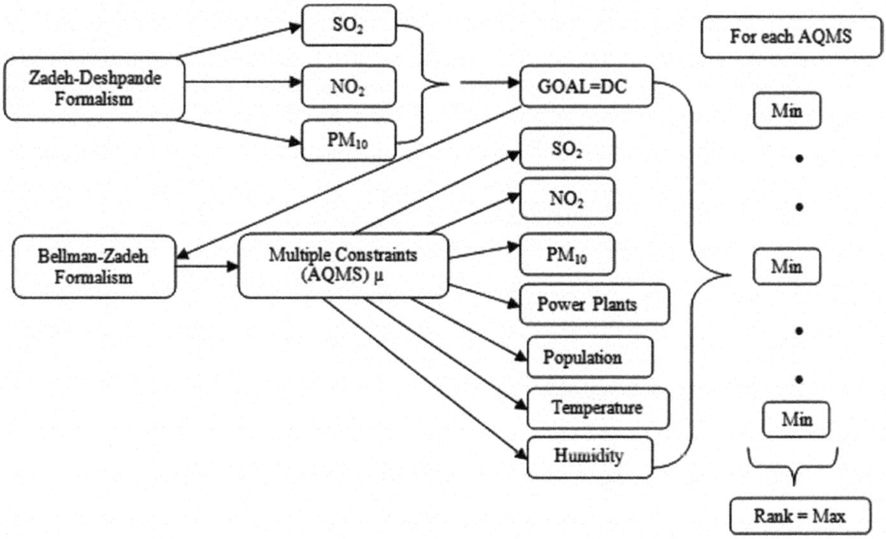

Fig. 14.3 Overall framework for the combined Zadeh-Deshpande and Bellman-Zadeh approach

linguistically using the concept of degree of certainty (DC). Like traditional AQI, ZD method used only three criteria pollutants to arrive at the air quality classification. This DC estimated in numeric terms using ZD formalism, is used as Goal in BZ formalism. BZ formalism along with the three criteria pollutants uses other constraints such as a number of power plants in the city, population, temperature and humidity to classify ambient air quality at the twelve AQMS. The general framework presented considers SO_2 as one of the pollution parameters because of the fact that the location of thermal power plants in the vicinity city population is not uncommon only in India but also in some other countries.

14.2.8 The Study

The unnoticed limitations of CAQI sound for devising ZD formalism based on fuzzy concepts (Jyoti et al. 2011, 2013, 2014), wherein air quality is linguistically classified and associated with linguistic degree of certainty to each classification. Fuzzy logic concepts thus help in modelling uncertainty of uncertainty. Epistemic uncertainty in expert's perception is modelled by defining expert fuzzy sets and aleotary uncertainty due to the degree of certainty is modelled by defining fuzzy sets for DC. Based on ZD formalism Aurangabad, Chandrapur, Jalna, Lote, Mumbai, Nagpur, Nanded, Nashik, Pune, Sangli, Solapur are the twelve AQMS with Very Unhealthy air quality status with varying degree of certainty potential from air quality viewpoint. Furthermore, the cities are also classified using FCM into five clusters with a varying weighting factor based on the reference group proposed by the second author of this paper. BZ formalism is then applied which ranks the cities as per their pollution levels taking into consideration supplementary parameters like wind velocity, temperature etc. for air quality classification. The phenomenon of increase in traffic could be one of the primary sources of air pollution. Nevertheless, the location of industrial installation in the vicinity of these cities has further compounded the problem of increasing pollution levels in some of these cities. The summary results are based on worst-case scenario which means that these cities are ranked with a very high risk if any one of the monitoring stations falls in this category (Ashok and Jyoti 2017).

14.3 Results and Discussion

14.3.1 Based on CAQI and Air Quality Using ZD Formalism

Table 14.1 presents the summary of the high-risk pollution levels with numeric DC using ZD formalism and the CAQI approach. The data for the three criteria pollutants has been fetched from the Maharashtra Pollution Control Board website for winter months from Nov. 2014 to Jan. 2015. The ZD formalism describes air quality first linguistically with numeric DC attached to each description, whereas CAQI gives a numeric value to air quality status and then describes it linguistically.

Table 14.1 Classification of air quality using CAQI and ZD methods

S. No.	Station	ZD Method with DC	CAQI
1	Aurangabad	(VP) 0.78	145.87 (P)
2	Chandrapur	(VP) 0.87	155.06 (VP)
3	Jalna	(VP) 0.85	228.48 (VP)
4	Lote	(VP) 0.72	101.00 (P)
5	Mumbai	(VP) 0.92	167.72 (VP)
6	Nagpur	(VP) 0.89	100.00 (P)
7	Nanded	(VP) 0.74	102.00 (P)
8	Nashik	(VP) 0.86	110.39 (P)
9	Pune	(VP) 0.76	114.35 (P)
10	Sangli	(VP) 0.84	152.46 (VP)
11	Solapur	(VP) 0.65	99.78 (P)
12	Thane	(VP) 0.87	99.62 (P)

14.3.2 Based on Fuzzy C Means Algorithm

The result of the FCM algorithm is computed using PM_{10} concentrations as it is the most important criteria pollutant in Indian Scenario. After running the algorithm through five iterations with varying values of weighting factor ($m = 2.2$) the following clusters were obtained:

Cluster 1 = *Good* (*G*) = {Lote}
Cluster 2 = *Moderate* (*M*) = {Aurangabad, Thane}
Cluster 3 = *Unhealthy* (*UH*) = {Nagpur, Nanded, Solapur, Nashik}
Cluster 4 = *Unhealthy for Sensitive Group* (*USG*) = {Pune, Sangli, Chandrapur, Mumbai}
Cluster 5 = *Very Unhealthy* (*VUH*) = {Jalna}.

From the clusters obtained using FCM, it is clear that Jalna, Pune, Sangli, Chandrapur and Mumbai should be given top priority to reduce pollution. In our view, CAQI and FCM methods could be used for reconfirming the results obtained using ZD formalism. We believe that unsupervised learning methods should precede supervised methods.

14.3.3 Based on Bellman-Zadeh Formalism

Assigning Membership Grade. In order to assign membership grade to various constraints following criteria has been applied. Though the three criteria pollutants: Sulphur Dioxide (SO_2), Oxides of Nitrogen (NO_x), Particulate Matter (PM_{10}), were considered in ZD approach, these three constraints and other constraints such as population density, wind velocity, wind direction, temperature, humidity and the capacity/number of thermal power plants in that particular city were considered in BZ method. Thermal power plants rank first in their pollution potential.

The concentration of the criteria pollutants for winter months has been considered. The maximum concentrations of the pollutants were considered for plotting the membership graph as shown in Fig. 14.4 for PM_{10}. A similar method was applied to SO_2 and NO_X pollutants.

Table 14.2 presents the city-wise membership grades for the first three constraints. The DC column obtained by application of ZD formalism now acts as a "Goal" for further computations in BZ formalism.

For power plants constraint, the city with maximum number of industries/ or industry with a very high air pollution potential is given highest membership grade, i.e. Chandrapur = 0.8 (because of highly polluting Super Power Thermal Power Plant), for the population constraint wind speed and population along the wind direction is given highest membership grade of 0.9, i.e. Mumbai. For the temperature constraint, null hypothesis was used with temp = 32 °C as 0.5 membership grade. For humidity constraint, humidity ranging from 70 to 79 was assigned $\mu = 0.7$, for

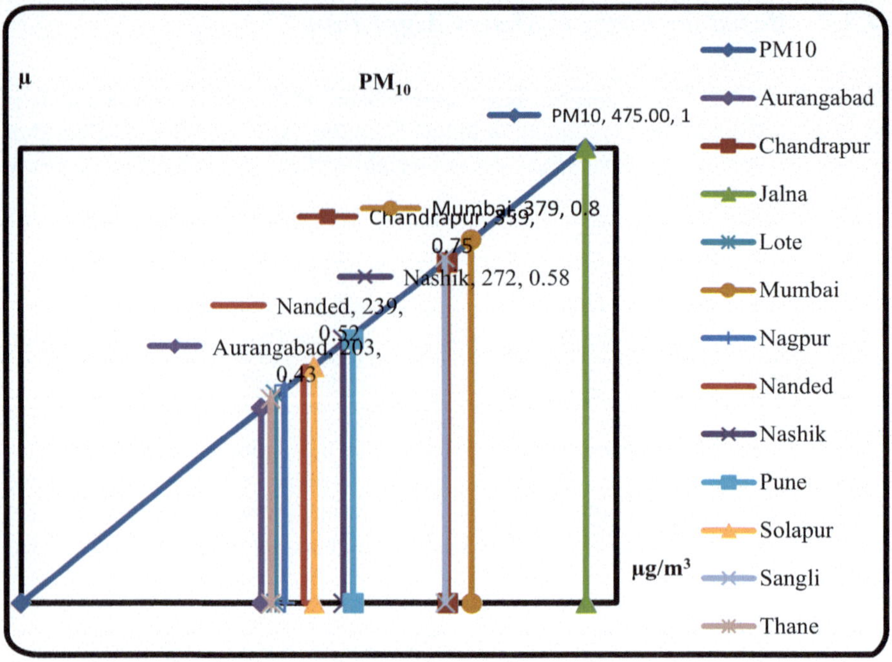

Fig. 14.4 Assigning membership grade to PM_{10} pollutant

Table 14.2 Assigning membership grades to SO_2, NO_X, PM_{10} constraints

S. No.	Monitoring station	Goal DC	Constraint 1 SO_2	Constraint 2 NO_x	Constraint 3 PM_{10}
1	Aurangabad	0.78	0.24	0.3	0.43
2	Chandrapur	0.87	0.47	0.48	0.75
3	Jalna	0.85	0.20	0.25	1.00
4	Lote	0.72	0.15	0.08	0.46
5	Mumbai	0.92	1.00	1.00	0.80
6	Nagpur	0.89	0.20	0.29	0.48
7	Nanded	0.74	1.00	0.47	0.52
8	Nashik	0.86	0.62	0.23	0.58
9	Pune	0.76	0.6	0.85	0.58
10	Sangli	0.84	0.30	0.49	0.76
11	Solapur	0.65	0.20	0.51	0.52
12	Thane	0.87	0.35	0.18	0.45

Table 14.3 Assigning membership grades to power plants, population, temperature and humidity constraints

S. No.	Monitoring station	Constraint 4 (Power plants)	Constraint 5 (Pop)	Constraint 6 (Temp)	Constraint 7 (Humidity)
1	Aurangabad	0.2	0.5	0.4	0.6
2	Chandrapur	0.8	0.4	0.5	0.8
3	Jalna	0.2	0.3	0.3	0.5
4	Lote	0.5	0.2	0.3	0.7
5	Mumbai	0.4	0.9	0.4	0.7
6	Nagpur	0.4	0.6	0.5	0.7
7	Nanded	0.5	0.5	0.3	0.8
8	Nashik	0.4	0.3	0.4	0.5
9	Pune	0.3	0.7	0.3	0.6
10	Sangli	0.2	0.3	0.4	0.7
11	Solapur	0.3	0.6	0.4	0.6
12	Thane	0.7	0.8	0.5	0.7

80 to 89 $\mu = 0.8$ and so on. Table 14.3 details the membership grades of these four constraints.

Further, the minimum membership grade for goal and the seven constraints combined was obtained city wise. Finally, the maximum of the all minimum value obtained for each AQMS was identified as the city with rank = 1, meaning the city with a high risk of pollution levels. Assignment of the membership grade is a subjective measure. We consulted domain expert (the second author is himself a domain expert). PM_{10} is the pollutant with very high concentration in air, some of the reason been dense vehicular traffic, construction activities and untarred roads. Additionally, minimum temperature and condensed perpendicular scattering of pollutants during winter months increase the concentration of PM_{10} in Mumbai city. Increase in vehicle density has gradually led to dangerous levels of emissions at various hotspots and in traffic sneers during peak hours. The industrial belts located near cities emit oxides of nitrogen, oxides of sulphur, and particulate matter. Chandrapur, Nagpur, Thane and Pune have a large number of vehicles plying on road thus resulting in high PM_{10} concentrations.

As per the BZ (Goal = DC and PM_{10}), Chandrapur and Mumbai cities should be given top priority to initiate pollution abatement measures. Constraint 4 (location of thermal power plants) is important in the case of Chandrapur. Nanded and Pune cities are ranked second due to hazardous industrial installation and excessive concentrations of PM_{10} due to the increase in vehicular traffic.

14.3.4 Optimal Ranking of AQMS Using ZD-BZ and FCM Methods

Table 14.4 finally gives the ranking of AQMS, comparing ZD and BZ (with Goal = DC and BZ with Goal = PM_{10}) formalism along with FCM results thus emphasizing that there is slight variation in the ranking of sites using the three methods as ZD formalism is based on concentrations of three pollutants only whereas ZD-BZ combined formalism takes into consideration a total of seven constraints to arrive at the ranking of cities. The last column of Table 14.4 also shows ranking results using FCM.

Perhaps to repeat, an attempt has been made to skillfully combine ZD formalism for describing air quality linguistically with certain DC specified numerically and BZ method wherein the concept of the intersection of goal and constraints are considered in multi-criteria decision-making problems. As we are interested in risk-based optimal ranking of AQMS, the DC in ZD formalism is considered as a Goal in BZ method to finally arrive at risk-based optimal ranking of AQMS for the Identification of Pollution Mitigation Measures in a Fuzzy Environment. Furthermore, the paper conclusively demonstrates the importance of supervised learning followed by unsupervised learning with a reference group approach in FCM clustering.

14.4 Concluding Remarks

It is our belief that ZD-BZ combination and reference level approach in fuzzy clustering—the two new concepts may have wider practical applications in some of the areas of science and technology wherein the goal and constraints are imprecise or fuzzy and the policymaker is confronted with estimating risk-based optimal strategy for a defined problem.

Table 14.4 Ranking of air quality monitoring stations

ZD formalism	BZ formalism with goal = DC	BZ formalism with goal = PM_{10}	Ranking using FCM
Mumbai	Chandrapur/Mumbai	Chandrapur/Mumbai	Jalna
Chandrapur	Nanded/Pune	Nanded/Pune	Pune, Sangli, Chandrapur, Mumbai
Nagpur/Thane	Nashik	Nashik	Nagpur, Nanded, Solapur, Nashik
Pune	Aurangabad/Jalna/Nagpur/Sangli/Solapur	Aurangabad/Jalna/Nagpur/Sangli/Solapur	Aurangabad, Thane
Jalna	Thane	Thane	Lote
Sangli	Lote	Lote	
Nashik/Nanded			
Lote			
Aurangabad			
Solapur			

References

Ashok D, Jyoti Y (2017) Risk based optimal ranking of polluted cities in a fuzzy environment: a case study. Grenze Int J Eng Technol 3(1):1–6

Bellman RE, Zadeh LA (1970) Decision making in a fuzzy environment. Manage Sci 17(4):B141–B164

Deshpande AW, Raje DV (2003) Fuzzy logic applications to environmental management systems: case studies. IEEE, pp 365—368

Jyoti Y, Kharat V, Deshpande A (2011) Fuzzy description of air quality: a case study. In: 6th international conference on rough sets and knowledge technology (RSKT). Banff, Canada, 9–12 Oct 2011, 420–427

Jyoti Y, Kharat V, Deshpande A (2013) Evidence theory and fuzzy relational calculus in estimation of health effects due to air pollution. Int J Intell Syst 22(1):9–22

Jyoti Y, Kharat V, Deshpande A (2014) Fuzzy description of air quality using fuzzy inference system with degree of match via computing with words: a case study. Int J Air Qual Atmos Health 7(3):325–334

Mckone T, Deshpande A (2005) Can fuzzy logic bring complex environmental problems into focus? Environ Sci Technol, 42A–47A

Chapter 15
Impact of Increasing Ozone on Agricultural Crop Yields

Sonal Kumari, Nidhi Verma, Anita Lakhani, and K. Maharaj Kumari

15.1 Introduction

Urbanization, industrialization, and increasing levels of anthropogenic emissions in the last two decades have resulted in an increase in tropospheric ozone (O_3) levels and substantial deterioration of air quality (Lal et al. 2008). The two main sources in the troposphere which determine the levels of surface O_3 are stratospheric input and photochemical O_3 formation (Sicard et al. 2009) (Fig. 15.1). O_3 is a secondary air pollutant formed by photochemical reactions of biogenic and anthropogenic volatile organic compounds (VOCs) and carbon monoxide (CO) in the presence of nitrogen oxides (NO_x) (Lin et al. 1988). Among biogenic VOCs, terpenes and isoprene are the most abundantly emitted species from deciduous trees (Atkinson and Arey 2003). Crops also contribute to biogenic VOCs emission but their emission rate is relatively lower as compared to deciduous trees (König et al. 1995; Redeker et al. 2003). Therefore, the contribution of biogenic VOCs emitted from crops is relatively less towards O_3 formation in the troposphere.

The rise in the concentration of O_3 negatively affects the growth of crops and is most likely to threaten food production in the future (Dingenen et al. 2009; Avnery et al. 2011). The influence of O_3 on crops is a function dependent on the plant phenotype and O_3 exposure dosage. In some phenotypes, O_3 enters in plants during normal gas exchange through stomata and results in impairment of plant metabolism while in certain other phenotypes exposure to O_3 results in interference with the plant hormone levels (Iriti and Faoro 2009). In some plants, O_3 exposure causes reactive oxygen stress (ROS), visible leaf injury, abscission, and senescence (Mills et al. 2011). Field experiment studies: National Crop Loss Assessment Network (NCLAN)

S. Kumari · N. Verma · A. Lakhani · K. Maharaj Kumari (✉)
Department of Chemistry, Faculty of Science, Dayalbagh Educational Institute, Dayalbagh, Agra 282005, India
e-mail: maharajkumari.k@rediffmail.com

S. M. Shiva Nagendra et al. (eds.), *Urban Air Quality Monitoring, Modelling and Human Exposure Assessment*, Springer Transactions in Civil and Environmental Engineering,
https://doi.org/10.1007/978-981-15-5511-4_15

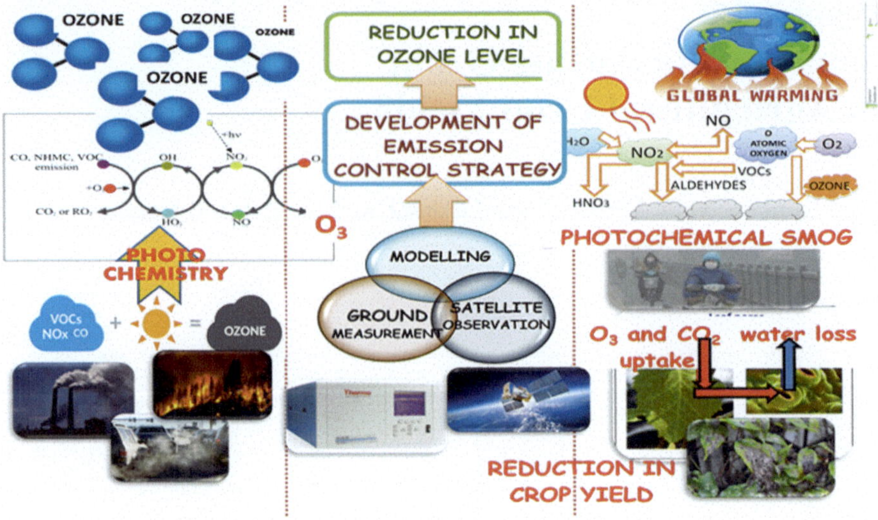

Fig. 15.1 Ozone production, its effects, and development of control strategies

in the USA and European Open Top Chamber Programme (EOTCP) in Europe have found 5–10% reduction in crop yield by O_3 exposure and reported that crop yield losses can increase in the future (Mauzerall and Wang 2001; Emberson et al. 2003). Global studies using M7 (mean 7-h ozone mixing ratio) and AOT40 (accumulated O_3 exposure over a threshold of 40 ppb) indices have shown that relative yield loss (RYL), crop production loss (CPL), and associated economic cost loss (ECL) are second highest in India (Dingenen et al. 2009; Avnery et al. 2011). Several studies have reported the effect of O_3 on crop yield for the Indian region but most of these are based on model simulations of ozone (Dingenen et al. 2009; Avnery et al. 2011; Ghude et al. 2014) or based on limited observations of O_3 (Emberson et al. 2009; Rai et al. 2010; Debaje 2014; Sinha et al. 2015; Lal et al. 2017). Dingenen et al. (2009) have estimated annual loss of 13–28% for wheat and 6–8% for rice in India using a global chemistry transport model. Annual loss estimated by Avnery et al. (2011) for wheat ranged from 9 to 30% based on simulations of ozone using Model for OZone And Related chemical Tracers-2 (MOZART-2). Ghude et al. (2014) using a Weather Research and Forecasting model coupled with chemistry (WRF-chem), have estimated annual loss of wheat and rice by about 5% and 2.1%, respectively. Sinha et al. (2015) have calculated crop losses for two states of India (Punjab and Haryana) in the range of 11–41% for wheat and 19–26% for rice. Lal et al. (2017) have reported reduction in crop yield using ground-based measurement of surface O_3 from a network of 17 sites covering different parts of India. To monitor more precisely the impact of O_3 on crops a long term study using ground-based observation is needed.

In northern India, the two main crop-growing seasons are Rabi and Kharif. Cotton, rice, sugarcane, maize, and vegetables are mainly grown in Kharif season while Rabi season mainly includes wheat. In India, wheat is mainly grown from January to April in Rabi season while rice is grown during July–October (Kharif season). In India, rice and wheat are the two most important staple crops, therefore the present study concerns only the crop-growing season of wheat and rice. The present study was conducted at a suburban site of Agra to determine the impact of O_3 on crop yield by determining the RYL, CPL, and associated ECL for rice and wheat during 2010–13 using M7 and AOT40 indices. In addition, a comparison of old and revised O_3 exposure relationships derived for Indian crop phenotypes was also done.

15.2 Site Description

Measurements for the present study were performed at Dayalbagh Educational Institute, Dayalbagh, Agra (27° 10′ N, 78° 05′ E) (suburban site) for the period 2010–13 (Fig. 15.2). The study site lies in the middle of Indo-Gangetic Plain (IGP) and is about 200 km southeast of Delhi. During summer (Mar–May) Agra is hot and dry while cool in the winter (Jan–Feb). 90% of the annual rainfall (650 mm) is observed during monsoon season (Jun–Sept). Strong surface winds are observed during summer and monsoon while winter is associated with a greater calm period. In winter maximum and minimum temperature ranges between 10 and 3 °C and in summer it ranges between 45 and 25 °C. Relative humidity at the site varies from 25 to 99%. There are no industrial activities in the surrounding area of the sampling site. Agricultural practices are carried out in the vicinity of the sampling site (~1 km). The seasonal crops that are mainly grown in this area are paddy, wheat, bajra, and vegetables.

Fig. 15.2 Site description map

15.3 Methodology

The O_3 levels were recorded using an online O_3 analyzer (Thermo Scientific 49i) which is based on UV absorption photometry (Beer–Lambert law). O_3 molecules absorb UV light at 253.7 nm and according to Beer–Lambert law concentration of O_3 is proportional to the absorbance. The detection limit of the analyzer is 0.5 ppb. Weekly calibration of the instrument was done by using an O_3 generator (built-in) and zero air generator (Kumari et al. 2018).

AOT40 and M7 are the two commonly used metrics for quantifying the impact of O_3 on crop yield (Aunan et al. 2000; Dingenen et al. 2009; Ghude et al. 2014). M7 index is the seasonal mean of O_3 concentration during 09:00–16:00 h and AOT40 is an O_3 exposure metric in which O_3 levels of more than 40 ppb during solar radiation hours (07:00–18:00) are considered. M7 metric gives equal importance to all measurements and quantifies the yield losses due to O_3 concentrations <40 ppb, while AOT40 gives more importance to O_3 concentration >40 ppb (Tuovinen 2000). Hence, M7 is a better metric for estimating yield loss at lower O_3 levels while at higher O_3 levels AOT40 performs better (Hollaway et al. 2012). Therefore, a comparison of both the indices will be helpful in determining better crop yield responses to O_3.

$$\mathrm{AOT40} = \sum_{i=1}^{n}([\mathrm{O_3}]_i - 40) \quad \text{for } [\mathrm{O_3}] > 40\,\text{ppb during daylight hours only} \tag{15.1}$$

$$\mathrm{M7} = \frac{1}{n}\sum_{i=1}^{n}[\mathrm{O_3}]_i \quad \text{for } 09:00 - 16:00\,\mathrm{LT} \tag{15.2}$$

The O_3-exposure dose-response functions for relative yield (RY) for rice and wheat crops on the basis of field studies are given in Table 15.1.

The data of crop production (CP) from 2010 to 2013 has been taken from Directorate of Economics and Statistics, Department of Agriculture and Cooperation (2014).

Table 15.1 O_3 exposure-relative yield equations for wheat and rice

Crop	O_3-exposure-relative yield relationship	Reference
Wheat	$\mathrm{RY} = e^{-(\mathrm{M7}/137)^{2.34}}/e^{-(25/137)^{2.34}}$	Lesser et al. (1990)
	$\mathrm{RY} = -0.0000161 * \mathrm{AOT40} + 0.99$	Mills et al. (2007)
Rice	$\mathrm{RY} = e^{-(\mathrm{M7}/202)^{2.47}}/e^{-(25/202)^{2.47}}$	Adams et al. (1989)
	$\mathrm{RY} = -0.0000039 * \mathrm{AOT40} + 0.94$	Mills et al. (2007)

$$RYL_i = 1 - RY_i \quad (15.3)$$

RYL_i is relative yield loss for the respective year.

$$CPL_i = (RYL_i * CP_i)/(1 - RYL_i) \quad (15.4)$$

CPL_i is crop production loss for the respective year.

$$ECL_i = CPL_i * MSP_i \quad (15.5)$$

ECL_i is economic cost loss and MSP_i is minimum selling price for the respective year.

15.4 Result

Figure 15.3 shows the monthly variation of hourly average O_3 for the period January 2010–December 2013. The highest values of O_3 were observed during summer with mean O_3 mixing ratios 48 ± 26.7 ppb. During summer season high temperature, high solar radiation, and low humidity favor the photochemical production of O_3.

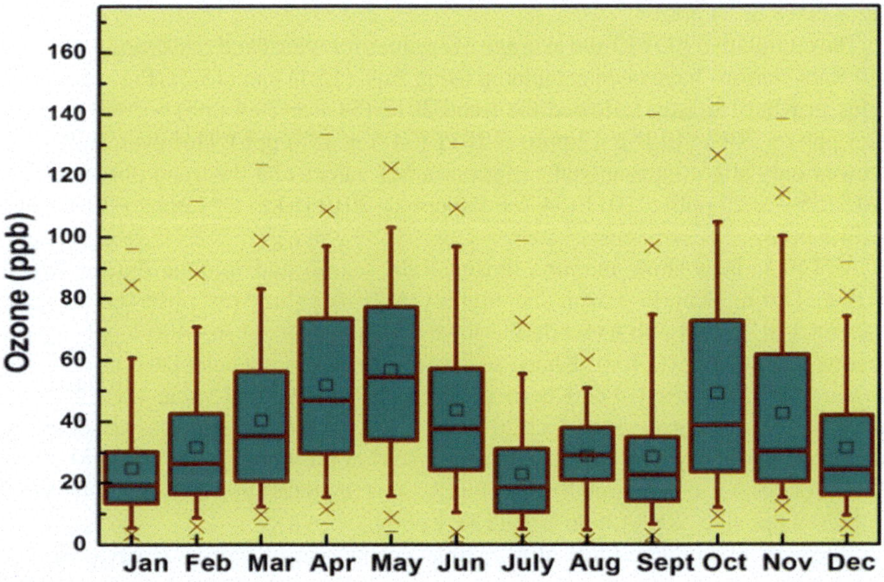

Fig. 15.3 Monthly mean values of O_3 during 2010–13

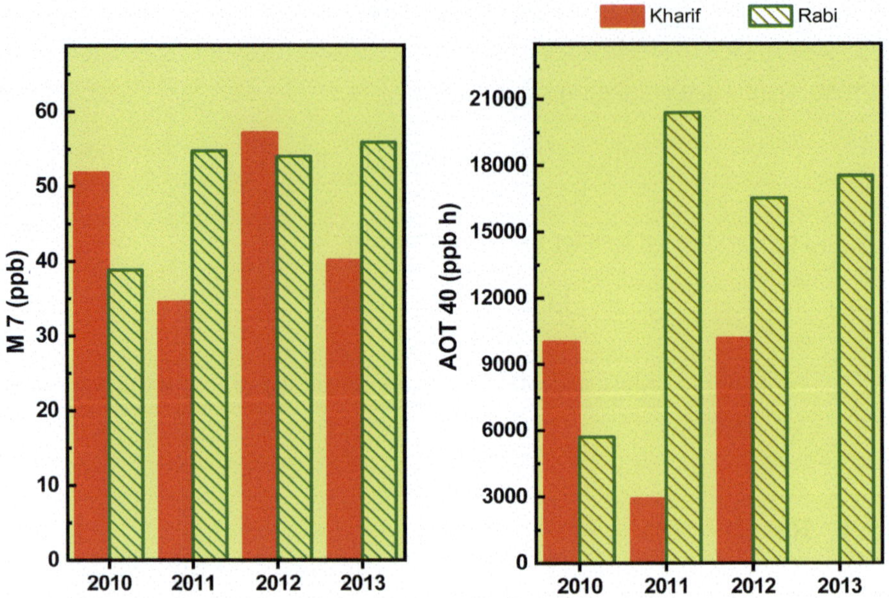

Fig. 15.4 Yearly M7 and AOT40 values for Kharif and Rabi crops

In winter O_3 levels were low (28.3 ± 19.3 ppb) due to low boundary layer and less solar radiation. The monsoon season is characterized by the lowest mean O_3 mixing ratios (19.8 ± 14 ppb).

The cumulative AOT40 and average M7 values for wheat and rice crops for Kharif and Rabi seasons have been calculated using Eqs. (15.1) and (15.2) (Fig. 15.4). M7 value in Kharif season followed the trend 2012 (57.1 ± 20.4 ppb) > 2010 (51.8 ± 22.8 ppb) > 2013 (40.1 ± 12 ppb) > 2011 (34.5 ± 19.6 ppb). However, Rabi crop showed only slight inter-annual variation in M7 values and the trend observed was 2013 (55.9 ± 25 ppb) > 2011 (54.7 ± 9.6 ppb) > 2012 (54 ± 18.1 ppb) > 2010 (38.8 ± 14.5 ppb).

AOT40 indices show maxima during Rabi season and minima during Kharif season. During Kharif season the highest AOT40 value was observed in 2012 (10,168.3 ± 3510.3 ppb h) while the lowest value (2896.44 ± 1406.1 ppb h) was observed in 2011. AOT40 of Kharif season for 2013 was not calculated as data was missing for that period. For Kharif crop, the lowest AOT40 value was observed in 2011 because during July–Oct 2011 relatively higher rainfall was observed as compared to other years. However, in Rabi season the trend in AOT40 values was 2011 (20,399.4 ± 6087.1 ppb h) > 2013 (17,533.4 ± 4821.5 ppb h) > 2012 (16,514.5 ± 4420.9 ppb h) > 2010 (5696.2 ± 1138.1 ppb h).

Independent samples *t*-test at a confidence interval of 95% was performed to ascertain the significant difference in inter-annual M7 and AOT40 values. For AOT40 value no statistically significant difference was observed between all the years in Kharif season. However, in Rabi season AOT40 values in 2010 were found to be statistically significantly different. M7 showed a significant difference for the year 2012 with 2010, 2011, 2013, and between 2010 and 2011 in Kharif season while in Rabi season all the years showed a significant difference in M7 values except between 2010 and 2011.

On comparing the observed AOT40 and M7 values with the values reported for other sites in India (Fig. 15.5), the observed value of AOT40 (15,036 ppb h) for wheat for the study site (average of 2010–13) is comparatively lower as compared to AOT40 value (27,553 ppb h) reported for an urban site of Pune by Beig et al. (2008) for the time period 2003–06. However, a study conducted at a suburban site Mohali (Sinha et al. 2015) for 2012 has reported 17,111 ppb h AOT40 which is higher than the value reported in the present study but the M7 value (51.7 ppb) reported at Mohali and study site are comparable. At two rural sites of India, Debaje et al. (2010) and Singh et al. (2015) have reported AOT40 values for wheat. Debaje et al. (2010) have reported 3745 ppb h for Ahmednagar considering Jan–Mar (2007) as the crop-growing season and Singh et al. (2015) have reported 11,600 ppb h for Varanasi considering Dec-Mar as crop-growing season. Both the studies report comparatively lower AOT40 values as compared to the present study.

Beig et al. (2008) have reported O_3-exposure-related study for the period 2003–06 at an urban site of Pune and reported a value of 1442 ppb h for rice while the study conducted at a rural site of Varanasi (Sarkar et al. 2015) reported a comparatively higher AOT40 value (11,200 ppb h). However, for the present study site, an average AOT40 7690 ppb h was observed (average of 2010–12) which is comparatively

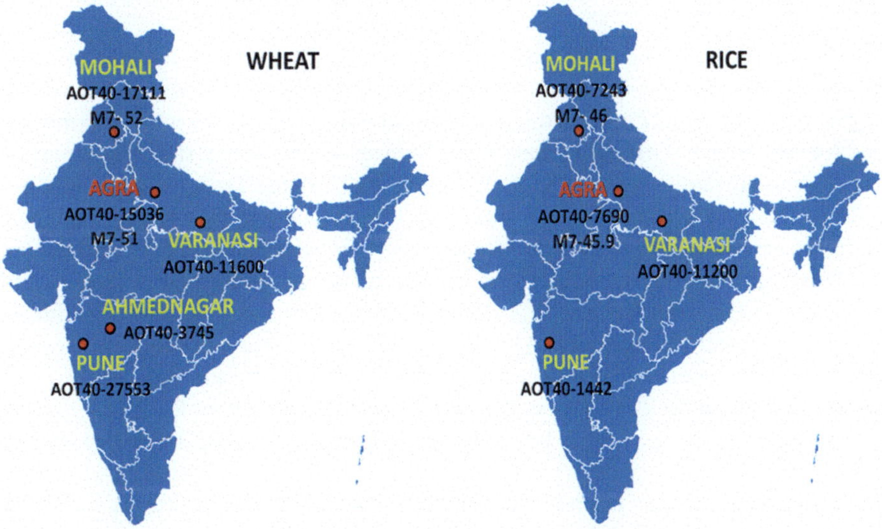

Fig. 15.5 Comparison of AOT40 (in ppb h) and M7 (in ppb) values for wheat and rice crop reported at different sites of India (Mohali (Sinha et al. 2015); Varanasi (Singh et al. (2015) for wheat), (Sarkar et al. (2015) for rice); Ahmednagar (Debaje et al. 2010); Pune (Beig et al. 2008) and Agra (present study))

lower than the values reported for Varanasi but higher than the value reported for Pune. Sinha et al. (2015) have tried to find the impact of O_3 exposure on rice using M7 indice and found value (M7-46 ppb) comparable to the present study (average 2010–13).

In the present study, AOT40 for the initial three months of agricultural crop growth period was determined (Kharif-July–Sept; Rabi-Jan–Mar) as a value of 3000 ppb h has been set as a critical value for exposure of agricultural crops to O_3 (Dingenen et al. 2009). It was found that AOT40 lies above the critical value in Rabi season in all years except 2010 while no such trend was observed in Kharif season. This observation indicates that the O_3 levels exceed the critical level during the Rabi crop-growing season.

To calculate relative yield AOT40 and M7 for the respective period have been used (Fig. 15.6). The relative yield obtained from AOT40 and M7 for wheat were highest in 2010 (0.9; 0.96) and lowest in 2011 (0.66) using AOT40 and in 2013 (0.9) using M7. For rice RY calculated from AOT40 was similar in 2010 and 2012 (0.9) and the highest value was observed in 2011 (0.93). For M7 metric also the highest value was observed in 2011 (0.99). However, the RY values obtained from M7 metric for both the crops were higher than the values obtained from the AOT40 metric.

A number of studies have reported the crop yield responses by using AOT40 and M7. Sinha et al. (2015) have used both the indices for estimating rice and wheat yield loss and found 5.2 and 4.1 times higher relative yield loss by AOT40. Similarly, Lal et al. (2017) have also reported higher RYL values derived from AOT40 metric than M7 for rice and wheat. In the present study, the RYL values obtained from AOT40 for the study period was 3.9 and 3.2 times higher than M7 values for rice and wheat, respectively.

Crop production loss (CPL) using AOT40 and M7 has been calculated using Eq. (15.4). The crop production loss for wheat during the study period 2010–2013 ranges from 0.06 to 0.27 million tons based on AOT40 metric and 0.02 to 0.05 million tons for M7 metric with the highest loss observed in the year 2011. For rice, crop production loss ranges from 993 to 1290 tons (AOT40) and 118 to 351 tons (M7).

To evaluate the economic cost loss (ECL) for wheat and rice (using Eq. 15.5) minimum selling prices (MSP) for respective years were taken from the database of Commission for Agriculture Costs and Prices (Directorate of Economics and Statistics 2014). For wheat, the ECL calculated on the basis of AOT40 was highest in 2011 (345.4 million rupees) and lowest in 2010 (64.8 million rupees) while on the basis of M7 the highest ECL was observed in 2013 (75.3 million rupees). ECL for rice was maximum in 2011 (1.4 million rupees) and minimum in 2010 (1.0 million rupees) by AOT40 while for M7 the highest value was observed in 2012 (0.44 million rupees).

Earlier studies in the US and Europe have tried to develop a crop-specific O_3-exposure-response relationship using NCLAN and EOTCP as an indicator of reduction in crop yield since the 1980s (Heck et al. 1982). Further Mills et al. (2007) derived O_3-exposure-relative yield functions using AOT40 on the basis of existing literature. But all these studies were conducted on European and US-specific crops and no specific relations have been developed for Asia specific crops. Therefore,

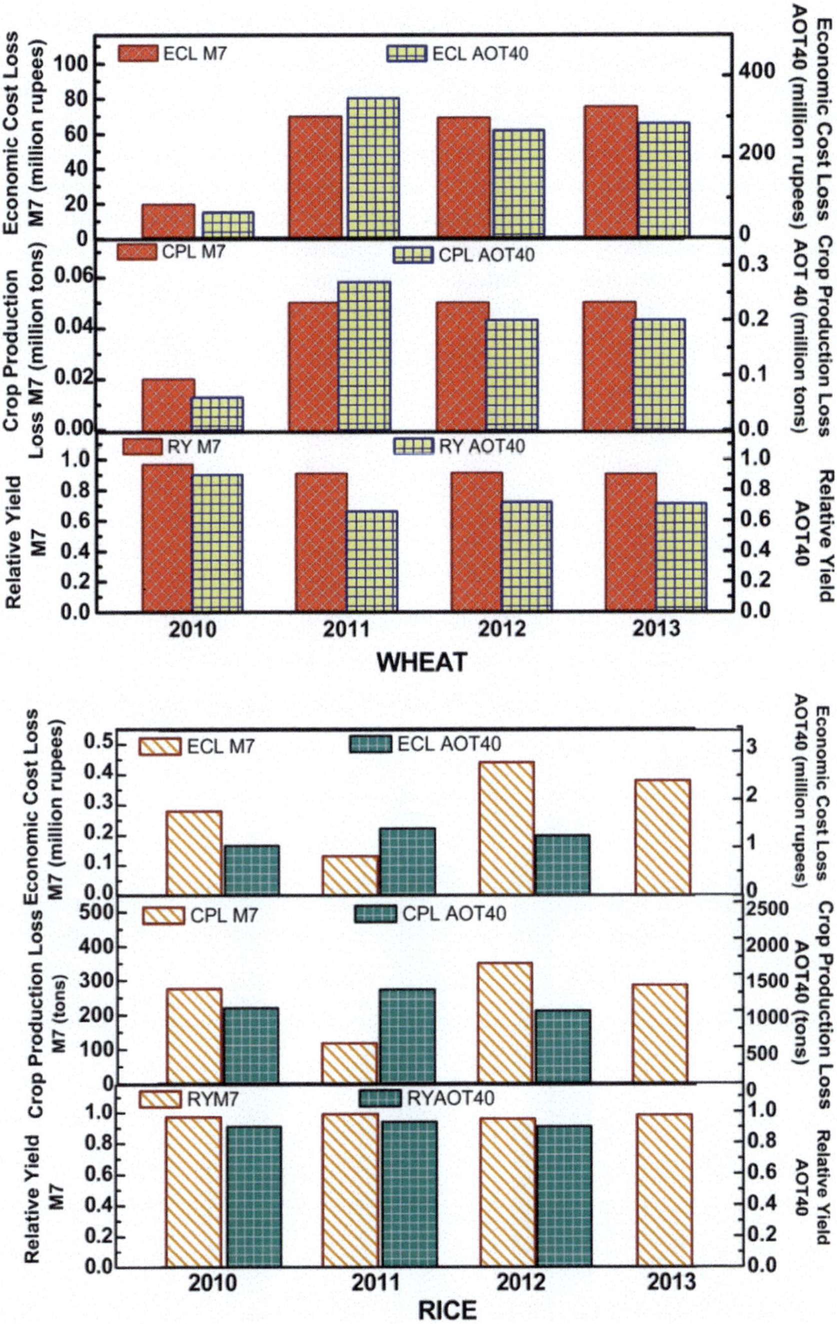

Fig. 15.6 Assessment of the impact of O_3 on wheat and rice crops during 2010–13

for determining the influence of O_3 exposure on Indian specific crop-exposure-yield relationships are needed.

Sinha et al. (2015) have tried to develop relative yield relationships for Indian crop species (Table 15.2). In the present study, a comparison of both the exposure-yield functions has been done to determine the variation in relative yield using old exposure-response relationships and revised relationships derived by Sinha et al. (2015) (Fig. 15.7). It was found that for wheat (AOT40) RY derived by Mills et al. (2007) overestimate the RY as compared to RY derived for Indian cultivars by Sinha et al. (2015). On comparing RY values of wheat on the basis of AOT40 a variation of 4–10% was found which is comparatively smaller than the variation found using M7 value (6–44%). The largest variation using M7 was observed in the year 2011.

Table 15.2 A comparison of old and revised O_3 exposure-related equations

Crop	Metric	Old relation	Revised relation
Wheat	M7	$RY = e^{-(M7/137)^{2.34}}/e^{-(25/137)^{2.34}}$	$RY = e^{-(M7/62)^{4.5}}/e^{-(25/62)^{4.5}}$
	AOT40	$RY = -0.0000161 * AOT40 + 0.99$	$RY = -0.000026 * AOT40 + 1.01$
Rice	M7	$RY = e^{-(M7/202)^{2.47}}/e^{-(25/202)^{2.47}}$	$RY = e^{-(M7/86)^{2.5}}/e^{-(25/86)^{2.5}}$
	AOT40	$RY = -0.0000039 * AOT40 + 0.94$	$RY = -0.00001 * AOT40 + 0.95$

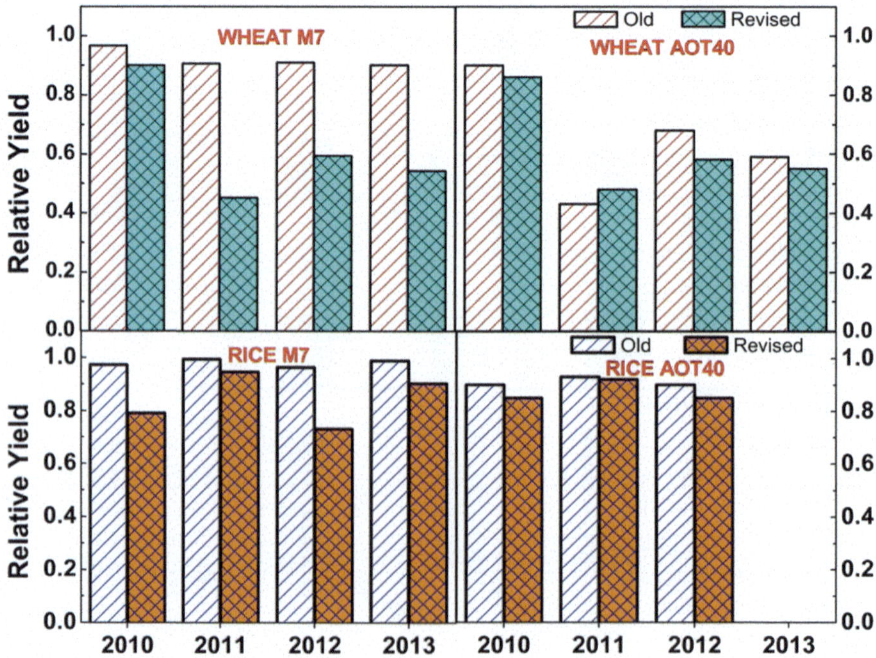

Fig. 15.7 Comparison of old and revised O_3 crop-exposure relationships

The magnitude of the overestimation of RY by earlier crop-exposure relations was comparatively less (AOT40-1–5% and M7-4–23%) for rice as compared to newly derived relations for both AOT40 and M7 metrics.

15.5 Conclusion

In this study, an estimation of O_3-induced crop yield losses for rice and wheat using two indices has been done. AOT40 indices showed maxima during Rabi season and minima during Kharif season. Similar variation has been found in M7 values but the magnitude of inter-annual variation was less as compared to AOT40 values. RYL for wheat and rice on the basis of AOT40 ranged 10–34% and 7–10%, respectively while on the basis of M7 relative yield losses ranged from 3 to 11% for wheat and 0.7 to 4% for rice. Using RYL an estimation of CPL has been done. CPL for wheat amounted to 0.02–0.05 million tons using M7 and 0.06–0.27 million tons using AOT40 was observed. CPL for rice was relatively smaller than wheat. For M7 it ranged from 118 to 351 tons and 993 to 1290 tons for AOT40. AOT40 and M7 metrics gave very different crop yield losses, as these two metrics are for US and European regions. Therefore, a comparison of US and European specific crops and Indian cultivars derived O_3 exposure relations has been done. It was found that old O_3 exposure-related relationships had over-estimated RY for both the crops in comparison to relationship derived for Indian cultivars. Studies on M7 and AOT40 metrics indicate that crop yield is potentially at a risk from rising O_3 levels. These studies can be helpful in better integration of the policies and abatement measures aimed at reducing air pollution. In addition, development of crop management practices for reducing the impact of O_3 like appropriate cultivar selection, withholding irrigation during the highest ozone episodes can be helpful for farmers to protect their crops from O_3.

Acknowledgements The authors are thankful to the Director of the institute and the Head of Chemistry Department for necessary help. The financial support for this work was provided by ISRO-GBP under AT-CTM project.

References

Adams RM, Glyer JD, Johnson SL, McCarl BA (1989) A reassessment of the economic effects of ozone on US agriculture. JAPCA 39(7):960–968

Atkinson R, Arey J (2003) Gas-phase tropospheric chemistry of biogenic volatile organic compounds: a review. Atmos Environ 37:197–219

Aunan K, Berntsen TK, Seip HM (2000) Surface ozone in China and its possible impact on agricultural crop yields. AMBIO 29(6), 294–301

Avnery S, Mauzerall DL, Liu J, Horowitz LW (2011) Global crop yield reductions due to surface ozone exposure 1: year 2000 crop production losses and economic damage. Atmos Environ 45:2284–2296

Beig G, Ghude SD, Polade SD, Tyagi B (2008) Threshold exceedances and cumulative ozone exposure indices at tropical suburban site. Geophys Res Lett 35(2)
Debaje SB (2014) Estimated crop yield losses due to surface ozone exposure and economic damage in India. Environ Sci Pollut Res 21(12):7329–7338
Debaje SB, Kakade AD, Jeyakumar SJ (2010) Air pollution effect of O_3 on crop yield in rural India. J Hazard Mater 183(1):773–779
Dingenen RV, Dentener FJ, Raes F, Krol MC, Emberson L, Cofala J (2009) The global impact of ozone on agricultural crop yields under current and future air quality legislation. Atmos Environ 43:604–618
Directorate of Economics and Statistics, Department of Agriculture and Cooperation (2014) Agricultural statistics at a glance 2014. Available at http://eands.dacnet.nic.in/PDF/Agricultural-Statistics-At-Glance2014.pdf
Emberson LD, Ashmore MR, Murray F (2003) Air pollution impacts on crops and forests: a global assessment. Imperial College Press, London
Emberson LD, Buker P, Ashmore MR, Mills G, Jackson LS, Agrawal M, Atikuzzaman MD, Cinderby S, Engardt M, Jamir C, Kobayashi K, Oanh NTK, Quadir QF, Wahid A (2009) A comparison of North American and Asian exposure-response data for ozone effects on crop yields. Atmos Environ 43, 1945–1953
Ghude SD, Jena C, Chate DM, Beig G, Pfister GG, Kumar R, Ramanathan V (2014) Reductions in India's crop yield due to ozone. Geophys Res Lett 41:51971
Heck WW, Taylor OC, Adams R, Bingham G, Miller J, Preston E, Weinstein L (1982) Assessment of crop loss from ozone. J Air Pollut Control Assoc 32(4):353-361
Hollaway MJ, Arnold SR, Challinor AJ, Emberson LD (2012) Intercontinental trans-boundary contributions to ozone-induced crop yield losses in the Northern Hemisphere. Biogeosciences 9:271–292
Iriti M, Faoro F (2009) Chemical diversity and defence metabolism: how plants cope with pathogens and ozone pollution. Int J Mol Sci 10(8), 3371–3399
König G, Brunda M, Puxbaum H, Hewitt CN, Duckham SC, Rudolph J (1995) Relative contribution of oxygenated hydrocarbons to the total biogenic VOC emissions of selected mid-European agricultural and natural plant species. Atmos Environ 29(8):861–874
Kumari S, Verma N, Lakhani A, Tiwari S, Kandikonda, MK (2018) Tropospheric ozone enhancement during post-harvest crop-residue fires at two downwind sites of the Indo-Gangetic Plain. Environ Sci Pollut Res 25(19):18879–18893
Lal S, Sahu LK, Gupta S, Srivastava S, Modh KS, Venkataramani S, Rajesh TA (2008) Emission characteristic of ozone related trace gases at a semi-urban site in the Indo-Gangetic plain using inter-correlations. J Atmos Chem 60(3), 189
Lal S, Venkataramani S, Naja M, Kuniyal JC, Mandal TK, Bhuyan PK, Kumari KM, Tripathi SN, Sarkar U, Das T, Swamy YV (2017) Loss of crop yields in India due to surface ozone: an estimation based on a network of observations. Environ Sci Pollut Res 1–10
Lesser VM, Rawlings JO, Spruill SE, Somerville MC (1990) Ozone effects on agricultural crops: statistical methodologies and estimated dose-response relationships. Crop Sci 30(1):148–155
Lin X, Trainer M, Liu SC (1988) On the nonlinearity of the tropospheric ozone production. J Geophys Res 93:15879–15888
Mauzerall DL, Wang X (2001) Protecting agricultural crops from the effects of tropospheric ozone exposure-reconciling science and standard setting. Annu Rev Energy Environ 26, 237–268
Mills G, Buse A, Gimeno B, Bermejo V, Holland M, Emberson L, Pleijel H (2007) A synthesis of AOT40-based response functions and critical levels for ozone for agricultural and horticultural crops. Atmos Environ 41:2630–2643
Mills G, Hayes F, Simpson D, Emberson L, Norris D, Harmens H, Buüker P (2011) Evidence of widespread effects of ozone on crops and semi–natural vegetation in Europe (1990–2006) in relation to AOT40—and flux–based risk maps. Glob Change Biol 17:592–613
Rai R, Agrawal M, Agrawal SB (2010) Threat to food security under current levels of ground level ozone: a case study for Indian cultivars of rice. Atmos Environ 44:4272–4282

Redeker KR, Meinardi S, Blake D, Sass R (2003) Gaseous emissions from flooded rice paddy agriculture. J Geophys Res Atmos 108(D13)

Sarkar A, Singh AA, Agrawal SB, Ahmad A, Rai SP (2015) Cultivar specific variations in antioxidative defense system, genome and proteome of two tropical rice cultivars against ambient and elevated ozone. Ecotoxicol Environ Saf 115:101–111

Sicard P, Coddeville P, Galloo JC (2009) Near-surface ozone levels and trends at rural stations in France over the 1995–2003 period. Environ Monit Assess 156:141–157

Singh P, Agrawal M, Agrawal SB, Singh S, Singh A (2015) Genotypic differences in utilization of nutrients in wheat under ambient ozone concentrations: growth, biomass and yield. Agric Ecosyst Environ 199:26–33

Sinha B, Singh Sangwan K, Maurya Y, Kumar V, Sarkar C, Chandra BP, Sinha V (2015) Assessment of crop yield losses in Punjab and Haryana using two years of continuous in-situ ozone measurements. Atmos Chem Phys Discuss 15(2):2355–2404

Tuovinen JP (2000) Assessing vegetation exposure to ozone: properties of the AOT40 index and modifications by deposition modelling. Environ Pollut 109:361–372

Chapter 16
Air Pollution Episode Analysis and Qualitative Evaluation of Proposed Control Measures in Delhi City

Sunil Gulia, S. K. Goyal, and Rakesh Kumar

16.1 Introduction

The Delhi city like other megacities of developing countries, is struggling to improve its urban air quality which is degraded rapidly in the last few years (Gulia et al. 2017; Pant et al. 2015; Gupta and Kumar 2006). Exposure to high pollutant concentrations associated with a wide range of health effects depends on the constituents of the pollutants (Heal et al. 2012). For example, Guttikunda and Goel (2013) have estimated that a high level of particulate matter can be responsible for thousands and millions of premature deaths and asthma attacks, respectively, each year in Delhi. Further, Kesavachandran et al. (2015) have reported morning walk in outdoor atmosphere in Delhi city may cause impairment in the functioning of due to $PM_{2.5}$ deposition in trachea and bronchia. However, these estimates are yet to be validated from hospital records. Delhi air pollution problem is complex with lots of unidentified, un-accounted, and unregulated sources that vary season to season, time of the day, and location of the city and also due to neighboring states. Numerous studies have been conducted to quantify the sources of pollution (Pant et al. 2015; Sharma and Dikshit 2016; CPCB (Central Pollution Control Board) 2010). The sources of air pollution in Delhi city are both natural as well as anthropogenic. The city is located in a tropical atmosphere, so metrological conditions are responsible for pollution

S. Gulia (✉) · S. K. Goyal
CSIR-National Environmental Engineering Research Institute, Delhi Zonal Centre, Naraina Industrial Area, New Delhi 110028, India
e-mail: s_gulia@neeri.res.in

S. K. Goyal
e-mail: sk_goyal@neeri.res.in

R. Kumar
CSIR-National Environmental Engineering Research Institute, Nehru Marg, Nagpur 440020, India
e-mail: r_kumar@neeri.res.in

S. M. Shiva Nagendra et al. (eds.), *Urban Air Quality Monitoring, Modelling and Human Exposure Assessment*, Springer Transactions in Civil and Environmental Engineering,
https://doi.org/10.1007/978-981-15-5511-4_16

caused by the transport of pollutants from far-field areas. In anthropogenic sources, the burning of fossil fuels and biomass are major causes of air pollution in addition to local sources such as garbage burning, re-suspension of road dust, bakeries, and crematoria (Sharma and Dikshit 2016; Nagpure et al. 2015). Further, the city is facing unique challenges of air pollution episodes which may occur for a short period (2–3 days or maximally a week). One such episode occurred in the months of October–November last year which might be due to combinations of burning of paddy crop residues in fields of neighboring states (Kumar et al. 2015) and burning of firecrackers during festivals and sudden drop of mixing height and wind speed, besides various other sources of air pollution in Delhi and nearby regions/states.

The frequency of air pollution episode is more during the winter season when pollutants exceed the specified standards. However, in months of April–May, particulates concentration was found exceeding many times which might be due transport of natural dust from the desert area of neighboring states (Sharma and Dikshit 2016). The Central and State Government have introduced and implemented many mitigation actions to manage the high level of air pollution in the city. Some of these actions are improvement in fuel quality in vehicles; low sulphur fuels, lead-free fuel; shifting of polluting industries outside Delhi city; use of CNG in public transportation instead of diesel; shifting of power plants from coal to natural gas; restricted entry of trucks in the city and implementation of the odd–even scheme in private cars for a specified period, etc. (CPCB (Central Pollution Control Board) 2010; Dholakia et al. 2013; Gulia et al. 2015; Kumar et al. 2017). Recently, Ministry of Environment, Forest and Climate Change (MoEF&CC 2017) has notified the graded response action plan (GRAP) to reduce the pollution level. The actions are defined based on the air pollution level such as severe/emergency, very poor, poor, and moderate as per Air Quality Index guidelines specified by Central Pollution Control Board (CPCB).

The present study is an attempt to evaluate the air pollution episode that happened during the first week of November 2016. Air quality and meteorology data have been analyzed for the period between 25th October and 15th November 2016. The study also discussed the possible preparedness/actions to reduce the air pollution level in the next winter and qualitative evaluations of action plans suggested by the MoEF&CC.

16.2 Study Area: Delhi City

The Delhi city is a megacity and surrounded by Aravalli hill from south side. The city is surrounded by major populated cities of Haryana (Sonepat, Bahadurgar, Gurugram, Fridabad) and Uttar Pradesh (Ghaziabad and Noida) states. There is no physical land use boundary that differentiates these cities from each other. The population of the city is increasing alarmingly with a growth rate of 47% in a decade. The total land area of the city is 1483 km^2 (NCR 2013). This exponential growth in population has resulted in rapid demand of transport, construction, industry and high energy consumption. The total registered vehicles were approx. 6.93 million in year 2011 which will be increased to be 25.6 million by 2030 (Kumar et al. 2017). This increasing vehicular population were plying on 33,198 km of road length which were not changed much from last 10 years. This leads to traffic congestions and huge emission of pollutants (NCR 2013; GoD 2016).

In Delhi city, Air quality monitoring is being carried out by total 49 stations. Out of them, 39 are CAAQMS (continuous ambient air monitoring station) and are being operated and maintained by three different agencies i.e., Central Pollution Control Board (CPCB, 7 stations), Delhi Pollution Control Committee (DPCC, 26 stations) and the Indian Institute of Tropical Meteorology Pune (8 stations). The pollutants being monitored by these stations are PM_{10}, $PM_{2.5}$ SO_2, NO_x, and O_3. Additionally, CPCB is being monitoring 10 more location through manual air quality monitoring under NAMP with support from CSIR-NEERI (03 out of 10).

16.3 Data Collection and Methodology

The hourly and daily average monitoring data of PM_{10}, $PM_{2.5}$, SO_2 and NO_x concentration were collected from continuous ambient air quality monitoring stations operated by Delhi Pollution Control Committee (DPCC) for the period of 25th October to 15th November 2016. Along with air quality data, hourly meteorological parameters of wind speed, wind direction, ambient temperature and relative humidity of each site were collected for the same period. The information on surrounding geographical and land use of each site was collected by conducting field survey and Google earth software. The collected monitoring data were analyzed by estimating the descriptive parameters, time series analysis, exploratory analysis between meteorological parameters and pollutant concentrations, pollution rose and correlation analysis of pollutants concentrations between sites. Finally, a qualitative analysis has been carried out to analyze the impact of recently suggested GRAP on the reduction of pollution level at these sites. This qualitative analysis is carried out based on the applicability of these actions on local sources.

16.3.1 Air Quality Monitoring Stations

Out of total CAAQMS, data of four stations was available in the form of Air Quality Index (AQI) and hourly concentration data. These monitoring stations are operated by DPCC and located at Anand Vihar, Mandir Marg, Punjabi Bagh and R.K. Puram. The monitoring sites are covering industrial, commercial and residential sectors of Delhi city. The details of each site including latitude and longitude, surrounding land use and direction w.r.t center of Delhi are described in Table 16.1 and shown in Fig. 16.1.

Table 16.1 Details of selected ambient air quality monitoring stations in Delhi city

Sr. No.	Site name	Features
1	Anand Vihar (AV)	It is located at Delhi-UP boarder on east of Delhi. The surrounding land use represents commercial, industrial, and residential areas. The station is located kerbside of the road corridor where approx. 2400 buses are being crossed (CNG + diesel buses)
2	R.K. Puram (RKP)	Located in south of Delhi on school premises along the arterial road. The major trafficked road is just ~120 m from. The station is surrounded by residential area
3	Punjabi Bagh (PB)	The station is surrounded by residential and commercial areas. It is located 30 m away from the kerbside of an arterial road in West Delhi
4	Mandir Marg (MM)	It is located at centre of Delhi and surrounding by green areas on one side and residential at the other side. Station is located on the premises of a school along the roadside. The traffic is moderate

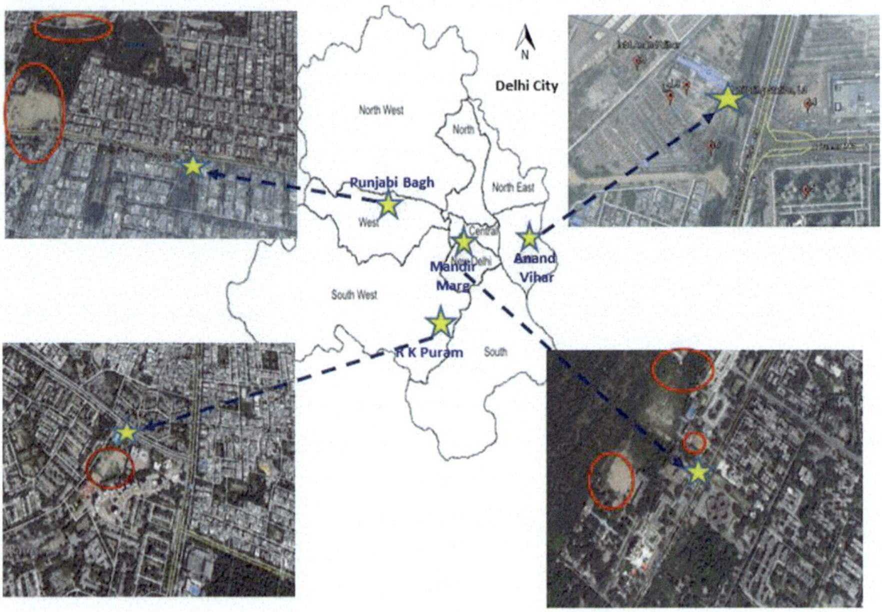

Fig. 16.1 Delhi map showing locations of selected AQM stations

16.4 Results and Discussion

16.4.1 Air Pollution Episode Analysis

As per AQI data analysis, air pollution conditions were found severe, most of the times of the study period at all four selected locations, and most prominent pollutant was observed to be either PM_{10} or $PM_{2.5}$. At AV station, air quality was observed severe having AQI value more than 400 and prominent pollutant was observed to be PM_{10} most of the days of the study period. However, at MM, severe was found 11 days, very poor was found 08 days and poor and moderate one day each. At PB, it was recorded as severe (16 days), very poor (4 days) and moderate (1 day) whereas at RKP station, air quality was found severe (16 days), very poor (4 days) and poor (1 day). The AQI value was recorded maximum (500) on 5–6th November 2016 at all locations for particulate matter. AQI category wise status of air quality is given in Table 16.2.

16.4.2 Air Pollutant Concentrations Analysis

The time series plot of 24 h averages of $PM_{2.5}$, PM_{10}, NO_x, and SO_2 concentration was plotted to analyze the trend and exceedances of respective national ambient air quality standards (NAAQS) during the study period (Fig. 16.2). It is observed that the $PM_{2.5}$ and PM_{10} concentrations have a more or less increasing trend from 25th October to 6th November (maximum of the study period) and then sharply decreased from 7th November onwards (Fig. 16.2a, b). However, the trend of NO_x and SO_2 was different then particulates. The NO_x was found maximum on 30–31st October (Diwali festival) at all locations except AV where peaks were observed on 6th November (Fig. 16.2c). In addition to that, concentrations were found higher on 12th Nov. but data were missing for 13th Nov. which make the trend incomplete and further explanation. The SO_2 concentration was observed maximum on 30–31st October 2016 during the study period (Fig. 16.2d). Higher levels of NO_x and SO_2

Table 16.2 Category wise status of AQI values during 25th Oct.–15th Nov. 2016 (21 days)

Air quality category	AQI values	No. of days			
		AV	RKP	PB	MM
Severe	401–500	21	16	16	11
Very poor	301–400	0	4	4	8
Poor	201–300	0	1	0	1
Moderate	101–200	0	0	1	1
Satisfactory	51–100	0	0	0	0
Good	0–50	0	0	0	0

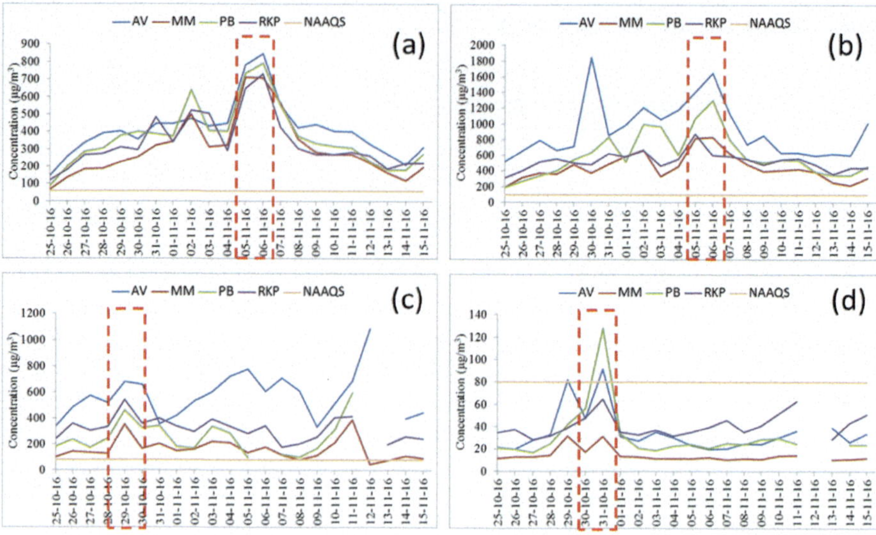

Fig. 16.2 Time series plot of 24 hourly $PM_{2.5}$ (**a**), PM_{10} (**b**), NO_x (**c**), and SO_2 concentration (**d**)

on 30–31st Oct. are expected due to bursting of firecrackers on the festival of Diwali whereas higher levels of NO_x on 12th Nov. is not found directly correlating to any particular activity. The $PM_{2.5}$ concentration exceeded in the range of 12–14 times (6th Nov.) of the NAAQS value of 60 $\mu g/m^3$ during the study period at all locations. This value for PM_{10} was observed 8–17 times (6th Nov.) of the NAAQS value of 100 $\mu g/m^3$ during the study period at all locations. The concentrations of $PM_{2.5}$ and PM_{10} were found maximum on 5–6th Nov. However, NO_x concentration exceeded up to 8 times of NAAQS values on 30th Oct. (10 times on 6th Nov. at AV station). The SO_2 exceeded up to 2 times of NAAQS values of 80 $\mu g/m^3$ on 30th Oct. The comparative trends between four locations indicate the highest concentration at AV and lowest at MM station except for SO_2 which is highest at RKP station (mainly a residential area).

It is inferred that particulate pollutant has a different trend than the gaseous pollutant both time-wise and location wise which may be due to the influence of local sources, especially at AV station. Further, correlation analysis has been carried out for a particular pollutant between different monitoring locations w.r.t MM station (Table 16.3). The correlation between monitoring stations is defined in qualitative

Table 16.3 Values of correlation coefficient between four locations

Pollutant	MM/AV	MM/PB	MM/RKP
$PM_{2.5}$	Excellent	Excellent	Excellent
PM_{10}	Good	Excellent	Excellent
NO_x	No correlation	Excellent	Excellent
SO_2	Excellent	Good	Average

forms by considering different ranges of correlation coefficient (r^2). The correlation is assumed to be Excellent (for $r^2 > 0.8$), Very Good (for r^2 between 0.6 and 0.8), Good (for r^2 values between 0.4 and 0.6), Average (for r^2 between 0.2 and 0.4) and No correlation (for $r^2 < 0.2$).

The correlation of different selected sites w.r.t. MM was found excellent for $PM_{2.5}$ which indicates a common cause of $PM_{2.5}$ level at all locations. However, the correlation between MM and AV station for PM_{10} and NO_x are found Good and no correlation, respectively. The correlation between MM and RKP is found average for SO_2. The average and no correlation indicate that local sources are influenced by pollutant concentration levels at the monitoring station.

16.4.3 Air Pollution Meteorology

Air pollution meteorology plays an important role in evaluating the fate and transport of air pollutants in the atmosphere. The winter months are known for poor dispersion of pollutants and the occurrence of air pollution episodes (high pollution level). In an urban area, except industries, most of the sources are at ground level, emission from which do not rise high and so accumulate more, in winter months when atmospheric inversion inhibits dispersion of pollutants in the upper atmosphere. This leads to formation of haze/smog. The time series plot of meteorological parameters such as wind speed, wind direction, ambient temperature, and relative humidity was drawn to further correlate with the trends of pollutant concentrations during the study period as shown in Fig. 16.3.

The ambient temperature reduced by 6 °C from 25th Oct. to 15th Nov. 2016. Further, it was observed that ambient temperature reduced by 2 °C on 5th and 6th Nov. compared to 4th Nov. 2016. However, relative humidity increased up to 74% on 6th Nov. compared to 51% on 25th Oct. The wind speed reduced by almost 0.5 m/s from 25th Oct. to 15th Nov. 2016 which is 45% w.r.t. average wind speed on 25th Oct. 2016 (1.1 m/s). Figure 16.3 c clearly indicates the reduction of wind speed on 5–6th Nov. compared to 4th and 7th Nov. 2016. Based on the above analysis, it can be concluded that meteorological conditions were worst for air pollution dispersion on these two days compared to other days of the study period at all the four locations.

16.4.4 Pollution Rose

The $PM_{2.5}$, PM_{10}, SO_2, and NO_x concentrations were further described in the form of polar plots (pollution rose) to know the location and types of sources. The smoothing technique has been applied in the adjacent bins in the polar plot for better visualization of sources (Kumar et al. 2017). Therefore, the actual concentrations are slightly different as mentioned according to the color bars (Fig. 16.4). At AV station, $PM_{2.5}$, PM_{10}, and NO_x were found higher when the wind was blowing from Northeast in

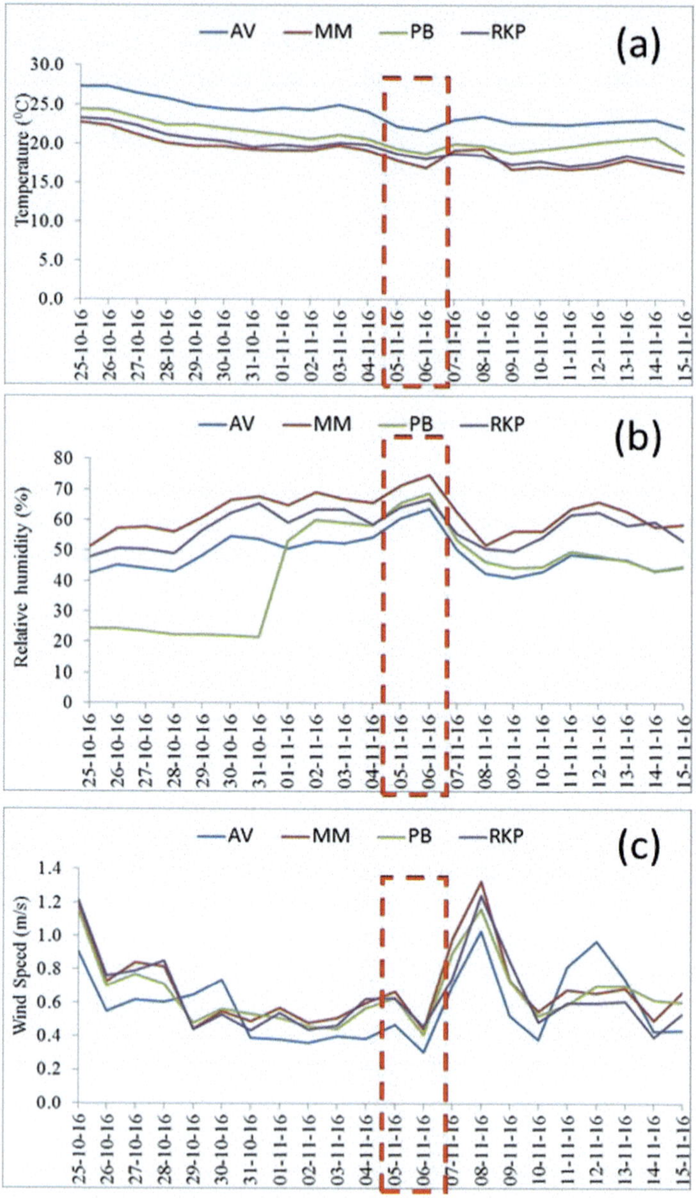

Fig. 16.3 Time series plot of meteorological parameters of (**a**) Temperature, (**b**) Relative Humidity, (**c**) Wind Speed during the study period at four AQM locations

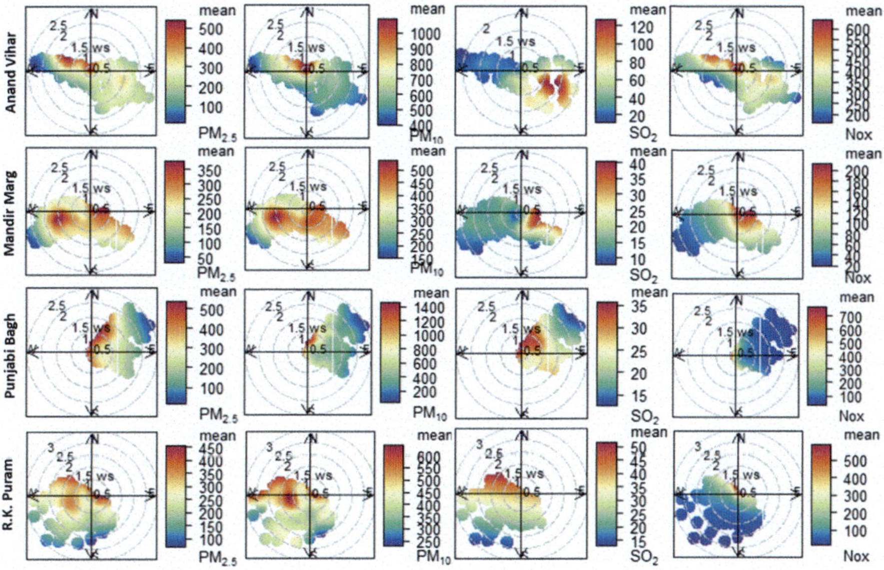

Fig. 16.4 Pollution rose diagram of four pollutants at four locations

wind speed up to 1.5 m/s whereas SO_2 was found higher when the wind was blowing from Southwest in wind speed up to 2 m/s. Further, this may indicate the presence of local sources located around monitoring stations which vary pollutant to pollutant. However, at MM station, concentrations of all pollutants were found higher when the wind was blowing from Southeast and Southwest directions in wind speed up to 2 m/s. Further, at PB location, pollution level was found higher irrespective of wind direction when wind speeds up to 1 m/s. The wind direction showed from East and Northeast. At RKP, pollution level was found higher when wind blows from West direction in wind speed range of 0–2 m/s. It observed that all four locations have different wind directions which might be due to local surrounding topography and urban structure. Further, this variation leads to variation spatial distribution of pollution level at an urban scale. Therefore, it may be concluded that local topography around the monitoring station affects the direction of the wind and also pollutant concentration.

16.4.5 Mitigation Measures for Next Winter

In the recent past, Central and Delhi state government introduced numerous action plans to manage the increasing air pollution levels in Delhi city. However, some actions are limited to Delhi state boundary and some go beyond it up to NCR. These management actions were long and short term as well as local and regional scale. The

MoEF&CC has recently notified the graded response action plan (GRAP) to combat air pollution in Delhi–NCR region which will be implemented by the concerned department/agency under the monitoring of Environment Pollution (Prevention and Control) Authority (EPCA). The GRAP defines the mitigation measures to be taken based on air quality level as per AQI, i.e., severe/emergency, very poor, poor, and moderate. These actions will be implemented in the whole NCR region including Delhi city irrespective of location-specific air pollution level which is a common issue in Delhi NCR. In the present study also, it is observed that the profile of pollution level is varying temporally and spatially in Delhi city. For example, pollutant behaves differently at AV station compared to the other three stations, while variation is seen in other stations also for SO_2 and NO_x. Given this, a qualitative matrix has been defined to evaluate the impacts of GRAP on the reduction of pollution levels at different selected monitoring locations. The pollutant concentration data of these monitoring stations may be used by policymakers to evaluate the GRAP efficacy (at least qualitatively) using pre, during and post-implementation scenario. The impacts of actions are defined in three categories, i.e., High (H), Moderate (M), and Low (L) based on the presence of sources within 500 m boundary around monitoring station (Table 16.4).

It is observed that impacts of a particular proposed action are varying from high to low at different sites. For examples, action on stop of entry of truck traffic into Delhi during "severe or emergency" may reduce pollution level at high rate at AV and RKP because first one is located at interstate border (one of the entry points of trucks) and the second is located near to ring road and outer ring road (way for movement of trucks in Delhi). However, it may have moderate impact at other locations because they are not so close to the movement of trucks.

16.5 Conclusion

The present study has analyzed the increased air pollution level data (episodic situation) and related causes happened in Delhi city on 5–6th November 2016 using air quality from four differently located monitoring stations. The following conclusion can be drawn from the present study:

- Air quality was found severe and very poor most of the time of the study period with a prominent pollutant of particulate matter at all seven locations except 2–3 days (after the episodic situation was over).
- $PM_{2.5}$ and PM_{10} concentrations exceeded 12–14 times and 8–17 times of NAAQS values of 60 $\mu g/m^3$ and 100 $\mu g/m^3$, respectively on 5–6th November in Delhi city.
- The meteorological conditions were found worst for pollution dispersion on 5–6th November 2016 days which may be one of the reasons for high air pollution build-up.
- The trend of NO_x and SO_2 concentration were found different than $PM_{2.5}$ and PM_{10} at particular locations.

Table 16.4 Graded response action plan impacts matrix for different locations in Delhi city

Sr. No.	Suggested actions[a]	Expected response from GRAP implementation			
		AV	PB	RKP	MM
1.	Stop entry of truck traffic into Delhi (except essential commodities)	H	M	H	M
2.	Stop construction activities	H	M	M	M
3.	Introduce odd and even scheme for private vehicles based on license plate numbers and minimize exemptions	H	M	H	H
4.	Taskforce to take a decision on any additional steps including shutting of schools	M	H	H	H
5.	Close brick kilns, hot mix plants, stone crushers	L	L	L	L
6.	Shut down Badarpur power plant	L	–	–	–
7.	Intensify public transport services	H	H	H	H
8.	Increase frequency of mechanized cleaning of road	H	H	M	H
9.	Stop use of diesel generator sets	H	H	H	M
10.	Enhance the parking fee by 3–4 times	M	M	M	M
11.	Increase bus and metro services by augmenting contract buses and increasing frequency of service	H	H	H	H
12.	Stop use of coal/firewood in hotels and open eateries	M	H	M	M
13.	Residential Welfare Associations and individual house owners to provide electric heaters during winter to security staff to avoid open burning by them	M	M	H	L
14.	Alert in newspapers/TV/radio	H	H	H	H
15.	Strictly enforce/stop garbage burning in landfill and other places and impose heavy fines on the person responsible	H	H	H	M
16.	Close/stringently enforce all pollution control regulations in brick kilns and industries	H	L	L	L
17.	Stringently enforce pollution control in thermal power plants through PCB monitoring	L	L	L	L

(continued)

Table 16.4 (continued)

Sr. No.	Suggested actions[a]	Expected response from GRAP implementation			
		AV	PB	RKP	MM
18.	Do periodic mechanized sweeping on roads with heavy water and traffic sprinkling on unpaved roads every two days	H	H	M	H
19.	Strict vigilance and no tolerance for visible emissions-stop plying of visibly polluting vehicles by impounding or heavy fine	M	M	M	H
20.	Strict vigilance and enforcement of PUC norms	M	L	L	M
21.	Stringently enforce rules for dust control in construction activities and close non-compliant sites	H	H	M	M
22.	Deploy traffic police for smooth traffic flow at identified vulnerable areas	M	M	M	H
23.	Strictly enforce Supreme Court order on diversion of non-destined truck traffic and ensure only trucks registered after 2005 are allowed entry in Delhi	H	M	H	L
24.	Strictly enforce Supreme Court ban on firecrackers	M	H	H	M
25.	Ensure fly ash ponds are watered every alternate day during summer months (March–May)	L	L	L	L
26.	Information dissemination in social media	H	H	H	H
27.	Mobile apps should be used to inform people about the pollution levels, contact details of the control room	H	H	H	H

[a]Severe + or emergency: all actions will be implemented; severe: all actions except 1–4; very poor: all actions except 1–8; moderate to poor: all actions except 1–14

- Local sources around monitoring station influences air quality significantly and accordingly local level action plans also needs to be developed.
- Qualitative evaluation of GRAP indicates that efficacy of a particular mitigation action on pollution reduction can vary from location to location.
- Each GRAP actions needs to be quantified in terms of expected pollution reduction through comprehensive local ward level emission inventory and air quality modeling.

References

CPCB (Central Pollution Control Board) (2010) Air quality monitoring, emission inventory and source apportionment study for Indian cities. National summary report. The Central Pollution Control Board, New Delhi

Dholakia HH, Purohit P, Rao S, Garg A (2013) Impact of current policies on future air quality and health outcomes in Delhi, India. Atmos Environ 75:241–248

GoD (2016) Government of Delhi. Chapter 12, transport, economic survey of Delhi, 2014–15. http://delhi.gov.in/wps/wcm/connect/DoIT_Planning/planning/economic+survey+of+dehli/economic+survey+of+delhi+2014+-+2015

Gulia S, Nagendra SMS, Khare M, Khanna I (2015) Urban air quality management—a review. Atmos Pollut Res 6:286–304

Gulia S, Nagendra SS, Khare M (2017) Extreme events of reactive ambient air pollutants and their distribution pattern at urban hotspots. AAQR. 17(2):394–405

Gupta I, Kumar R (2006) Trends of particulate matter in four cities in India. Atmos Environ 40(14):2552–2566

Guttikunda SK, Goel R (2013) Health impacts of particulate pollution in a megacity—Delhi, India. Environ Dev 6:8–20

Heal MR, Kumar P, Harrison RM (2012) Particles, air quality, policy and health. Chem Soc Rev 41:6606–6630

Kesavachandran CN, Kamal R, Bihari V, Pathak MK, Singh A (2015) Particulate matter in ambient air and its association with alterations in lung functions and respiratory health problems among outdoor exercisers in National Capital Region, India. Atmos Pollut Res 6:618–625

Kumar P, Khare M, Harrison RM, Bloss WJ, Lewis AC, Coe H, Morawska L (2015) New directions: air pollution challenges for developing megacities like Delhi. Atmos Environ 122:657–661

Kumar P, Gulia S, Harrison RM, Khare M (2017) The influence of odd–even car trial on fine and coarse particles in Delhi. Environ Pollut 225:20–30

MoEF&CC (2017) Graded response action plan to combat air pollution in Delhi and National Capital Region. Ministry of Environment, Forest and Climate Change, Govt. of India

Nagpure AS, Ramaswami A, Russell A (2015) Characterizing the spatial and temporal patterns of open burning of municipal solid waste (MSW) in Indian cities. Environ Sci Technol 49:12904–12912

NCR (2013) National Capital Region Regional Plan. Draft revised regional plan 2021, National Capital Region (approved in 33rd meeting of the NCR planning board held on 1st July, 2013), July 2013. National Capital Region Planning Board, Ministry of Urban Development, Government of India

Pant P, Shukla A, Kohl SD, Chow JC, Watson JG, Harrison RM (2015) Characterization of ambient $PM_{2.5}$ at a pollution hotspot in New Delhi, India and inference of sources. Atmos Environ 109, 178–189

Sharma M, Dikshit O (2016) Comprehensive study on air pollution and green house gases (GHGs) in Delhi. Final report. IIT Kanpur

Chapter 17
Stabilizing Different Subgrade Soils with Pond Ash to Lower Greenhouse Gas Emissions for Bituminous Pavements in India

Gaurav Gupta, Hemant Sood, and Pardeep Kumar Gupta

17.1 Introduction

The increase in the population and associated pollution has severely deteriorated the air quality of India, specifically in the urban areas. India stands at number three position across the globe for the amount of greenhouse gas (GHG) emissions released by any nation, trailing ranked two the USA and ranked one China (World Resource Institute 2014). In the year 2014, New Delhi, the capital of India was rated as the world's highest polluted city by the World Health Organization (Report: Ambient Air Pollution Database-Update 2014). The year 2015 reported a higher death toll in India (3283) than China (3233) (Greenpeace Study 2016). India scored 141st place among 180 participating nations in the 2016 Environmental Performance Index, citing the country's performance on top-rated environmental concerns (Yale University 2016). In the year of 2016, India attained a diminished 110th position on sustainable development index (Ryan Swaney 2016), accounting the performance and actions taken by countries to achieve a sustainable environment. In order to curb GHG emissions, engineers urgently need to practice innovative green methodologies for a sustainable environment.

The rapid rise in urbanization has resulted in increased population and traffic, to counter which, construction of new pavements and widening of the existing pavements are required. During the construction of pavements, GHG emissions are

G. Gupta (✉) · H. Sood
National Institute of Technical Teachers' Training and Research, Chandigarh, India
e-mail: gaurav007gupta@gmail.com

H. Sood
e-mail: sood_hemant@yahoo.co.in

P. K. Gupta
Punjab Engineering College, Chandigarh, India
e-mail: p_gupta_2000@yahoo.com

S. M. Shiva Nagendra et al. (eds.), *Urban Air Quality Monitoring, Modelling and Human Exposure Assessment*, Springer Transactions in Civil and Environmental Engineering,
https://doi.org/10.1007/978-981-15-5511-4_17

released in the atmosphere causing a detrimental effect on the environment. Pavements can be either flexible (bituminous pavement) or rigid (concrete pavement) in their behavior. In India, the majority of the pavements constructed are of bituminous type, comprising granular and bituminous layers built over the subgrade layer. As compared to rigid pavements, bituminous pavements are economical to construct but require higher maintenance. This study is limited to discussions on bituminous pavements. This study aims to reduce GHG emissions associated with bituminous pavements by reducing the thickness of the layers overlying the subgrade, due to the better subgrade strength through soil stabilization with pond ash. Six soils, comprising clay, silt and sand found predominantly in the northern part of India, were dealt with in this study.

17.2 Materials Used

Materials used in this study were pond ash, and six types of soils of different plasticity are classified as intermediate plasticity clay (CI), low plasticity clay (CL), low plasticity silt and clay (ML-CL), low plasticity silt (ML), non-plastic silt (MNP) and sand with significant silt content (SM). The soils were classified according to the Indian standards (BIS 1970).

17.2.1 Pond Ash

Coal is the mainstay for the generation of electricity in India and results in the production of an enormous volume of coal ash residue. One hundred thirty million tonnes of coal ash was produced in the year 2006–2007 which is estimated to rise to 600 million tonnes by the year 2020 (Chatterjee 2011). The thermal power plants generate three types of coal ashes viz. fly ash, bottom ash and pond ash. Fly ash refers to the lighter ash particles that rise with the flue gases, while bottom ash refers to the more massive ash particles that settle down in the boiler. Being lighter in weight, fly ash gets carried away with air causing air pollution, whereas bottom ash causes land pollution in the form of leaching. A slurry of water mixed with bottom ash and fly ash is disposed of in nearby areas leading to the formation of ash ponds. A significantly higher amount of pond ash is generated than fly ash and bottom ash (Subbarao and Ghosh 1997). Pond ash occupies more than 200 million m^2 of valuable land in India (Chand and Subbarao 2007). The extensive generation of pond ash causes severe disposal inconvenience alongside numerous ecological and environmental problems, over and above large engaging areas for its storage. Utilization of pond ash would reduce the burden on the environment in addition to reducing the thickness of pavement layers turning out to be an approach toward sustainable construction techniques.

In the current study, gray-colored class-F pond ash (ASTM C 1993) was used, and its geotechnical properties are presented under Table 17.1.

17.2.2 Soils

With an aim to include the variable soil strata encountered in north India, a total of six soils, i.e., CI, CL, ML-CL, ML, MNP and SM were testing for performance evaluation of pond ash stabilization, and their geotechnical properties are reported in Table 17.2.

17.3 Study Methodology

This study aims to curb GHG emissions emitted from the construction and maintenance of bituminous pavement by treating its subgrade with pond ash. The study was performed in two stages, the first being the laboratory studies performed on treated and non-treated soils and the other being the assessment of the potential of pond ash stabilization to lessen the GHG emissions.

17.3.1 Experimental Investigations

A number of experiments were carried out on the soils above mixed with 10, 20, 30 and 40% of pond ash. The contents mentioned are the optimum contents derived from the literature (Kumar and Gupta 2016; Kolay et al. 2011). For assessment of the strength characteristics, CBR tests were performed and compaction characteristics, i.e., optimum moisture content (OMC) and maximum dry density (MDD), were determined (BIS 1983, 1987). The CBR specimens were prepared at OMC, and 97% MDD were soaked in water for four days.

17.3.2 Assessment of GHG Emissions Release from Pavement Construction

17.3.2.1 Production of Construction Materials

The materials necessary for building bituminous pavements are the chief cause of GHG emissions. Embodied carbon dioxide (CO_2) factor for a particular material denotes the life cycle quantity of GHG emissions in equality to CO_2. Despite the fact

Table 17.1 Geotechnical properties of pond ash

Grain size distribution (%)				Plastic limit (%)	Liquid limit (%)	Plasticity index	Specific gravity	MDD (kN/m^3)	OMC (%)
Clay	Silt	Sand	Gravel						
–	45.9	54.1	–	–	–	Non-plastic	2.0	13.1	21.2

Table 17.2 Geotechnical properties of subgrade soils

Properties		CI	CL	ML-CL	ML	MNP	SM
Grain size distribution	Clay (%)	18.7	15.1	9.4	4.3	–	–
	Silt (%)	72.4	71.0	71.9	62.5	59.7	34.9
	Sand (%)	8.9	13.9	18.7	33.2	40.3	65.1
	Gravel (%)	–	–	–	–	–	–
Consistency limits	Plastic limit (%)	24.9	18.6	18.5	15.8	–	–
	Liquid limit (%)	37.3	28.8	24.8	19.1	–	–
	Plasticity index (%)	12.4	10.2	6.3	3.3	–	–
Specific gravity		2.78	2.70	2.68	2.67	2.64	2.66

that the consumption of materials used in construction does not release GHG emissions on the construction site, the progression for their creation tends to emit GHG emissions citing the exhaustive energy procedure. Being aligned with the ongoing research in the construction industry of India, deprived vigilance and un-trustworthy data record form the chief obstacles to achieving decisive embodied CO_2 factors, leading to a small collection of outstanding research. After a rigorous review, the embodied CO_2 factors suggested and determined by Auroville Earth Institute (Maini and Thautam 2009).

The factors emerged relevant and comprehensive, where an extensive collection of materials was studied and their India-specific embodied energy and CO_2 values were determined. The study accounted extraction of raw materials and their processing to generate the end product for the determination of embodied energy and CO_2 emissions. As the embodied CO_2 coefficient for bitumen was not available, the value was obtained from the international literature (Hammond and Jones 2008). The embodied CO_2 factors used in the current study are listed in Table 17.3.

A generalized design mix of bitumen and aggregates (coarse and fine) for different pavement layers (granular sub-base (GSB), wet mix macadam (WMM), dense bituminous macadam (DBM) and bituminous concrete (BC), meeting the specifications of Ministry of Road Transport and Highways, (MoRTH) is reported in Table 17.4 (Ministry of Road Transport and Highways (MORTH) 2013).

The embodied CO_2 factors are worked out mathematically using the standard embodied CO_2 of materials used in pavement construction and the quantities of respective construction materials consumed in pavement construction as shown in equation one below.

Table 17.3 Embodied CO_2 factors for different construction materials

Materials	Coarse aggregate	Fine aggregate	Bitumen
Embodied CO_2 (kg/kg material)	0.0216 (Maini and Thautam 2009)	0.002 (Maini and Thautam 2009)	0.48 (Hammond and Jones 2008)

Table 17.4 Proportioning of different construction materials for the layers of bituminous pavement

Properties	Pavement layers			
	Bituminous concrete	Dense bituminous macadam	Wet mix macadam	Granular sub-base
Density (kg/m^3)	2400	2300	2300	2300
Bitumen (% by mass)	5.5	4.5	0	0
Aggregates (% by mass)	94.5	95.5	100	100
Coarse fraction of total aggregates (% by mass)	55	60	70	80
Fine fraction of total aggregates (% by mass)	45	40	30	20

$$\begin{aligned} &\text{Embodied } CO_2 \text{ for 1 m}^3 \text{ of pavement layer} = \text{Density of the layer}\left(\text{kg/m}^3\right) \\ &* \text{ Volume of layer}\left(1 \text{ m}^3\right) * \left[\text{Bitumen by mass (\%)} * \text{ Embodied } CO_2 \text{ of bitumen (0.48)}\right. \\ &+ \text{ coarse aggregates by mass (\%)} * \text{Embodied } CO_2 \text{ of coarse aggregates (0.0216)} \\ &\left.+ \text{ fine aggregates by mass (\%)} * \text{Embodied } CO_2 \text{ of fine aggregates (0.002)}\right]\text{kg} \end{aligned} \tag{17.1}$$

The sample calculation for the embodied CO_2 released during laying of the BC layer, prepared as per the mix proportioning stated under Table 17.4, has been illustrated in the following section.

Density of BC mix = 2400 kg/m^3
Calculating quantities for a 1 m^3 BC mix,
Mass of BC mix = 2400 kg
Mass of bitumen (5.5% by weigh of total mix) = 132 kg
Mass of aggregates (94.5% by weight of total mix) = 2268 kg
Mass of coarse aggregates (55% by weight of aggregates (2268 kg)) = 1247.6 kg
Mass of fine aggregates (45% by weight of aggregates (2268 kg)) = 1020.4 kg
Multiplying the quantities with their standard embodied CO_2 coefficients

$$\begin{aligned} \text{Embodied } CO_2 \text{ for 1 m}^3 \text{ of BC layer} &= 132 * 0.48 + 1247.4 * 0.0216 + 1020.4 \\ &* 0.002 \text{ kg} = 92.4 \text{ kg Embodied } CO_2 \end{aligned}$$

Similarly, the embodied CO_2 related to other pavement layers is worked out and reported in Fig. 17.1. The reduction in the thickness of pavement layers can be directly related to the reduction in embodied CO_2 for the respective pavement section.

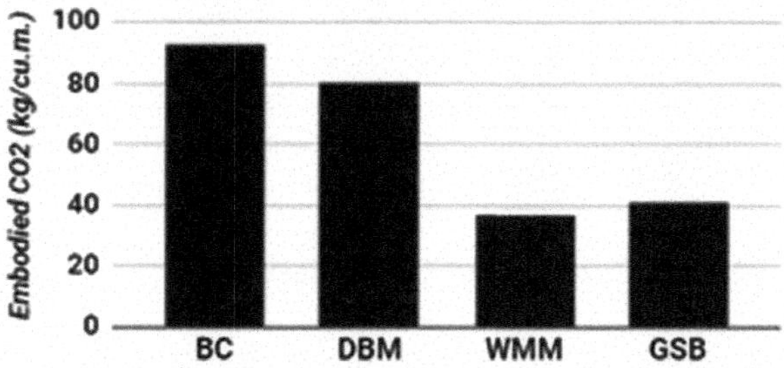

Fig. 17.1 Factors for the embodied CO_2 for different pavement layers

17.3.2.2 Transportation of Construction Materials

The transportation of aggregates (coarse and fine) from quarry sites and bitumen from refineries to the construction site results in the release of GHG emissions. Transportation sector forms one of the significant sources of air pollution due to emissions by vehicular operations. A study conducted by the India GHG Program released India Specific Road Transport Emission Factors (2015) for different categories of vehicles in India (India Specific Road Transport Emission Factors for Stakeholder Consultation 2015). The report illustrates the methodology to arrive at country (India)-specific emission factors to help the Indian corporation to strengthen its GHG accounting process based on fuel efficiency and gross weight of freight vehicles. The freight vehicles were classified into three categories of low-, medium- and high-duty vehicles based on their payload capacity. Table 17.5 denotes emission factors for freight vehicles.

17.4 Results

17.4.1 Experimental Investigation on Pond Ash-Treated Subgrade Soils

Proctor (modified) and CBR tests were carried out to determine the compaction and strength parameters of different subgrade soils. Further, the bituminous pavement

Table 17.5 Emission factors of freight vehicles

Category	kg CO_2/km
Low-duty vehicles (payload capacity <3.5 tonnes)	0.3070
Medium-duty vehicles (payload capacity <12 tonnes)	0.5928
Heavy-duty vehicles (payload capacity >12 tonnes)	0.7375

layer's thickness was designed to rest on pond ash-admixed subgrade soils. GHG emissions associated with these pavements were presented and compared on the grounds of consumption of materials and their haulage requirements. The following sections detail the results for afore-mentioned laboratory tests.

17.4.1.1 Compaction Characteristics

Figures 17.2 and 17.3 depict the results of Proctor (modified) tests that were carried out on different subgrade soils mixed with varying proportions of pond ash. The MDD falls with the gain in pond ash dosage as a result of the replacement of soil particles with lighter (lower specific gravity) pond ash particles. The OMC increased with the increasing contents of pond ash due to the presence of unburnt carbon in the pond ash. As pond ash particles are replacing the soil particles, the total mass of the mixture falls, resulting in a lightweight mixture. These lightweight blends are preferred to be used in the building structures such as retaining walls, due to the reduced earth pressure on the retaining walls.

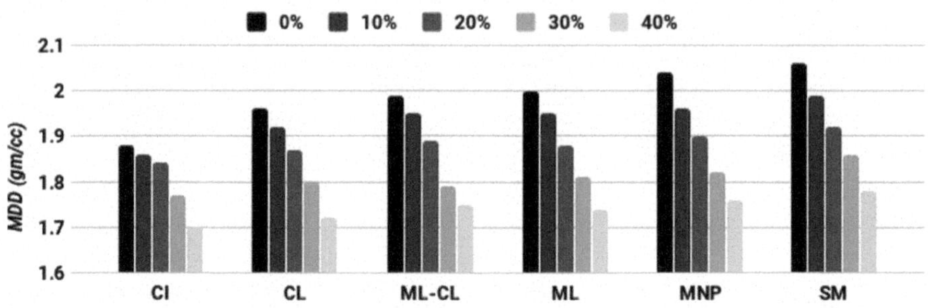

Fig. 17.2 Change in MDD with the contents of pond ash

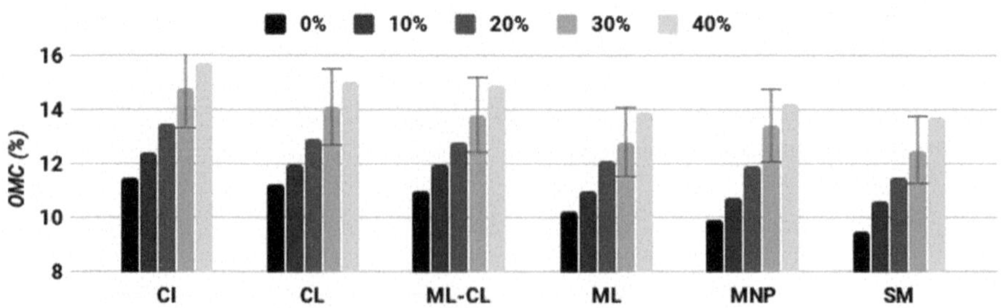

Fig. 17.3 Change in OMC with the contents of pond ash

17.4.1.2 Strength Characteristics

In order to evaluate the strength of the subgrade soil, CBR tests were performed on variable plasticity subgrade soils. Figures 17.4 depicts the results of CBR tests that were performed on different soils mixed with varying dosages of pond ash. The results depict that with the rise of pond ash dosage, the CBR increases up to its optimum content is attained, further than which a decrease in CBR was reported. The optimized pond ash content for different soils was 30% for all the soils.

The higher improvement was observed in the plastic soils (CI, CL, ML-CL and ML soils) than the non-plastic soils (MNP and SM soils). The CBR got enhanced as the non-plastic particles of pond ash substituted the clay particles of soil, eventually reducing the plasticity of the plastic soils. Regarding the non-plastic soils, the role of pond ash was to improve the gradation of the mix due to a better-graded particle size distribution, though the improvement was lower than in case of plastic soils.

17.4.2 GHG Emissions for Bituminous Pavements

GHG emissions were estimated for a 20-year service life of the pavement comprising initial construction stage followed by three periodic maintenance phases which may fall due every fifth, tenth and fifteenth year of pavement service. The fourth periodic maintenance at the end of twenty years was omitted from the study as the pavement would be reconstructed after 20 years. Construction of pavement primarily requires material, their transportation and construction machinery. This study addresses the material, and its transportation issues as the GHG emissions from the construction machinery were found negligible (<1%) in the literature (Kar et al. 2015).

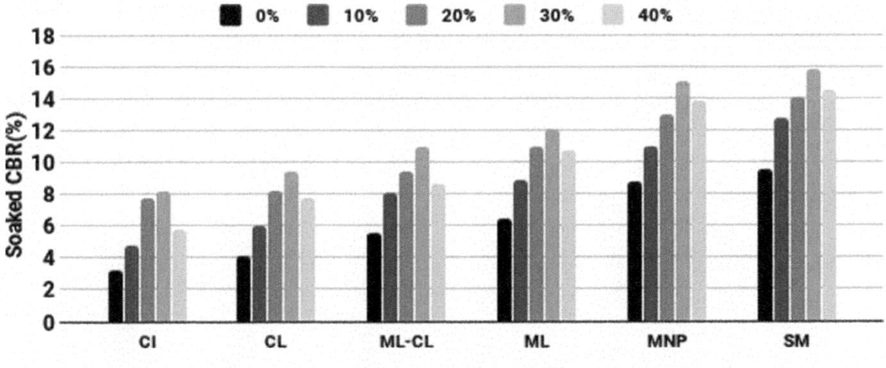

Fig. 17.4 Change in soaked CBR with the contents of pond ash

17.4.2.1 Construction of Bituminous Pavements

The bituminous pavements were designed following the guidelines laid down by the Indian Road Congress (Congress 2012). Pond ash-stabilized subgrade pertained higher CBR leading to diminished pavement thickness. The road section was expected to carry 150 million standard axles traffic, having four lanes, each of 21-m width, running the length of a unit km. Figures 17.5, 17.6, 17.7, 17.8 and 17.9 depict the thickness of BC, DBM, WMM, GSB layers and total pavement with pond ash-stabilized and -non-stabilized soils.

Figure 17.10 represents the embodied CO_2 from construction materials during the construction phase for a unit-kilometer-length road on pond ash-treated and -non-treated subgrades. Results depict that pond ash-stabilized soils had reduced GHG emissions due to the minimized need of construction materials. It could be inferred that the reductions were maximum for CI soil (25%) and minimum (7%) for MNP and SM soil.

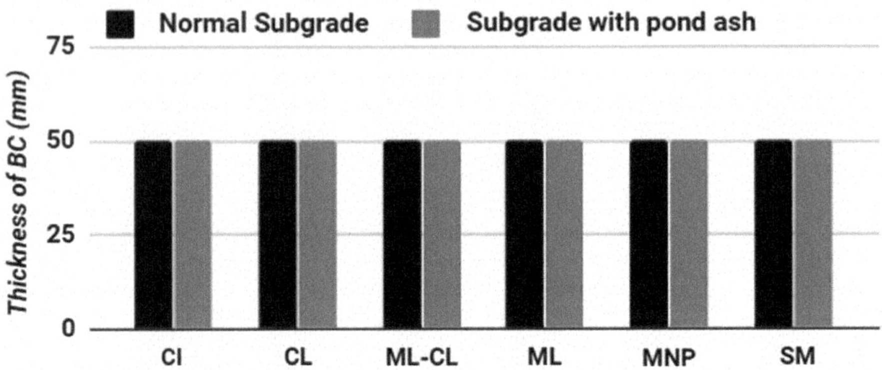

Fig. 17.5 Thickness of the BC layer for natural and pond ash-stabilized subgrade

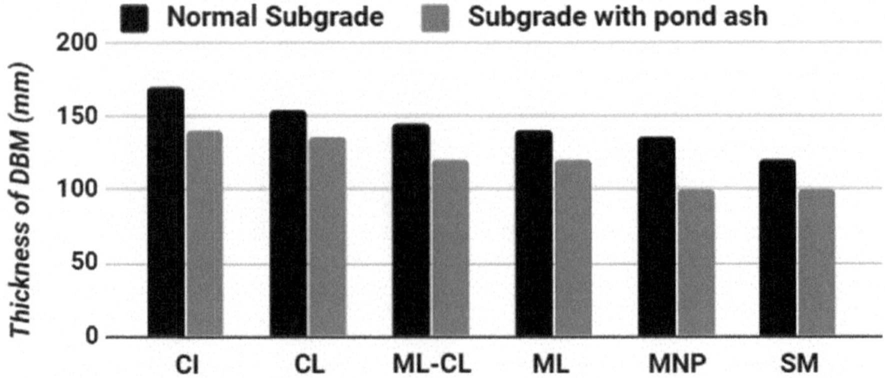

Fig. 17.6 Thickness of the DBM layer for natural and pond ash-stabilized subgrade

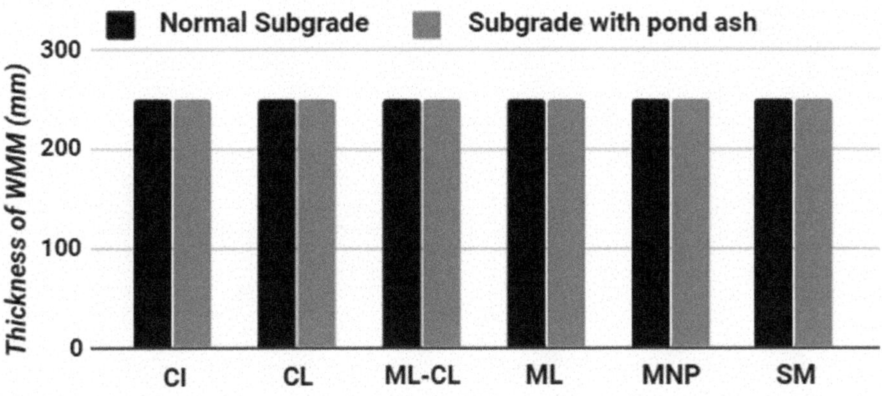

Fig. 17.7 Thickness of the WMM layer for natural and pond ash-stabilized subgrade

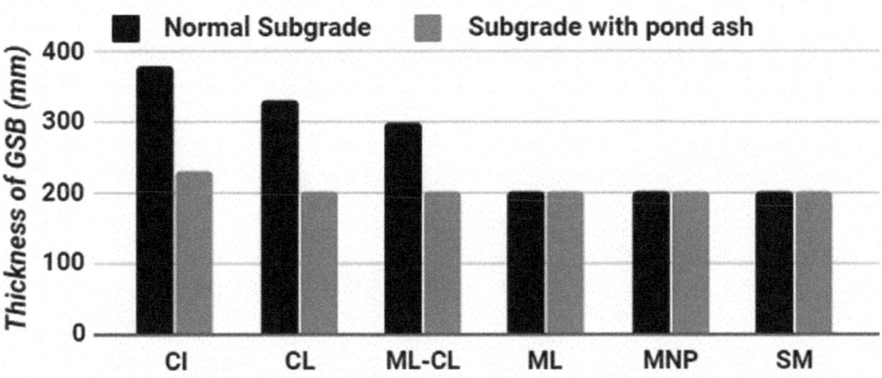

Fig. 17.8 Thickness of the GSB layer for natural and pond ash-stabilized subgrade

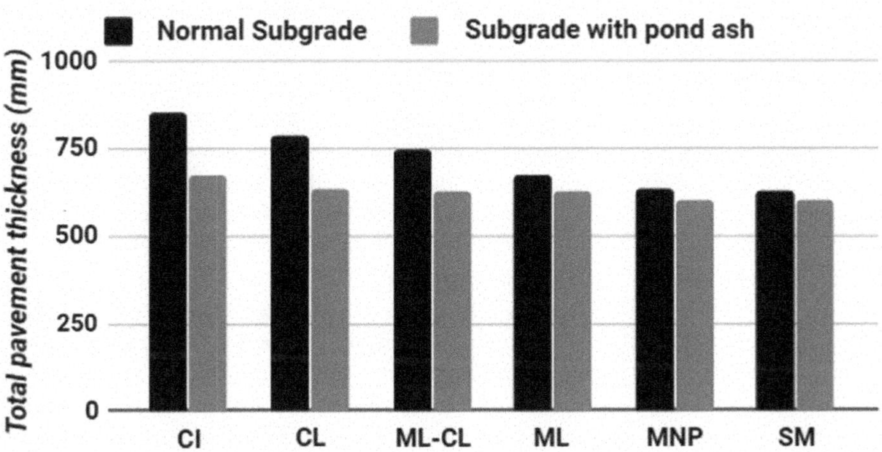

Fig. 17.9 Total pavement thickness for natural and pond ash-stabilized subgrade

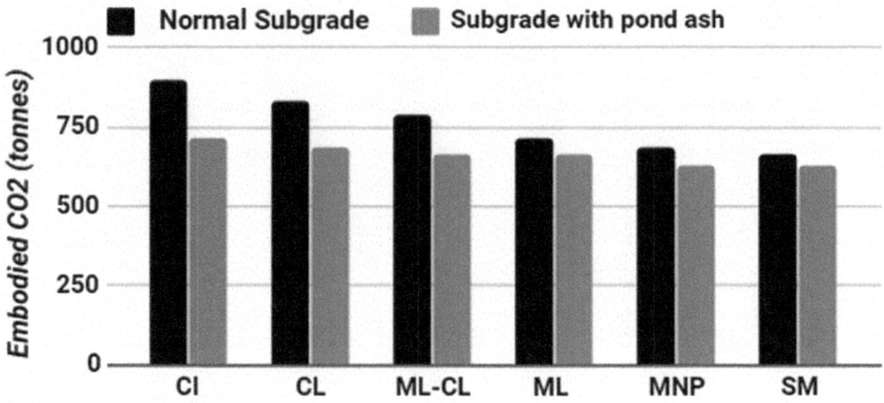

Fig. 17.10 Embodied CO_2 from construction materials during the construction phase

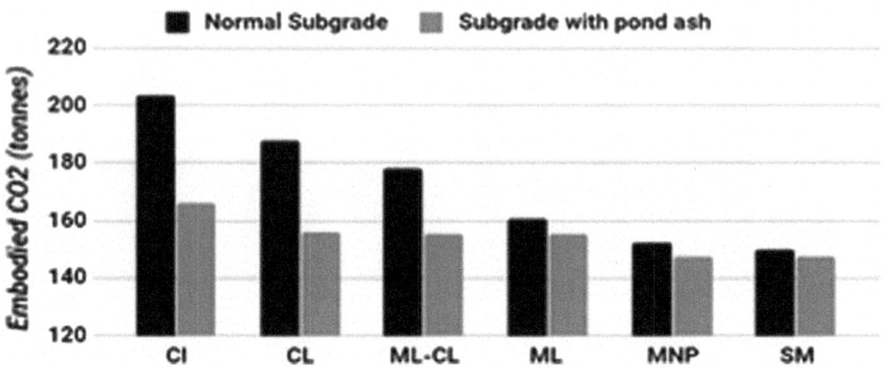

Fig. 17.11 Variation in embodied CO_2 due to the transportation for the construction stage

Figure 17.11 depicts the GHG emitted by the transportation of bitumen and aggregates from their source to the construction site for the construction project described above. Results depict that pond ash-stabilized soils had diminished need for the transportation of construction materials along with added haulage requirement for the transportation of pond ash. The GHG emissions reduced moderately for plastic soils and insignificantly for non-plastic soils as the reduced haulage for materials were counterbalanced by increased haulage for pond ash. As this study is of a general nature, a constant haulage distance of 50 km is assumed for all the materials including pond ash. The lead distance would be highly variable and site-specific for real-time construction projects. The effect of transportation would be significant in case of the construction sites lying in the close vicinity of thermal power plants. The authors suggest the practice of pond ash stabilization in construction projects which have distant quarries and refineries with thermal power plants located nearby.

17.4.2.2 Maintenance of Bituminous Pavements

A bituminous pavement necessitates periodic maintenance for sustained serviceability. The periodic maintenance schedule for bituminous roads comprises laying of a renewal coat to the wearing surface at a predetermined frequency that is usually determined based on the present deteriorated condition of pavement due to traffic movement and temperature changes. Based on literature review and inputs from site engineers and contractors, a bituminous concrete overlay of 40 mm thickness was proposed with a periodicity of 5 years.

Figures 17.12 and 17.13 represent the GHG emission from consumption of construction materials and their transportation for laying a bituminous concrete overlay of 40 mm thickness after 5, 10 and 15 years of pavement construction. As the thickness of overlay is independent of the pavement subgrade characteristics, the GHG emissions due to the transportation of bitumen and aggregates used in the overlay construction and its transportation are the same for all soil types. The

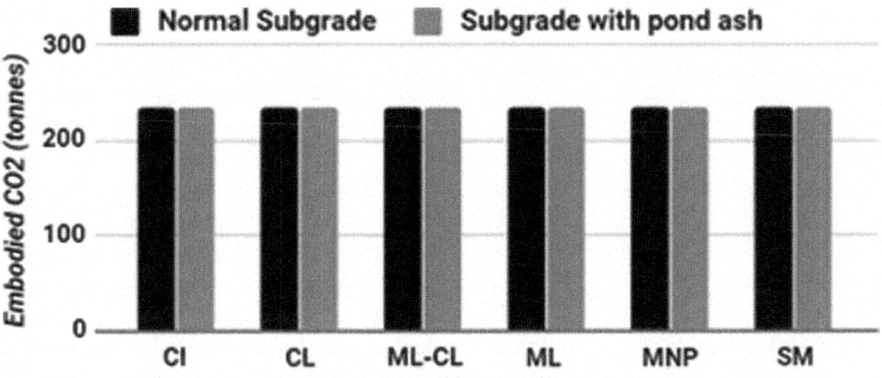

Fig. 17.12 Embodied CO_2 from construction materials during the maintenance phase

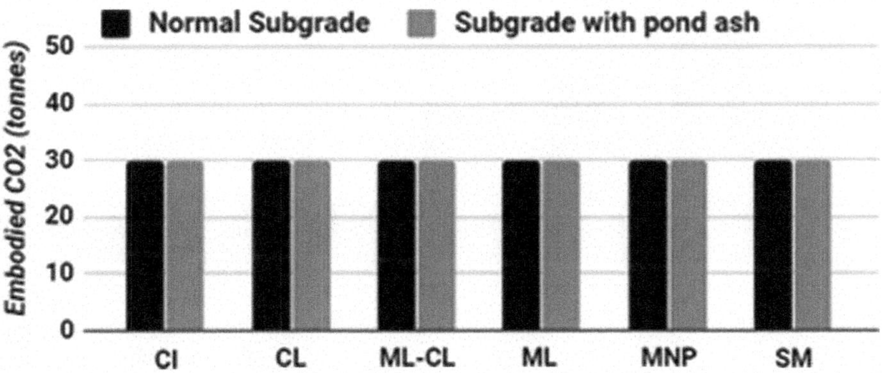

Fig. 17.13 Embodied CO_2 from transportation of materials during the maintenance phase

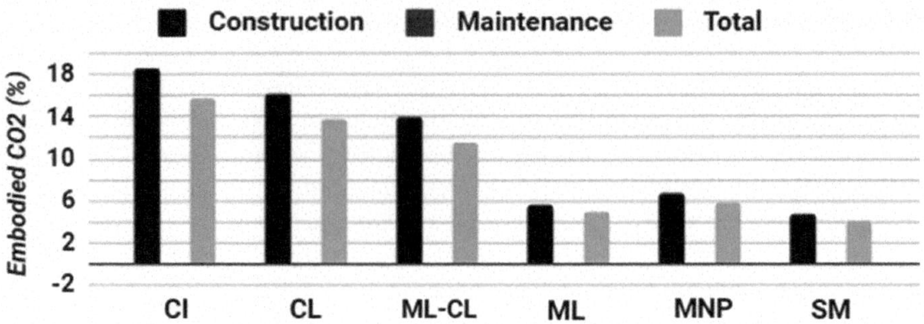

Fig. 17.14 Percentage reduction in CO_2 emissions for the construction, maintenance and total service life of bituminous pavements

percentage reduction of GHG emissions shown in Fig. 17.14 depicts the sustainable benefits of pond ash stabilization.

17.4.3 Carbon Credits

Kyoto Protocol (Protocol 1997) was agreed upon with the consent of assorted countries to mitigate the release of CO_2 emissions for a sustainable environment. Diminution of a unit tonne of CO_2 earned one carbon credit, accepted as a worldwide exchange and has an equivalent international value of 24 € (Shende and Jadhao 2014). Enumerations of the carbon credits earned from pond ash stabilization were calculated to assess the effect of the proposed stabilization on the environment. Figure 17.15 depicts the accretion of carbon credits from pond ash stabilization, leading toward constructing a sustainable environment.

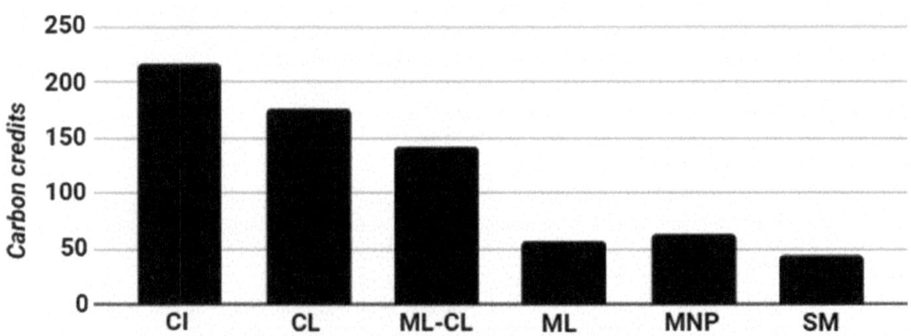

Fig. 17.15 Carbon credits amassed along the life cycle of the bituminous pavement

17.5 Conclusions

The present study determined the potential of pond ash to mitigate GHG emissions from the construction of pavement through stabilization of variable plasticity soils found typically in north India. The result from laboratory tests shows that as the pond ash increases, the MDD decreases and OMC increases for all type of soils. The CBR of the plastic soils treated with the optimum content of pond ash showed a noteworthy gain in the CBR, whereas the consequence was comparatively restrained for the non-plastic soils. The enhanced CBR resulted in minimized thickness of the pavement layers, saving construction material and associated haulage requirements. The mitigation in GHG emissions from materials and transportation during the initial construction stage was highest for the CI soil and lowest for the SM soil. The overall maintenance period released equal GHG emissions. The GHG emissions released during service life were curbed by 15.8% for CI soil, 13.8% for CL soil, 11.6% for ML-CL soil, 4.9% for ML soil, 5.8% for MNP soil and 4.1% for SM soil.

References

ASTM C 618 (1993) Specification for fly ash and raw or calcined natural pozzolana for use as a mineral admixture in portland cement concrete. Annual Book of ASTM Standards, Section, 4, pp 310–312

BIS I (1970) 1498 Classification and identification of soils for general engineering purposes

BIS I (1983) 2720 methods of test for soils: Part 8 Determination of water content-dry density relation using heavy compaction

BIS I (1987) 2720 methods of test for soils: Part 16 Laboratory determination of CBR of soil (second revision)

Chand SK, Subbarao C (2007) Strength and slake durability of lime stabilized pond ash. J Mater Civ Eng 19(7):601–608

Chatterjee AK (2011) Indian fly ashes: their characteristics and potential for mechanochemical activation for enhanced usability. J Mater Civ Eng 23(6):783–788

Congress IR (2012) Guidelines for the design of flexible pavements. Indian code of practice, IRC, p 37

Greenpeace Study (2016) Deaths due to outdoor pollution in India and China. In: Greenpeace India Report. Available online http://bit.ly/AirPollutionGreenpeaceIndiaReportNov2016. Accessed 26 Jan 2017

Hammond G, Jones C (2008) Inventory of carbon and energy: ICE. Sustainable Energy Research Team, Department of Mechanical Engineering, University of Bath, Bath

India Specific Road Transport Emission Factors for Stakeholder Consultation (2015) In: India GHG Program, Mumbai, India. Available online http://indiaghgp.org/sites/default/files/Road%20Transport%20Technical%20Paper.pdf Accessed 26 Jan 2017

Kar SS, Behl A, Shukla A, Jain PK (2015) Estimation of carbon footprints of bituminous road construction process. J Civil Environ Eng 5(198):2

Kolay PK, Sii HY, Taib SNL (2011) Tropical peat soil stabilization using class F pond ash from coal fired power plant. Int J Civil Environ Eng 3(2):79–83

Kumar A, Gupta D (2016) Behavior of cement-stabilized fiber-reinforced pond ash, rice husk ash–soil mixtures. Geotext Geomembr 44(3):466–474

Maini S, Thautam V (2009) Embodied energy of various materials and technologies. Auroville Earth Institute (AEI), Tamil Nadu
Ministry of Road Transport and Highways (MORTH) (2013) MORTH specifications for road and bridge works, 5th revision, India
Protocol K (1997) United Nations framework convention on climate change. Kyoto Protocol, Kyoto, p 19
Report: Ambient Air Pollution Database-Update (2014) Available online: http://www.who.int/phe/health_topics/outdoorair/databases/AAP_database_results_2014.pdf Accessed 23 Jan 2017
Ryan Swaney (2016) A global report: SDG index and dashboards. Available online https://www.dropbox.com/s/gy2zmh9065v0mr5/SDG_Index_Dashboard_full.pdf?dl=0%20Sustainable%20Development%20Solutions%20Network Accessed 29 Jan 2017
Shende BR, Jadhao RS (2014) Carbon credit science and business. Sci Rev Chem Commun 4(2)
Subbarao C, Ghosh A (1997) Fly ash management by stabilization. J Solid Waste Technol Manag 24(3):126–130
World Resource Institute (2014) Climate analysis indicators tool, Version 2.0. Available online http://www.wri.org/resources/data-sets/cait-historical-emissions-data-countries-us-states-unfccc.Accessed 02 Jan. 2017
Yale University (2016) Report: global metrics for the environment. Available online https://issuu.com/2016yaleepi/docs/epi2016_final. Accessed 21 January 2017

Chapter 18
Strategic Approach for Emission Reduction from Coal-Fired Thermal Power Plants in India

Shipra and Asim Kumar Pal

18.1 Introduction

Secure, reliable and affordable energy supplies are fundamental to global economic stability and growth. Energy consumption is directly related with the development of the society. Being a developing country, consumption of the energy in India is increasing very fast. India is the third largest producer of electricity.[1] In India per person per year electricity consumption is 957 kWh which is very less compared to world average electrical energy per capita that is about 2700 kWh.[2] As the country is developing the energy requirement of the society will enhance which needs to be met in upcoming years. Since consumption of energy is increasing, there is a need to give more stress on development, thereby consuming more energy with associated higher emission of carbon. India needs economic development that is conducted without damaging the natural resources which in turn requires energy efficiency, i.e., to use the limited resources in such a manner that energy demand is fulfilled. More than 61.44% of India's energy demand is met through the coal-based thermal power plant (Central Electricity Authority Report 2016).

[1] **Businessstandard.** http://www.business-standard.com/article/economy-policy/now-india-is-the-third-largest-electricity-producer-ahead-of-russia-japan-118032600086_1.htm.

[2] **User guide for India's 2047 energy calculator coal and gas power stations.**

Shipra (✉)
Electricity Supply Sub Division Motipur, North Bihar Power Distribution Company Ltd., Muzaffarpur, Bihar 843111, India
e-mail: nehushipra@gmail.com

A. K. Pal
Department of Environmental Science and Engineering, IIT(ISM), Dhanbad, India
e-mail: palasim2003@yahoo.co.in

S. M. Shiva Nagendra et al. (eds.), *Urban Air Quality Monitoring, Modelling and Human Exposure Assessment*, Springer Transactions in Civil and Environmental Engineering,
https://doi.org/10.1007/978-981-15-5511-4_18

Table 18.1 Regionwise total Capacity of coal-based thermal power plant in (MW) in India as on 31.03.2016

Regions	Sector	Coal-Based
Northern Region	State private central (total)	**45644.50**
Western Region	State private central (total)	**72153.01**
Southern Region	State private central (total)	**36442.50**
Eastern Region	State private central (total)	**30622.87**
North Eastern Region	State private central (total)	**310.00**
All India	State	64320.50
	Private	69462.38
	Central	51390.00
	Subtotal	185172.88

Source CEA Report (2016)

National Thermal Power Corporation (NTPC), Public sector undertaking and several other state level power generating companies are engaged in operating coal-based thermal power plants. Apart from NTPC and other state level operators, some private companies are also operating the power plants. All India installed capacity of coal-based thermal power plants as on 31.03.2016 is shown in Table 18.1. India had estimated coal reserves of 306.6 billion metric tons, the fifth largest coal reserves in the world (EnergyStatistics 2016). The production of coal in the country has grown from a little over 341 million tons in 2002–03 to over 557.5 million tons in 2012–13. The growth in production of coal in the 10th plan was 5.6% and in the 11th Plan was 4.6% and in the year 2014–15, the production of coal was 612.44 million tons in a growth of 8.25% over the previous year (Prasad and Anand 2013).

However, the rapid growth in coal production has not been sufficient to meet the demand for coal, which has resulted in demand gap. These gaps are met through imports of coal, as per the economics Times report India's coal import increase to 35% in September from previous fiscal year. A Summary of coal production and consumption in India from 2001 to 2013 in (million tons) is shown in Table 18.2.

As per Annual Report 2015–16 Ministry of Coal, Govt. of India, 70% of the coal is consumed in power sector, 7% in steel industry, 3% in cement industry and remaining 20% in other area, most of these are high ash content coal in the calorific value range of 3000 kilo calorie per kilogram to 4500 kilo calorie per kilogram and ash content in the range of 30–45%. Using the high ash coal for the power sector is a major challenge, from the point of view of achieving high level of efficiency of consumption, and more particularly, from the point of view of environmental management due to fly ash emissions. Release of high quantum of pollutants from coal-based power plants contributes significant impacts to health and environment. Coal combustion during electricity generation releases various pollutants into the

Table 18.2 Summary of coal production and consumption (million tons) in India

Year	Production	Consumption
2001	385.39	406.17
2002	401.18	422.68
2003	425.97	435.68
2004	452.38	484.18
2005	482.00	511.05
2006	509.40	543.71
2007	541.30	592.56
2008	578.91	639.54
2009	624.03	716.76
2010	628.79	746.26
2011	641.86	782.86
2012	665.69	853.86
2013	675.47	886.05

Source http://www.indexmundi.com/

environment. As a consequence, emissions of green house gases and other pollutants are increasing in India with the increasing demand for electricity. Though there are different sources of electricity generation major part of electricity generated from coal-fired power plants and result a tremendous produce the highest rate of CO_2 per kilowatt hour (Department of Energy and Environmental Protection Agency, Washington DC 2000). Evidence has shown that the climate change will affect the natural resources in different ways, which will ultimately threaten the livelihoods of the most poor and marginalized sector of the population who are closely tied to India's natural resource base. During the combustion of coal in thermal power plant mercury present in the coal is released. Mercury is one of the most dangerous heavy metal for human being as well as ecosystem. Mercury is invisible hazardous pollutants; whose effects are visible after a long time so it becomes very difficult to identify its effects.

The important problem in this world is the reduction of non renewable energy source and increase in the pollution level due to power plant. Both situations can be controlled if the efficiency of the energy system is improved. Every power plant losses their efficiency due to its continuous operations, age and many more reasons. This efficiency deterioration causes an increase in CO_2 emission. Economics of power generation does not only require designing an efficient power plant, but also following proper operation and maintenance strategy such that the energy conversion efficiency of the plant throughout its life cycle remains high.

In terms of carbon emission the contribution of Coal-based power plants is approximately 60% of the total carbon emissions in India. In 2014, a study carried out by Guttikunda and Jawahar (2014), about the status of operational coal-fired thermal power plants in India. The key findings from that study are as summarized below:

- The total coal consumption at the 111 coal-fired power plants was 500 million tons for the year 2010–11, leading to significant emissions of PM, SO_2, NO_2, CO_2, and VOC.
- **The higher $PM_{2.5}$ concentration resulted deaths of more than 80,000 people.**
- **The emissions standards practiced here were at-least 5–10 times worse than those practiced in developed countries like US, China, Australia.**

Technology used in the power plants are mostly subcritical type so emissions from coal-based thermal power plants in the country are increasing day by day and it will continue to go up in the near future. These plants are inefficient and release considerable air pollution in the surrounding atmosphere. In order to reduce air pollution by these coal-fired thermal power generation units, Ministry of Environment Forests and Climate Change (MoEF and CC) has formulated stricter pollution control norms for these units. As per the guideline released by MoEF and CC Indian power plants will have to cut the emission of particulate matter by 25%, Sulfur dioxide by 90%, oxide of nitrogen by 70% and mercury by 75% to improve air quality status in and around thermal power plant. As per this standard, power plant is categories into three different section.

- Started before December 31, 2003.
- Started after 2003 up to December 31, 2016.
- Going to be start after December 31, 2016.

In the earlier rules particulate matter was allowed up to 150 mg per normal cubic meter for units having generation capacity of 210 MW or more and 350 mg per Nm^3 for units less than 210 MW capacity. Now for the first category the permissible limit is 100 mg per Nm^3, 50 mg for the second and 30 mg for new units. The main purpose of imposing the rule is to control emissions of particulate matter (PM), sulfur dioxide, nitrogen oxides and mercury and also cut water use by coal-based thermal power plants. The proposed standard of Ministry of Environment and Forest and Climate Change is shown in Table 18.3.

There is no standard limit of CO_2 emission from thermal power plant. Major contributor of CO_2 to the environment is thermal power plants. To improve the

Table 18.3 MOEF CC proposed standard for coal-fired thermal power plants

Pollutants	Older units		Older new		Future
	Before December 31, 2003		After 2003–2006		From January, 2017
	<500 MW	≥ 500 MW	<500 MW	≥ 500 MW	
PM	100 mg/Nm^3		50 mg/Nm^3		30 mg/Nm^3
SO_2	600 mg/Nm^3	200 mg/Nm^3	–	200 mg/Nm^3	100 mg/Nm^3
NO_X	600 mg/Nm^3		300 mg/Nm^3		100 mg/Nm^3
Hg	–	0.03 mg/Nm^3	0.03 mg/Nm^3		0.03 mg/Nm^3

Source www.moef.nic.in

condition of coal-fired plant, measurements of these pollutants are very much essential so that a necessary major can be taken for the reduction of such harmful gases. In developing country like India Power generating facilities are quite old and very poorly maintained due to which the carbon emissions are increasing. The need to mitigate these environmental impacts has driven the development of a variety of technologies designed to burn coal more cleanly and efficiently. From this scenario Clean Coal Technologies comes into picture. The term Clean Coal Technologies (CCT) is used to mean every option capable of reducing emissions upstream, downstream, or within the power generation process. Adoption of different clean coal technology such as supercritical (SC) stream cycle, ultra-supercritical stream cycle (USC), integrated gasification combined cycle (IGCC) will increase the efficiency of the plant. These technologies are more environmentally benign and would help meet increased energy demands. These are highly efficient technologies and consequently use less coal hence released less amount of pollutant to the environment.

18.2 Materials and Methods

Population growth (existing as well as projected) of the country for the year 2010–2020 was collected from statistical portal of India. A research analysis carried out by Energy and Resources Institute (TERI) about the Indian energy scenario and they have suggested a tactical approach for sustainable development. As per their research, the electricity demand, growth rate was taken as 5.7%/year during the period 1997–2019 and 3.9%/year during the period 2020–2047. The same growth rate was used in this study to forecast electricity demand. Population of India from 2010 to 2020 (existing as well as projected) and projected electricity demand is shown in Table 4.15. This table shows the trend of annual electricity demand in India, nearly 63% of the total electricity demand met through coal-based thermal power plant. Heat rate (i.e., heat required to produce one unit of electricity), calorific value of coal and efficiency of coal-based thermal power plant were used to calculate the amount of coal required to produce 1 kW of electricity and on the basis of this, total annual coal requirement of coal-fired plant was calculated. Indian thermal power plants use bituminous/sub-bituminous coal with gross calorific value ranging in between 3000 and 5000 kcal.

$$\begin{aligned} &\text{Average calorific value taken in this study} \\ &= (3000 + 5000)/2 \\ &\approx \mathbf{4000 kcal/kg} \\ &= 16800\ \text{kJ/kg} \quad (1\ \text{kcal/kg} = 4.2\ \text{kJ/kg}) \end{aligned} \tag{18.1}$$

In this study it was assumed that energy conversion in power plant is 100% efficient hence one unit of electricity generation requires 3600 kJ of energy. Average efficiency of coal-based thermal power plants of different combustion technology is

Table 18.4 Average efficiency of coal-based thermal power plant of different technology

Combustion technology	Efficiency
Subcritical	32.8[a]
Supercritical	38.5[b]
Ultra-supercritical	43.3[b]

Source [a]Centre for Science and Environment (February 21, 2015)
[b]"The Future of Coal", MIT (2007)

Table 18.5 Heat rate and per kWh coal requirement of different technology

Combustion technology	Heat rate	Coal requirement in kg per kWh
Subcritical	10975.60976	0.653310105
Supercritical	9350.649351	0.556586271
Ultra-supercritical	8314.08776	0.494886176

shown in Table 18.4.

$$\text{Heat rate} = \frac{3600\frac{\text{kJ}}{\text{h}} = 10975.6\ \text{kJ/h}}{32.8} \tag{18.2}$$

To produce 1 kWh of electricity coal requirement in kg/kWh is

$$= \frac{10,975.6\ \text{kJ/h}}{16,800\ \text{kJ/kg}}$$
$$= 0.653\ \text{kg of coal per kWh} \tag{18.3}$$

As per the above information heat rate and amount of coal required per kWh are calculated as shown in Tables 18.5 and 18.6.

Table 18.6 Total projected electricity demand in India from 2010 to 2020

Year	Population (billion)	Electricity demand (billion kWh/year)
2010	1.19	927.01
2011	1.21	979.85
2012	1.24	1035.70
2013	1.26	1094.73
2014	1.28	1157.13
2015	1.29	1223.08
2016	1.31	1292.79
2017	1.32	1366.47
2018	1.34	1444.35
2019	1.36	1526.67
2020	1.38	1583.20

Table 18.7 Amount of coal used in thermal power plants

Electricity generated from coal-based thermal power plant (MW)	Year	Coal used (million tons/year)
66668.53	2010	381.36
70468.66	2011	403.10
74485.27	2012	426.08
78730.58	2013	450.36
83218.25	2014	476.03
87961.23	2015	503.16
92974.62	2016	531.84
98273.53	2017	502.15
103874.50	2018	594.19
109794.80	2019	628.06
113860.30	2020	651.31

Amount of coal used in the coal-fired plant was calculated based on above information as shown in Table 18.7.

18.3 Result and Discussion

Emission rate of CO_2 was calculated using the formula given by Coal Swarm coverage of India and coal (2016). As per this formula

$$\text{Emissions per kWh} = CO_2 \text{ emissions per unit of energy in coal (expressed in Btu)} \times (\text{Btu to kWh})/(\text{Plant efficiency}) \tag{18.4}$$

In India most of the coal are of sub-bituminous rank, approximately 214.3 lb of CO_2 is emitted per million British Thermal units (Btu) of energy for sub-bituminous coal (US Energy Information Administration).

For 2010 to 2020, emission rate and the estimated annual CO_2 emissions are evaluated as presented in Tables 18.8 and 18.9.

The annual emission data in Table 18.9 shows that CO_2 emissions are increasing at rate of 5% per year on all India basis in these power plants during the period 2010–2020. A research estimate total annual CO_2 emissions from all the coal-fired power plants in India as 395 million tons in 1997–98 (Mittal and Sharma 2003).

Table 18.8 Emission rate for different technology (kg of CO_2/kWh)

	Subcritical	Supercritical	Ultra-supercritical
Emission rate	1.01	0.85	0.76

Table 18.9 Summary of predicted annual CO_2 emission (million tons CO_2/year) in different technology at the power plants in India in 2010–20

Year	Subcritical	Supercritical	Ultra-supercritical
2010	589.86	496.41	443.85
2011	623.48	524.71	469.15
2012	659.02	554.62	495.89
2013	696.58	586.23	524.16
2014	736.28	619.64	554.03
2015	778.25	654.96	585.61
2016	822.60	692.29	618.99
2017	869.48	731.74	654.27
2018	919.04	773.45	691.55
2019	971.42	817.53	730.97
2020	1007.39	847.80	758.04

These estimates were based on the installed plant capacity, since the actual electricity generation data was not available at that time. Plant having subcritical technology releases heavy CO_2 but penetration of supercritical and ultra-supercritical technology helps in 19 and 33% emission reduction respectively.

Indian coals have less sulfur content (0.3–0.55%) compared to coal import from Indonesia and other country. Emission of SO_2 mostly depend upon the sulfur content in the coal whereas emissions of CO_2 and NO depends on the operating conditions and the design type of the plant (Mittal et al. 2012). Since sulfur contains of coal is very low so there is of no practical significance in reducing SO_2 emissions to the atmosphere (Rees et al. 1966). Hence all the sulfur in the coal is considered to have been converted to SO_2. Table 18.10 shows the estimated SO_2 and NO emissions from thermal power plants in India using the emission factor of SO_2 (0.00521 kg/kWh) and NO (0.00154 kg/kWh) (Chakraborty et al. 2008). Annual SO_2 emission has

Table 18.10 Summary of SO_2 and NO emissions (million tons) during 2010 to 2020 from coal-fired plant

Year	SO_2	NO
2010	3.04	0.90
2011	3.22	0.95
2012	3.40	1.00
2013	3.59	1.06
2014	3.80	1.12
2015	4.01	1.19
2016	4.24	1.25
2017	4.49	1.33
2018	4.74	1.40
2019	5.01	1.48
2020	5.20	1.54

Table 18.11 Estimated Hg emissions (tons) during 2010 to 2020 from coal-fired plant

Year	Coal used (million tons/year)	Hg emission (tons/year)
2010	381.36	202120.8
2011	403.10	213643.0
2012	426.08	225822.4
2013	450.36	238690.8
2014	476.03	252295.8
2015	503.16	266674.8
2016	531.84	281875.2
2017	502.15	266139.5
2018	594.19	314920.7
2019	628.06	332871.8
2020	651.31	345194.3

increased from 3.04 million tons in 2010 to 5.20 million tons in 2020 at an average annual rate of 0.22 million tons per year and annual NO emission has increased from 0.9 million tons in 2010 to 1.54 million tons in 2020 at an average annual rate of 0.06 million tons per year.

Coal-fired thermal power plants are the 2nd largest mercury emission in India (Rai et al. 2013). The residence time of Hg in elemental state varies between 0.5 and 2 years whereas in oxidized and particulate form much less (Schoeder and Munthe 1988). Hg contents in coal varies from 0.01 to 0.5 mg/kg depends upon the type of coal used in combustion (Pacyna et al. 2006). In this study emission factor of mercury is considered as 0.53 mg/kg (Rai et al. 2013) for estimating the total Hg emission from coal-fired thermal power plants.

Mercury estimation from the Indian Thermal power plants was done on the basis of annual consumption of coal and emission factor (EF) of Hg used in this sector.

$$(\text{EHg})\text{Year} = (\text{coal consumption})\text{Year} \times (\text{EF})\text{Hg} \quad (18.5)$$

where

$(E_{Hg})_{Year}$ = Annual emission of mercury (million tons) from thermal power plants.

Annual Hg emission has increased from 202.12 tons in 2010 to 345.19 tons in 2020 at an average annual rate of 14.3 tons per year. Total estimated emission of mercury from coal-fired thermal power plants as shown in Table 18.11.

18.3.1 Cost Analysis

To compare the cost of electricity production of different technologies, the total specific cost of each technology is calculated over the entire lifetime of the power plant and an average levelized cost of energy (LCOE) is calculated. Levelized cost of energy measure the lifetime cost involves in energy production. It allows the comparison of different technologies of unequal life spans and capacities, so that the minimum price at which energy can be sold will be determined. The cost for electricity produced by different power generation technologies is calculated under different assumptions. For simplification the methodology applied is based on a single input parameter overall years considered.

$$\text{LCOE} = \frac{\text{Capital cost} \times \text{CRF}(1 - \text{TDpv})}{8760 \times \text{CF} \times (1 - T)} + \frac{\text{Fixed OM}}{8760 \times \text{CF}} + \text{Variable OM} + \text{Fuel price} \times \text{Heat rate} \quad (18.6)$$

Source: Transparent cost database (4/14/2016**) where,**
Capital cost = Cost of the plants CRF = Capital expenditure.

$$\text{CRF} = D(1 + D)N/((1 + D)N-1) \quad (18.7)$$

Source: Transparent cost database (4/14/2016) where,
D = Discount rate, i.e., used to attribute a value to future cash flows.
N = Lifetime of the investment, T = Tax rate, Dpv = Depreciation, CF = Capacity factor (The ratio between average power and peak capacity), In this study it was assumed that plant is running at full capacity for exactly 80% time in a year.

Fixed OM = Fixed operation and maintenance cost includes staffing and other costs that are independent of operating hours.

Variable OM = Variable operation and maintenance cost, which are a function of operating hours Assumption, Source: Transparent cost database (4/14/2016).

Dpv = 0.59, D = 7%, N = 30 Years, 8760 = Number of hours in a year, Fixed OM = 0.0048 $/kWh

Variable OM = 0.022 $/kWh, Fuel price = 2.34 $/mmBTU, CF = 80%, T = 33%, Heat rate = 0.01008 mmBTU CRF = 0.08 (Calculated)

There are different modes of levelized cost calculation but this study is based on normalized mode in which single discount rate and other cost for all technologies are considered. Capital cost of subcritical, supercritical and ultra-supercritical is 1150, 1350 and 1190 $/kw.[3]

Calculated levelized cost of estimation for different technology is shown in Table 18.12.

[3]**Business and public administration studies**: https://www.bpastudies.org/bpastudies%20/article/view/170/318.

Table 18.12 Calculated power generation cost in different combustion Technology

Technology	LCOE (Rs/kWh)
Subcritical	4.03
Supercritical	4.32
Ultra-supercritical	4.06

18.4 Conclusion

Population of India is increasing rapidly with respect to population, electricity demand is also increasing and to overcome the demand electricity generation will also increase. Total generation from coal-fired plant is expected to increase two folds from 66,668 MW in 2010 to 113,860 MW in 2020 and total coal consumption is estimated to increase 1.5–2 times from 380 million tons/year to 651 million tons/year, accordingly the CO_2 emissions will also increase from 589 to 1007 million tons/year. This study thus provides a methodology for the development of emission inventory of different pollutants from coal combustion in thermal power plants in India and the role of supercritical and ultra-supercritical technology in emission reduction. CO_2 is the major problem of thermal power plant in India, so if the clean coal technology like supercritical or ultra-supercritical technology are used to, up to 33% of emission reduction can be achieved. Though, these technologies are costly in nature but it reduces the emission significantly.

References

Annual Report (2015–16) Ministry of Coal, Government of India. https://coal.nic.in/content/annual-report-2015-16

Central Electricity Authority Report (2016). http://cea.nic.in/reports/monthly/executivesummary/2016/exe_summary-03.pdf

Chakraborty N, Mukheriee I, Santra AK, Chowdhury S, Chakraborty S, Bhattacharya S, Mitra AP, Sharma C (Feb 2008) Atmos Environ 42:1073–1082

Coal Swarn Coverage of India and Coal (2016). https://www.sierraclub.org/sites/www.sierraclub.org/files/uploads-wysiwig/Final%20Boom%20and%20Bust%20report_0.pdf

Department of Energy and Environmental Protection Agency, Washington DC, (2000). https://www.eia.gov/energyexplained/energy-and-the-environment/where-greenhouse-gases-come-from.php

Disha (2001) Green India 2047, TERI. http://www.dae.gov.in/node/128

"EnergyStatistics 2016" (PDF)

Guttikunda SK, Jawahar P (2014) Atmospheric emissions and pollution from the coal-fired thermal power plants in India. Atmos Environ 92:449–460

India's first environmental rating of coal-based power plants by Centre for Science and Environment under CSE's Green Rating Project (GRP), 21 Feb 2015

Levelized cost calculations/Transparent cost database (14/4/2016)

Ministry of Environment, Forest & Climate Change notification (emission standards) dated April 2015

MIT (2007) The future of coal, Table 3.1, p 19

Mittal M, Sharma C (2003) Anthropogenic emissions from energy activities in India: generation and source characterization, Part I
Mittal ML, Sharma C, Singh R (2012) Estimates of emissions from coal fired thermal power plants in India. Environmental Protection Agency
Pacyna EG, Pacyna JM, Steenhwsen F, Wilson S (2006) Global anthropogenic mercury emission inventory for 2000. Atmos Environ 37:109–117
Prasad DN, Anand MR (2013), Recent trends in production and import of coal in India. Ministry of Coal Occasional Working paper Series, 13 October 2013
Rai VK, Raman NS, Chaudhary SK (2013) Mercury in thermal power plants—a case study. Int J Pure App Biosci 1(2):31–37
Rees OW, Shimp NF, Beeler CW, Kuhn JK, Helfinstine RJ (1966) Sulphur retention in bituminous coal ash. Illinois State Geological Survey, Circula 396
Schoeder MH, Munthe J (1988) Atmospheric mercury an overview. Atmos Environ 32:809–822
User guide for India's 2047 energy calculator coal and gas power stations
U.S. Energy Information Administration: https://www.eia.gov/tools/faqs/
http://www.indexmundi.com

Chapter 19
Plasma-Based Hybrid Technique for Abatement of Pollutants (NO_x and CO) from Stationary Engine Exhaust-Effect of Loading Condition and Flow Rate

A. D. Srinivasan, N. Jagadisha, and R. Rajagopala

19.1 Introduction

In industries and automobile sector, diesel engine plays very important role. It is a type of manmade pollution source. Emission control from this lean-burnt engine from the conventional techniques' available catalytic reduction and SCR techniques in present scenario is less successful because of its environment and continues to emit large amount of pollutants. Most toxic and main pollutants present in the engine exhausts are NO_x and CO which are regulated by standards, and it is mandated to meet the regulations which is a challenging task and which needs for suitable after treatment techniques (Mohapatro and Rajanikanth 2011; Rajanikanth and Ravi 2002).

The EDP is a non-conventional technique, which promote various chemical reactions in exhaust by producing various chemically active radicals which will facilitate the removal of NO_x and other pollutants in lean-burnt engine exhaust (Penetrante 1997; Harano et al. 1998; Penetrante et al. 1999; Hackam and Akiyama 2000; Chae 2003).

Further, plasma promotes catalysis and adsorption when it is cascaded with a catalyst and adsorbent. Plasma-aided catalysis and adsorption are gaining lot of importance, and many researchers across the globe have been working on this (Rajanikanth et al. 2003; Mizuno et al. 1996; Shimizu and Oda 1997). Further, majority of the research work is carried out at exhaust temperatures more than 150 °C to make use of catalytic property and addition of agents in the reactors.

A. D. Srinivasan · N. Jagadisha (✉) · R. Rajagopala
Department of EEE, S.J.C.E, Mysuru, India
e-mail: jagga86@gmail.com

A. D. Srinivasan
e-mail: adsrinivasan.jce@gmail.com

S. M. Shiva Nagendra et al. (eds.), *Urban Air Quality Monitoring, Modelling and Human Exposure Assessment*, Springer Transactions in Civil and Environmental Engineering,
https://doi.org/10.1007/978-981-15-5511-4_19

The results reported by the literatures have limitations with regard to energy consumption, operating temperature window, by-product formation and pollutant removal efficiency (Srinivasan and Rajanikanth 2007).

In the present work, a study on the pollutants removal from the stationary lean-burnt engine exhaust was carried out using EDP hybrid techniques. The main objective of the study is to explore the effect of loading and exhaust flow rate on the pollutant removal efficiency of electric discharge plasma associated catalysis and adsorbent configurations. For the purpose of study, the exhaust treatment was done with two steps. In the first step, the engine exhaust was treated with single-step plasma-assisted catalyst and in the second step with single-step plasma-assisted adsorbent process. To study the effect of loading and flow rate, experiments were carried out at no load and partial load and at different flow rate. The effectiveness of the technique with regard to NO_x, CO removal and by-product reduction was discussed.

19.2 Experimental Setup

The schematic laboratory setup is shown in Fig. 19.1. A high-voltage rotating spark gap pulse source of 30 kV was used, and frequency of the pulses was kept constant at 100 pps (pulses per second). A 150 MHz digital oscilloscope (DL1540, 200MS/s, Yokogawa) was used to measure the voltage applied to the plasma reactor through a 2000:1 EP-50K, 50 MHz, voltage divider.

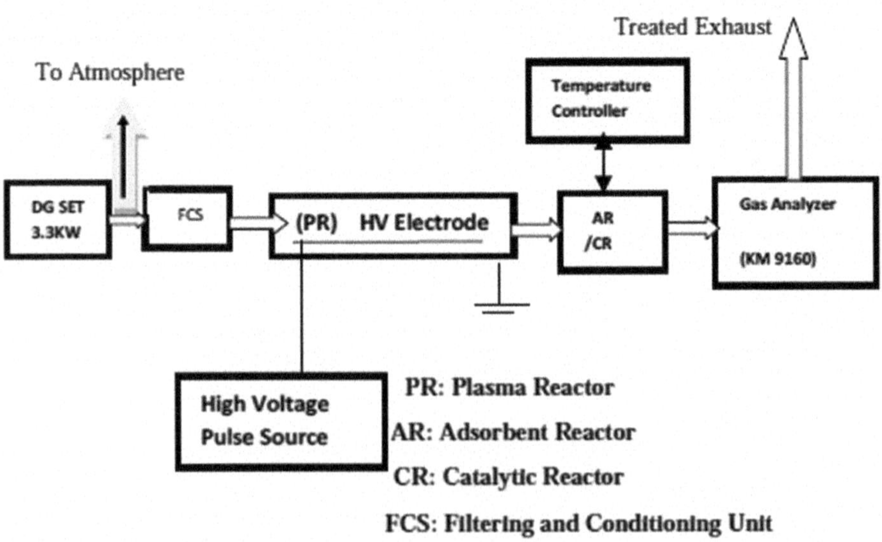

Fig. 19.1 Diesel engine exhaust treatment using single-step plasma hybrid catalysis/adsorbent technique

A 3.3 kW diesel engine was used in the laboratory as exhaust source, and only a part of the main exhaust from the engine was treated because of limitation in the laboratory infrastructure. Further, the exhaust flow rate was controlled through the flow meter to vary the exhaust from 2 to 8 lpm.

A cylindrical glass tube referred as DBD reactor (inner diameter: 15 mm and outer diameter: 17 mm) was employed in the present studies which is consisting of a SS wire with a thickness 1 mm used as the inner electrode and aluminium foil wrapped over the glass tube as the outer electrode. The effective discharge occurs at 30 cm. The experiments with plasma reactor were carried out at room temperature.

A waste by-product commercially available red mud has been used as catalyst, and MS13X molecular sleeves and activated alumina balls (AAB) were used as adsorbent. Both catalyst and adsorbent in the form of pellets were placed inside reactor referred as catalytic/adsorbent reactor (CR/AR).

In the plasma hybrid process, the catalytic/adsorbent reactor was placed at the outlet of the plasma reactor where the PR and AR were operated at room temperature, whereas the CR operated at 400 °C.

The exhaust gas made to pass through FCS which is shown in Fig. 19.1 was used for filtering of the exhaust in order to remove oil partials and particulate matter up to 10 μm and then allowed to enter the treatment zone.

A QUINTOX KM 9160 gas analyzer is used for the measurement of NO_x, CO and other gaseous pollutants present in engine exhaust.

19.3 Methodology

To cram the effect of flow rate on NO_x and CO removal process by plasma hybrid system, the experiments were conducted at different flow rates of 2, 4 and 8 lpm using flow meter. Further, to investigate the effect of loading condition on the pollutant removal process, experiments were conducted on no load and partial load using generator electric loading system.

19.4 Results and Discussion

Table 19.1 shows initial concentrations of pollutants under no load and partial load (27.27%) conditions.

Table 19.1, shows different pollutant concentrations where NO and NO_2 were measured independently and then added to get the NO_x concentration.

In the present paper, the effect of flow rate on the pollutant removal is first discussed, and then effect of loading condition on pollutant removal is discussed.

Table 19.1 Initial concentration of pollutants/components present in diesel engine exhaust

Main pollutants	No load	27.7% load
H_2O	1.0% vol	1.0% vol
CO_2	1.7%	3.2%
CO	410 ppm	311 ppm
NO	127 ppm	284 ppm
NO_2	93 ppm	262 ppm
NO_x	220 ppm	546 ppm
Aldehydes	50 ppm	80 ppm
O_2	18.6%	15.8%

19.4.1 Effect of Flow Rate

1. **On NO_x removal by adsorbent process**

From Fig. 19.2, it can be observed that flow rate has little effect on NO_x removal efficiency. This is because the NO_x removal by an adsorbent depends on the size of the pollutant molecule (NO_x and CO) and pore size of adsorbent material (Mohapatro and Rajanikanth 2010). Hence, for a given concentration of the pollutant and adsorbent material, the flow rate has less effect on the NO_x removal efficiency.

2. **On NO_x removal by plasma-assisted adsorbent process**

From Fig. 19.3, it is observed that the flow rate effect on NO_x removal depends on the plasma voltage. At low plasma voltages, the flow rate affects the NO_x removal considerably, and at higher plasma voltages, the effect is quite small. This can be explained as below.

At lower plasma voltage, the conversion of NO to NO_2 by the plasma is not effective, and further, higher flow rate would decrease the resident time of the exhaust in the plasma reactor. In this condition, the exhaust entering the adsorbent reactor contains more NO molecules which are not effectively adsorbed by the adsorbent material that has been used in our work. However, at higher plasma voltage, the

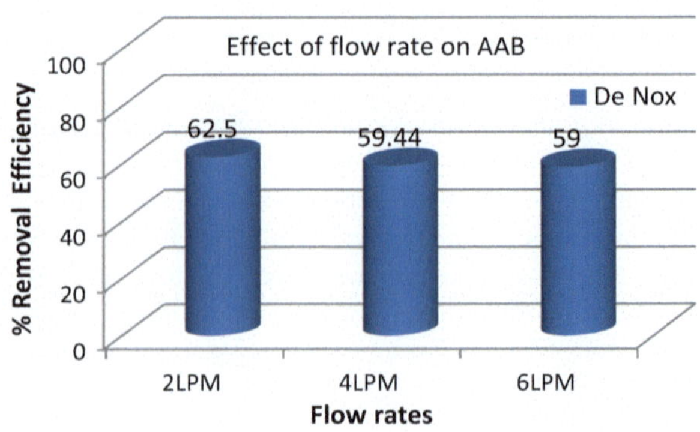

Fig. 19.2 Effect of flow rate on NO_x removal by adsorbent process

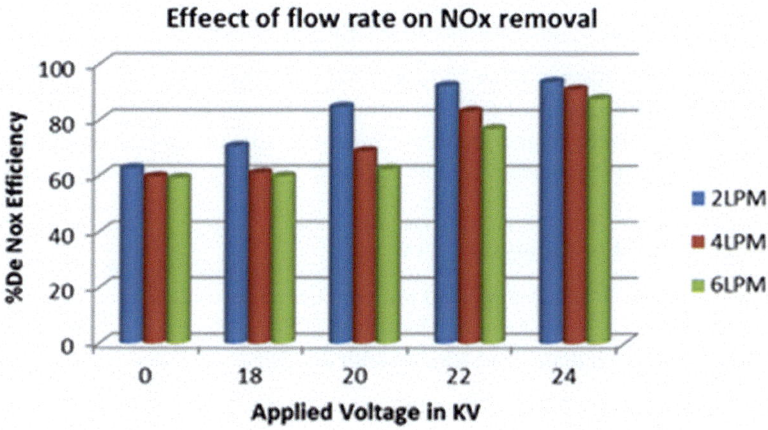

Fig. 19.3 Flow rate effect on NO_x removal by plasma-assisted adsorbent process

plasma will be more active in converting NO to NO_2 molecules even at higher flow rates, and hence, exhaust treated by plasma will have more NO_2 molecules. When this exhaust enters adsorbent reactor, the adsorption process will be more active as NO_2 molecules get adsorbed more effectively by the adsorbent material that has been used in our work. Hence, the flow rate matters little at higher voltages. Further, at any flow rate, plasma-assisted adsorption process shows superior NO_x removal process compared to adsorption process alone (Broer and Hammer 2000; Hoard 2001; Yoon et al. 2002; Rajanikanth et al. 2004).

3. **On NO_x removal by catalyst process**

From Fig. 19.4, it is noted that NO_x removal rate decreases with increase in flow rate which can be attributed to the low resident time of the exhaust in the CR (Mohapatro and Rajanikanth 2012).

Fig. 19.4 Flow rate effect on NO_x removal by catalyst process

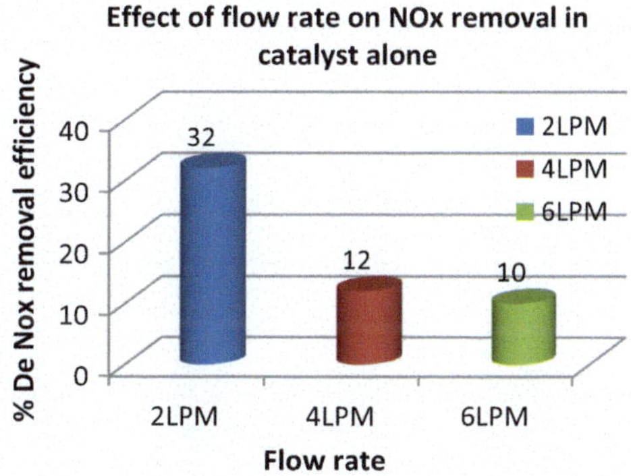

4. **On NO_x removal by plasma-assisted catalytic process**

From Fig. 19.5, it is observed that NO_x removal decreases with increase in flow rate which is due to the decrease of resident time as explained above. However, at given flow rate, NO_x removal is higher than catalyst alone which is due to combined action of plasma and catalyst process (Rajanikanth and Srinivasan 2007; Srinivasan et al. 2012a).

5. **On CO removal by catalyst process**

From Fig. 19.6, it is noted that the CO removal decreases with increase in flow rate which is again due to decrease in resident time of the pollutant inside the reactor at higher flow rates (Mohapatro and Rajanikanth 2012).

6. **On CO removal by plasma-assisted catalytic process**

From Figs. 19.6 and 19.7, it is observed that the plasma-assisted catalytic technique shows better CO removal compared to catalytic technique alone at given flow rate. This is because the plasma being an oxidizing environment converts CO to CO_2, thus reducing the CO concentration. Further, when the exhaust containing low CO

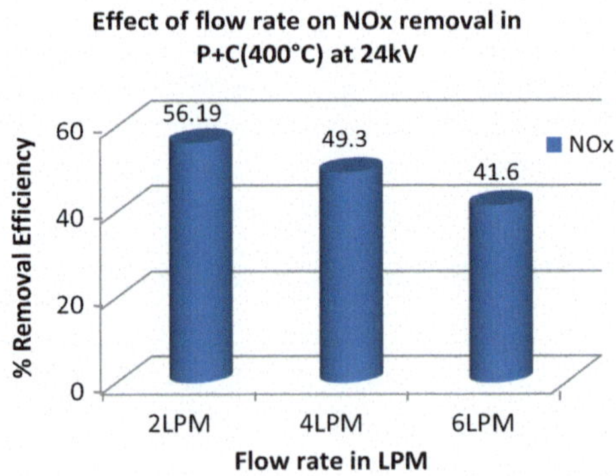

Fig. 19.5 Effect of flow rate on NO_x removal by plasma-assisted catalytic process

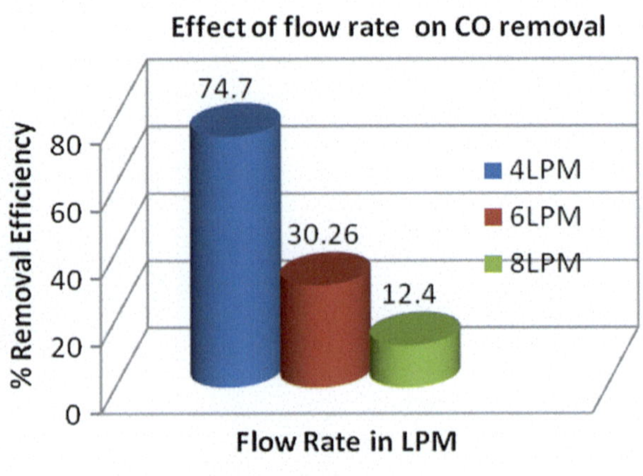

Fig. 19.6 Flow rate effect on CO removal by catalyst process

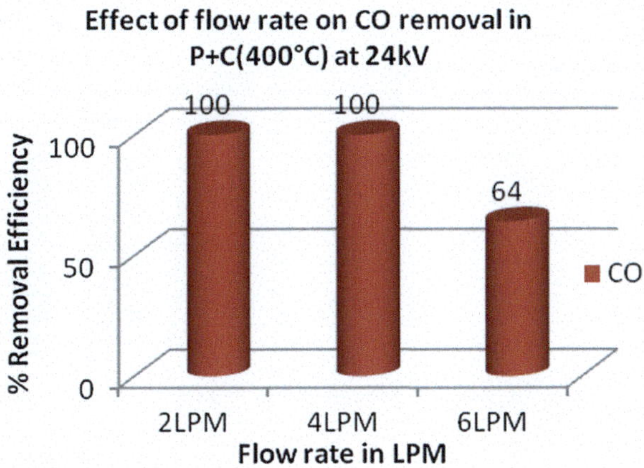

Fig. 19.7 Flow rate effect on CO in removal by plasma-assisted catalytic process

concentration enters the catalyst, the catalytic activity will not be much affected by the flow rate. Hence, the flow rate is not much affected on CO removal in plasma-assisted catalytic process (Srinivasan et al. 2012a).

19.4.2 *Effect of Loading*

1. **On NO_x removal by adsorbent process**

From Fig. 19.8, it is observed that when adsorbent alone is used, the loading effect on the NO_x removal is less pronounced for each of the adsorbents used in our work. Further, under any load, MS 13X shows better NO_x removal compared to activated

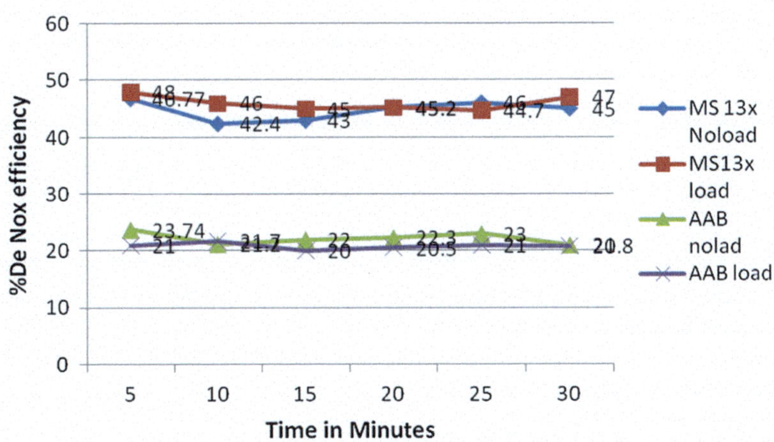

Fig. 19.8 Comparison between adsorbents on NO_x removal

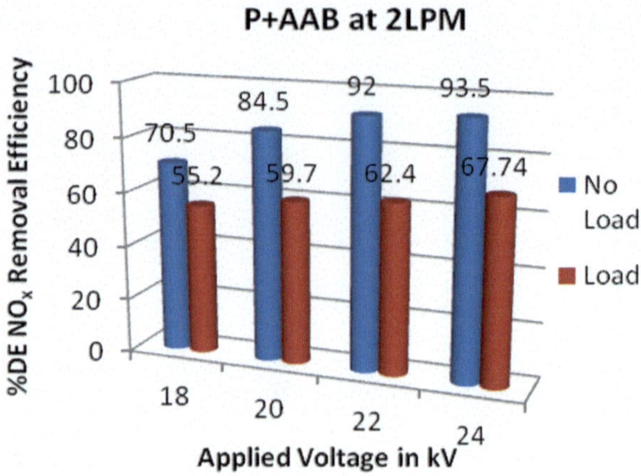

Fig. 19.9 Effect of load on NO_x removal by plasma-assisted adsorbent process

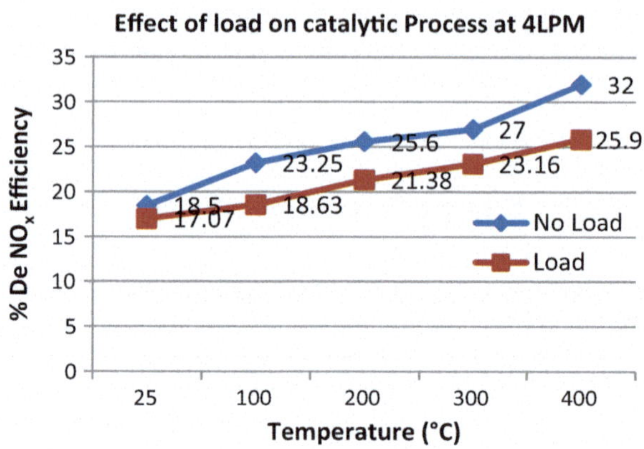

Fig. 19.10 Effect of loading on NO_x removal by catalyst process

alumina. This can be attributed to the appropriate pore dimension and large surface area of the MS 13X (Srinivasan et al. 2012b).

2. **On NO_x removal by plasma-assisted adsorbent process**

From Fig. 19.9, it is observed that plasma-assisted adsorbent (particularly at higher voltages) shows better NO_x removal performance compared to adsorbent alone both at no load and load condition. This is attributed to effective production of NO_2 molecules by the plasma process which subsequently gets adsorbed effectively by the adsorbent. However, the effect of loading is more pronounced in case of plasma-assisted adsorbent process compared to adsorbent alone. This can be due to high initial concentration under load condition which decreases NO_x removal performance of plasma process (Srinivasan et al. 2012b).

3. **On NO_x removal by catalytic process**

From Fig. 19.10, it is observed that at a given catalyst temperature, the NO_x removal

Fig. 19.11 Effect of load on NO_x removal by plasma-assisted catalytic process

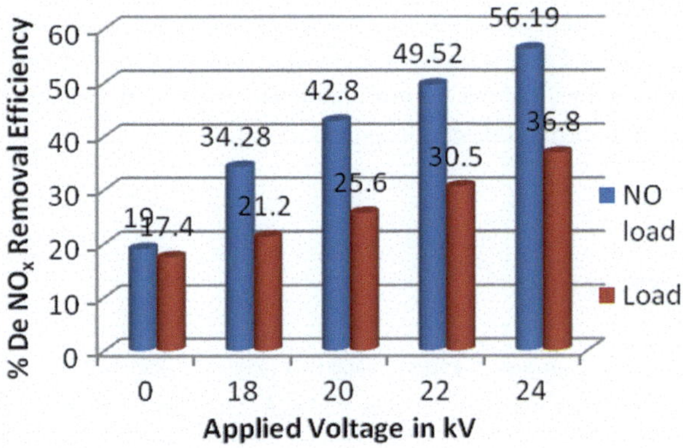

by the catalyst decreases under load condition. This is because as loading increases, the initial concentration of NO_x increases which will decrease the catalytic activity, thus decreasing the NO_x removal (Mohapatro and Rajanikanth 2012).

4. **On NO_x removal by plasma-assisted catalytic process**

From Fig. 19.11, it is observed that plasma-assisted catalyst shows better NO_x removal performance compared to catalyst alone both at no load and load condition particularly at higher voltages. This is owing to the fact that in this condition, plasma significantly contributes to NO_x removal. Further, it is also noted that the NO_x removal gets much affected by the loading in case of plasma-assisted catalyst compared to catalyst alone. This is because initial concentration of NO_x increases under loading condition which significantly lowers the activity of the plasma process affecting NO_x removal (Rajanikanth and Srinivasan 2007).

19.5 Conclusion

The following conclusions are listed from the present work:

- For a given concentration of the pollutant and adsorbent material, the flow rate has less effect on the NO_x removal efficiency.
- The effect of flow rate on NO_x removal in case of plasma-assisted adsorption depends on the plasma voltage. At low plasma voltages, the flow rate affects the NO_x removal considerably, and at higher plasma voltages, the effect of flow rate on NO_x removal is quite small.
- The CO removal by plasma-assisted catalytic process is not much affected by the flow rate.
- The plasma-assisted adsorbent process (particularly at higher voltages) shows better NO_x removal performance compared to adsorbent alone both at no load and load condition. However, the effect of loading is more pronounced in case of plasma-assisted adsorbent process compared to adsorbent alone.

- The plasma-assisted catalyst shows better NO_x removal performance compared to catalyst alone both at no load and load condition particularly at higher voltages. Further, the NO_x removal gets much affected by the loading in case of plasma-assisted catalyst compared to catalyst alone.

References

Broer S, Hammer T (2000) Selective catalytic reduction of nitrogen oxides by combining a non-thermal plasma and a V_2O_5-WO_3/TiO_2 catalyst. Appl Catal B Environ 28:101–111

Chae JO (2003) Non-thermal plasma for diesel exhaust treatment. J Electrostat 57:251–262

Hackam R, Akiyama H (2000) Air pollution control by electrical discharges. IEEE Trans Dielectr Electr Insul 7(5), 654–683

Harano A et al (1998) Oxidation of carbonaceous particles in silent discharge reactor. J Chem Eng Japan 31(5):700–705

Hoard J (2001) Plasma catalysis for diesel exhaust treatment: current state of art. In: Society of automotive engineers, 2001-01-0185

Mizuno A, Shimizu K, Yanagihara K, Kinoshita K, Tsunoda K, Kim HH, Katsura S (1996) Effect of additives and catalysts on removal of nitrogen oxides using pulsed discharge. In: IEEE-IA Annual Meeting, pp 1806–1812

Mohapatro S, Rajanikanth BS (2010) Cascaded cross flow DBD-adsorbent technique for NO_x abatement in diesel engine exhaust. IEEE Trans Dielectr Electr Insul 17(5)

Mohapatro S, Rajanikanth BS (2011) Portable HVAC and pulsed plasma sources for control of NO_x in diesel engine exhaust. IEEE Trans Dielectr Electr Insul 18(6)

Mohapatro S, Rajanikanth BS (2012) Dielectric barrier discharge cascaded with red mud waste to enhance NO_x removal from diesel engine exhaust. IEEE Trans Dielectr Electr Insul 19(2), 641–647

Penetrante BM (1997) Removal of NO_x from diesel generator exhaust by pulsed electron beams. In: IEEE IA annual meeting, pp 91–96

Penetrante BM, Brusasco RM, Merrit BT, Pitz WJ, Vogtlin GE (1999) Feasibility of plasma after treatment for simultaneous control of NO_x and particulates. In: SAE, 1999-01-3637, pp 2056–2061

Rajanikanth BS, Ravi V (2002) Removal of nitrogen oxides in diesel engine exhaust by plasma assisted molecular sieves. Plasma Sci Technol 4

Rajanikanth BS, Srinivasan AD (2007) Pulsed plasma promoted adsorption/catalysis for NO_x removal from stationary diesel engine exhaust. IEEE Trans Dielectr Electr Insul 14(2):302–311

Rajanikanth BS, Srinivasan AD, Nandiny BA (2003) A cascaded discharge plasma adsorbent technique for engine exhaust treatment. Plasma Sci Technol 5(3):1825–1833

Rajanikanth BS, Das S, Srinivasan AD (2004) Unfiltered diesel engine exhaust treatment by discharge plasma: effect of soot oxidation. Plasma Sci Technol 6(5):2475–2480

Shimizu K, Oda T (1997) $DeNO_x$ process in flue gas combined with non-thermal plasma and catalyst. Presented at IEEE-IAS annual meeting, New Orleans, pp 1942–1949

Srinivasan AD, Rajanikanth BS (2007) Pulsed plasma treatment for NO_x reduction from filtered/unfiltered stationary diesel engine exhaust. In: IEEE IA annual meeting, pp 1893–1900

Srinivasan AD, Rajagopala R, Jagadisha N, Bhargavi A (2012a) Laboratory investigation of pulse discharge based techniques for engine exhaust treatment—effect of exhaust nature and operating conditions. Int J Plasma Energy Sci Technol 6(3), 215–221

Srinivasan AD, Rajagopala R, Jagadisha N, Bhargavi A (2012b) A laboratory investigation of pulsed discharge plasma associated catalysis for engine exhaust treatment. In: International symposium on plasmas for catalyses and energy materials (ISPECEM)-2012, Tanjin

Yoon S, Panov AG, Tonkyn RG, Ebeling AC, Barlow SE, Balmer L (2002) An examination of the role of plasma treatment for lean NO_x reduction over sodium zeolite Y and gamma alumina: part 1. Plasma assisted NO_x reduction over Na-Y and Al_2O_3. Catal Today 72, 243–250

Chapter 20
Evaluation of Carbon Foot Print Reduction in Aviation Infrastructure

D. M. M. S. Dissanayaka and W. W. A. S. Fernando

20.1 Introduction

20.1.1 Background

According to the International Air Transport Association (IATA) records, 3.8 billion air passengers were recorded in 2016 and anticipated that there would be 7.2 billion in 2035 (IATA 2016). Aircraft transported about 35% of world trade by value, although it was only 0.5% by volume (Gençsü and Hino 2015). Aviation has brought benefits to society by connecting places, people and products with a range of services compared with any other modes of transport. Thus, aviation needs to improve its reputation in environmental efficiency and lower emission levels when heading towards sustainability.

Transport sector predominantly contributed to greenhouse gas (GHG) emissions by 14% of the total GHG contribution. It was 13% by aviation from the total GHG contribution in the transport sector (IPCC 2015). Even though the contribution of aviation was relatively smaller comparing to the other modes, the Intergovernmental Panel on Climate Change (IPCC) declared that aviation was a fast-growing source of CO_2 emission (Sims et al. 2014).

D. M. M. S. Dissanayaka (✉) · W. W. A. S. Fernando
Department of Transport and Logistics Management, Faculty of Engineering, University of Moratuwa, Moratuwa, Sri Lanka
e-mail: mano90diss@gmail.com

W. W. A. S. Fernando
e-mail: anujas@uom.lk

S. M. Shiva Nagendra et al. (eds.), *Urban Air Quality Monitoring, Modelling and Human Exposure Assessment*, Springer Transactions in Civil and Environmental Engineering,
https://doi.org/10.1007/978-981-15-5511-4_20

20.1.2 Environmental Impact

Aviation sector recognized that the rapid growth in aviation is influencing the climate change. International aviation accounted for 65% of total fuel consumption in 2010, contributed 448 MT of CO_2 emissions. That emission level was 185 MT in 1990. ICAO anticipated international aviation emissions will increase up to 682–755 MT CO_2 by 2020 (Gençsü and Hino 2015).

Blocking the outgoing energy radiating from the surface of the earth, GHGs influence climate change. The IPCC said that the surface temperature of the earth would rise by 1.0 °C and the world should immediately find solutions in limiting total warming to less than 2.0 °C (OECD 2012).

20.1.3 Current Applications Towards Carbon Emission Reduction by Aviation Bodies

ICAO goals towards emission reduction in aviation

1. Improving aircraft fuel efficiency by 1.5% per year until 2020
2. To cap net emissions from 2020 following a carbon-neutral growth
3. To halve the net aviation carbon emissions of the year 2005 by 2050.

Governments reached a global agreement on targets addressing aviation emissions at the 37th ICAO Assembly 2010 (ICAO 2010). ICAO analysed the potential economic impact of introducing a market-based method (MBM) to maintain a specific level of net carbon emission globally (ICAO 2010). MBMs were to fill the gap which was the failure amount of carbon emission reduction when achieving carbon-neutral growth with operational measures. MBMs were expected to follow until the arrival of standard biofuel. Biofuels can be made using algae, Jatropha and Camelina had been proved to reduce the carbon footprint of aviation fuel by 80% over their full lifecycle (ATAG 2011). MBMs must be cost-efficient to receive global acceptance. MBMs for aviation must be a bridge towards the carbon neutrality.

Aviation infrastructure and air traffic management should be effectively guided towards carbon neutrality. Achieving the second goal of ICAO by 2020, aviation would cap its net carbon emissions following a carbon-neutral growth. Accomplishing the third goal by 2050, it aimed to trim down the net emissions compared to 2005 levels as agreed by IATA, the Airports Council International (ACI), the International Coordinating Council of Aerospace Industries Associations (ICCAIA) and the Civil Air Navigation Services Organization (CANSO) (ATAG 2011).

20.1.4 Current Applications in Emission Reduction in Airport Operations

Carbon emission in aviation can be categorized into two sections such as emission from aircraft operation and emission from airport operation. Airport operations emitted only 5% of carbon, whereas the movement of aircraft was responsible for the remaining 95% (Štimac et al. 2013).

Very few aviation bodies pay attention to carbon emission in airport operation. This research aimed to focus on the requirement of reducing carbon emission in airport operation. Some airports are voluntarily involved in carbon emission reduction programmes at airport operation. Airport Carbon Accreditation is one of the methods that strives in managing, reducing and ultimately neutralizing their carbon footprint. Achieving the level of neutrality, there were 28 accredited airports in 9 countries, representing 19.6% of European air passenger traffic (ACI 2009a).

20.1.5 Benefits of Measuring and Monitoring the Carbon Footprint

- Improve air quality—The emission reduction methods can be initiated by measuring the current carbon emission of different sources. The air quality can be improved due to emission reduction practices.
- Cost reduction—An airport operator can make significant cost savings by reducing its emissions, mainly through energy efficiency.
- Improve international and national image—Reducing GHG emissions with a carbon management plan, the image of the airport will be promoted. The better image can achieve a competitive advantage over the other competitive airports within the region.
- Support for future GHG reduction policies and regulations preparation—With the emerging impact of climate changes, there is an emergence of new environmental regulations and policies designed to reduce GHG emissions. Emission inventory can help the business operation to prepare for any future policies or regulations relating to GHG emissions.
- Customer service relationship (CSR) communication strategy—Passengers and future investors can be attracted to green concepts.
- Guide for sustainable airport operation—Assessing carbon emission shows the level of emission that aviation stakeholders are responsible for. Showing the level of emission, it is guiding to reduce it by forcing and motivating for new methods of emission reduction.

20.2 Problem Statement

Green aviation involves activities like improving aircraft fuel efficiency, developing efficient air traffic control, evolving new technologies and system engineering processes to achieve carbon neutrality in aviation. Since ICAO sets goals in the environmental aspect which are not even mandatory, many parties ignore the environmental impact of aviation. ICAO has been unable to enforce this as a rule, due to the objections from the giants of the industry. Due to the high level of operations, it is difficult for industry giants to control their emission. Thus, ICAO has to be satisfied with setting goals.

A few airports are voluntarily involved in reducing carbon emission at airport operations. Many airports ignore its airport operational emission as it is lower compared to the aircraft emission. Therefore, existing literature sparsely addressed in the area of emission in airport operation. However, the smaller portion should not be ignored when achieving green aviation. Every source of emission should be accounted for when calculating aviation emission. Sustainability measures should be considered in every source of carbon emission to achieve green aviation.

20.2.1 Case Study Airport—Bandaranaike International Airport (BIA)

Bandaranaike International Airport (IATA: CMB, ICAO: VCBI) is the main international gateway of Sri Lanka. Its design capacity is 6 million annual passenger movements. 28 scheduled airlines are operating in BIA from 45 cities. 25 movements per hour are the aircraft handling capacity in the runway. Its cargo handling capacity is 25,000 MT per annum (AASL 2014). This research was based on the BIA airport operation.

Most functions of an international airport are outsourced. Those functions are catering services, ground handling functions, aircraft maintenance, fuelling, etc. Since the airport inventory was owned by many parties, measuring its total emission is difficult. Some airlines follow methods to reduce emission in its aircraft. Sustainable aviation cannot be achieved only through reducing aircraft emission. All the emission sources should be considered to achieve green aviation. The absence of the proper mechanism to identify the emission sources, its emission levels and techniques to mitigate the emission dragged away from the aviation sector from its sustainability.

20.2.2 Research Aims and Objectives

The ultimate aim of this study was to identify whether an international airport should have measures to mitigate Carbon emission at its operation. To achieve the aim, the research was based on the following objectives;

- To identify the current carbon emission sources at an international airport
- To evaluate the current carbon emission levels of those sources.

20.3 Methodology

The research methodology was comprised of a literature review, an interview survey of industry experts, a collection of primary data from Airport and Aviation Services Limited, Civil Aviation Authority, Central Environmental Authority and Sri Lankan Airlines, the calculation of carbon emission from the sources, a statistical analysis on secondary data and a quantitative analysis on the data gathered from the survey (Fig. 20.1).

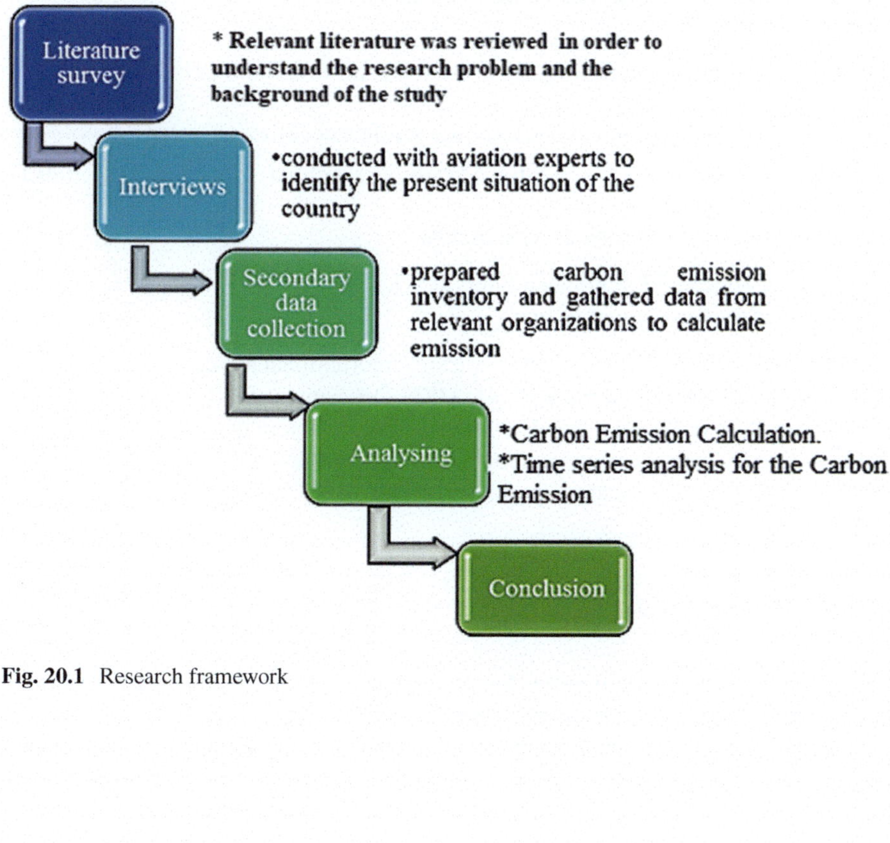

Fig. 20.1 Research framework

20.3.1 Data Collection

Relevant literature was reviewed to achieve background knowledge of the study. As the first objective of the study was to identify the emission inventory of the airport, Standard International Airport Carbon Emission Inventory list was requested from Airport Council International (ACI). It was used to gain the idea of emission inventory in other international airports. Then, interviews were carried out to identify the carbon emission sources at BIA.

The population for the interview was the personnel in the industry of different sectors such as regulatory body, airport service provider and airlines. The convenient sampling method was used when interviewing industry experts since less personnel in the Sri Lankan aviation industry were interested in the area of carbon emission. Then, the list of carbon emission sources was prepared. Due to the involvement of several outside parties in airport operation, the carbon emission sources had not been identified and listed before at BIA. Thus, carbon emission and sources at BIA were recognized in this research through the interviews and literature review.

20.3.2 Carbon Footprint Data Calculation

The basic equation for calculating a carbon footprint is (Wilding et al. 2009):

$$\begin{aligned} CO_2\text{ Emissions} &= \text{activity data} \times \text{relevant emission factor} \\ &\quad \times \text{Global Warming Potential} \end{aligned} \tag{20.1}$$

Activity data quantifies an activity such as:

- Electricity consumed in kilowatt hours (kWh)
- Amount of fuel consumed in litres.

Emission factors were published by various international agencies such as IPCC, the International Energy Agency (IEA), UK's Department for Environment, Food and Rural Affairs (DEFRA), etc. The latest emission factors provided by the 'GHG Protocol 2012' which were updated on a regular basis and based on improvements and upgrades in international methodologies are used in this analysis.

20.3.3 Assumptions

All carbon footprint calculations were inevitably involved in data/information limitations and made certain assumptions (Table 20.1).

Standard and country-specific emission factors for Sri Lanka have not been established so far for many sources of carbon emission. Thus, published emission factors

Table 20.1 Specific emission factors (IPCC 2006)

Emission factors used	Unit and source
Grid electricity, Sri Lanka	**0.46011 kg$CO_{2\text{-eq}}$/kWh** (International Energy Agency, 2009/GHG Protocol, 2012)
Diesel (stationary combustion)	**2.691 kg$CO_{2\text{-eq}}$/litre** (IPCC, 2006 in GHG Protocol, 2012)
Kerosene (stationary combustion)	**2.533 kg$CO_{2\text{-eq}}$/litre** (IPCC, 2006 in GHG Protocol, 2012)
Petrol vehicle (mobile combustion)	**2.272 kg$CO_{2\text{-eq}}$/litre** (IPCC, 2006 in GHG Protocol, 2012)
Diesel vehicle (mobile combustion)	**2.676 kg$CO_{2\text{-eq}}$/litre** (IPCC, 2006 in GHG Protocol, 2012)

Source IPCC Emissions Factor Database | Greenhouse Gas Protocol (IPCC 2006)

from the GHG Protocol 2012 were used in this emission calculation to maintain some degree of uniformity and reliability of GHG emissions estimates. The GHG Protocol 2012 emission factors were mainly based on the IPCC, which was the main international body that provided most of the research and scientific information relating to global warming and climate change.

20.3.4 Global Warming Potential (GWP)

It is a ratio between one unit mass of CO_2 over a specified time and one unit mass of the greenhouse gas (GHG) to that. Different greenhouse gases have different GWP. GWP for the CO_2 is 1. GWP was developed to allow comparisons among different GHGs with their global warming impact (Wilding et al. 2009).

20.3.5 Units and Carbon Dioxide Equivalents ($CO_{2\text{-eq}}$)

All emissions were calculated as **$CO_{2\text{-eq}}$**, as required by the GHG Protocol to put everything to a common scale and CO_2 is the highest contributor of all GHGs. $CO_{2\text{-eq}}$ was the 'universal currency' for a carbon footprint measurement. All emissions in this carbon footprint analysis were presented as metric tonnes of carbon dioxide equivalents (t$CO_{2\text{-eq}}$) and kg of carbon dioxide equivalents (kg$CO_{2\text{-eq}}$).

20.3.6 Analysis Method

Descriptive analysis of the secondary data was done and the entire emission sources at BIA were recognized. With the use of the equation, the primary data was converted to carbon emission. Monthlywise data was obtained from the year 2011 to 2015.

The emission levels were used for time series analysis and Minitab software has been used. Autoregressive integrated moving average (ARIMA) (p, q, d) was used for modelling and considered autoregressive (AR) (p), the degree of differencing (d) and moving average (MA) (q). The time series analysis was conducted for the overall distribution of emission in airport operations in the case study.

20.4 Results and Findings

Airport carbon emission inventory could be categorized into three sections (Table 20.2).

Scope 1—inventory owned and controlled by the airport service provider
Scope 2—inventory purchased by the airport service provider
Scope 3—inventory not owned or directly controlled by airport service provider (ACI 2009b).

Under each scope, the carbon emission sources at BIA were identified. It was classified in Table 20.3. Scope 1 and 2 emission sources were focused on this research. Only those areas which were managed by the airport operator—Airport and Aviation Services (Sri Lanka) Ltd—were selected.

According to objective 1 of this research, above carbon emission inventory at BIA was identified. Carbon emission of the selected inventory was calculated with the use of the previously mentioned equation. Monthly emission levels were identified from the year 2011 to 2015.

Considering the monthly emission levels from the year 2011 to 2015, time series analysis was done. It was said that total tons of carbon emission in airport operation was increasing with years and it was confirmed in the time series analysis. ARIMA model for the time series was obtained with the use of Minitab software.

Table 20.2 Airport carbon emission sources categorized according to the scope

Scope 1 (Owned and controlled by AASL)	*Scope 2* (Purchased by AASL)	*Scope 3* (Not owned or directly controlled by AASL)
Stationary	Electricity facility	Aircraft
Mobile vehicle	Steam and hot water facility	Ground service equipment/vehicles
Waste		Ground access vehicles

Table 20.3 Carbon emission sources at BIA and its emission level in 2015

	Carbon emission source	Scope	Emission of $kgCO_{2\text{-eq}}$
1	*Waste incinerators (diesel)*	1	180,572
2	Owned by the airport	1	572,525
	Not owned by the airport	3	Not measured
3	*Ground handling machines, equipment*		
	Owned by the airport	1	102,915
	Not owned by the airport	3	Not measured
4	*Fire training-kerosene*	1	10,637
5	*CO2 extinguisher*	1	13.6
6	*Wastewater processing (diesel)*	1	77,388
7	*Generators (diesel)*	1	102,915
8	*Electricity*	2	17,349,803

The model was checked using diagnostic plots for standardized residuals, ACFs of residual, and *p*-value of Ljung–Box statistics. Since there was no significant autocorrelation of the residuals and the *p*-value of the Ljung–Box statistic was high, deriving that the model was adequate for forecasting emission at the airport (Fig. 20.2).

The ARIMA (1, 1, 3) model was fitting for this series. The following were the Minitab output for this model (Tables 20.4 and 20.5).

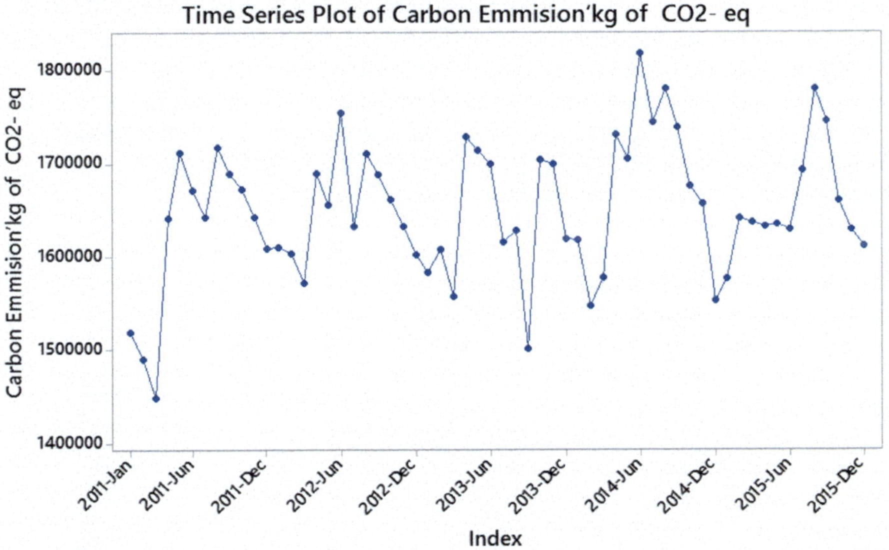

Fig. 20.2 Carbon emission in $kgCO_{2\text{-eq}}$/month

Table 20.4 Final estimates of parameters

Type	Coefficient	SE coefficient	*T*	*P*
AR 1	0.1586	0.4276	0.37	0.712
MA 1	0.6379	0.4290	1.49	0.143
MA 2	0.0364	0.2604	0.14	0.889
MA 3	0.2604	0.1849	1.41	0.165
Constant	3.024	1.271	2.38	0.021

Differencing: 1 regular difference
Number of observations: original series 60, after differencing 59
Residuals: SS = 299,537 (back forecasts excluded)
MS = 5547 DF = 54

Table 20.5 Modified Box-Pierce (Ljung–Box) chi-square statistic

Lag	12	24	36	48
Chi-square	20.9	31.4	40.3	47.0
DF	7	19	31	43
P-value	0.004	0.037	0.122	0.312

The *p* values were considered only as significant at the 10% level for the first-order coefficient of the autoregressive part of the model and the third-order coefficient of the moving average part of the model. Furthermore, the Ljung–Box chi-square statistics suggested that there may be a seasonal effect of at least order 1. Therefore, the model could be constructed as an ARIMA (1, 1, 3) (1, 0, 0) 12 model. The first set of parentheses explained the lags for the autoregressive (AR) and integrated (I) parts of the model will be 1, while the moving average (MA) would be based on lag 3. The second set of parentheses indicates the seasonal effect, which is supposed to follow a 12-period (annual) cycle around AR (1) (Tables 20.6 and 20.7).

Therefore, the equation for the fitted model for the time series was

Table 20.6 Final estimates of parameters

Type	Coefficient	SE coefficient	*T*	*P*
AR 1	0.1948	0.3654	0.53	0.596
SAR 12	0.6007	0.1219	4.93	0.000
MA 1	0.5927	0.3628	1.63	0.108
MA 2	0.0486	0.2222	0.22	0.828
MA 3	0.2986	0.1856	1.61	0.114
Constant	1.3399	0.5610	2.39	0.021

Differencing: 1 regular difference
Number of observations: original series 60, after differencing 59
Residuals: SS = 217,122 (back forecasts excluded)
MS = 4097 DF = 53

Table 20.7 Modified Box-Pierce (Ljung–Box) chi-square statistic

Lag	12	24	36	48
Chi-square	7.0	15.5	27.8	32.2
DF	6	18	30	42
P-value	0.318	0.631	0.583	0.862

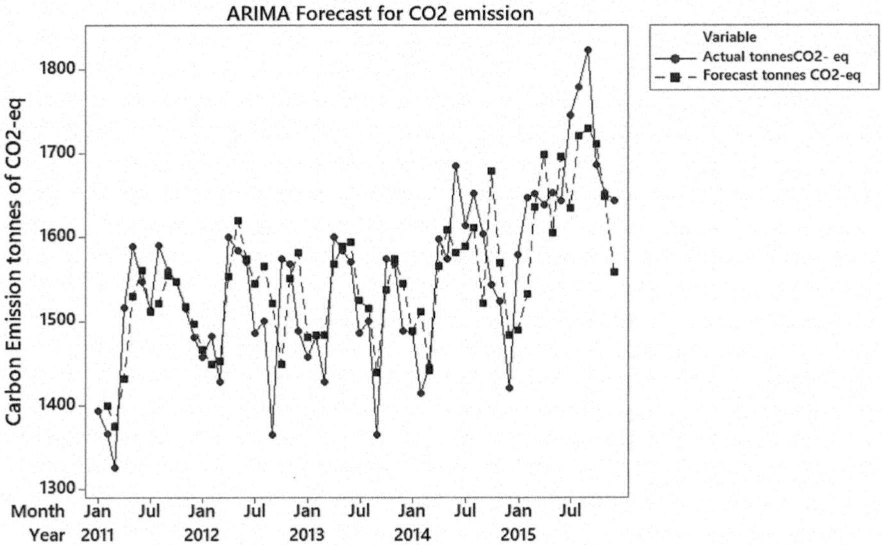

Fig. 20.3 ARIMA forecast for carbon emission

$$Y_t = 1.34 + Y_{t-1} + 0.19(Y_{t-1} - Y_{t-2}) + 0.60(Y_{t-12} - Y_{t-13})$$
$$-0.12(Y_{t-13} - Y_{t-14}) + 0.59e_{t-1} - 0.05e_{t-2} - 0.30 * e_{t-3} \quad (20.2)$$

Y_t—current values for emission
$Y_{t-1} \ldots Y_{t-14}$—emission at period $t-1 \ldots t-14$
e—random errors due to moving averages
$e_{t-1} \ldots e_{t-3}$—random errors at the periods $t-1, \ldots, t-3$ (Fig. 20.3).

20.5 Conclusion

According to the time series analysis, it was obvious that carbon emission from airport operation is increasing with years. Its emission level for the year 2015 was 18,294 tones and analysis showed it would be increasing. Thus, emission from airport operation cannot be negligible when calculating total carbon emission in the aviation sector. The emission from airport operation at BIA showed a significant level of emission. Reporting on carbon emission level was one of the best ways to reduce emission. The organization can identify the emission level that they were responsible for, once it measures emission from all the sources.

With the growth of aviation demand, it leads to more aviation operation. More operations lead to more emissions until sustainable solutions would be initiated. Airport development should aim to become carbon neutral in the long term to achieve sustainable aviation. Reducing carbon emission only from aircraft cannot achieve this sustainable target. Thus, reducing emission from airport operation is obviously to be achieved on the way to sustainability.

Acknowledgements The authors would like to thank Ms. Chanika Mannawaduge, Assistant Guidance Material, and Civil Aviation Authority of Sri Lanka for providing guidance to gather data from different parties. The authors would also like to thank Mr. Sumith Tennakoon, Air traffic controller at Airport and Aviation Services (Sri Lanka) Limited for the briefing on the current situation in the industry. Then we would be grateful to Mr. R. W. Wijesinghe, Head of Mechanical Engineering at Airport and Aviation Services (Sri Lanka) Ltd and Mr. S. M. R. Rafeek, Head of Projects at Airport and Aviation Services (Sri Lanka) Ltd, for helping to collect relevant data.

References

AASL (2014) Airport & Aviation Services (Sri Lanka) Ltd annual report

ACI (2009a) Airport carbon accreditation—EUROPE. [Online]. Available: http://www.airportcarbonaccreditation.org/airport/participants/europe.html. Accessed 17 Aug 2017

ACI (2009b) Guidance manual: airport greenhouse gas emissions management

ATAG (2011) Beginner's guide to aviation biofuels

Gençsü I, Hino M (2015) Raising ambition to reduce international aviation and maritime emissions

IATA (2016) IATA—IATA forecasts passenger demand to double over 20 years. [Online]. Available: http://www.iata.org/pressroom/pr/Pages/2016-10-18-02.aspx. Accessed 10 Apr 2018

ICAO (2010) Resolutions adopted at the 37th session of the assembly. Montreal

IPCC (2006) IPCC emissions factor database | greenhouse gas protocol. [Online]. Available: http://www.ghgprotocol.org/Third-Party-Databases/IPCC-Emissions-Factor-Database. Accessed 30 Aug 2017

IPCC (2015) Climate change 2014 synthesis report. Geneva

OECD (2012) Green growth and the future of aviation

Sims R, Schaeffer R, Soares Moreira Cesar Borba B, Schaeffer R, Creutzig F, Cruz-Núñez X, Dimitriu D, Figueroa Meza MJ, Fulton L, Kobayashi S, Lah O, McKinnon A, Newman P, Ouyang M, Schauer JJ, Sperling D, Tiwari G, Sokona Y, Farahani E, Kadner S, Seyboth K, Adler A, Baum I, Brunner S, Eickemeier P, Kriemann B, Savolainen J, Schlömer S, von Stechow C, Zwickel T, Minx J (2014) Transport. Paul Peeters

Štimac I, Vince D, Jakšić B, El MI (2013) Model of environment-friendly aircraft handling—case study: Zagreb airport

Wilding J, Barker JB, Authority T, Biehler AD, Dot P, Brown LL, Dot M, Clark WAV, Angeles L, Ekern DS, Dot V (2009) Guidebook on preparing airport greenhouse gas emissions inventories. Washington

Chapter 21
Structural Analysis of Interactions Between Airborne Pollutants and Chemically Modified RNAs

Kannan Krishnamurthi, Pravin K. Naoghare, Saravana S. Devi, Amit Bafana, and Patrizio Arrigo

21.1 Introduction

The occurrence of drastic climate changes is a consolidated knowledge. These events have an effect on land modification and alteration of precipitation level. The changes of geophysical factors are indicted to lead to biodiversity decline and a loss of functionality of ecosystems. The human quality of life and health is clearly affected by these environmental changes. The combination of climate changes with environmental pollution, due to human activities, has become a critical factor for a sustainable development and to human quality of life. The air pollution, foremost associate with mankind activities, is a global menace for whole environment and, in particular, for human well-being. The air pollutants have been classified into two main classes: biotic and abiotic. The first group contains different elements as pollen, virus and bacteria. The second group contains physical and chemical pollutants as radioactivity or organic compounds. Taking their origin into account the air pollutants are classified as primary and secondary. The primary air pollutants are those generated by a natural events or anthropic activities. The secondary air pollutants are originated by

K. Krishnamurthi · P. K. Naoghare · S. S. Devi · A. Bafana
CSIR NEERI Environmental Health Division, Nagpur, India
e-mail: k_krishnamurthi@neeri.res.in

P. K. Naoghare
e-mail: pk_naoghare@neeri.res.in

S. S. Devi
e-mail: ss_devi@neeri.res.in

A. Bafana
e-mail: a_bafana@neeri.res.in

P. Arrigo (✉)
CNR Institute for Macromolecular Studies, Genoa, Italy
e-mail: arrigo@ge.ismac.cnr.it

S. M. Shiva Nagendra et al. (eds.), *Urban Air Quality Monitoring, Modelling and Human Exposure Assessment*, Springer Transactions in Civil and Environmental Engineering,
https://doi.org/10.1007/978-981-15-5511-4_21

chemical reactivity of primaries ones induced by physical or other chemical factors. The contribution of air pollution to the whole environmental stress is very high taking also into account the complete cycle of air pollutant (production, spread in the atmosphere, fall out in water and soil, contamination of drinking water and food). The technological advancement continuously increases the number of air pollutants originating more complex mixtures of substances spread also far away from their emission source. The regulation of pollutant emission is changed according to the availability of new scientific information. The Clean Air Act in the 1970 was the first law enacted to protect the atmosphere.[1] The Environmental Protection Agency (E.P.A) has formally defined the hazardous air pollutants (HAP) as chemical entities, conveyed by air flow, that are clearly identified or suspected to be a critical cause of much severe pathology.[2] The HAP includes inorganic, organic compounds and, more recently, nanomaterials. The composition of an HAP blend varies in different environmental contexts as urban, rural, outdoor or indoor. The damage, induced by an acute exposure, is widely demonstrated; conversely, the knowledge about long term exposure is not enough. This limitation is attributable to the complexity of data and long time required for epidemiological studies. In the last decade, it is emerged a new 'holistic' approach to investigate the effects of pollution on human health. This new way of thinking is commonly defined 'exposome' (Rappaport 2011). This new analytical perspective is based on the integration of the results obtained from each particular source of stress, separately screened. The simultaneous evaluation of information, obtained from each specific analysis, allows to improve the exposure risk prediction. It is important to underline that exposome takes into account both exogenous and endogenous toxicants (Wild 2012). The application of exposome approach is strongly dependent on availability of new high-throughput techniques, as transcriptomics and metabolomics that allow to deeply investigate multiple facets of molecular mechanisms underlying the arising of a severe chronic pathology. The environmental pollution burden is, in an exposome viewpoint, a pivotal element acting on the whole life of an individual. The impact of environmental chemical stressors on genomes is demonstrated by many studies mainly focused on DNA damage (Jiang et al. 2017). Only more recently, the effects of chemical pollutants on post-transcriptional and translational mechanism have been supported by a sufficiently large number of searches. Several studies on non-coding RNAs have corroborated the hypothesis that these molecules are capable to activate a rapid response against environmental changes. These findings have encouraged further deep investigation about molecular mechanisms of environmental stress response in which non-coding RNAs are involved. MicroRNAs are, among the all non-coding RNAs, the more extensively studied, nevertheless other non-coding RNAs can also be affected by pollutants (Karlsson O. and Baccarelli A. A. 2016).

The advancement of molecular technology has highlighted a relatively new possible target for contaminants in the post-transcriptional and translational phase of mRNAs. The centrality, in gene expression, of RNA processing is known for

[1]https://www.epa.gov/clean-air-act-overview.

[2]https://www.epa.gov/.

long time, only thanks to high sensitivity experimental methods is possible to deeply investigate different steps of post-transcriptional and translational processes (Mannin and Cooper 2017). The experiments unravel a complex system of enzymatic activities controlling the quality of a pre-mRNA (non-sense mediated decay) and the translation of a checked mRNA into a protein. The non-sense mediated decay is the mechanism controlling the 'quality' of genetic information carried by an pre-mRNA (Baker and Parker 2004). The 'half-life' of a mRNA is depending by its stability. The stability is under control of a combination of recognition elements located in mRNA, mainly in the upstream (5'-UTR) and downstream (3'-UTR) untranslated region. The 5'-UTR appears to be a pivotal element in ribosome recruitment (Hinnebusch et al. 2016; Fraser 2015). The 3'-UTR region of a mRNA is particularly relevant in the regulation of mRNA fate. (Geissler and Grimson 2016). This region contains several types of nucleotide motif for regulatory signals as poly-A (Skolnik-David et al. 1987) or AU-rich (Chen and Shyu 1995). A gene can possess different isoforms of 3'-UTR with different lengths. Several studies indicate that the length of 3'-UTR seems to be related to subcellular localization (Di Liegro et al. 2014). A different length suggests a different combination of regulatory signals. Recently, a relatively new regulatory mechanism has been identified in RNA: the chemical modifications. The RNA appears to be prone to a large number of chemical modification (Boschi-Muller and Motorin 2013). NGS techniques permit to screen modified nucleoside in a RNA sequence. This kind of modification affects coding and non-coding RNAs(Song and Yi 2017). The ensemble of chemical modification affecting RNA is called 'epitranscriptome' (Dominissini 2014). The pseudouridine (Ψ) is probably the first type of chemically modified nucleobase, however, also other modifications, such as the inosine derived by A-to-I (RNA editing), are well known. Among different chemical modification of RNAs, the methylation is presumably supported by a large number of experimental studies. Initially, the methylation was identified on DNA and hit a cytosine that becomes methyl-cytosine. The same modification has been also discovered on RNAs. The level of DNA methylation is commonly correlated with mutagenesis (CpG motif), while in RNA, the role of this modification is meanwhile under investigation. There are many different possibilities to methylate a nucleobase in RNA, however, one of them is supported by a substantial set of experimental data: the methylation of adenosine in position N6 (m6A) (Pan 2013) (Fig. 21.1).

At the present, the biological role of this modification is under investigations. Currently, available literature underlines the multiplicity of processes in which the m6A appears to be involved as pre-mRNA splicing, mRNA stability, nuclear export of mRNAs, translational efficiency and stem-cell differentiation. This modification seems also having a role in obesity, neuronal disorders, innate immune response, cancer and substance abuse. Nowadays, little information is available about the effect of xenobiotics, principally, those associated with environmental pollution on epitranscriptome. It is possible that a pollutant can influence the different kinetics, required to methylate and de-methylate an adenosine, interacting with a specific enzyme or directly interact with chemically modified nucleic acid. In this paper, we show the

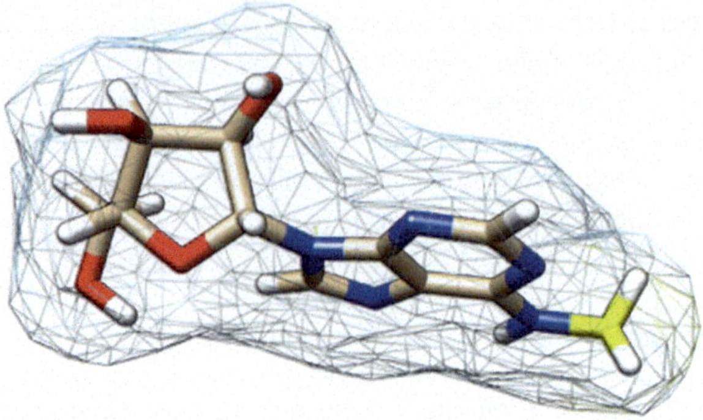

Fig. 21.1 N6-methyladenosine (CID 102175). The methyl group is marked by yellow stick. This figure has been obtained using UCSF Chimera (https://www.cgl.ucsf.edu/chimera/)

results of an '*in silico*' analysis focused on the prediction of interactions between hazardous air pollutants and RNA containing a methylated adenosine. This study is centered on the N6-methyladenosine modification of RNA independent by the type of RNA (mRNA, ncRNAs and so on).

21.2 Method

The inadequacy of available information about the role played N6-methyladenosine on different biological processes, under environmental stress, encouraged us to develop and apply an '*in silico*' approach to estimate the potential harmful effects of HAP on this chemical modification of RNAs. The proposed computational workflow is essentially based on the estimation of interaction energy (docking) between each hazardous air pollutant and a structure of a m6A methylated RNA. The preliminary step was based on the preparation of a hazardous air pollutant dataset. They were selected from EPA HAP list[3] and subsequently downloaded from PubChem database.[4] In this analysis, we did not considered the type of source and if they are preferentially indoor or outdoor pollutants. The HAP data set is summarized in Table 21.1. The chemicals were selected according the accessibility to 'omics' information because this knowledge was essential to associate the interaction strength with possible alteration of mRNA processing.

We extracted from PubChem, if available, the 3D conformer of each selected HAP. Those chemicals lacking of 3D conformer were processed by Frog2 system (Miteva et al. 2010). This tool, starting from SMILE (Weininger 1988) coding, is capable to generate a 3D conformation useful for the forthcoming docking phase. The little structural information about chemically modified RNAs convinced us to

[3]https://www3.epa.gov/ttn/atw/pollsour.html.

[4]https://pubChem.ncbi.nlm.nih.gov/.

Table 21.1 EPA hazardous air pollutants selected for this analysis

PubChem code (CID)	Chemical name	PubChem code (CID)	Chemical name
6115	Aniline	996	Phenol
5897	2-Acetamidofluorene	6338	Vinyl chloride
568	Dibenzofuran	5569	Trifluralin
5943	Carbon tetrachloride	991	Parathion
5558	Bromoform	7416	Nitrobenzene
31369	Chloroprene	7276	Styrene oxide
300	Chloroacetic acid	7261	2,4-Diaminotoluene
177	acetaldehyde	7243	O-Phenylenediamine
10967	Chloromethyl ether	7242	O-toluidine
10757	2-chloroacetophenone	7114	4-Nitrobiphenyl
10522	Chlorobenzilate	7047	Quinoline
10039	Carbonyl sulfide	720	2-amino-6-[[4,5-dihydroxy-6-(hydroxymethyl)-3-oxooxan-2-yl]amino]hexanoic acid
6581	Acrylic acid	6658	Methyl methacrylate
1140	Toluene	6564	1,2-Dichloropropane
1443	15-kete	6569	2-Butanone
11641	Vinyl bromide	6544	Isophorone

(continued)

Table 21.1 (continued)

PubChem code (CID)	Chemical name	PubChem code (CID)	Chemical name
12228	Methyl isocyanate	6378	Propylene oxide
12669	Methane-D1	6377	2-Methylaziridine
15625	Dioxin	6366	Vinylidene chloride
31373	Tetrachloroethylene	6278	1,1,1-Trichloroethane
4115	Methoxychlor	6124	N-Nitrosodimethylamine
4650	p-benzoquinone	6061	Methylhydrazine
4944	Propoxur	6046	N-nitrosomorpholine

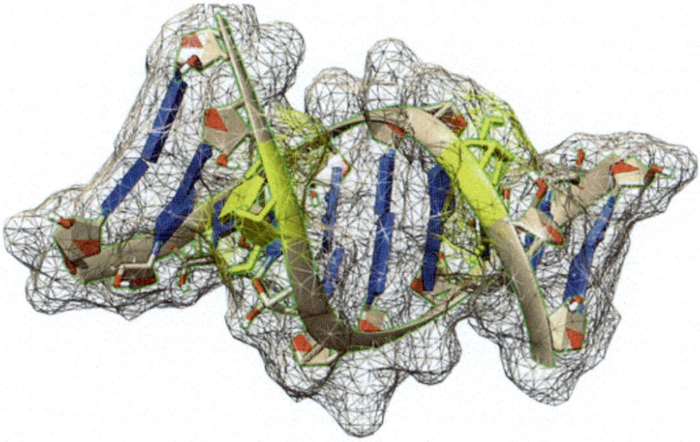

Fig. 21.2 Synthetic N6-methylated RNA. The PDB entry 2MVS containing a methylated adenosine (yellow labeled). This figure has been obtained using UCSF Chimera (https://www.cgl.ucsf.edu/chimera/)

use, for the docking, a synthetic oligomers containing a N6-methyladenosine (PDB code:2MVS) (Roost et al. 2015) (Fig. 21.2).

The PDB entry contains two synthetic chains, for our interaction analysis, we have taken only one chain. This choice was suggested to focus the analysis on the interaction with modified nucleobase without a possible interference with sugar backbone in the double-strand oligomer. The docking was performed using the PatchDock-FireDock suite Web server (Mashiach et al. 2010). This tool is based on a rigid docking method, in which the conformational modifications, due to molecular flexibility, are not considered. In the case of a small oligomer of nucleic acid, this methodology is the more ease to apply. The docking phase produced a set of numerical features describing the ligand-target binding efficiency. The whole set of features is tabulated below. The docking parameters represent the total interaction energy (GCE) and the Van der Waals contribution. In addition, the system estimates if the interaction occurs in the internal region of the target (insideness) and it evaluated the contribution of desolvation energy (ACE) to ligand-target binding. The final set of features was used to apply an unsupervised classification method. This information was integrated with some chemical-physics characteristics of air pollutants extracted from PubChem database regarding the hydrogen acceptor and donor atoms, the number of heavy atoms, and other characteristics of each selected HAP. The complete set is summarized in the table (Table 21.2).

For classification phase, a synthetic $\begin{bmatrix} chem_{1,1} & chem_{1,m} \\ chem_{n,1} & chem_{n,m} \end{bmatrix}$ data matrix was generated where N is the number of HAP in the dataset and M is number of features. The classification was initially carried out by a k-means clustering (Everitt et al. 2011). The cluster analysis was applied three times: (1) using only docking features, (2) using only PubChem features and (3) using the whole set of features. Cluster

Table 21.2 Set of HAP features used for classification

Variable acronym	Description	Origin
GCE	Global contact energy	PatchDock
aVdW	Van der Waals attractive contribution	PatchDock
rVdW	Van der Waals repulsive contribution	PatchDock
Insid	Degree of interaction internality	PatchDock
ACE	Atomic contact energy (desolvation energy)	PubChem
H accept	Number of hydrogen bond acceptor atoms	PubChem
H donor	Number of hydrogen bond donor atoms	PubChem
RB	Number of rotable bonds	PubChem
HA	Number of heavy atoms (atoms different from hydrogen)	PubChem
TPS	Topological surface area (spatial characteristics of chemical)	PubChem
XlogP3	Estimated partition coefficient octanol/water	PubChem

validity has been assessed by davies-bouldin index (Davies and Bouldin 1979). The final steps of data analysis has been centred on estimate the relative variable relevance by machine learning. The relative dependency, among the more relevant features, has been estimated by regression tree. The regression trees were cross-validated using k-fold method with $k = 10$ (Fushiki 2011).

21.3 Results

This analysis outlined two interesting aspects of hazardous air pollutants. The insufficiency, for a large part of HAP, of 'omics' information was a constrain for the generalization of HAP–RNA interactions. Taking this notice into account, it is interesting to note the differences emerging information from different cluster analysis. It is remarkable how the classification based on PubChem information allows to find a larger number of cluster compared to the result with docking parameters and the overall set of features. Table 21.3 shows the minimal value of Davies–Bouldin index

Table 21.3 Summary of cluster validation (Davies–Bouldin index)

Type of features	Number of features	Optimal number of clusters	Minimal value of Davies–Bouldin index
Docking features	5	2	0.5840
PubChem features	6	4	0.4950
Overall features	11	3	0.6230

Table 21.4 Hazardous air pollutants minimizing the distance from cluster centroid

Cluster	Docking features	PubChem features	Complete set of features
Cluster 1	2-Methyaziridine (cid 6377)	Nitrobenzene (cid 7416)	N-nitrosomethyl amine (cid 6124)
Cluster 2	Nitrobenzene (cid 7416)	Bis(chloro)methyl ether (cid10967)	P-phenylendiamine (cid 7814)
Cluster 3		O-toluidine (cid 7242)	Bis(chloro)methyl ether (cid 10967)
Cluster 4		Trifluralin (cid 5569)	

for the three different cluster analysis. The cluster analysis was performed using not normalized data using a city-block metric (Larson and Sadiq 1983).

The compounds, minimizing the distance from cluster centroid are, for each cluster analysis, are summarized in the following table (Table 21.4).

The cluster analysis has given information about the groups of chemicals but not about the dependency (relevance) of chemical-RNA interaction by predictive features. In order to achieve this goal, a regression analysis was carried out. The analysis of regression tree performed on each feature set separately has originated the following optimized trees. The features were normalized by using interquartile normalization approach generally used for microarray data processing (Bolstad et al. 2003). We performed different regressions: (a) using only docking feature; (b) using only PubChem features and (c) using only the relevant parameters identified.

The second regression tree was used to estimate the octanol/water partition coefficient (XLogP3) dependency on chemical-physics characteristics reported in PubChem database This analysis underlined the relevance of heavy atoms number (HA) and topological surface area (TPS) on the XlogP3 estimation. The final regression analysis was focused to estimate the dependency of ligand-target interaction energy on the base of all features. This analysis has outlined the role of attractive Van der Waals contribution on HAP-RNA binding.

The same behavior emerged from the analysis carried out only on docking features (Table 21.5).

The estimation of relative relevance of PubChem feature to predict the XlogP3 value has indicated the number of heavy atoms as more important parameter (Table 21.6).

The relevance of these feature has been confirmed using only those variables having a relevance parameter >0. The features that satisfy this condition were: 1) Attractive Van der Waals contribution; 2) Atomic contact energy; 3) Heavy atoms;

Table 21.5 Feature obtained from docking and used for classification

aVdW	rVdW	Inside	ACE
0.866	0	0	0.02

Table 21.6 Physico-chemical feature extracted from PubChem and used for classification

H acceptor	H donor	Rotable bonds	Heavy atoms	Topological surface
0	0.0032	0	**0.2036**	0.046

Table 21.7 Set of more informative features (docking and PubChem) used for classification

aVdW	ACE	Heavy atoms	H donor	XlogP3
0.022	**0.30**	0	0	0

Table 21.8 Set of more informative feature (docking and PubChem) without ACE

aVdW	H donor	H acceptor	Heavy atoms	Rotable bonds	XlogP3
0.2406	0	0.038	0.0005	0	0

4) XlogP3. The results have underlined their influence on HAP-RNA interaction (Table 21.7).

In order to investigate the influence of other features except ACE, we trained a tree with the following parameters (Table 21.8).

These different regressions highlighted the role of Van der Waals contribution, as predictable, on the HAP-m6A_RNA. The desolvation energy (ACE) seems to mask the role of physico-chemical parameters. The last regression, based on a set of feature without ACE, has confirmed aVdW as more discriminative feature followed by H acceptor and heavy atom.

21.4 Discussion

The fundamental role of N6-methyladenosine (m6A) in the mRNA translational process has been demonstrated by several studies. The evolutionary conservation (Sergiev et al. 2016) suggests that it is a pivotal regulator of mRNA processing, and the reversibility of N6-adenosine methylation of RNA is a clue of a rapid response of a biological system, not only eukaryote, to endogenous and exogenous signals (endogenous metabolite drugs or hazardous air pollutants). The current information about a possible direct interaction between an hazardous air pollutant and an RNA is quite scarce. This analysis shows the results of an 'in silico' analysis of HAP–RNA interaction. The application of docking is limited by the availability of solved 3D structure of chemically modified RNA. It is important to underline that interactions with inorganic pollutants (ions) has been neglected because this approach requires too much computational time. Our survey, even if carried out on a limited set of HAP, underlined the possibility that these compounds could affect the mechanisms involved in mRNA quality control and translation by a direct interaction with chemically modified nucleotides. The interaction between an HAP and N6-methylated

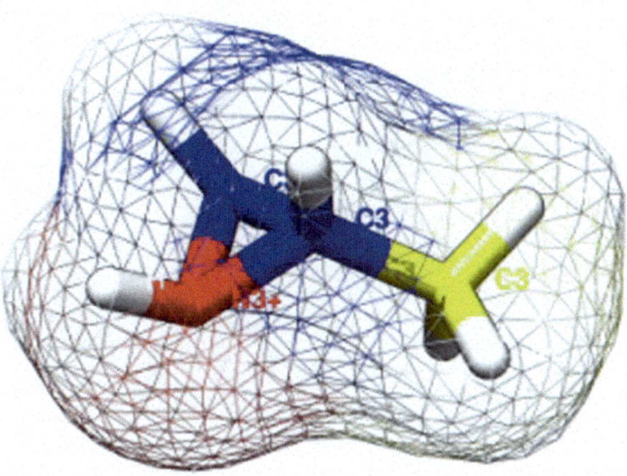

Fig. 21.3 Structure of methylaziridine (cid 6377). The yellow stick indicated the CH_3 group and the red residue is an NH_3^+. This figure has been obtained using UCSF Chimera (https://www.cgl.ucsf.edu/chimera/)

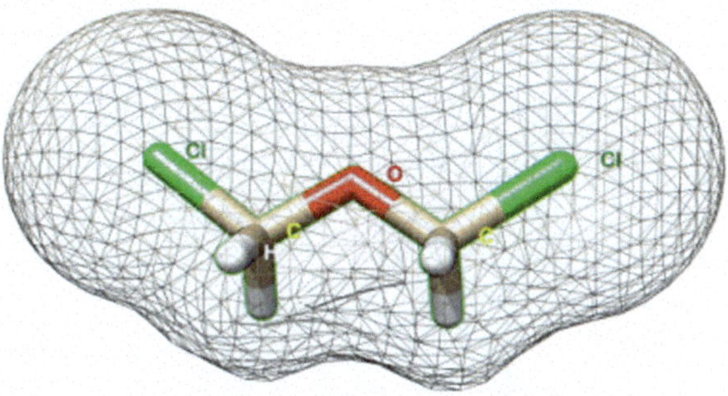

Fig. 21.4 Bis(chloro)methyl ether (cid10967). The green atoms mark the Cl and the red the O. This figure has been obtained using UCSF Chimera (https://www.cgl.ucsf.edu/chimera/)

adenosine RNA is exemplified by the complexes obtained by docking. The figures below exemplify two different complexes of hazardous air pollutants and a single chain of N6-methylated RNA oligomer deposited in PDB database (Figs. 21.3, 21.4, 21.5 and 21.6).

The two figures highlight the different interacting behavior of each HAP with modified methylated RNA. The first pollutant, *Bis(chloro)methyl ether*, contains two Cl^- substituents in methyl groups. In this case, the pollutant seems to recognize the methylated adenosine. The second example, the *2-methylaziridine*, contains a CH_3 and an N^+ atom. In this case, the interaction does involve the methylated adenosine. Both HAPs selected as example are defined as carcinogens.

The 2-methylaziridine appears to be involved in several renal diseases and its potential gen target is HEXB. Excluding genetic diseases an abnormal function of these proteins is related to metabolic diseases affecting neurons or lipid accumulation. The estrogen receptor 1 (ESR1) is, taking the information extracted from PubChem, the potential target for Bis(chloro)methyl ether. The other HAPs (*nitrobenzene*, *O-toluidine*, t*rifluralin*, *N-nitrosomethylamine*, *P-phenylendiamine*), highlighted by cluster analysis, do not interact with modified adenosine. This compound contains a *N* characterized by a partial positive charge. The nitrobenzene has ALDH1A1 gene as

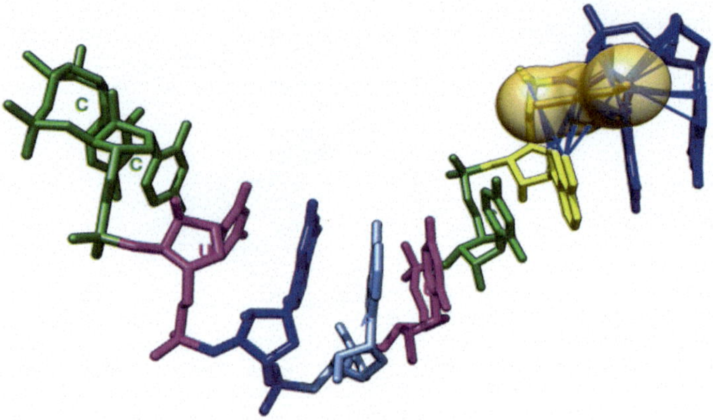

Fig. 21.5 Complex of 2MVS chain A with Bis(chloro)methyl ether (cid10967). The yellow residue indicates the position of m6A

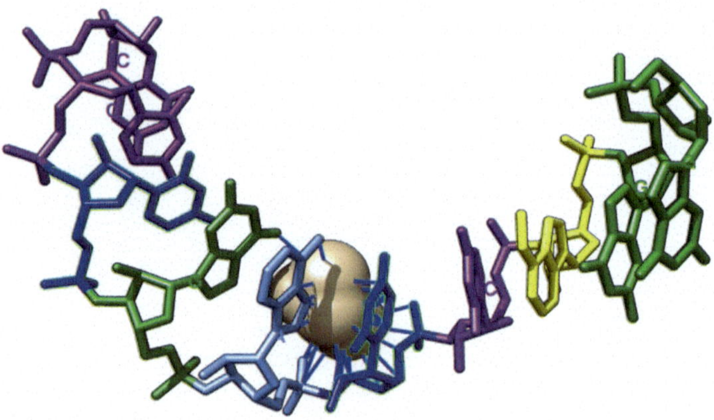

Fig. 21.6 Complex of 2MVS chain A with 2-methylaziridine (cid 6377). The yellow residue indicates the position of m6A. This figure has been obtained using UCSF Chimera (https://www.cgl.ucsf.edu/chimera/)

target and it is a potential carcinogen. The O-toluidine is a human carcinogen acting on *thyroid stimulating hormone receptor* (TSHR). O-toluidine also contains a N as $-NH_2$ group. The trifluralin is also a carcinogen and it can target, as o-.toluidine, the THSR. However this chemical is also capable to target another protein: the *nuclear receptor subfamily 1 group I member 3* (NRI3). The *N-nitrosomethylamine* also contains N and its carcinogenicity is well established. At the moment, biological studies are not yet identified a human target on which this compound is active. The last chemical, the *P-phenylendiamine* contains also N atoms in the NH_2 groups in para position. In this case, the N does not have a partial charge. No evidences support its inclusion as human carcinogen, despite biological test has identified a human protein targeted by this compound: the *microtubule associate protein tau* (MAPT). These proteins are involved in several neurodegenerative diseases.

21.5 Conclusion

The investigation of molecular mechanisms, underlying the RNA methylation, is becoming more and more important in order to elucidate the fine tuning in involvement of translational process regulation. The chemical modifications affecting different types of RNA both coding and regulatory suggest a possible direct interaction between small molecular pollutants with ribonucleic acid. The interactions can occur on specific enzymes involved in RNA modification (methyltransferases), in the RNA editing (ADAR) or decay (poly-A binding proteins). It is important to underline that specific non-coding RNAs are also capable to methylate other RNAs; for instance, the role played by o C/D box snoRNA in rRNA methylation is well established. The role of adenosine methylation in position N6 is emerging as pivotal element in the regulation of mRNA stability. The physiological reversibility of this chemical modification suggests that any factor capable to modify the reversibility of the process can indirectly affect the mRNA stability.

Our analysis also suggests that interference, between HAP, due to a long term exposure, and m6A methylated RNAs could be a cause of alteration of reversibility (Jia et al. 2013) in the N6-methylation enzymatic cascade. A modification of this enzymatic cascade can affect several other processes involved in cell reprogramming. It is important also to underline that N6-methyladenosine occur also in miRNA and other non-coding RNAs supporting the hypothesis that environmental stress, acting on N6-methyladenosine process is a critical path to induce unapparent damages in the regulatory mechanism controlling the physiological response to environmental changes. These small variations can, on long term, accumulate and originate a blown pathology (Chandola et al. 2015). We guess that compared to the investigation of interaction between HAP and specific enzymes, the information about the potential interaction with N6-methylated RNA could be valuable to estimate the proneness of RNAs to environmental stress factors.

Acknowledgements This study has been supported by EU TECO project mobility grant.

References

Baker KE, Parker R (2004) Nonsense-mediated mRNA decay: terminating erroneous gene expression. Curr Opin Cell Biol 16:293–299

Bolstad BM, Irizarry RA, Astrand M, Speed TP (2003) A comparison of normalization methods for high density oligonucleotide array data based on variance and bias. Bioinformatics 19:185–193

Boschi-Muller S, Motorin Y (2013) Chemistry enters nucleic acids biology: enzymatic mechanisms of RNA modification. Biochemistry (Mosc) 78:1392–1404

Chandola U, Das R, Panda B (2015) Role of the N6-methyladenosine RNA mark in gene regulation and its implications on development and disease. Brief Funct Genomics 14:169–179

Chen CY, Shyu AB (1995) AU-rich elements: characterization and importance in mRNA degradation. Trends Biochem Sci 20:465–470

Davies LD, Bouldin DW (1979) A cluster separation measure. IEEE Trans Pattern Anal Mach Intell PAMI-1 2:224–227

Di Liegro CM, Schiera G, Di Liegro I (2014) Regulation of mRNA transport, localization and translation in the nervous system of mammals (review). Int J Mol Med 33:747–762

Dominissini D (2014) Genomics and proteomics. Genomics and proteomics. Science 346:1192

Everitt B, Landau S, Leese M, Stahl D (2011) Cluster analysis. Wiley, Chichester, United Kingdom

Fraser CS (2015) Quantitative studies of mRNA recruitment to the eukaryotic ribosome. Biochimie 114:58–71

Fushiki T (2011) Estimation of prediction error using by *K*-fold cross-validation. Stat Comput 21:137–146

Geissler R, Grimson A (2016) A position-specific 3'UTR sequence that accelerates mRNA decay. RNA Biol 13(11):1075–1077

Hinnebusch AG, Ivanov IP, Sonenberg N (2016) Translational control by 5'-untranslated regions of eukaryotic mRNAs. Science 17(352):1413–1416

Jia G, Fu Y, He C (2013) Reversible adenosine methylation in biological regulation. Trends Genet 29:108–115

Jiang CL, He SW, Zhang YD, Duan HX, Huang T, Huang YC, Li GF, Wang P, Ma LJ, Zhou GB, Cao Y (2017) Air pollution and DNA methylation alterations in lung cancer: a systematic and comparative study. Oncotarget 8:1369–1391

Karlsson O, Baccarelli AA (2016) Environmental health and long non-coding RNAs. Curr Environ Health Rep 3:178–187

Larson RC, Sadiq G (1983) Facility location with the Manhattan distance in presence of barrier to travel. Oper Res 31:652–699

Mannin KS, Cooper TA (2017) The roles of RNA processing in translating genotype to phenotype. Nat Rev Mol Cell Biol 18:102–114

Mashiach E, Schneidman-Duhovny D, Peri A, Shavit Y, Nussinov R, Wolfson HJ (2010) An integrated suite of fast docking algorithms. Proteins 78:3197–3204

Miteva MA, Guyon F, Tufféry P (July, 2010) Frog2: efficient 3D conformation ensemble generator for small compounds. Nucleic Acids Res 38 (web server issue): W622-7

Pan T (2013) N6-methyl-adenosine modification in messenger and long non-coding RNA. Trends Biochem Sci 38:204–209

Rappaport SM (2011) Implications of the exposome for exposure science. J Expo Sci Environ Epidemiol 21:5–9

Roost C, Lynch SR, Batista PJ, Qu K, Chang HY, Kool ET (2015) Structure and thermodynamics of N6-methyladenosine in RNA: a spring-loaded base modification. J Am Chem Soc 137:2097–2115

Sergiev PV, Golovina AY, Osterman IA, Nesterchuk MV, Sergeeva OV, Chugunova AA, Evfratov SA, Andreianova ES, Pletnev PI, Laptev IG, Petriukov KS, Navalayeu TI, Koteliansky VE, Bogdanov AA, Dontsova OA (2016) N6-methylated adenosine in RNA: from bacteria to humans. J Mol Biol 428:2134–2145

Skolnik-David H, Moore CL, Sharp PA (1987) Electrophoretic separation of polyadenylation-specific complexes. Genes Dev 1:672–682

Song J, Yi C (2017) Chemical modifications to RNA: a new layer of gene expression regulation. ACS Chem Biol 12:316–325

Weininger D (1988) SMILES a chemical language and information system. 1. Introduction to methodology and encoding rules. J Chem Inf Comp 28:31–36

Wild CP (2012) The exposome: from concept to utility. Int J Epidemiol 41:24–32

Chapter 22
Characterization of PM_{10} and Its Impact on Human Health During Annual Festival of Lights (Diwali) in Northeast India

Rajyalakshmi Garaga and Sri Harsha Kota

22.1 Introduction

One of India's most visually spectacular festivals being celebrated is Diwali. It is the five-day festival of lights (deep = light and avali = a row, i.e., a row of lights) in which Diwali is traditionally celebrated on the third day. People of all age groups celebrate this holy festival with the theme of triumph of light over darkness and good over evil. They offer expression to their joy by lighting earthen "divas" (lights), enhancing the houses, and welcoming precious ones to their homes for sharing in a grand gala. This festival falls in October or November every year and includes the burning of huge amounts of fireworks. Display of fireworks burning starts from the evening and continues till late at night.

However, apart from sharing happiness and goodness on this auspicious occasion, contamination of air due to the discharge of large amounts of harmful gases and toxic substances into the environment due to the burning of these firecrackers prompts adverse effects to human health. Sharp increment in the concentrations of constituents of firecrackers was reported constantly over the years (Gouder and Montefort 2014). Determination of PM (<10 μm) in the ambient air is essential in light of the fact that they carry a mixture of toxins from firecrackers which enter the respiratory system causing pulmonary effects (Verma and Deshmukh 2014), and its impact is highly acknowledged in the case of infants, women and elderly people (Ambade and Ghosh 2013). PM_{10} creates thick clouds of smoke that contain potassium nitrate, charcoal, and sulfur (Ambade and Ghosh 2013). Health effects commonly include short-term (acute) or can be extended to chronic in case of long-term exposure. Pollutants released from fireworks at a higher elevation can be scattered in a substantial volume of air and in this manner may not have a severe health danger (Betha

R. Garaga · S. H. Kota (✉)
Department of Civil Engineering, Indian Institute of Technology Guwahati, Guwahati, India
e-mail: harshakota@iitg.ernet.in

S. M. Shiva Nagendra et al. (eds.), *Urban Air Quality Monitoring, Modelling and Human Exposure Assessment*, Springer Transactions in Civil and Environmental Engineering,
https://doi.org/10.1007/978-981-15-5511-4_22

and Balasubramanian 2013). However, the ground level firecrackers display has an immediate implication towards human health. Acute eosinophilic pneumonia (AEP) has been tested positive for a patient due to inhalation of smoke continuously for three nights from the burning of fireworks (Hirai et al. 2000). While fever, cough, and dyspnoea were reported initially as an acute effect. Exposure to particulate matter causes chronic respiratory and cardiovascular diseases, pulmonary effects, premature death, and cancer (Nasir and Brahmaiah 2015; Barman et al. 2008). Patients with wheezing, respiratory diseases, exacerbation of bronchial asthma and bronchitis increased to 30–40% reported during the Diwali celebration in India (Gouder and Montefort 2014).

Several studies were carried out in different parts of India to assess the impact of the fireworks burning on PM_{10} related to air quality. A short-term study (Ravindra et al. 2003) reported 2–3 times elevation in PM_{10} concentrations during the Diwali festival in Hisar city in the north-west part of India. PM_{10} during Diwali day in Lucknow in northern India was reported as 7.53 times higher than normal days (Barman et al. 2008). Five times increase in PM_{10} concentration was observed in Kolkata in eastern India during Diwali (Chatterjee et al. 2013). An uncommon increment in PM_{10} concentration to 35 times during Diwali day in comparison to a normal day, due to abrupt pollution load caused by burning of fireworks was reported in Gujarat in western India (Nasir and Brahmaiah 2015). In Jharkhand, PM_{10} concentration during the Diwali period has raised to nearly 3 times than normal days (Ambade and Ghosh 2013). 4–10 times increase in PM_{10} concentrations were observed during Diwali in Nagpur in central India (Khaparde et al. 2012). Owing to already high PM_{10} concentrations, several studies were conducted in the national capital, New Delhi. For example, remarkably huge concentrations of PM_{10}, i.e., 767 and 620 $\mu g/m^3$ were observed during the Diwali festivals of 2008 and 2009, respectively (Perrino et al. 2011). Also, a community-based health survey (Sharma et al. 2015) conducted during Diwali in 2013 revealed an increase in number of patients in Delhi with problems related to respiratory diseases, hearing issues, irritation in eyes, and headache.

Only two investigations till now were reported from northeastern India. In Tezpur, mean PM_{10} concentration observed was 87.45 $\mu g/m^3$ which was 2.13 times more than the normal day concentrations (Deka and Hoque 2014). Another study from Dibrugarh (Pathak et al. 2015), reported that PM_{10} and $PM_{2.5}$ concentrations during Diwali were 168 and 160 $\mu g/m^3$ respectively which was 5.33 and 5.74 times higher than the normal day concentrations. However, no health correlation and risk level study was done.

Over the years, these celebrations during Diwali have only gained more importance. However, studies in a residential campus with a goal of understanding the direct environmental impact of fireworks on the health of residents are rare. Moreover, no investigation was documented in the biggest city of northeast, i.e., Guwahati, which has severe air quality issues. Thus, the main aim of this study is to study PM_{10} and its associated health risk on the residents during Diwali in 2015. The chemical composition of PM_{10} was studied and then used for source identification using the United States' Environmental Protection Agency's Positive Matrix Factorization (PMF) model. Additionally, noise levels and biological characteristics of PM_{10} during Diwali, which is the first of its kind analysis till date, were also investigated.

22.2 Methodology

22.2.1 Site and Sample Collection

Guwahati is one of the fast-growing cities in India and the largest city of northeast India. The present study was conducted in the Indian Institute of Technology Guwahati (IITG), which covers 285 hectares plot of land on the north bank of the river Brahmaputra and 20 km from the heart of the city. Surrounded by mountains, the campus is located in a valley. It is geographically positioned around 26 °11′ 14″ N and 91° 41′ 30″ E. The sampling location, included in the Fig. 22.1, is at an industrial region surrounded by export promotion industrial park, Guwahati biotech park, skin and health care industry, a pharmaceutical company, a machine industry, and a liquid petroleum gas bottling plant. Moreover, construction activities prevail throughout the year as a part of capacity expansion within the premises of residential areas.

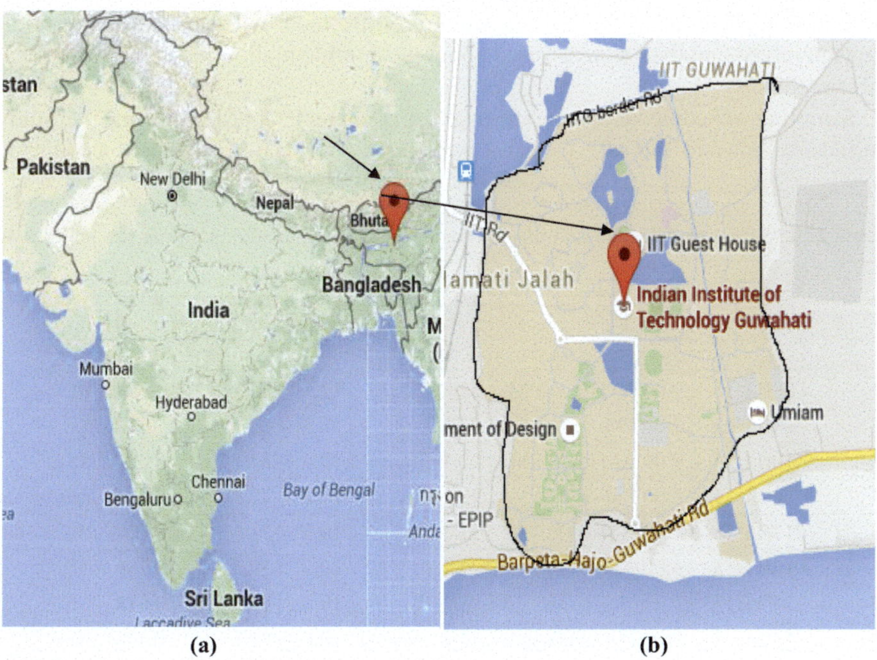

Fig. 22.1 **a** Guwahati city in India. **b** Study area—IIT Guwahati

Vehicle distribution mainly includes motorcycles and passenger cars with few heavy commercial vehicles used in the construction sites.

For quality check, filter papers were sterilized by autoclaving at 121 °C (250 °F) and 100 kPa (15 psi) above atmospheric pressure for 15 min to ensure zero microorganisms before using them in the sampler. Filters were carried to and fro from the site in sealed polyethylene bags and utmost care was taken to avoid handling errors. Also, filter papers were desiccated for 24 h before and after the sampling to minimize the errors due to moisture and achieve accurate measurements. To prevent manual contamination, few drops of ethanol was applied to hands during filter placement and retrieval.

Noise level was measured using a sound level meter (Amprobe SM—20 A) at a height of 3–4 m from ground level, following the Central Pollution Control Board guidelines (CPCB 2003).

Weather data was collected using a weather station (Vantage Pro 2). Sampling started from 8th and continued till 17th of November 2015, to predict the variation of air quality. Samples collected during 10th, 11th, and 12th, were treated as Diwali samples. The samples before and after Diwali period are considered pre- and post-Diwali samples. A health-based survey on the patients attending the institute's hospital during those days was conducted. Information regarding age, body weight, and various diseases such as headache, fatigue, irritation, coughing, sneezing, and sinusitis, with which the patients were suffering were documented. This information was utilized for further analysis of the study such as risk level calculation. The need and importance of the present study are informed beforehand to the participants involved and they were assured about the confidentiality of disclosed personal information.

Sampling schedule and meteorological parameters recorded during the study period are shown in Table 22.1. Vector average of wind speed and direction collected

Table 22.1 Sampling schedule and meteorological parameters recorded during the monitoring period

S. no.	Sampling date	Sampling time	Temperature (°C)	Dew point (°C)	Relative humidity (%)	Wind speed (m/s)	Wind direction
1	08/11/15	9:00 AM–9:00 PM	22.6	19.5	83.7	1.084	SE
2	09/11/15	9:00 PM–9:00 AM	22.6	19.3	82.9	1.312	SE
3	10/11/15	9:00 AM–5:00 PM	22.3	19.4	84.1	0.682	SSE
4	11/11/15	5:00 PM–1:00 AM	23.1	19.2	79.5	0.624	SE
5	12/11/15	1:00 AM–9:00 AM	23.7	19.5	78.5	0.572	SE
6	13/11/15	9:00 AM–9:00 PM	23.1	19.2	79.5	0.786	SE
7	14/11/15		23.2	19.2	79.2	0.618	SE
8	15/11/15	9:00 PM–9:00 AM	23.3	19.6	81.2	0.436	SSE
9	16/11/15		22.8	19.9	84.6	0.592	SSE
10	17/11/15		21.7	19.8	89.1	0.867	SE

every 5 min during the analysis period by the weather station were averaged to find the dominant wind direction as Southeast with wind speed being 0.5–1 m/s for 33% of hours.

22.2.2 Source Identification

Receptor models are applied in many studies (Kim and Hopke 2004; Polissar et al. 1998; Begum et al. 2004; Bove et al. 2014), have proven capability in identifying the accurate and potential emission sources of PM_{10} at a receptor location (Waked et al. 2014). The present study used the US EPA's PMF (Norris 2014) to quantify the source contributions to PM_{10} mass. PMF, a multivariate factor analysis tool, divides a matrix of speciated sample data into two matrices denoting factor contributions and factor profiles (Paatero and Tapper 1994; Paatero 1997). The detailed descriptions of EPA PMF v5.0 are described in EPA PMF User Guide (2014), and briefly description is given here. If "X_{ij}" is the concentration of specie "i" in sample "j", PMF identifies the number of sources "p", the species profile "g" of each source and the amount of mass "f" contributed by each source to each individual sample as expressed in Eq. (22.1). Minimization of the residual e_{ij}, in Eq. (22.1) is done using a least-squares approach (Paatero and Tapper 1994; Paatero 1997).

$$X_{ij} = \sum_{p=1}^{l} g_{ip} f_{pj} + e_{ij} \tag{22.1}$$

PMF allows the results to be constrained to have non-negative source contributions and profiles. PMF values each data point to be individually weighted, based on the level of confidence in the measurement. For example, higher uncertainty relates to species with below detection limit concentrations than with those of the above detection limit. The object function "Q", minimized depending on the uncertainties (u) associated with each sample using Eq. (22.2).

$$Q = \sum_{i=1}^{m} \sum_{j=1}^{k} \left[\frac{X_{ij} - \sum_{p=1}^{l} g_{ip} f_{pj}}{u_{ij}} \right]^2 \tag{22.2}$$

where, "m" and "k" are total number of samples and species, respectively.

In this study, 23 samples and 13 species were used as input data for PMF model. Categorization of quality of data was based on the signal to noise ratio (S/N) and the percentage of sample method detection limit (MDL). Method detection limit (MDL) for all the samples and species was calculated using Eq. (22.3), using species-specific detection limit (Harrison and Shallcross 2011), volume of air sampled (V_i)

and volume of sample used for analysis (V_a) which is 20 mL in the present study.

$$\mathrm{MDL}_{ij} = \frac{\mathrm{DL}_j \times V_a}{V_i} \quad (22.3)$$

Thereafter, uncertainties "UC_{ij}" was calculated using Eq. (22.4), unless C_{ij} is less than MDL_{ij}, where UC_{ij} is taken as 5/6th of MDL_{ij}.

$$\mathrm{UC}_{ij} = \sqrt{(0.1 \times C_{ij})^2 + (0.5 \times \mathrm{MDL}_{ij})^2} \quad (22.4)$$

As this study focuses on a particular episode only 23 samples were collected. It is not uncommon to carry out PMF analysis in such studies. For example, Tian, Wang (Tian et al. 2014) studied the impacts of fireworks in Chinese New Year using 19 species analyzed from 26 samples. However, in such small data sets, high concentrations could potentially drive the solution. To estimate the uncertainty of the estimated profiles, 100 bootstrap runs with a minimum co-relation of 0.6 were performed.

22.2.3 Chemical Analysis

PM_{10} collected on each sample filter was divided into aliquot parts, i.e., one half and two-quarter halves for various purposes. For water-soluble ions, an aliquot of PM_{10} sample was immersed in ultrapure water and ultrasonicated for 20 min. Residual is filtered and the filtrate volume is made up to 15 mL. Samples were stored in prewashed polyethylene bottles and kept at 4 °C until the analysis in Ion Chromatograph (Metrohm 792 basic IC).

Anion column (Metrosep A Supp 5-250/4.0) with a suppressor was used for anions analysis. A solution mixture of 3.2 mM Na_2CO_3 and 1 mM $NaHCO_3$ was used as the eluent and the flow rate was maintained at 0.7 mL/min. 50 mM H_2SO_4 was used for regeneration.

Cations were analyzed by a cation column (Metrosep C 4 150/4.0). The eluent was prepared in ultrapure water with 1.7 mmol/L nitric acid and 0.7 mmol/L dipicolinic acid. The eluent flow rate was maintained at 0.9 mL/min. 20 μL of sample was measured using an inbuilt loop and injected into the IC system. Before injecting, all the samples were filtered through Millipore 0.22 μm PTFE filters.

For metals, digestion of aliquot of PM_{10} sample in 9 mL HNO_3 using hot plate under the operating conditions of 100 °C for 2 h was done. After filtration, final the volume was diluted to 50 mL using ultrapure water, and storage was made in polyethylene vials at 4 °C maintaining pH ~ 2. Atomic Absorption Spectroscopy (AAS) (Varian Spectra AA-55) was used to analyze the heavy metals. Following are the metals and ions quantified in the present study: Cd, Co, Fe, Ni, Sr, Zn and Ca^{2+}, NH_4^+ , Na^+, K^+, F^-, Cl^-, NO_3^-, and SO_4^{2-} (Table 22.2).

Table 22.2 Comparison of PM_{10} ($\mu g/m^3$) and metal concentration monitored during various firework episodes across the world along with present study

Event, year	Diwali, 2015[a]	Las Fallas, 2005[b]	FIFA World Cup victory, 2006[c]	Diwali, 2009[d]	Diwali, 2009[e]	Yanshui 2009[f]	Diwali, 2009[g]	Lantern festival, 2013[h]
Location	Guwahati, India	Valencia, Spain	Milan, Italy	Delhi, India	Delhi, India	Kaohsiung, Taiwan	Tezpur, India	Tianjin, China
PM_{10}	311	79	63.9	507.2	620	NA	40.88	212.95
Cd	0.14	0.0006	NA	NA	0.0043	0.02	NA	NA
Co	0.52	0.0004	NA	NA	0.0016	NA	0.007	0.01
Fe	3.52	0.4	1.731	0.82	3.900	1.19	2.46	7.5
Ni	3.41	0.003	0.005	6.30	0.013	NA	0.14	NA
Sr	9.22	0.03	0.139	0.03	1.6	0.31	NA	NA
Zn	0.71	0.08	0.190	0.08	1.6	1.29	NA	0.55

Note
NA not available
[a]Present study: 12-h sampling during the Diwali period
[b]Moreno et al. (2007): 24-h sampling duration
[c]Vecchi et al. (2008): 4-h sampling
[d]Sarkar et al. (2010): 24-h average sampling of three sites during Diwali day
[e]Perrino et al. (2011): 24-h sampling
[f]Tsai et al. (2012): 4-h sampling during Diwali day
[g]Deka and Hoque (2014): 12-h sampling, whole monitoring campaign
[h]Tian et al. (2014): 24-h sampling during the study period

22.2.4 Biological Analysis

Aliquot of PM_{10} sample was dipped in the reagent bottle having sterilized distilled water to transfer the bacteria. Serial dilution technique was followed and sample was poured on the petri plate with agar. Further, bacteria was allowed to grow on nutrient agar medium for 2–3 day incubated at 37 °C. Finally, number of bacteria was quantified and reported as CFU/m^3.

22.3 Results and Discussion

22.3.1 PM_{10} Concentrations

Figure 22.2 shows the variation of PM_{10} concentrations during pre-Diwali, Diwali, and post-Diwali days. As seen in the figure, PM_{10} concentrations always exceeded NAAQS limit of 100 $\mu g/m^3$. Mean concentrations during non-Diwali days were

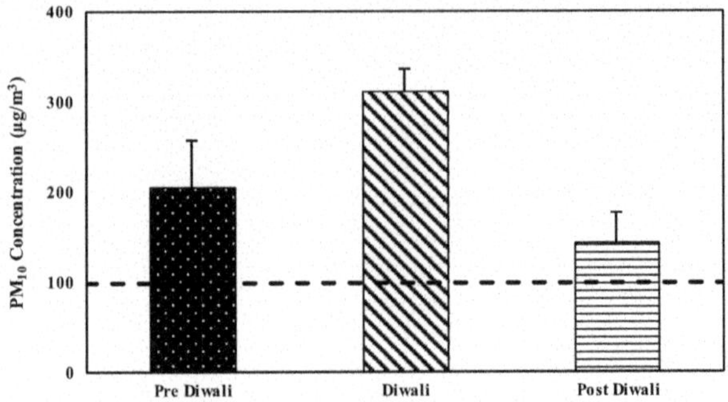

Fig. 22.2 Observed mean and standard deviation of PM_{10} concentrations ($\mu g/m^3$) during pre-Diwali, Diwali, and post-Diwali days. Dashed line indicates Indian NAAQS of 100 $\mu g/m^3$

172 $\mu g/m^3$. The mean PM_{10} concentration during Diwali days was 311 $\mu g/m^3$ (minimum = 294, maximum = 328 $\mu g/m^3$), which is 1.81 times higher than non-Diwali day. The mean concentration during pre-Diwali days was 203 $\mu g/m^3$ (minimum = 163, maximum = 264 $\mu g/m^3$), and post- Diwali days was 141 $\mu g/m^3$ (minimum = 106, maximum = 295 $\mu g/m^3$).

22.3.2 *Noise Levels and Biological Analysis*

Ambient noise levels measured during Diwali day was found to be 101 dBA. The mean noise level of non-Diwali day was 61 dBA which is 65% less than Diwali day. In both cases, ambient noise level has crossed over the NAAQS limit of 55 dBA, in residential areas during day time.

Variation in concentrations of bacteria is shown in Fig. 22.3. Panel (a) indicates that bacteria concentration was lower during Diwali compared to pre and post-Diwali days. A similar conclusion can be achieved from the panel (b), which shows the ratio

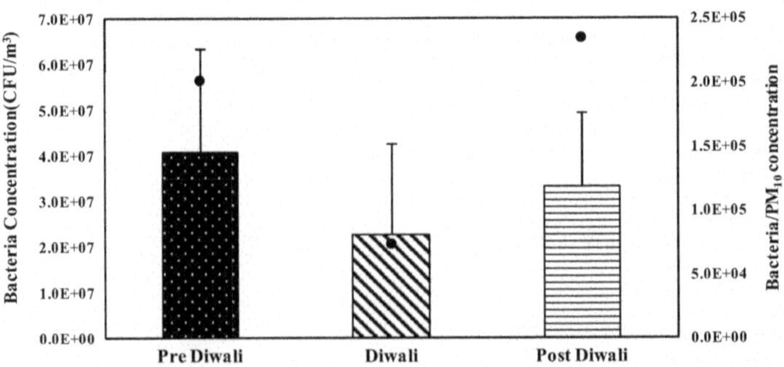

Fig. 22.3 Variation in bacteria concentration (CFU/m^3) during pre-Diwali, Diwali, and post-Diwali days. Ratio (CFU/μg) of change in concentrations of bacteria and PM_{10} during those days is shown as filled circle using right *y*-axis

of concentrations of bacteria and PM_{10}. This could be due to excess heat produced in the atmosphere due to the burning of fireworks.

22.3.3 Metals and Ionic Components of PM_{10}

The report shows the comparison of concentrations of PM_{10} and metals during days on which fireworks are burnt in this study and other studies around the world. The metal concentrations in this study, except for Fe and Zn, are often more than the other studies reported.

Concentrations of six metals during Diwali, pre- and post- Diwali days are shown in Fig. 22.4. There is an increasing trend observed between all the six metals specifically during Diwali period when compared to pre and post-Diwali periods, similar to the findings of earlier studies (Perrino et al. 2011; Kulshrestha et al. 2004; Wang et al. 2007; Sarkar et al. 2010; Pervez et al. 2016). Diwali/non-Diwali day ratios were greater than 1 for all the metals. Gradual increase of metals concentrations from pre to Diwali days and then decrease in post-Diwali days indicated the certainty of fireworks burning as main emission source. This is also supported by a 2.3 times increase in Sr, a tracer metal of fireworks burning (Sarkar et al. 2010), which is a greater increase of than other metals. Altogether, the metals concentrations during Diwali period were found to be increased 1–2.5 folds when compared to non-Diwali period. Fireworks are made up of various organic and inorganic chemicals such as charcoal, sulfur, potassium, lead, aluminium, iron and barium nitrate (Kulshrestha et al. 2004; Steinhauser et al. 2008).

Chemical reactions propel and burst them into special shapes. While Na and K are used as metal oxidizers, Zn is used to produce smoke effects, Sr is used to stabilize fireworks mixtures (Kulshrestha et al. 2004). Moreover, chloride, nitrate,

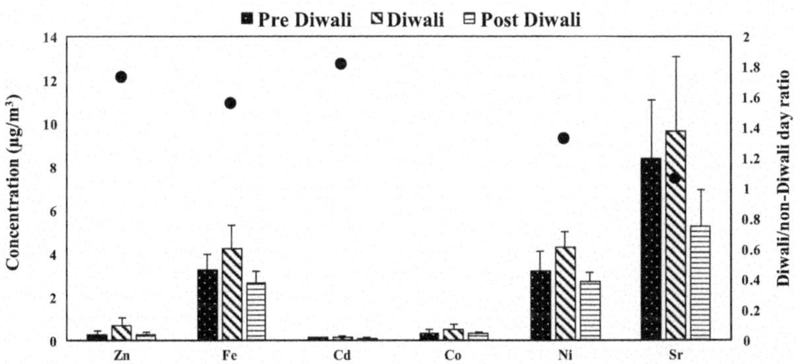

Fig. 22.4 Concentrations (μg/m^3) of metals, Zn, Fe, Cd, Co, Ni, and Sr, during pre-Diwali, Diwali, and post-Diwali days. Diwali/non-Diwali day ratios of those metals are also shown as filled dots using right *y*-axis

and sulphate are used in different metal salts to produce different colors (Tandon et al. 2008; Moreno et al. 2007). For example, Calcium chlorides and sulphates produce orange flames, and potassium nitrate, potassium chlorate, and potassium perchlorate imparts violet-pink color to the sparks which are the component of black powder, the combustible material (Moreno et al. 2007). Therefore, these metals are bound to increase when fireworks are burnt in large amounts. Table 22.3 shows the mean concentrations of water-soluble ions measured during fireworks in the current study and some other studies in the world. In this study, decreasing order of anions and cations were $Cl^- > SO_4^{2-} > F^- > NO_3^-$ and $K^+ > NH_4^+ > Na^+ > Ca^{2+}$, respectively, during whole monitoring campaign. For example, in comparison with the other study in this region, concentrations of all ions are higher. This is mainly due to the study location which is a city with more industries which leads to higher PM_{10} concentrations than the other study.

Figure 22.5 depicts the percentage of variations of water-soluble ionic species collected during pre-Diwali, Diwali, and post-Diwali periods. Concentrations of all the ionic species peaked during the Diwali period than on normal ambient days. Concentrations of Ca^{2+}, NH_4^+, Na^+, K^+, F^-, Cl^-, NO_3^- and SO_4^{2-} increased by 1.1, 2.3, 1.3, 2.2, 1.8, 1.6, 1.7 and 2 times, respectively, which were similar to the findings by Wang et al. (2007). Significant increase in concentrations of K^+ and Cl^- was observed during the Diwali period stressing that fireworks mixture comprises of potassium salts and chlorates and perchlorates which are used as oxidizers.

Table 22.3 Comparison of concentrations of ionic components of PM_{10} ($\mu g/m^3$) monitored during various firework episodes across the world along with present study

Event, year	Diwali, 2015[a]	New Year, 2005[b]	Diwali, 2009[c]	Diwali, 2009[d]	Diwali, 2011[e]	Lantern festival, 2013[f]
Location	Guwahati, India	Mainz, Germany	Delhi, India	Tezpur, India	Chhattisgarh, India	Tianjin, China
Cl^-	20.72	5.13	16	6.28	11	7.0
NO_3^-	0.63	6.48	11	0.84	10	20.5
SO_4^{2-}	13.08	36.17	44	8.05	15	40.0
Ca^{2+}	2.69	NA	6.2	1.26	13	NA
NH_4^+	9.93	2.39	0.64	1.13	14	NA
Na^+	6.19	NA	1.6	4.43	13	1.5
K^+	13.06	33.15	NA	0.15	13.75	9.0

Note
NA not available
[a]Present study: 12-h sampling during the Diwali period
[b]Drewnick et al. (2006): 8-h sampling duration
[c]Perrino et al. (2011): 24-h sampling
[d]Deka and Hoque (2014): 12-h sampling, whole monitoring campaign
[e]Ambade and Ghosh (2013): 24-h sampling. Approximate values, interpreted from graphs
[f]Tian et al. (2014): 24-h sampling during the study period

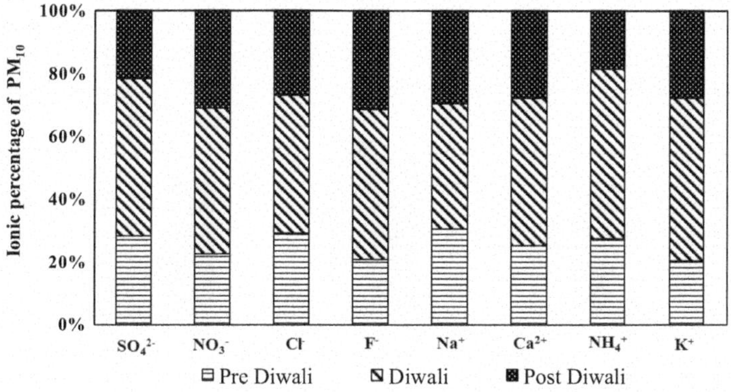

Fig. 22.5 Percentage variation of ions in PM_{10} during pre-Diwali, Diwali and post-Diwali days

In the present study, NO_3^-/SO_4^{2-} declined by 95% indicating high concentration of SO_4^{2-} which can be attributed to release of huge amount of SO_2 during burning of fireworks. This SO_2 could be transformed to particulate SO_4^{2-} to gas-phase oxidation by hydroxyl radicals or aqueous phase oxidation involving H_2O_2, O_3, and other oxidants (Chatterjee et al. 2013; Seinfeld and Pandis 2016). Also, sulphate salts are mainly used as coloring agents in crackers which can also be the reason behind higher concentration of sulphate in PM_{10}.

22.3.4 *Acidity/Basicity of PM_{10}*

Anion to cation (A/C) ratio gives an insight to the acidic or basic nature of the aerosols. In order to calculate ion balance, mass concentrations of ions are converted to microequivalents (μeq), as shown in Eq. (22.5), proposed by Tao et al. (2013)

$$\frac{\text{A (Anion}\ \mu\text{eq/m}^3}{\text{C (Cation}\ \mu\text{eq/m}^3)} = \frac{\left(\frac{Cl^-}{35.5} + \frac{NO_3^-}{62} + \frac{SO_4^{2-}}{48}\right)}{\left(\frac{Na^+}{23} + \frac{NH_4^+}{18} + \frac{K^+}{39} + \frac{Ca^{2+}}{20}\right)} \tag{22.5}$$

Figure 22.6 shows the graphical representation of A/C ratios for the Diwali period along with pre- and post-Diwali periods. A/C ratio greater than 1 indicates the aerosols' acidic nature whereas slightly less than 1 indicates the contribution of unmeasured carbonate ion (Tao et al. 2013; Kerminen et al. 2001). Too low A/C ratio indicates the basic nature of aerosols (Cao et al. 2005). In the present study, during non-Diwali days, the average A/C ratio was found to be 0.45, which was half of 0.9 found on Diwali days. Similar observations were made by Pervez et al. (2016). This indicates that fireworks burning made the particle composition more acidic.

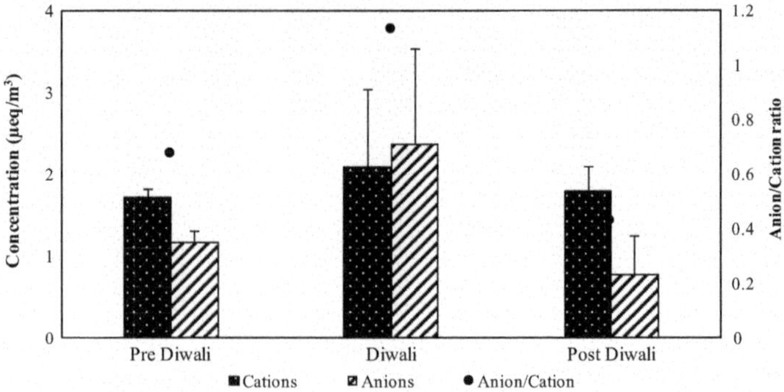

Fig. 22.6 Cation and anion concentrations (μeq/m^3) and their anion/cation ratios (secondary *y*-axis) during pre-Diwali, Diwali, and post-Diwali days

22.3.5 Source Apportionment of PM_{10}

For the entire monitoring campaign, PMF analysis was performed to identify the contributing sources of PM_{10}. All the species had *S*/*N* ratio of greater than 1 and were treated as strong by PMF (USEPA 2014). Performance of the model was studied using regression diagnostics, as shown in Table 22.4. For example, most of the species,

Table 22.4 Coefficient of determination (R^2), standard error (S_e), slope

S. no.	Species	Species-specific detection limit (μg/L)	No. of samples below the MDL	No. of samples above the MDL	Model performance		
					Slope	S_e	R^2
1	SO_4^{2-}	20	0	23	0.85	1.12	0.95
2	NO_3^-	20	0	23	0.91	0.06	0.96
3	Cl^-	20	0	23	0.91	2.32	0.92
4	Na^+	5	0	23	0.98	0.33	0.97
5	Ca^{2+}	7	0	23	1.04	0.33	0.85
6	NH_4^+	5	0	23	0.81	0.63	0.96
7	K^+	8	0	23	0.96	1.15	0.96
8	Zn	1	0	23	1.07	0.04	0.99
9	Fe	6	0	23	0.71	0.44	0.71
10	Cd	2	0	23	0.92	0.02	0.82
11	Co	5	0	23	0.84	0.05	0.91
12	Ni	10	0	23	1.00	0.26	0.93
13	Sr	2	0	23	0.93	0.70	0.95

Species-specific detection limit used for uncertainty calculations and no. of samples below/above the MDL's are also included except for Fe, Ca^{2+}, and Cd, had a correlation coefficient (R^2) greater than 0.9, indicating the good performance of the model.

Six sources were identified from the analysis. Figure 22.7 shows the source profiles and percentage contributions of species apportioned to each source. The first source was named dust emissions as it contributed to high concentrations of Ca^{2+} (30%), Fe (23%), and Ni (27%). These species are usually the tracer elements of construction and soil dust emissions (Zhang et al. 2013; Wen et al. 2016; Wang et al. 2008; Tan et al. 2016; Sharma et al. 2016) and usually present in cement and clay in higher concentrations. As the study site is a residential region, construction activities take place through out the year in many areas within the campus.

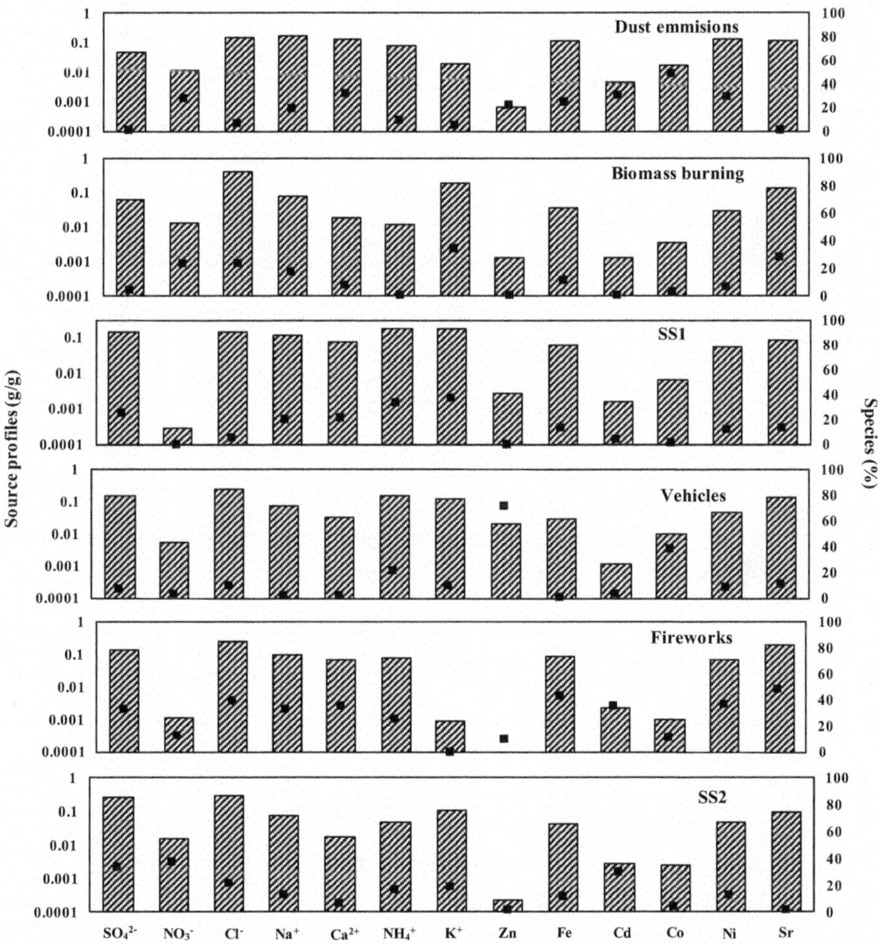

Fig. 22.7 Predicted source profiles and percentage of species contributed from dust emissions, biomass burning, secondary sources 1 (SS1), vehicular sources, fireworks, and secondary sources 2 (SS2). Median values estimated using the bootstrap analysis are only shown

35% of K^+ was attributed to source 2. Previous studies showed that K^+ is a tracer for biomass burning (Zhang et al. 2013; Wang et al. 2008; Tan et al. 2016; Sharma et al. 2016; Police et al. 2016). Owing to socio-economic reasons biomass is used as an energy source in this region. Moreover, as the sampling happened during post-monsoon where temperatures are low, this activity is prevalent, and hence this source is expected.

The third source was treated as secondary sources 1 and was characterized by contributions of 26% of SO_4^{2-} and 33% of NH_4^+. Possible reasons could be fertilizers used in agricultural activities in this region (Qiao et al. 2015). Zn and Co were 70% and 37%, respectively from the fourth source. These species are mainly emitted from fuel burning, lubrication oil, brake linings, and tyres of vehicles (Zhang et al. 2013; Sharma et al. 2016; Police et al. 2016). 34% of SO_4^{2-} and 36% of NO_3^-, are from a single source. These secondary ions are derived from gas to particle conversion processes involving photochemical reaction of gaseous precursors such as SO_2 and NO_x (Tao et al. 2013; Police et al. 2016; Chakraborty and Gupta 2010). Since the event took place in post-monsoon, stagnant conditions along with NO_x and SO_x emissions are highly favorable for the secondary aerosol generation. Thus, this fifth source is treated as secondary source-2 in this study. 49% of Sr, 36% of Cl^-, 32% of SO_4^{2-}, 42% of Fe, 33% of Na^+, 37% of Ni and 36% of Ca^{2+} were attributed to source 6. This indicates an association with fireworks burning (Perrino et al. 2011; Kulshrestha et al. 2004; Wang et al. 2007; Moreno et al. 2007; Licudine et al. 2012; Sijimol and Mohan 2014; Do et al. 2012). High mass fractions of Sr and Cl^- were due to the presence of strontium compounds, chlorates, and perchlorates in major quantities in raw materials of fireworks. Cl^-, Sr, and SO_4^{2-} dominated the source profile by 25, 20 and 15%, respectively.

Eventhough % of species attributed to a factor helped in identifying the source, most of the source profiles were dominated by species that are not generally emitted from the source. For example, only 8% of Cl^- was from vehicles, Cl^- was the most dominant species (20%) in that source profile. This error in source profiles could be due to higher concentrations of Cl^- and less number of samples used for PMF analysis.

22.3.6 Health Survey and Assessment of Health Risk Due to Heavy Metals

Panel (a) in Fig. 22.8. indicates that the number of patients suffering from headache, fatigue, irritation in eyes, coughing, sneezing, and sinusitis, increased by 3.8, 3, 3.3, 3.3, 3.5, and 6.5 times, respectively, on Diwali days compared to non-Diwali days. Similar findings have been reported by (Sharma et al. 2015) at Delhi.

Also, in this study, five major metals (Cd, Co, Fe, Zn, and Ni) were used to assess the possible non-carcinogenic human health risks due to this pollution episode. Two scenarios were taken into consideration. In Scenario I, cumulative Hazard Index (HI)

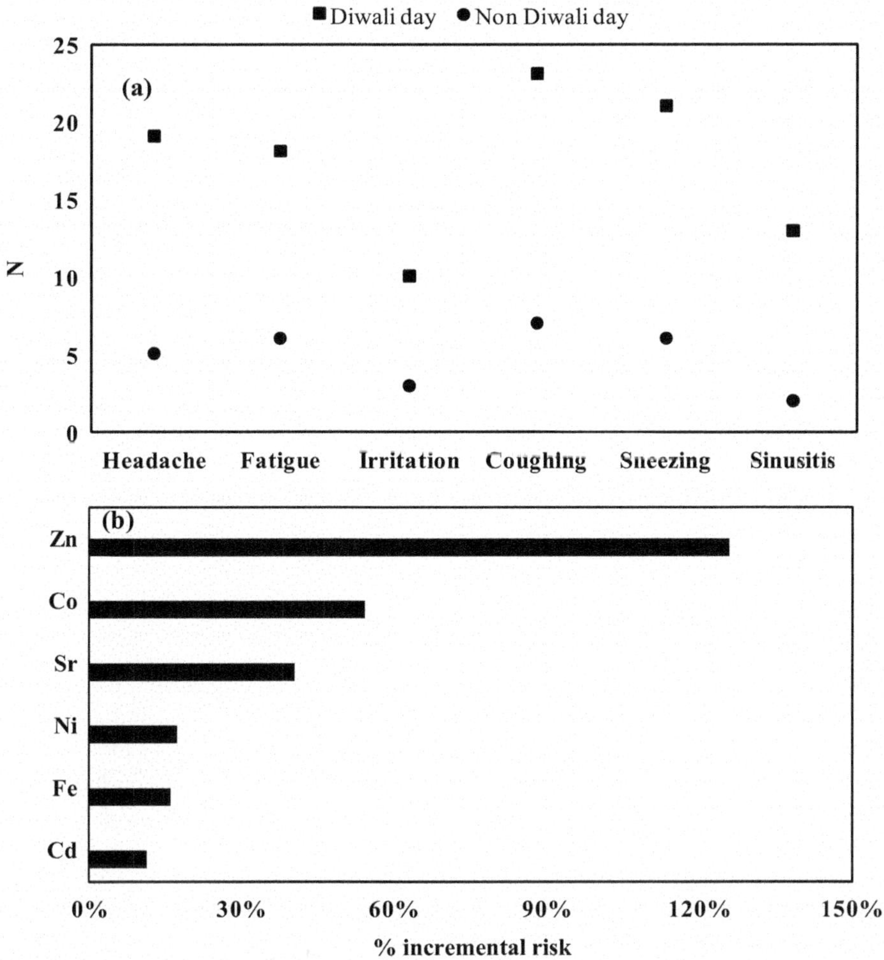

Fig. 22.8 **a** Number of patients suffering from headache, fatigue, irritation, coughing, sneezing and sinusitis during Diwali and Non-Diwali days, and **b** incremental risk (%) due to Diwali from met

due to metals (Cd, Co, Fe, Zn, and Ni) for an annual period (350 days) was estimated using average concentrations during non-Diwali days. Scenario II involved two parts (a) and (b). In II (a), HI was calculated for all the metals for 347 days, excluding Diwali period, by utilizing concentrations as in Scenario I. In II (b), HI for the Diwali period was calculated. Summation of II (a) and (b) was compared with HI from Scenario I to get the excess risk due to Diwali.

Non-carcinogenic health risk to an individual due to exposure to all metals "*j*" is given by Eq. (22.6):

$$HI = \sum_{j=1}^{k} \frac{M_j \times EF \times ED}{AT \times RfC_j} \tag{22.6}$$

where RfC is the inhalation chronic reference concentration (mg/m^3) obtained from The Risk Assessment Information System (www.rais.ornl.gov), M is a concentration of a metal (mg/m^3), EF is exposure frequency (days/year), ED is Exposure duration (years), and AT indicates averaging time [age (years) × 365 (days/years)]. Using data collected from patients, and concentrations measured during the analysis, the percentage incremental HQ due to each metal is estimated and shown in panel (b) of Fig. 6.14. Estimated Incremental risk was highest for Zn, followed by Co and Sr. Results indicated that, cumulative HI for Scenario II which includes Diwali period showed a risk level increase of ~0.5% in health effects due to chronic lifetime exposure to metals alone. It should be noted that this risk level rise was only due to metals, and in ambient settings, the risk level will be higher due to chronic exposure to many other pollutants like SO_2, NO_x, and EC, etc., released by fireworks.

22.4 Conclusions

In this study, concentrations of metals and ions in PM_{10} were measured for a period of 10 days in 2015, coinciding with fireworks festival in northeast India. Results indicated that while the concentration of metals increased by 51%, ions increased by 74%. PM_{10} showed the condition of cation deficiency during fireworks burning, indicating acidic nature of particles. Additionally, ambient noise level measured during Diwali days was 101 dB, 65% higher than non-Diwali days mean noise level and exceeds the NAAQS standards by 1.84 times. Furthermore, the mean concentrations of bacteria during Diwali days reduced to 61% of non-Diwali days.

Six dominant sources, i.e., biomass burning, two secondary sources, vehicular sources, dust emissions, and fireworks were contributing to PM_{10} during the analysis period as suggested by the USEPA'S PMF model. Analysis indicated that 49% of Sr and 36% of Cl^- in this study were due to fireworks burning.

Increased level of particulate matter resulted in 67% increase of number of patients attending the hospital during Diwali period. The chronic lifetime exposure to metals alone, during Diwali, increased the health risk level by 0.5%. This study stresses the importance of regulated and monitored practice of burning of fireworks in regions with high population density.

However, this health study showed only the possible impact of fireworks on human health. In future, a time series analysis [for example, see Atkinson et al. (2012) and Bhaskaran et al. (2013)], using data collected from more fireworks burning events, is to be carried out to quantify the relation between PM_{10} concentrations and health risks in a region.

Acknowledgements The authors would like to thank the Indian Institute of Technology, Guwahati for providing facilities to carry out this research. The authors would also like to thank medical professionals in the institute hospital and patients for the co-operation in the health survey. The authors also thank the anonymous reviewers for their inputs.

References

Ambade B, Ghosh S (2013) Characterization of PM_{10} in the ambient air during Deepawali festival of Rajnandgaon district, India. Nat Hazards 69(1):589–598

Atkinson RW et al (2012) Systematic review and meta-analysis of epidemiological time-series studies on outdoor air pollution and health in Asia. Air Qual Atmos Health 5(4):383–391

Barman SC et al (2008) Ambient air quality of Lucknow City (India) during use of fireworks on Diwali Festival. Environ Monit Assess 137(1–3):495–504

Begum BA et al (2004) Investigation of sources of atmospheric aerosol at urban and semi-urban areas in Bangladesh. Atmos Environ 38(19):3025–3038

Betha R, Balasubramanian R (2013) Particulate emissions from commercial handheld sparklers: evaluation of physical characteristics and emission rates. Aerosol Air Qual Res 13(1):301–307

Bhaskaran K et al (2013) Time series regression studies in environmental epidemiology. Int J Epidemiol 42(4):1187–1195

Bove MC et al (2014) An integrated $PM_{2.5}$ source apportionment study: positive matrix factorisation vs. the chemical transport model CAMx. Atmosp Env 94:274–286

Cao JJ et al (2005) Characterization of airborne carbonate over a site near Asian dust source regions during spring 2002 and its climatic and environmental significance. J Geophys Res D Atmos 110(3):1–8

Chakraborty A, Gupta T (2010) Chemical characterization and source apportionment of submicron (PM 1) aerosol in Kanpur Region, India. Aerosol Air Qual Res 10(5):433–445

Chatterjee A et al (2013) Ambient air quality during Diwali Festival over Kolkata—a mega-city in India. Aerosol Air Qual Res 13(3):1133–1144

CPCB (2003) Guidelines for ambient air quality monitoring. C.P.C. Board, Editor, Delhi

Deka P, Hoque RR (2014) Diwali fireworks: early signs of impact on PM10 properties of rural Brahmaputra Valley. Aerosol Air Qual Res 14:1752–1762

Do T-M et al (2012) Metals present in ambient air before and after a firework festival in Yanshui, Tainan, Taiwan. Aerosol Air Qual Res 12(5):981–993

Drewnick F et al (2006) Measurement of fine particulate and gas-phase species during the New Year's fireworks 2005 in Mainz, Germany. Atmosp Env 40(23):4316–4327

Gouder C, Montefort S (2014) Potential impact of fireworks on respiratory health. Lung India Off Organ Indian Chest Soc 31:375–379

Harrison T, Shallcross D (2011) Smoke is in the air: how fireworks affect air quality. Sci Sch 21:47–55

Hirai K et al (2000) Acute eosinophilic pneumonia associated with smoke from fireworks. Intern Med 39(5):401–403

Kerminen V-M et al (2001) Ion balances of size-resolved tropospheric aerosol samples: implications for the acidity and atmospheric processing of aerosols. Atmos Environ 35(31):5255–5265

Khaparde VV et al (2012) Influence of burning of fireworks on particle size distribution of PM_{10} and associated Barium at Nagpur. Environ Monit Assess 184(2):903–911

Kim E, Hopke PK (2004) Source apportionment of fine particles in Washington, DC, utilizing temperature-resolved carbon fractions. J Air Waste Manag Assoc 54(7):773–785

Kulshrestha UC et al (2004) Emissions and accumulation of metals in the atmosphere due to crackers and sparkles during Diwali festival in India. Atmos Environ 38(27):4421–4425

Licudine JA et al (2012) Hazardous metals in ambient air due to New Year fireworks during 2004–2011 celebrations in Pearl City, Hawaii. Public Health Rep 127(4):440
Moreno T et al (2007) Recreational atmospheric pollution episodes: Inhalable metalliferous particles from firework displays. Atmos Environ 41(5):913–922
Nasir UP, Brahmaiah D (2015) Impact of fireworks on ambient air quality: a case study. Int J Environ Sci Technol 12(4):1379–1386
Norris G et al (2014) EPA positive matrix factorization (PMF) 5.0 fundamentals and user guide. DC EPA/600/R-14/108. Prepared for the US Environmental Protection Agency Office of Research and Development, Washington, DC
Paatero P (1997) Least squares formulation of robust non-negative factor analysis. Chemometr Intell Lab Syst 37(1):23–35
Paatero P, Tapper U (1994) Positive matrix factorization: a non-negative factor model with optimal utilization of error estimates of data values. Environmetrics 5(2):111–126
Pathak B et al (2015) Short term introduction of pollutants into the atmosphere at a location in the Brahmaputra Basin: a case study. Atmosp Pollut Res 6(2):220–229
Perrino C et al (2011) Chemical characterization of atmospheric PM in Delhi, India, during different periods of the year including Diwali festival. Atmosph Pollut Res 2(4):418–427
Pervez S et al (2016) Chemical speciation of aerosols and air quality degradation during the festival of lights (Diwali). Atmosp Pollut Res 7(1):92–99
Police S, Sahu SK, Pandit GG (2016) Chemical characterization of atmospheric particulate matter and their source apportionment at an emerging industrial coastal city, Visakhapatnam, India. Atmosp Pollut Res 7(4):725–733
Polissar AV et al (1998) Atmospheric aerosol over Alaska: 2. Elemental composition and sources. J Geophys Res Atmosp 103(D15):19045–19057
Qiao X et al (2015) Modeling dry and wet deposition of sulfate, nitrate, and ammonium ions in Jiuzhaigou National Nature Reserve, China using a source-oriented CMAQ model: part I. Base case model results. Sci Total Env 532:831–839
Ravindra K, Mor S, Kaushik CP (2003) Short-term variation in air quality associated with firework events: a case study. J Environ Monit 5(2):260–264
Sarkar S et al (2010) Chemical speciation of respirable suspended particulate matter during a major firework festival in India. J Hazard Mater 184(1–3):321–330
Seinfeld JH, Pandis SN (2016) Atmospheric chemistry and physics: from air pollution to climate change, 3rd edn. Wiley
Sharma S, Nayak H, Lal P (2015) Post-Diwali morbidity survey in a resettlement colony of Delhi. Indian J Burns 23(1):76–80
Sharma S et al (2016) Source apportionment of $PM_{2.5}$ in Delhi, India using PMF model. Bull Env Contam Toxicol 97:1–8
Sijimol M, Mohan M (2014) Environmental impacts of perchlorate with special reference to fireworks—a review. Env Monit Assess 186(11):7203–7210
Steinhauser G et al (2008) Heavy metals from pyrotechnics in New Years Eve snow. Atmos Environ 42(37):8616–8622
Tan J et al (2016) Long-term trends of chemical characteristics and sources of fine particle in Foshan City, Pearl River Delta: 2008–2014. Sci Total Environ 565:519–528
Tandon A, Yadav S, Attri AK (2008) City-wide sweeping a source for respirable particulate matter in the atmosphere. Atmos Environ 42(5):1064–1069
Tao J et al (2013) Chemical composition of $PM_{2.5}$ in an urban environment in Chengdu, China: importance of springtime dust storms and biomass burning. Atmosp Res 122:270–283
Tian YZ et al (2014) Estimation of the direct and indirect impacts of fireworks on the physicochemical characteristics of atmospheric PM_{10} and $PM_{2.5}$. Atmosp Chem Phys 14(18):9469–9479
Tsai H-H et al (2012) Influences of fireworks on chemical characteristics of atmospheric fine and coarse particles during Taiwan's Lantern Festival. Atmos Environ 62:256–264
USEPA (2014) EPA Positive Matrix Factorization (PMF) 5.0 Fundamentals and User Guide. Office of Research and Development, US EPA, Washington, DC

Vecchi R et al (2008) The impact of fireworks on airborne particles. Atmos Environ 42(6):1121–1132
Verma C, Deshmukh DK (2014) The ambient air and noise quality in India during Diwali festival: a review. Recent Res Sci Technol 6(1):203–210
Waked A et al (2014) Source apportionment of PM 10 in a north-western Europe regional urban background site (Lens, France) using positive matrix factorization and including primary biogenic emissions. Atmos Chem Phys 14(7):3325–3346
Wang Y et al (2007) The air pollution caused by the burning of fireworks during the lantern festival in Beijing. Atmos Environ 41(2):417–431
Wang H et al (2008) Long-term monitoring and source apportionment of $PM_{2.5}$/PM_{10} in Beijing. China. J Env Sci 20(11):1323–1327
Wen W et al (2016) Source apportionment of $PM_{2.5}$ in Tangshan, China—hybrid approaches for primary and secondary species apportionment. Front Env Sci Eng 10(5):1–14
Zhang R et al (2013) Chemical characterization and source apportionment of $PM_{2.5}$ in Beijing: seasonal perspective. Atmosp Chem Phys 13(14):7053–7074

Chapter 23
Toxicological Study of Nanoparticles: An Attempt to Relate Physicochemical Characters with Toxicity

A. Seenivasan, M. Muthuraj, and T. Panda

23.1 Introduction

Nanoparticles (NPs) have been gaining significant attention in different aspects of engineering and medicine attributed to their exquisite optical, magnetic, thermal, electrical and biological properties which are strongly related to their physicochemical characteristics. Thus, physicochemical characteristics, such as, dimensions, structure (shape), state of agglomeration, surface chemistry, higher surface area to volume ratio, charge and lipophilicity, chemical nature, ionization status, and porosity are the potential factors that govern the application of NPs. Among several applications, NPs are employed in real time imaging, bio-sensing systems, targeted drug and/or gene delivery systems etc. According to the recent reports, increased utilization of NPs has been evidenced over the past years with a frequency of three to four NPs associated products per week in the market (Gajewicz et al. 2012) resulting in an unregulated environmental release and human exposures. These miniscule molecules having properties as that of biomolecules may interact, react or alter the biological systems (human beings, plants, aquatic systems, etc.,) thereby causing unpredictable changes to life. Exposure to nickel NPs while painting the bushes of turbine bearings with nickel aerosols, resulted in respiratory distress syndrome leading to the death

A. Seenivasan (✉)
Department of Biotechnology, National Institute of Technology Andhra Pradesh, Tadepalligudem 534101, Andhra Pradesh, India
e-mail: ayothiraman@nitandhra.ac.in

M. Muthuraj
Department of Bioengineering, National Institute of Technology Agartala, Agartala 799046, India
e-mail: msrpmsiva@gmail.com

T. Panda
MSB140A, Biochemical Engineering Laboratory, Department of Chemical Engineering, Indian Institute of Technology Madras, Chennai 600036, India
e-mail: pandaiitm@gmail.com

S. M. Shiva Nagendra et al. (eds.), *Urban Air Quality Monitoring, Modelling and Human Exposure Assessment*, Springer Transactions in Civil and Environmental Engineering,
https://doi.org/10.1007/978-981-15-5511-4_23

of a person (Phillips et al. 2010). Similarly, nanoparticles exhibit toxic effects to workers and environment, who involved in the preparation and production processes under laboratory conditions and on industrial scales. Hence, nanotoxicology has emerged which targeted to evaluate the hazards and risks involved in handling NPs while framing the management techniques required to minimize their release to environment.

NPs can cause numerous toxic effects through several mechanisms in the body such as oxidative stress, genetic mutations, immunomodulatory activities, etc., leading to hepatotoxicity, pulmonary toxicity, nephrotoxicity, splenic toxicity, hematological toxicity, cytotoxicity effects and many more. For instance, a 38 years old adult died due to the respiratory distress syndrome and acute tubular necrosis after the continuous exposure to nickel NPs for a period of 13 days. The pathology reports clearly witnessed the presence of elevated levels of translocated nickel NPs in urine sample and in the lung macrophages (Phillips et al. 2010). Carbon nanotubes also caused epithelioid granuloma and interstitial inflammations (Lam et al. 2003; Srivastava et al. 2015). Ryman-Rasmussen et al. (2009) reported in the animal studies that fibrosis and scarring were caused in the walls of thorax attributed to the exposure to multiwalled carbon nanotubes. Mueller and Nowack (2008) discussed capturing the effects of titanium oxide NPs exposed to the different environment in a model system as these NPs are utilized widely in different applications. Toxicity, biological activity, interaction, translocation, and excretion of the NPs are determined by their physicochemical characteristics. The major physicochemical characteristics believed to determine the toxicity levels of NPs are enlisted in Oberdörster et al. (2005a, b), Luyts et al. (2013) and Beddoes et al. (2015). For instance, the agglomeration state of the particles in the aerosol determines the localization, subsequent retention and clearance of inhaled nanomaterials in the lungs as reported by Creutzenberg (2012), which supports the theory. Apart from these parameters, molecular topology is also an important phenomenon that determines solubility, binding and permeability characteristics of any molecules (Yang et al. 2012). It was found in many cases that the slight change in the charge, dimensions, structural and surface attributes etc., of same material results in huge variation in their interactions and toxicological assessments. Thus, it depicts the significant affiliation between the physicochemical properties of the NPs and their toxicity. Significant research is being conducted to examine the physical and chemical characteristics of NPs that modulates their toxicity (Luyts et al. 2013; Beddoes et al. 2015). However, there is a huge gap in realizing their complete potentials and hazards. Thus, the present review has tried to capture and relate the physical and chemical properties of NPs with their toxicity.

23.2 Physicochemical Characteristics Determine Toxicity of Nanoparticles

23.2.1 Toxicity Based on the Dimensions of Nanoparticles

Among the different parameters, size is a pivotal parameter that regulates the overall efficiency and the toxicity of NPs. The dimension limits are not well defined for the NPs, however, their size is often mentioned. Auffan et al. (2009) reported 30 nm as the critical diameter, below which the particles exhibit the properties of NPs while it is also reported that <50 nm, <10 nm and <100 nm as NPs leading to confusions (Anastasio and Martin 2001; Kumar et al. 2010; Standard and British 2005; Morris et al. 2007). Kumar et al. (2010) termed the particles less than 300 nm as NPs in their study since 99% of particles belong to the range. Fubini et al. (2010) reported that all particles measuring diameters from 1 to 100 nm and do not possess the properties which non-nanosized particles own are termed as NPs and this has been accepted widely by research community.

The absorption, distribution, involvement in metabolic activities and excretion efficiency of nanoparticles are dependent on their dimensions, which in turn modulates their toxic effects. For instance, size-dependent absorption and accumulation of nanoparticles was reported by Kumar et al. (2010) in alveolar and tracheobronchial regions of lungs. The study also showed an inverse correlation between size and toxicity of NPs as also supported by Luyts et al. (2013) which states size and toxicity as two mutually exclusive parameters in many cases. Similarly size-dependent half-life period of NPs in blood and immunotoxicity were reported by Kiessling et al. (2014). The absorption or cellular internalization takes place through several mechanisms such as phagocytosis, pinocytosis collectively termed as endocytosis or semi-endocytosis leading to cytotoxicity. The smaller particles of size less than or equal to 4.5 nm were reported to penetrate vesicles directly while the large particle of size >4.5 nm was reported to cross the vesicular membrane through three different pathways which includes direct penetration, chain like co-operative penetration, and inverted micelle-like penetration in a vesicle of 60 nm diameter (Chen et al. 2013). It was also mentioned that the large particles form clumps or clusters which do not enter the cells directly. The other mechanism is the receptor-mediated cellular uptake of NPs which is based on the dimensions and other surface attributes. For example, gold NPs when incubated with HeLa cells, found to absorb by receptor-mediated endocytosis attributed to the surface coating by serum proteins of the HeLa cells (Chithrani et al. 2006). It was also found that the NPs of 50 nm trapped more inside the vesicles of the HeLa cells than smaller and bigger particles when a range of NPs of size 1 to 100 nm were tested (*cf*. Fig. 23.1). In this study, the atoms of gold obtained by ICP-AES were converted into gold (Au) NPs using the following expression (Eqs. 23.1 and 23.2) (Chithrani et al. 2006).

$$N = \frac{\text{No. of Au atoms that can fit to each vol. of AuNPs } (U)}{\text{No. of Au atoms measured using ICP - AES } (M)} \tag{23.1}$$

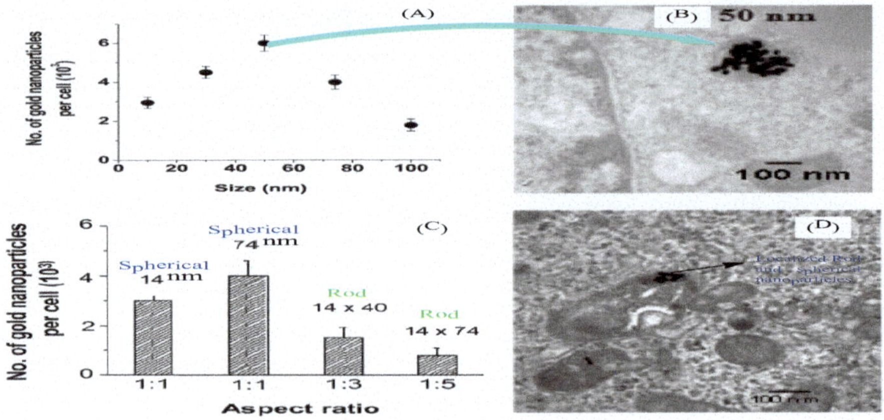

Fig. 23.1 Size (**a**) and shape (**c**) dependent uptake of gold nanoparticles on HeLa cells. **b** Transmission electron microscopic capture of trapped 50 nm gold nanoparticles in vesicles of cells. Similarly, **d** represents TEM image of localized spherical and rod shaped gold nanoparticles. Reproduced with permission from Chithrani et al. (2006)

$$U = \frac{2}{3} \times \pi \times \left(\frac{D}{a}\right)^3 \quad (23.2)$$

where D and "a" refer to the diameter of particles (spherical) and the boundary of a unit cell, respectively and "a" equals 4.0786 Å on the edge. Each unit cell consists of four atoms of gold.

Comparable study was conducted by Meng et al. (2007) to find the difference in the cytotoxicity of particles from bulk to nano size. In an in vivo analysis carried out in mice the morphological changes in the stomach was captured for different sizes of copper particles 23.5 nm, 17 μm, ionic copper and a control (without any copper exposure). The study showed significant change in the morphology of stomach for the 23.5 nm particle whereas the micron sized and ionic copper particles resembled the control. Chen et al. (2006) have shown a distinct modulation in the toxicity levels when the size of the copper particles varied from micron to nano size. The copper nanoparticles (23.5 nm) resulted in the grave nanotoxicity than micron (17 μm) particles and copper ions as found in the in vivo studies in mice and induced heavy damage to spleen, liver and kidney. Size specific cytotoxicity of gold NPs on connective tissue fibroblast, macrophages, and epithelial cells was observed by Pan et al. (2007). The NPs sized of 1.4 nm observed to exhibit more cytotoxicity as compared to the bigger (1.8 and 15 nm) and smaller (0.8 and 1.2 nm) sizes. Further, they have also reported the size-dependent cellular response. Carlson et al. (2008) have shown the size-dependent toxicity of silver NPs of 15 and 30 nm. They have studied the viability of mitochondria and cell membrane, and the production of reactive oxygen spices (ROS) as an indication of nanotoxicity of silver NPs. Surprisingly, there was a tenfold raise in the production of ROS with 15 nm silver nanoparticles. Apart from these, ultrafine pollutants measuring <2.5 μm produced from different processes such as combustion of fuels, welding, volcanic activities etc., were reported to induce cardiopulmonary toxicity (Oberdörster 2001). These particles make them a suitable carrier for other adsorbed particles, pollutants, gases and metal particles due to the large surface area compared their volume. These ultrafine particles occupy more than 50% of the total particles (13.4 $\mu g/m^3$) in the particle group of size <10 μm in industrial locations (Dominici et al. 2006). An increase in the concentration of particulate matters (of size <2.5 μm) to 10 $\mu g/m^3$ was reported to increase the health risk to 8–18% (Dockery et al. 2005). Size-dependent induction of lung tumor was identified by testing a low dosage of 10 mg/m^3 of 20 nm particles and a high dose of 250 mg/m^3 of 300 nm particles. It was determined that the 20 nm sized particles induced high rates of lung tumor even at the low dose (10 mg/m^3) which could be due to aggregated form of small sized particles (Hoet et al. 2004). On the contrary, mesoporous (ranges from 100 to 600 nm) and amorphous (ranges from 2–335 nm) particles of silica showed an increase in the cytotoxicity as the particle size increases which affected the human red blood cells membrane integrity (Zhao et al. 2011; Rabolli et al. 2010). Hence, these conflicting data makes the generalization of concepts and categorization difficult.

Karlsson et al. (2009) attempted to compare the toxicity (cytotoxicity, mitochondrial depolarization (damage), DNA damage and oxidative DNA lesion) of most commonly used NPs of copper oxide (CuO), titanium dioxide (TiO_2), iron oxides (Fe_2O_3 and Fe_3O_4), at micron and nanometer (<100 nm) scales. Among the different NPs tested CuO was observed to be more cytotoxic and genotoxic as compared to other compounds studied at their nanoscales attributed to the mitochondrial damage of cells. On the contrary, TiO_2 was found to be more toxic at micron level caused by heavy damage to the DNA as compared to TiO_2NPs which showed oxidative DNA damage and mitochondrial damage. Iron(III) oxide also caused DNA damage at micron level than nanometer level. However, the overall toxicity of iron oxides was considerably low than other two types, even though it could cause oxidative DNA and mitochondrial damage to the A549 cell lines. Similarly, Heinlaan et al. (2008) have studied the toxicity of various oxides of zinc (ZnO), copper (CuO) and titanium (TiO_2) at their bulk and nano levels. Zinc in different forms $ZnSO_4$, ZnO NPs, and bulk ZnO showed high toxicity toward *Daphnia magna*, *Thamnocephalus platyurus*, and *Vibrio fischeri*, whose lethal concentration were also reported. It was reasoned that the solubilized zinc ions caused the toxicity as observed from recombinant zinc sensor bacteria. The CuO has also exhibited toxicity toward the test organism and reasoned due to the release of soluble copper ions. TiO_2 did not show any toxic effect even at the concentration of 20 g/L.

As mentioned above, size is a key parameter that controls the distribution and elimination of NPs that are absorbed. Many reports have shown that the NPs of size greater than 6 nm are not generally eliminated by kidneys and waits until the mononuclear phagocyte system breaks down the NPs in to smaller ones. It was also reported that these molecules gets accumulated in the respective organs leading to numerous toxic effects (Shin et al. 2015). Quantum dots of cadmium selenide remained for eight months in the hepatic tissue leading to hepatotoxicity (Ballou et al. 2004). Even though there were many controversies, smaller NPs are highly toxic as compared to the larger ones. The conflicts in the data can also be explained by the involvement of other factors which includes concentration of NPs, surface chemistry and other material properties that have also influenced the toxicity when larger NPs are used.

23.2.2 *Toxicity Based on the Surface Reactivity and Surface Charge*

The surface charge and reactivity have been crucial parameters that determine clearance and distribution of the particles inside the cell. With reduction in the particle dimension (size), the reactivity increases due to the thermodynamic instability of smaller particles. Hence, an immense consideration must be paid on the particle surface than the core region (Nel et al. 2009). The surface area greatly influences the toxicity of the particles than its mass (Beddoes et al. 2015; Driscoll et al. 1997). To add a point to the existing discussion, Meng et al. (2007) have reported that ultrahigh surface reactivity provoked the grave cytotoxicity of copper NPs. Due to its ultrahigh reactivity and surface area, these particles facilitated the accumulation of alkalescent and heavy copper ions, which simultaneously resulted in metabolic alkalosis and overloading of copper (Cu) ions at the end. Positively charged lipid NPs activated immunological responses (Solmesky et al. 2011). The cationic component of the Acramin F utilized in textile industries led to lethal effect over a broad range of cell lines. Agglutination of red blood cells was induced by the positively charged amine group in poly L-glutamic acid (Dekie et al. 2000). This cationic charge density and its surface orientation were found to determine the membrane accessibility which in turn determines the toxicity and biocompatibility.

Toxicological assessment of different sized gold nanoparticles (2, 4 and 6 nm) with varying surface charges i.e., positive, negative and zwitterion forms in HeLa cells was carried out by Jiang et al. (2015). The cellular internalization efficiency was found to be high for positively charged NPs as compared to the zwitterions and anions as shown in Fig. 23.2, and the maximum uptake was found for the 6 nm particles as compared to other sized NPs (2 and 4 nm). The uptake of positively charged particles increased with raising particle size while the internalization of neutral and anionic particles reduced with the increase in the size of NPs (Jiang et al. 2015).

Cells possess a negative charge on their surface and therefore a positively charged nanoparticle may get easily attracted toward the cellular environment thus leading to maximum toxicity as compared to the anionic and the neutral NPs (Fig. 23.3). The gold nanoparticles with positive charge were actively up taken by various cell lines that includes human airway epithelial cells, ovarian cancer (CP70 and A2780) cell lines, and human bronchial smooth myocytes leading to changes in the membrane potential of the cell lines. The positively charged NPs modulated the membrane potential by altering the Ca^{2+} influx in all cell lines thereby leading to apoptosis and

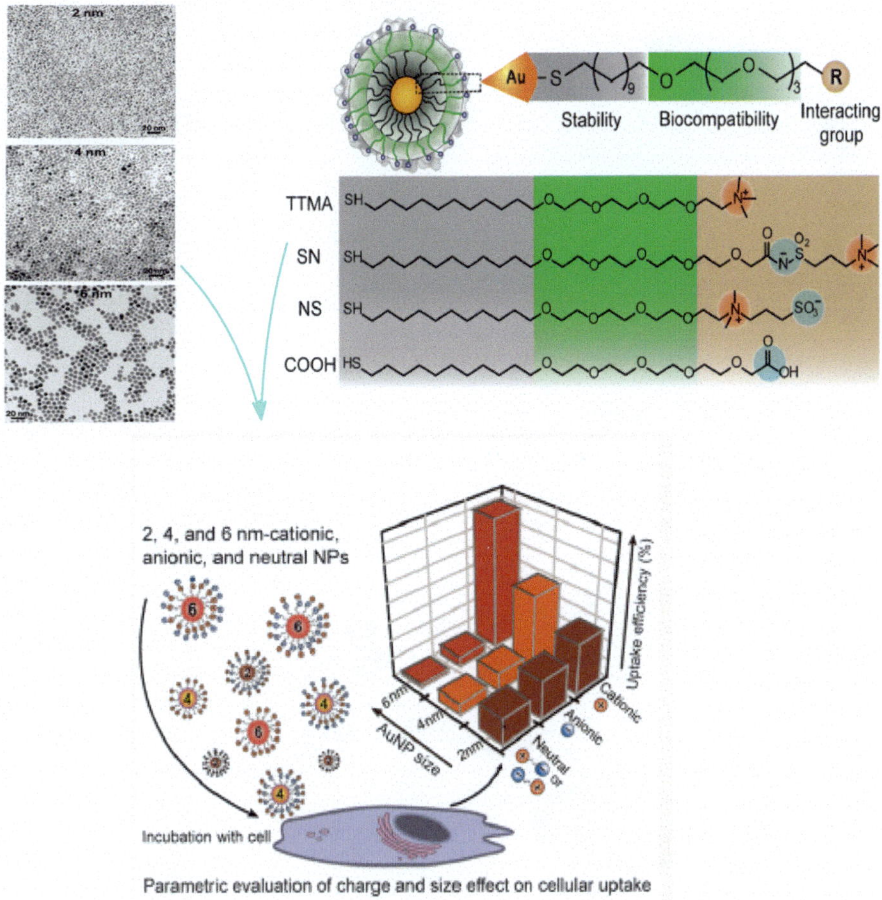

Fig. 23.2 Parametric evaluation of dimension (size) and surface charge on the cellular internalization of AuNPs. Various sizes (2, 4, 6 nm) and charges (positive (+), negative (−), and zwitterionic (− and +) ligands has been tested on cellular internalization. The cellular uptake of zwitterionic and anionic ligands was decreased with increase in particle size, whereas, the positive particles has internalized more as its size increased. Reproduced with permission from (Jiang et al. 2015)

inhibition of proliferation in the normal cell lines (Arvizo et al. 2010). Polymers with a low positive charge density used in globular or linear rigid structure were reported to have less cytotoxicity as compared to the branched and flexible materials (Hoet et al. 2004; Singh et al. 1992). Surface modified NPs are found to be relatively safer with less cytotoxicity as compared to the native molecules as that in quantum dots. In general quantum dots are surface modified to reduce its toxicity and membrane diffusivities (Lovric et al. 2005; Cho et al. 2007). Detailed examination of the NPs

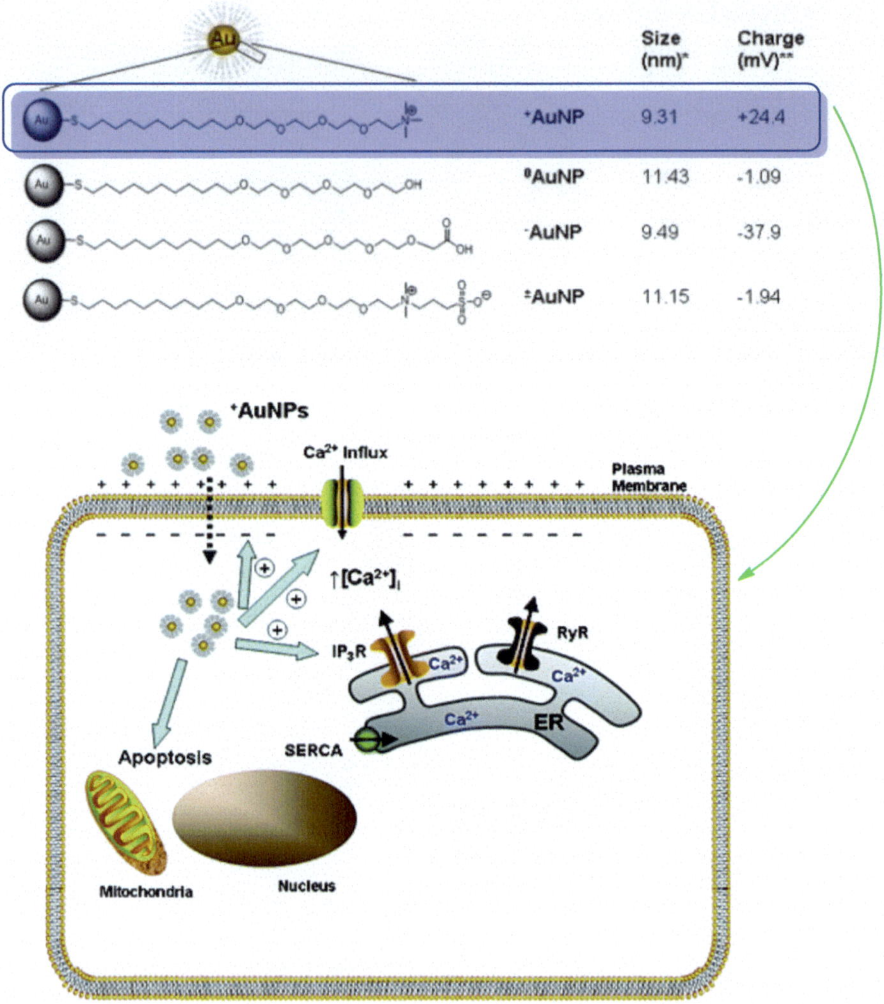

Fig. 23.3 The modulation of membrane potential and intracellular activities in different cancerous and normal cell types by the surface modified AuNPs. A rapid and considerable membrane depolarization was induced by the cationic AuNPs at 1.2 μM and it was also influenced by its concentration. The particles uptake by the cell followed by depolarization of membrane that induced influx of calcium ion from surrounding as well as from endoplasm reticulum via IP_3R channels, which leads to augmented apoptosis and reduced cellular proliferation. Reproduced with permission from Arvizo et al. (2010)

and their behavior in the different environment has to be performed to understand the properties. The surface properties are altered as a response to its interaction between the NPs and their environments such as soil, air, and water (Handy et al. 2008; Nowack 2009).

23.2.3 Toxicity Based on Status of Agglomeration

The NPs tend to attract each other at nanometer range forming agglomeration as it exhibits a highly thermodynamically unstable structure. Hence, many people use suitable surfactants or any other dispersing agent to keep the particles in dispersed form. The dispersion of the particles highly depends on the different parameters, such as surface ionic strength, adsorption with dispersant, solution pH and size of the particles. Still, there are many questions regarding the aggregation status of the particles and its translocation behavior. The other question is that what would be the effect of mono-dispersive and poly dispersive particles on the cell. The bioavailability of the particles is strongly affected by both aggregation and agglomeration of the NPs (Luyts et al. 2013). The agglomeration status of NPs also determined the toxicity level, deposition, retention, and clearance rates of the inhaled NPs (Creutzenberg 2012). This finding gained attention in the scientific community to determine the interaction between the NPs and the environmental factors to understand their behavior in different environmental constraints (Nowack 2009; Elsaesser and Howard 2012).

23.2.4 Toxicity Based on Shape

Stoehr et al. (Stoehr et al. 2011) reported that the spherical silver NPs are less toxic than other forms showing the effect of shape on the particle toxicity. Similarly, the fiber form of NPs was found to be more toxic than any other forms attributed to their low clearance rate in lungs leading to respiratory syndromes when inhaled (Hoet et al. 2004). Fiber particles with a width to height ratio of 3:1 penetrate the lungs based on their aerodynamic properties (Lippmann 1990). The length to width is known as aspect ratio, which has more roles in the cellular uptake of the particles. Absorption of spherical and rod shaped NPs with different aspect ratio was captured in the study conducted by Chithrani et al. (Chithrani et al. 2006) which showed maximum uptake of spherical NPs by the HeLa cells as compared to rod shaped (Fig. 23.1). Clearance of any foreign particles from lungs occurs with the help of mucociliary escalators from the upper ways while in the alveoli, the clearance was performed through phagocytosis. In the case of large fibers the clearance rate through phagocytosis is too slow as compared to other shapes (Hoet et al. 2004). At low inhalation concentrations or at low deposition rates with high clearance rate through phagocytosis, the half retention time is around 1680 h (70 days) which increases to the maximum when the deposition rate is higher than the clearance rate. The perseverance of fibers like asbestos or silica in lungs for a longer period was reported to induce cancer (Hoet et al. 2004; Donaldson et al. 2010). Carbon nanotubes were also found to cause similar effect and are more toxic than localized quartz and carbon blocks (Lam et al. 2003; Donaldson et al. 2010; Warheit et al. 2003). The longer fibers are broken down into shorter fibers in the lungs which were then easily removed through phagocytosis performed by macrophages. Such longitudinal broken down pieces of long asbestos fibers with reduced diameter were witnessed in the lungs which proves the phenomena. Amorphous fibers found to break perpendicular to their long axis (Searl 1994). These findings indicate that the retention of such particles creates a burden to the tissues thus, increasing the adverse effects.

Stanton et al. (Stanton and Wrench 1972; Stanton et al. 1977) observed that the fibers with the aspect ratio of 8:1.3 μm (length:diameter) induced mesothelioma in rats which is known as Stanton hypothesis. Thus, the increase in the length of the fiber increases the chances for cancer induction and it was found that the frequency of cancer was higher with the NPs of length more than 40 μm (Luyts et al. 2013; Schins 2002). Carbon nanotubes are an excellent example of such phenomena and they also show toxicity by inducing platelet aggregations (Hoet et al. 2004). Poland et al. (2008) reported the distorted macrophages caused by the failure of phagocytosis of long fibers of multiwalled carbon nanotubes. The macrophages were unable to uptake the whole length of multiwalled fibers thereby leading to its own destruction. There are models that deals with the internalization of the particles based on the geometry of the particles. There are two parameters used in the model. One is the contact angle at initial contact between the particles and cells/macrophages. Another parameter is the volume ratio between the particles to macrophage or cell. It is observed as the contact angle exceeds 45°; it abridged the uptake rate of NPs by macrophages. Similarly, the cells or macrophages can uptake equal or less than its volume, but it could not engulf bigger particles as the particle exceed its volume (Champion and Mitragotri 2006). These models are valid for the particles with size less than 50 nm. The contact area is the key factor that governs the uptake rate of phagocytosis (Luyts et al. 2013).

23.2.5 *Toxicity Based on Surface Coating/Surface Chemistry*

As similar to the effect of structural attributes and dimensions of NPs on their toxicity, the chemical composition also affects the degree of toxicity (Beddoes et al. 2015). The surface coating on the particles affects their uptake and excretion process. The enhanced surface area obtained in the zinc oxide NPs served as matrix and played a role in their cytotoxicity as observed in rats (Ho et al. 2011). As compared to the anionic coated particles with a zeta (ζ) potential of -11 ± 2.0 mV, chitosan coated poly (D, L, -lactide-co-glycolide) NPs with a zeta (ζ) potential of $+36.8 \pm 0.4$ mV has served as a better-sustained vehicle for gene delivery (Luyts et al. 2013; Baoum et al. 2010). However, the positively charged NPs showed more toxicity than the negative charges as enters lysosome, where it activates oxidase activity of nanoceria (Asati et al. 2010). These positively charged particles adversely affect the DNA and cell cycle.

Gold was served as food colorant and additives as safe material certified by Joint FAO/WHO Expert Committee (De Jong et al. 2008). However, the use of stabilizers like CTAB stabilizes the shape and size of the particles but make them more toxic even at low concentrations of CTAB (Niidome et al. 2006). Hence, this surface coating can either influence positively or negatively that demand through a study about the relationship between the surface charges and toxicity.

23.2.6 *Toxicity Based on Dosage and Exposure Period*

Influence of dosage and exposure period are well known as critical parameters that are directly related to the toxicity of NPs. However, obtaining a direct relation between the dosages required and its respective toxicity remains complex. In general, a logical dosage metric was utilized which relates the number of particles in the cytoplasm of each cell with the number of particles in the intracellular compartments (Wittmaack 2007; Oberdörster 2010). The quantification and prediction of the number of particles in the subcellular environment are also laborious (Elsaesser et al. 2010). Thus, the surface area along with size and shape of the particles were used as parameters instead of determining the number or mass of particles inside the cell (Oberdörster 2010). Lin et al. (2006) described the impact of dosage and exposure time of silica NPs (15 and 46 nm) on cytotoxicity in *in vitro* condition. The study revealed that the cells undergo for an oxidative stress as observed from the increased levels of nascent oxygen species and reduction in the glutathione levels. The dosage and exposure time were also reported to disturb the cell multiplication and biochemistry of the cell membranes (Chang et al. 2007; Jin et al. 2007). Elsaesser and Howard (2012) explained the direct relationship between the dose of the chemical and its toxicity.

23.3 Toxicity of Inhaled NPs

Many investigations have been targeting to understand the translocation behavior of the NPs and to determine the appropriate route for administration with minimal translocation and reduced toxicity in pharmaceutical practices (Warheit and Sayes 2015). Nasal uptake of the NPs through inhalation has gained interest as it was determined to be the best way to administer drugs with minimal translocation as how in the case of xenobiotic administration (Florence and Hussain 2001). Several studies were targeting to understand the effects of nasal administration of micro-, nano- and ultrafine particles and their toxicity (Elsaesser and Howard 2012; Donaldson et al. 2002). Translocation of NPs was reported to induce cardiovascular arrhythmia and coagulation (Yeates and Mauderly 2001), failure of autonomic nervous systems, etc. (Liao et al. 1999). The failure of cardiovascular system may be due to the initiation of inflammatory reaction in pulmonary system followed by compulsion in the heart muscles. A similar induction of inflammatory reaction in the pulmonary systems was reported by Chen et al. (2006) which exhibited the translocation of lipopolysaccharide coated NPs into other secondary organs through micro environmental changes. The other hypothesis may be due to the translocation of NPs to the systemic circulation which in turn affecting the cardiovascular functions (Oberdörster et al. 2005a; Hoet et al. 2004; Radad et al. 2012).

Hackenberg et al. (2011) reported immediate translocation of zinc oxide NPs from nasal mucosa into the cytoplasm (10%) and nucleus (1.5%) of nasal mucosa cells in an in vitro study. These translocated zinc oxide NPs induced increased secretion of interleukins (IL-8) and caused damage to the host DNA. In similar to this, nickel nanomaterial was reported to induce leakage of LDH in the epithelial cells of lungs and damage to mitochondria of those cells. In addition to that, it also induced oxidative stress and associated lipid degradation, followed by apoptosis in the cells as determined by their increased caspase-3 activity (Ahamed 2011).

Most worrying is the internalization and translocation of NPs into the central nervous system (CNS). A translocation rate of 2.5 mm per hour into the CNS via olfactory nerves was reported by Hoet et al. (2004). Similar translocation of colloidal gold NPs into olfactory bulb through olfactory nerves was observed under in vivo conditions in a monkey (Elder and Oberdorster 2006). Simko and Mattsson (2010) observed the transport of NPs via olfactory, trigeminal, and sensory nervous systems into the CNS. Similarly, the quantum dots of titanium oxide NPs translocated into the central nervous system resulting in its failure (Radad et al. 2012). Translocation of copper NPs adversely affected organs such as liver, kidney, and testicles (Yang et al. 2010a, b) resulting in increased expression of various enzymes such as, alanine aminotransferase, aspartate aminotransferase. Thus, there are several ways the translocation of NPs may affect the host system in which they are injected. In order to avoid such toxicity, several attempts were made to predict the toxicological behavior and to assess the positive and negative impacts of the NPs which are reported in the following paragraphs.

23.3.1 Destruction of Cellular Membranes

Cell to cell interaction and their interaction with the surrounding materials are guarded by the membranes which allow selective uptake and rejection through membrane channels and receptors (Vasir and Labhasetwar 2008). In similar to the outer environment, the organelles in the intracellular environment of Eukaryotes are guarded by the membrane systems. These are the susceptible zones targeted by NPs leading to physical or chemical damages. The binding properties of the NPs, the surface pressure it imposes on the membrane and adhesion are determined by the structural attributes of the NPs (Peetla and Labhasetwar 2008). There are several ways a nanoparticle can damage the membrane integrity which includes physical damage and chemical oxidation that modulates selectivity and sensitivity of the membranes (Ginzburg and Balijepalli 2007). This membrane damage varied significantly based on the type of nanomaterial, physicochemical characteristics and concentration as observed from gene expression analysis (Gou et al. 2010). They also target the intracellular membrane-bound organelles such as nucleus by diffusion or through receptor-mediated channel proteins causing damage to the nuclear DNA and its functions (Elsaesser et al. 2010).

23.3.2 Non-specific Interaction with Biomolecules

Once the NPs enter the intracellular environment they are suddenly opened up to several possibilities of reactions with several biomolecules. The major problem is with the proteins in which these NPs mimic as chaperons which guide them in the folding process. The intervention of NPs with protein leads to the refolding of proteins or misfolding of proteins thereby leading to the termination of certain process or pathways from functioning (Dobson 2003). Such miss folding of proteins was reported to cause functional loss, neurodegenerative diseases due to the amyloid structure formation if occurs in the nervous system. Similar changes in the protein folding were reported on exposure to NPs acting as molecular chaperons (Akiyoshi et al. 1999; Takahashi et al. 2011; Marano et al. 2011). Thus, misfolding and overproduction of proteins and other macromolecules have been considered to be the most important concern in the evaluation of toxicology of NPs (Elsaesser and Howard 2012).

NPs interaction with DNA has been studied by several researchers, however, a complete understanding of the toxicology effects were still lagging (Barnes et al. 2008). The release of ROS as a response to oxidative stress may be one of the reasons which might have involved in damaging the DNA (Elsaesser and Howard 2012). However, a complete understanding of the phenomena involved won't be possible only with an in vitro study as suggested by Gil et al. (Gil et al. 2010). In a complex system like human beings with several organ systems, a wholistic approach with in vivo studies will be required. In vivo studies provide help in understanding real time modulations, evaluation of chronic exposures, convenient trial numbers, with proper evaluation of retention and clearance rate of NPs (Radad et al. 2012). A strategic approach and trials in *in vivo* will be essential to evaluate the hazardous properties of NPs employed in the industrial environments (Elsaesser and Howard 2012).

23.4 Conclusions

There are attempts to investigate the impact of physicochemical parameters on the toxicity of exposed NPs. It is highly complex to compare and generalize the phenomena, as there were more conflicts. Sometime, the presence of trace materials and solvents during the preparation of NPs also caused high toxicity. Hence, finding sources of toxicity and removal of trace materials from NPs is mandatory to enhance the accuracy of the prediction. However, there is much scope for the researcher to perform experiment and model the behavior of the particles as it interact with biologicals and environmental factors.

References

Ahamed M (2011) Toxic response of nickel nanoparticles in human lung epithelial A549 cells. Toxicol In Vitro 25(4):930–936

Akiyoshi K, Sasaki Y, Sunamoto J (1999) Molecular chaperone-like activity of hydrogel nanoparticles of hydrophobized pullulan: thermal stabilization with refolding of carbonic anhydrase B. Bioconjug Chem 10:321–324

Anastasio C, Martin ST (2001) Atmospheric nanoparticles. In: Banfield JF, Navrotsky A (eds) Nanoparticles and the environment, vol 44. Mineralogical Society of America, Washington, DC, pp 293–349

Arvizo R, Miranda O, Thompson MA, Pabelick CM, Bhattacharya R, Robertson JD, Rotello VM, Prakash YS, Mukherjee P (2010) Effect of nanoparticle surface charge at the plasma membrane and beyond. Nano Lett 10(7):2543–2548

Asati A, Santra S, Kaittanis C, Perez JM (2010) Surface-charge-dependent cell localization and cytotoxicity of cerium oxide nanoparticles. ACS Nano 4:5321–5331

Auffan M, Rose J, Bottero JY, Lowry GV, Jolivet JP, Wiesner MR (2009) Towards a definition of inorganic nanoparticles from an environmental, health and safety perspective. Nat Nanotechnol 4:634–641

Ballou B, Lagerholm BC, Ernst LA, Bruchez MP, Waggoner AS (2004) Noninvasiveimaging of quantum dots in mice. Bioconjug Chem 15:79–86

Baoum A, Dhillon N, Buch S, Berkland C (2010) Cationic surface modification of PLG nanoparticles offers sustained gene delivery to pulmonary epithelial cells. J Pharm Sci 99:2413–2422

Barnes CA, Elsaesser A, Arkusz J, Smok A, Palus J, Lesniak A, Salvati A, Hanrahan JP, Jong WH, Dziubaltowska E, Stepnik M, Rydzynski K, McKerr G, Lynch I, Dawson KA, Howard CV (2008) Reproducible comet assay of amorphous silica nanoparticles detects no genotoxicity. Nano Lett 8:3069–3074

Beddoes CM, Case CP, Briscoe WH (2015) Understanding nanoparticle cellular entry: a physicochemical perspective. Adv Colloid Interface Sci 218:48–68

Carlson C, Hussain SM, Schrand AM, Braydich-Stolle LK, Hess KL, Jones RL, Schlager JJ (2008) Unique cellular interaction of silver nanoparticles: size-dependent generation of reactive oxygen species. J Phys Chem B 112(43):13608–13619

Champion JA, Mitragotri S (2006) Role of target geometry in phagocytosis. Proc Natl Acad Sci USA 103:4930–4934

Chang JS, Chang KLB, Hwang DF, Kong ZL (2007) In vitro cytotoxicity of silica nanoparticles at high concentrations strongly depends on the metabolic activity type of the cell line. Environ Sci Technol 41:2064–2068

Chen Z, Meng H, Xing G, Chen C, Zhao Y, Jia G, Wang T, Yuan H, Ye C, Zhao F, Chai Z (2006a) Acute toxicological effects of copper nanoparticles in vivo. Toxicol Lett 163(2):109–120

Chen J, Tan M, Nemmar A, Song W, Dong M, Zhang G, Li Y (2006b) Quantification of extrapulmonary translocation ofintratracheal-instilled particles in vivo in rats: effect of lipopolysaccharide. Toxicology 222:195–201

Chen X, Tian F, Zhang X, Wang W (2013) Internalization pathways of nanoparticles and their interaction with a vesicle. Soft Matter 9:7592–7600

Chithrani BD, Ghazani AA, Chan WC (2006) Determining the size and shape dependence of gold nanoparticle uptake into mammalian cells. Nano Lett 6(4):662–668

Cho SJ, Maysinger D, Jain M, Röder B, Hackbarth S, Winnik FM (2007) Long-term exposure to CdTe quantum dots causes functional impairments in live cells. Langmuir 23:1974–1982

Creutzenberg O (2012) Biological interactions and toxicity of nanomaterials in the respiratory tract and various approaches of aerosol generation for toxicity testing. Arch Toxicol 86(7):1117–1122

De Jong WH, Hagens WI, Krystek P, Burger MC, Sips AJ, Geertsma RE (2008) Particle size-dependent organ distribution of gold nanoparticles after intravenous administration. Biomaterials 29(12):1912–1919

Dekie L, Toncheva V, Dubruel P, Schacht EH, Barrett L, Seymour LW (2000) Poly-L-glutamic acid derivatives as vectors for gene therapy. J Control Rel 65(1–2):187–202

Dobson CM (2003) Protein folding and misfolding. Nature 426:884–890

Dockery DW, Luttmann-Gibson H, Rich DQ, Link MS, Mittleman MA, Gold DR, Koutrakis P, Schwartz JD, Verrier RL (2005) Association of air pollution with increased incidence of ventricular tachyarrhythmias recorded by implanted cardioverter defibrillators. Environ Health Perspect 113:670–674

Dominici F, Peng RD, Bell ML, Pham L, McDermott A, Zeger SL, Samet JM (2006) Fine particulate air pollution and hospital admission for cardiovascular and respiratory diseases. JAMA 295(10):1127–1134

Donaldson K, Brown D, Clouter A, Duffin R, MacNee W, Renwick L, Tran L, Stone V (2002) The pulmonary toxicology of ultrafine particles. J Aerosol Med 15:213–220

Donaldson K, Murphy FA, Duffin R, Poland CA (2010) Asbestos, carbon nanotubes and the pleural mesothelium: a review of the hypothesis regarding the role of long fibre retention in the parietal pleura, inflammation and mesothelioma. FibreToxicol 7:5

Driscoll KE, Deyo LC, Carter JM, Howard BW, Hassenbein DG, Bertram TA (1997) Effects of particle exposure and particle-elicited inflammatory cells on mutation in rat alveolar epithelial cells. Carcinogenesis 18:423–430

Elder A, Oberdorster G (2006) Translocation and effects of ultrafine particles outside of the lung. Clin Occup Environ Med 5:785–796

Elsaesser A, Howard CV (2012) Toxicology of nanoparticles. Adv Drug Deliver Rev 64:129–137

Elsaesser A, Taylor A, de Yanes GS, McKerr G, Kim EM, O'Hare E, Howard CV (2010) Quantification of nanoparticle uptake by cells using microscopical and analytical techniques. Nanomedicine 5:1447–1457

Florence AT, Hussain N (2001) Transcytosis of nanoparticle and dendrimer delivery systems: evolving vistas. Adv Drug Deliv Rev 50:S69–S89

Fubini B, Ghiazza M, Fenoglio I (2010) Physico-chemical features of engineered nanoparticles relevant to their toxicity. Nanotoxicology 4(4):347–363

Gajewicz A, Rasulev B, Dinadayalane TC, Urbaszek P, Puzyn T, Leszczynska D, Leszczynski J (2012) Advancing risk assessment of engineered nanomaterials: application of computational approaches. Adv Drug Deliv Rev 64(15):1663–1693

Gil PR, Oberdorster G, Elder A, Puntes V, Parak WJ (2010) Correlating physicochemical with toxicological properties of nanoparticles: the present and the future. ACS Nano 4:5527–5531

Ginzburg VV, Balijepalli S (2007) Modeling the thermodynamics of the interaction of nanoparticles with cell membranes. Nano Lett 7:3716–3722

Gou N, Onnis-Hayden A, Gou AZ (2010) Mechanistic toxicity assessment of nanomaterials by whole-cell-array stress genes expression analysis. Environ Sci Technol 44:5964–5970

Hackenberg S, Zimmermann FZ, Scherzed A, Friehs G, Froelich K, Ginzkey C, Koehler C, Burghartz M, Hagen R, Kleinsasser N (2011) Repetitive exposure to zinc oxide nanoparticles induces dna damage in human nasal mucosa mini organ cultures. Environ Mol Mutagen 52(7):582–589

Handy RD, Von der Kammer F, Lead JR, Hassellöv M, Owen R, Crane M (2008) The ecotoxicology and chemistry of manufactured nanoparticles. Ecotoxicology 17(4):287–314

Heinlaan M, Ivask A, Blinova I, Dubourguier HC, Kahru A (2008) Toxicity of nanosized and bulk ZnO, CuO and TiO_2 to bacteria *Vibrio fischeri* and crustaceans *Daphnia magna* and *Thamnocephalus platyurus*. Chemosphere 71(7):1308–1316

Ho M, Wu KY, Chein HM, Chen LC, Cheng TJ (2011) Pulmonary toxicity of inhaled nanoscale and fine zinc oxide particles: mass and surface area as an exposure metric. Inhal Toxicol 23:947–956
Hoet PH, Brüske-Hohlfeld I, Salata OVJ (2004) Nanoparticles—known and unknown health risks. J Nanobiotechnol 2(1):12
Jiang Y, Huo S, Mizuhara T, Dar R, Lee YW, Hou S, Moyano DF, Duncan B, Liang XJ, Rotello VM (2015) The interplay of size and surface functionality on the cellular uptake of sub-10 nm gold nanoparticles. ACS Nano 27:9986–9993
Jin Y, Kannan S, Wu M, Zhao JX (2007) Toxicity of luminescent silica nanoparticles to living cells. Chem Res Toxicol 20:1126–1133
Karlsson HL, Gustafsson J, Cronholm P, Möller L (2009) Size-dependent toxicity of metal oxide particles—a comparison between nano-and micrometer size. Toxicol Lett 188(2):112–118
Kiessling F, Mertens ME, Grimm J, Lammers T (2014) Nanoparticles for imaging: top or flop? Radiology 273(1):10–28
Kumar P, Robins A, Vardoulakis S, Britter R (2010) A review of the characteristics of nanoparticles in the urban atmosphere and the prospects for developing regulatory controls. Atmos Environ 44(39):5035–5052
Lam CW, James JT, McCluskey R, Hunter RL (2003) Pulmonary toxicity of single-wall carbon nanotubes in mice 7 and 90 days after intratracheal instillation. Toxicol Sci 77:126–134
Liao D, Creason J, Shy C, Williams R, Watts R, Zweidinger R (1999) Daily variation of particulate air pollution and poor cardiac autonomic control in the elderly. Environ Health Perspect 107:521–525
Lin W, Huang YW, Zhou XD, Ma Y (2006) In vitro toxicity of silica nanoparticles in human lung cancer cells. Toxicol Appl Pharmacol 217:252–259
Lippmann M (1990) Effects of fiber characteristics on lung deposition, retention, and disease. Environ Health Perspect 88:311–317
Lovric J, Bazzi HS, Cuie Y, Fortin GR, Winnik FM, Maysinger D (2005) Differences in subcellular distribution and toxicity of green and red emitting CdTe quantum dots. J Mol Med 83:377–385
Luyts K, Napierska D, Nemery B, Hoet PH (2013) How physico-chemical characteristics of nanoparticles cause their toxicity: complex and unresolved interrelations. Environ Sci Process Impacts 15(1):23–38
Marano F, Hussain S, Rodrigues-Lima F, Baeza-Squiban A, Boland S (2011) Nanoparticles: molecular targets and cell signaling. Arch Toxicol 85:733–741
Meng H, Chen Z, Xing G, Yuan H, Chen C, Zhao F, Zhang C, Zhao Y (2007) Ultrahigh reactivity provokes nanotoxicity: explanation of oral toxicity of nano-copper particles. Toxicol Lett 175(1):102–110
Morris J, Willis J, Gallagher K (2007) Nanotechnology white paper. US Environmental Protection Agency, Washington, DC, www.epa.gov/osa/pdfs/nanotech/epa-nanotechnology-whitepaper-0207.pdf
Mueller N, Nowack B (2008) Exposure modeling of engineered nanoparticles in the environment. Environ Sci Technol 42:4447–4453
Nel AE, Madler L, Velegol D, Xia T, Hoek EM, Somasundaran P, Klaessig F, Castranova V, Thompson M (2009) Understanding biophysicochemical interactions at the nano-bio interface. Nat Mater 8:543–557
Niidome T, Yamagata M, Okamoto Y, Akiyama Y, Takahashi H, Kawano T, Katayama Y, Niidome Y (2006) PEG-modified gold nanorods with a stealth character for in vivo applications. J Control Rel 114:343–347
Nowack B (2009) The behavior and effects of nanoparticles in the environment. Environ Pollut 157:1063–1064
Oberdörster G (2001) Pulmonary effects of inhaled ultrafine particles. Int Arch Occup Environ Health 74(1):1–8
Oberdörster G (2010) Safety assessment for nanotechnology and nanomedicine: concepts of nanotoxicology. J Intern Med 267:89–105
Oberdörster G, Oberdörster E, Oberdörster J (2005a) Nanotoxicology: an emerging discipline evolving from studies of ultrafine particles. Environ Health Perspect 113:823–839

Oberdörster G, Maynard A, Donaldson K, Castranova V, Fitzpatrick J, Ausman K, Carter J, Karn B, Kreyling W, Lai D, Olin S, Monteiro-Riviere N, Warheit D, Yang H (2005b) Principles for characterizing the potential human health effects from exposure to nanomaterials: elements of a screening strategy. Part FibreToxicol 2:8–43

Pan Y, Neuss S, Leifert A, Fischler M, Wen F, Simon U, Schmid G, Brandau W, Jahnen-Dechent W (2007) Size-dependent cytotoxicity of gold nanoparticles. Small 3:1941–1949

Peetla C, Labhasetwar V (2008) Biophysical characterization of nanoparticle–endothelial model cell membrane interactions. Mol Pharm 5:418–429

Phillips JI, Green FY, Davies JC, Murray J (2010) Pulmonary and systemic toxicity following exposure to nickel nanoparticles. Am J Ind Med 53:763–767

Poland CA, Duffin R, Kinloch I, Maynard A, Wallace WA, Seaton A, Stone V, Brown S, Macnee W, Donaldson K (2008) Carbon nanotubes introduced into the abdominal cavity of mice show asbestos-like pathogenicity in a pilot study. Nat Nanotechnol 3:423–428

Rabolli V, Thomassen LC, Princen C, Napierska D, Gonzalez L, Kirsch-Volders M, Hoet PH, Huaux F, Kirschhock CE, Martens JA, Lison D (2010) Influence of size, surface area and microporosity on the in vitro cytotoxic activity of amorphous silica nanoparticles in different cell types. Nanotoxicology 4:307–318

Radad K, Al-Shraim M, Moldzio R, Rausch WD (2012) Recent advances in benefits and hazards of engineered nanoparticles. Environ Toxicol Pharmacol 34(3):661–672

Ryman-Rasmussen JP, Cesta MF, Brody AR, Shipley-Phillips JK, Everitt JI, Tewksbury EW, Moss OR, Wong BA, Dodd DE, Andersen ME, Bonner JC (2009) Inhaled carbon nanotubes reach the subpleural tissue in mice. Nat Nanotechnol 4:747–751

Schins RP (2002) Mechanisms of genotoxicity of particles and fibers. Inhal. Toxicol 14:57–78

Searl A (1994) A review of the durability of inhaled fibers and options for the design of safer fibers. Ann Occup Hyg 38:839–855

Shin SW, Song IH, Um SH (2015) Role of physicochemical properties in nanoparticle toxicity. Nanomaterials 5(3):1351–1365

Simko M, Mattsson MO (2010) Risks from accidental exposures to engineered nanoparticles and neurological health effects: a critical review. Part FibreToxicol 21:42–57

Singh AK, Kasinath BS, Lewis EJ (1992) Interaction of polycations with cell surface negative charges of epithelial cells. Biochim Biophys Acta 1120:337–342

Solmesky LJ, Shuman M, Goldsmith M, Weil M, Peer D (2011) Assessing cellular toxicities in fibroblasts upon exposure to lipid-based nanoparticles: a high content analysis approach. Nanotechnology 22(49):494016

Srivastava V, Gusain D, Sharma YC (2015) A critical review on the toxicity of some widely used engineered nanoparticles. Ind Eng Chem Res 54(24):6209–6233

Standard, BSI British (2005) Vocabulary: nanoparticles. PAS 71, no. 2005:32

Stanton MF, Wrench C (1972) Mechanisms of mesothelioma induction with asbestos and fibrous glass. J Natl Cancer Inst 48:797–821

Stanton MF, Laynard M, Tegeris A, Miller E, May M, Kent E (1977) Carcinogenicity of fibrous glass: pleural response in the rat in relation to fiber dimension. J Natl Cancer Inst 58:587–603

Stoehr LC, Gonzalez E, Stampfl A, Casals E, Duschl A, Puntes V, Oostingh GJ (2011) Shape matters: effects of silver nanospheres and wires on human alveolar epithelial cells. Part FibreToxicol 8:36

Takahashi H, Sawada S, Akiyoshi K (2011) Amphiphilic polysaccharide nanoballs: a new building block for nanogel biomedical engineering and artificial chaperones. ACS Nano 5:337–345

Vasir JK, Labhasetwar V (2008) Quantification of the force of nanoparticle–cell membrane interactions and its influence on intracellular trafficking of nanoparticles. Biomaterials 29:4244–4252

Warheit DB, Sayes CM (2015) Routes of exposure to nanoparticles: hazard tests related to portal entries. In: Dolez PI (ed) Nanoengineering: global approaches to health and safety issues, pp 41–54. Elsevier, New York

Warheit DB, Laurence BR, Reed KL, Roach DH, Reynolds GA, Webb TR (2003) Comparative pulmonary toxicity assessment of single wall carbon nanotubes in rats. Toxicol Sci 77:117–125

Wittmaack K (2007) In search of the most relevant parameter for quantifying lung inflammatory response to nanoparticle exposure: particle number, surface area, or what? Environ Health Perspect 115:187–194

Yang BH, Wang QJ, Lei RH, Wu CQ, Shi C, Wang Q, Yuan Y, Wang Y, Luo Y, Hu Z, Ma H, Liao M (2010a) Systems toxicology used in nanotoxicology: mechanistic insights into the hepatotoxicity of nano-copper particles from toxicogenomics. J Nanosci Nanotechnol 10:8527–8537

Yang Z, Liu ZW, Allaker RP, Reip P, Oxford J, Ahmad Z, Ren G (2010b) A review of nanoparticle functionality and toxicity on the central nervous system. J R Soc Interface 4:S411–S422

Yang Y, Engkvist O, Llinàs A, Chen H (2012) Beyond size, ionization state, and lipophilicity: influence of molecular topology on absorption, distribution, metabolism, excretion, and toxicity for drug like compounds. J Med Chem 55(8):3667–3677

Yeates DB, Mauderly JL (2001) Inhaled environmental/occupational irritants and allergens: mechanisms of cardiovascular and systemic responses: introduction. Environ Health Perspect 109:479–481

Zhao Y, Sun X, Zhang G, Trewyn BG, Slowing II, Lin VS (2011) Interaction of mesoporous silica nanoparticles with human red blood cell membranes: size and surface effects. ACS Nano 5:1366–1375

Chapter 24
Comparative Study of Gas and Particulate Phase Concentrations of Polycyclic Aromatic Hydrocarbons (PAHs) at Two Sites in Agra

Puneet Kumar Verma, Dinesh Sah, K. Maharaj Kumari, and Anita Lakhani

24.1 Introduction

Polycyclic Aromatic Hydrocarbons (PAHs) are organic environmental contaminants containing fused aromatic rings. They are ubiquitous and persistent in nature and have been a subject of concern due to their elevated concentration, carcinogenic and mutagenic properties (Wei et al. 2015). These are generally formed and released from incomplete combustion of fossil fuels and biomass (Albinet et al. 2007; Lima et al. 2005). Anthropogenic sources of these contaminants include emissions from industries, vehicular exhaust and domestic combustion.

PAHs are found in atmosphere in both the particulate and gas phase (Harrison et al. 1996). Due to their semi-volatile character, they get partitioned between the two phases; the partition of these compounds in both the phase governs their occurrence in the atmosphere, fate and toxicity (Ringuet et al. 2012; Schnelle-Kreis et al. 2007). Atmospheric partitioning of these compounds follows mainly two mechanisms i.e., adsorption onto the surface of particles and absorption into the organic layer on the particle surface (Ringuet et al. 2012). Removal of PAHs from the atmosphere through dry and wet deposition processes, their fate and transport are strongly influenced by their partitioning behavior (Bidleman 1988).

Gas and particle phase PAHs were monitored at a traffic site in Eskisehir from January to October 2006. During the sampling period, 66 and 69% of the total PAHs were found in gaseous formin heating and non-heating period respectively (Gaga and Ari 2011). The gas and particle phase PAHs were determined concurrently in a forest and a clearing region at a rural site in southern Ontario, Canada from October 2001 to November 2002. Gas-particle partitioning was not found significantly different

P. K. Verma · D. Sah · K. Maharaj Kumari · A. Lakhani (✉)
Department of Chemistry, Faculty of Science, Dayalbagh Educational Institute, Dayalbagh, Agra 282005, India
e-mail: anita.lakhani01@gmail.com

S. M. Shiva Nagendra et al. (eds.), *Urban Air Quality Monitoring, Modelling and Human Exposure Assessment*, Springer Transactions in Civil and Environmental Engineering,
https://doi.org/10.1007/978-981-15-5511-4_24

between the forest and the clearing region (Su et al. 2006). The gas-particle partitioning behavior of PAHs has been evaluated at various other sites like in Dalian (Zhou et al. 2013), sites near the Yellow sea (Wang et al. 2013), sites in the Chinese Northern plains (Yang et al. 2010) Guangzhou City (Liu et al. 2009) in China by simultaneous monitoring of PAHs in both gas and particulate phases.

Several studies have been carried out in India like in Nagpur (Etchie et al. 2018), Lucknow (Bhargava et al. 2004; Pandey et al. 2013), Ahmedabad (Aggarwal et al. 1982), Delhi (Gadi et al. 2012), Chennai (Mohanraj et al. 2011), Tiruchirappalli (Mohanraj et al. 2011), Coimbatore (Mohanraj et al. 2012), Agra (Rajput and Lakhani 2012; Singla et al. 2012) focusing only on the particulate phase PAHs to evaluate the carcinogenic risk from PAH exposure. In the past, the only study reported on gas-particle partitioning of PAHs was in Mumbai (Venkataraman et al. 1999). Most recently we have conducted simultaneous measurement of gas and particulate phase PAHs in Agra to justify physical adsorption of PAHs using Pankow model (Verma et al. 2017; Kumari and Lakhani 2018).

Agra lies in the Indo-Gangetic plain and is located in the state of Uttar Pradesh in India (27.1° N, 78.0° E). It has a population of about 1.275 million and area of about 4028 km^2. Despite their environmental importance, only few studies have been conducted on PAH levels in Agra. In the present study, PAHs have been measured both in the gas and particulate phases to determine their concentrations and their distribution between both the phases. Health risk assessment was also performed for all the 16 priority PAHs using Toxic equivalent approach.

24.2 Methodology

24.2.1 Sampling Sites

Sampling was performed simultaneously at a traffic site and a rural site. Bhagwan Talkies crossing situated on the National highway—19 (old National Highway—2) was the traffic site; this site observes mixed vehicular traffic throughout the day and night. In addition, it is also one of the major residential and commercial areas of the city. Sikandarpur, a small village in the northern part of Agra was selected as the rural site (Verma et al. 2017). Both the sampling sites are shown in Fig. 24.1.

24.2.2 Sample Collection

Samplers were placed on the roof of double storey buildings (approximately 12 m above the ground level) at both the sites. Sampling was performed during November and December 2015. Samples were collected using air samplers (TISCH International, USA Model TE-1000X) on quartz fiber filters (QFFs) (10.16 cm in diameter)

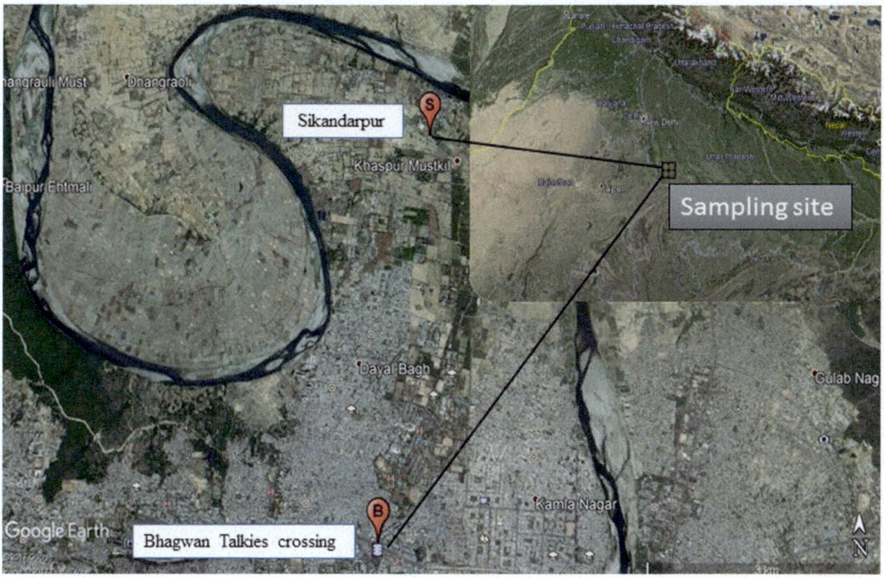

Fig. 24.1 Sampling sites in Agra (https://earth.google.com)

and on poly-urethane foam (PUF) plugs (8 cm long and 6 cm in diameter) for the particulate phase and gas phase respectively at a flow rate of 0.2 m^3 min^{-1}. Two samples each of 24 h duration were collected weekly at both the sites; the total sampled volume of air ranged from 317 to 321 m^3 for individual samples. Before sampling, QFFs were baked in a muffle furnace at 750 °C overnight to eliminate background contamination and packed in aluminum foil packages until used. QFFs were also desiccated at room temperature before and after sample collection for 24 h and weighed by a microbalance (Mettler AJ 150). Total suspended mass concentration (TSP) was determined from the difference of the post-and pre-sampling weights. PUFs plugs were pre-cleaned in dichloromethane (HPLC grade) by soxhlet extraction. Cleaned PUF were wrapped in aluminum foil and excessive DCM were removed by baking them in an oven at 70 °C for 40 min. PUFs were packed in zip locked packets and transported to the field without allowing exposure to ambient air. After sampling, PUFs and QFFs were stored at -20 °C until extraction (Verma et al. 2017).

24.2.3 Sample Extraction and Clean Up

PUFs were extracted in a mixture of DCM and n-hexane (4:1 ratio) by a soxhlet extraction for 8 h while QFFs were cut in small pieces and extracted ultrasonically at room temperature for 60 min. The organic extract obtained from the QFFs (particulate PAHs) was filtered using Sartorius (393) filter papers. Both the QFFs (Particulate phase PAHs) and PUF (Particulate phase PAHs) extracts were further cleaned by using a silica gel column and further concentrated to 1.5 ml by rotavapor (BUCHI). Concentrated PAHs extracts were stored at −20 °C in glass vials until analysis (Verma et al. 2017).

24.2.4 *Analysis*

Gas Chromatograph-Mass Spectrometer (GC-MS, BRUKER SCION SQ) was used in selected ion monitoring mode to analyze 16 priority PAHs. Helium was used at a flow rate of 1 mL min^{-1} as carrier gas. Separation of 16 PAHs was achieved on Rtx-5 MS capillary column (30 m in length, 0.25 mm internal diameter). Injector temperature was maintained at 270 °C. Initially the temperature of oven was set at 60 °C for 4 min and then it was programmed to increased to 290 °C at a rate of 8 °C min^{-1} and then finally held for 30 min. For every analysis 1 μl sample was injected in GC-MS in split less mode and the duration for entire run was of 62 min. The mass scanning ranged from 50 to 500 (Verma et al. 2017).

24.2.5 *Quality Control*

Recovery of compounds during the complete extraction procedure was evaluated by spiking known concentration of standard in samples. Recoveries for all the particulate and gas phase PAHs were between 97 ± 6.1% and 95 ± 2.3%, respectively. Same analytical procedure was used to extract and analyze field blanks. Limit of detection (LOD) were determined for all the PAHs (Miller and Miller 2010). The LOD values of individual PAHs are shown in Table 24.1 along with their abbreviation and *m*/*z* ratio (Verma et al. 2017).

24.3 Results and Discussion

24.3.1 *Atmospheric Concentrations of Polycyclic Aromatic Hydrocarbons (PAHs)*

The mean and standard deviation (SD), for all the particulate and gas phase concentration of PAHs at both the sites are given in Table 24.2. At rural site, dominant PAHs in the particle phase were BkF, IP, BghiP, Chy and BbF whereas in the gasphase the dominant PAHs were Phen, Anth, Fluo, Pyr and Ace. At traffic site, the dominant PAHs in the particulate phase were Chy, BkF, BbF, BghiP and IP whereas Phen, Anth, Pyr, Chy were the dominant compounds in the gas phase. The total concentrations

Table 24.1 Limit of detection (LOD)

PAHs	Abbreviation	*m*/*z* ratio	LOD (μg ml^{-1})
Napthalene	Nap	128	0.84
Acenapthylene	Acy	152	1.85
Acenapthene	Ace	154	0.05
Fluorene	Fluo	166	0.51
Phenanthrene	Phen	178	0.78
Anthracene	Anth	178	0.02
Fluoranthene	Fla	202	0.01
Pyrene	Pyr	202	0.01
Benzo(a)anthracene	BaA	228	0.15
Chrysene	Chy	228	0.02
Benzo(b)fluoranthene	BbF	252	0.18
Benzo(k)fluoranthene	BkF	252	0.18
Benzo(a)pyrene	BaP	252	1.42
Indeno[*1,2,3-cd*]pyrene	IP	276	0.12
Dibenz[a,h]anthracene	DbA	278	0.11
Benzo[g,h,j]perylene	BghiP	276	0.06

of PAHs were 4080 ng m^{-3} in the particle phase and 6888 ng m^{-3} in the gas phase at traffic site, whereas they were 1866.2 ng m^{-3} and 2782.7 ng m^{-3} in the gas and particle phase respectively, at the rural site. The volatile PAHs (Ace, Phen, Pyr and Fla) contributed maximum to the total gas phase PAH concentrations, in particular, Phen was found as the most abundant PAH in this study, approximately 37.4% of total PAHs at traffic site and 28.3% at the rural site. These results are in agreement with the previously reported studies (Akyüz and Çabuk 2010; Odabasi et al. 1999). Total PAHs concentration (gas + particulate) at traffic site (10,968 ng m^{-3}) was approximately 2.5 times higher than the PAHs concentration at rural site (4649 ng m^{-3}). Individual PAHs concentration (gas + particulate) were also found higher at the traffic site with a traffic/rural ratio varying between 0.9 and 3.9 except for Ace which dominated at rural site.

The low molecular mass PAHs (Nap, Acy, Ace, Fluo, Phen and Anth) were found to be higher in the gas phase and high molecular mass PAHs (BbF, BkF, BaP, IP, DbA and BghiP) were found mainly in the particle phase at both sites (Fig. 24.2). At the rural site, 3 and 4-ring PAHs were associated primarily with the gas phase except Ace and Fluo whereas BaA and Chy dominated in the particulate phase at both the sites, 5 and 6-ring PAHs (BbF, BkF, BaP, DbA, IP and BghiP) were found almost completely in the particle phase.

The distribution of PAHs between the particle and gas phase is consistent with other studies (Akyüz and Çabuk 2010; Li et al. 2006; Cincinelli et al. 2007; Tasdemir and Esen 2007). Based on number of rings, the 3-ring PAHs comprised approximately 75% of the total PAHs at both the sites in the gaseous phase as shown in Fig. 24.3,

Table 24.2 Concentration of PAHs at rural and traffic site

PAHs	Rural site			Traffic site			Gas + particulate						
	Particulate (ng m^{-3})	Gas (ng m^{-3})	G/P^a	Particulate (ng m^{-3})	Gas (ng m^{-3})	G/P^a	Rural (ng m^{-3})	Traffic (ng m^{-3})	T/R^b	Rural BaP-TEQ	Rural % BaP-TEQ Contribution	Traffic BaP-TEQ	Traffic % BaP-TEQ Contribution
Nap	1.9 ± 2.7	18.2 ± 21.4	9.1	29 ± 39.1	49.5 ± 66.7	1.7	20.2	78.5	3.8	0.02	0.005	0.08	0.008
Acy	7 ± 5.4	11.2 ± 7.8	1.5	11.4 ± 11.4	27 ± 22.9	2.3	18.3	38.4	2.1	0.02	0.004	0.04	0.004
Ace	112 ± 64.8	148.8 ± 106.2	1.3	174.3 ± 158.9	83.5 ± 57.1	0.4	260.8	257.8	0.9	0.26	0.061	0.26	0.026
Fluo	174.4 ± 126.4	325.6 ± 293.2	1.8	396.9 ± 383.1	257.5 ± 218.9	0.6	500	654.4	1.3	0.50	0.117	0.65	0.067
Phen	18.5 ± 14.5	1298.2 ± 653.9	70.1	387.1 ± 643.9	3724.9 ± 3251.9	9.6	1316.70	4112	3.1	1.32	0.308	4.11	0.418
Anth	11.3 ± 0.8	350.5 ± 95.8	30.8	141.4 ± 157.1	989.4 ± 867	6.9	361.8	1130.9	3.1	0.36	0.085	1.13	0.115
Fla	3.6 ± 3.4	47.8 ± 24.4	13.2	30.7 ± 25.5	151.4 ± 94.5	4.9	51.4	182.1	3.5	0.05	0.012	0.18	0.019
Pyr	36.7 ± 31.3	305.9 ± 149.3	8.3	196.5 ± 191.4	975.1 ± 542.2	4.9	342.7	1171.7	3.4	0.34	0.080	1.17	0.119
BaA	118.7 ± 120.2	53.9 ± 28.2	0.4	283.6 ± 152.2	152.1 ± 109.5	0.5	172.6	435.7	2.5	17.26	4.037	43.57	4.430
Chy	220.7 ± 154	166.5 ± 61.6	0.7	467.2 ± 214.5	446.7 ± 460.4	0.9	387.2	913.9	2.3	3.87	0.906	9.14	0.929
BbF	211.2 ± 175.7	15.9 ± 11.5	0.07	413.6 ± 198.3	7.4 ± 3.4	0.01	227.2	421.1	1.8	22.71	5.312	42.10	4.281
BkF	270.5 ± 238.7	7.6 ± 3.8	0.02	449.3 ± 263.0	3.8 ± 2.9	0.008	278.2	453.2	1.6	27.81	6.505	45.31	4.607
BaP	148 ± 140	13.3 ± 4	0.09	292.8 ± 155.4	8.8 ± 3.1	0.03	161.4	301.7	1.8	161.30	37.728	301.60	30.666
IP	266.1 ± 218.8	6.5 ± 3.3	0.02	322.6 ± 223.5	4.9 ± 1.6	0.01	272.6	327.6	1.2	27.26	6.376	32.75	3.330
DbA	29.7 ± 25.3	2.7 ± 0.3	0.09	96.6 ± 90.0	2.9 ± 0.4	0.03	32.5	99.6	3.0	162.00	37.892	497.50	50.585
BghiP	235.1 ± 204.4	9.4 ± 3.0	0.04	386.3 ± 287.3	2.9 ± 1.7	0.007	244.6	389.3	1.5	2.45	0.572	3.89	0.396
∑PAHs	1866.2	2782.7		4080.1	6888.6		4648.9	10,968.7		427.5		983.4	

G/P^a—Gas/Particulate ratio
T/R^b—Rural/Traffic ratio

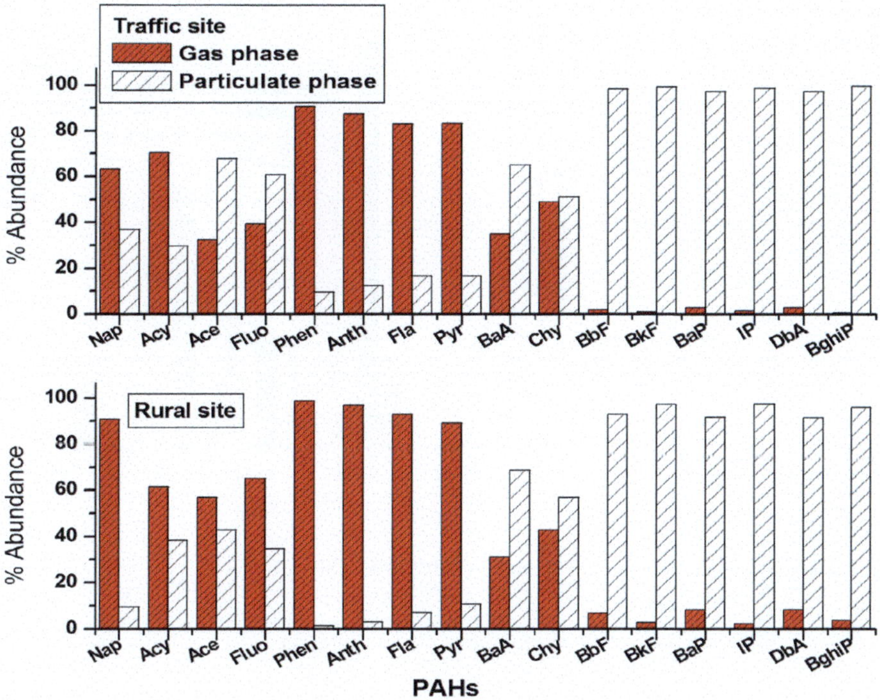

Fig. 24.2 Percentage abundance of PAHs in gas and particle phase at both sites

while the 5 and 6-ring PAHs accounted for 48% and 63% of total particulate phase PAHs at the traffic and rural site respectively. The carcinogenic PAHs mainly BaA, BaP, BbF, BkF, Chy, DbA and IP dominated in the particulate phase being 81% at the rural site and 66% at the traffic site. The greater abundance of these compounds at the rural site might be due to emissions from domestic combustion involving fuels like cow dung cakes, crop residue, coal and wood.

Health risk assessment of individual PAHs can be assessed by calculating toxicity equivalent (TEQ) relative to BaP. BaP-TEQ (ng m^{-3}) of individual PAHs are represented as the product of toxic equivalent factor (TEF) (Nisbet and Lagoy 1992) of each PAH with its atmospheric concentration. The BaP-TEQ of all the 16 priority PAHs (gas + particulate) are listed in Table 24.2. The levels of BaP-TEQ of all the non-carcinogenic PAHs is less than 1 (ng m^{-3}) at both the sites except for Phen, Anth and Pyr at traffic site because of their elevated concentration (Table 24.2). The % relative contribution of individual PAHs to Bap-TEQ clearly shows that DbA and BaP both contributed maximum (37% each) at the rural site where as at traffic site the BAP-TEQ concentration is dominated by DbA (50%) followed by BaP (30%).

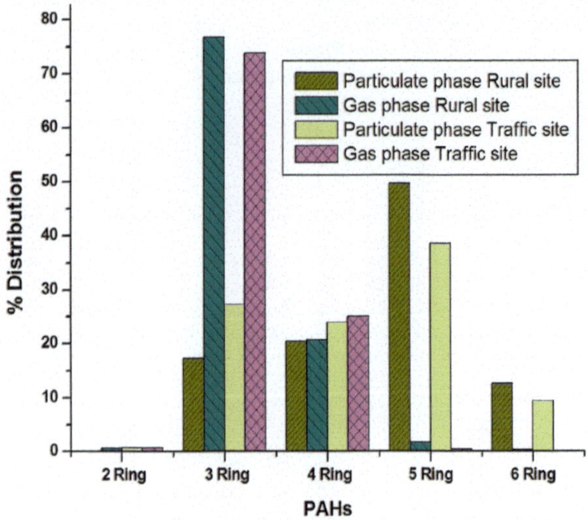

Fig. 24.3 Percentage variations of PAHs

Table 24.3 Mean values of $K_p \times$ TSP and K_pTSP/1 + K_pTSP (Φ)

PAHs	$K_p \times$ TSP		Φ	
	Rural site	Traffic site	Rural site	Traffic site
Nap	1.2	0.9	21.3	32.6
Acy	0.7	0.2	40.5	16
Ace	0.9	1.5	43	57.6
Fluo	0.9	0.7	36.9	38.2
Phen	0.01	0.8	1.2	16.9
Anth	0.03	0.18	3.2	11.5
Fla	0.07	0.5	6.8	21.7
Pyr	0.11	0.5	10	20.2
BaA	2.4	4.4	60.1	59.9
Chy	1.4	2.9	50.8	51.7
BbF	11.6	46.9	77.3	97.2
BkF	72.4	121.4	87.1	98.5
BaP	9.5	26.4	80.6	94.5
IP	40.5	56	95.4	97.1
DbA	15.8	25.3	91.7	94.1
BghiP	36.8	117.9	94.4	98.5

24.3.2 Gas–Particle Partitioning analysis of PAHs

Atmospheric fate and transport of PAHs are strongly influenced by gas–particle partitioning. Partitioning of PAHs depends on their vapor pressures, nature of particulate matter, ambient temperature and relative humidity (Chirico et al. 2007). Partitioning of atmospheric PAHs is determined in terms of gas–particle partition coefficient (K_p) ($m^3\ \mu g^{-1}$), which can be calculated by using equation proposed by Yamasaki (Yamasaki et al. 1982):

$$K_p = (F/\mathrm{TSP})/A \tag{24.1}$$

where, F is PAHs concentration in particle phase (ng m^{-3}), A is PAHs concentration in gas phase (ng m^{-3}), and TSP is total suspended particulate matter ($\mu g\ m^{-3}$) in ambient air. Particle to gas phase ratio can be obtained on multiplying K_p by TSP:

$$K_p \times \mathrm{TSP} = F/A \tag{24.2}$$

The product of $K_p \times$ TSP is > than 1 indicates partitioning of PAHs primarily into the particle phase and values <1 indicates towards the dominance of gas phase concentration (Volckens and Leith 2003). The mean values of $K_p \times$ TSP for individual PAHs for both the sites are presented in Table 24.3. The mean values of $K_p \times$ TSP for Acy, Ace, Fluo, Phen, Anth, Pyr and Fla were less than 1 at both sites, implying their tendencies to exist in the gas phase while, PAHs like Chy, BbF, BkF, BaP, IP, DbA and BghiP show their predominance in the particulate phase as their values for $K_p \times$ TSP were higher than 1.

The particle-bound fraction, ϕ (%), of each PAH was determined using the following equation:

$$\Phi = F/(F + A) = K_p\mathrm{TSP}/\left(1 + K_p\mathrm{TSP}\right) \tag{24.3}$$

The particle bound fraction of low molecular mass compounds exhibited significant variability at both the sites varying between 1 and 60% while low variability in the φ values was observed for the high molecular mass PAHs, namely BbF, BkF, BaP, IP, DbA and BghiP (Table 24.3). At both the sites high molecular mass PAHs were present in particle-bound state varying between 80 and 98% but low volatile PAHs showed their predominance in the gas phase.

24.4 Conclusion

Significant variations in gas and particulate phase concentrations of PAHs were observed with higher levels at the traffic site due to automobile exhaust and residential fuel combustion. Carcinogenic PAHs were abundant in the particulate phase and their percent contribution with respect to other PAHs were higher at rural site. Partitioning of PAHs in both the phases clearly shows the dominance of 2–3 ring PAHs in the gas phase and 5–6 ring PAHs in the particulate phase while on the other hand 4 ring PAHs are present almost equally in both the phases. Health risk assessment of PAHs shows that DbA-TEQ contributes maximum cancer risk followed by BaP-TEQ at the traffic site whereas both DbA-TEQ and BaP-TEQ shows equal cancer risk at the rural site. The results of present study show alarming concentration of PAHs in Agra. People living in the vicinity of traffic dominated site may be associated with various health problems. So, to overcome the problem of atmospheric emissions of PAHs, high fuel-efficient engines should be promoted in automobiles. The use of non-conventional energy and green gas should be promoted in both the residential and commercial areas discouraging wood and coal combustion.

Acknowledgements Authors are grateful to the Head, Department of Chemistry, Dayalbagh Educational Institute, Agra and are also thankful to Department of Science and Technology, New Delhi (Project No.: SB/S4/AS-150/2014), CSIR-SRF (09/607(0047)/2019-EMR-1) and DST-FIST (SR/FST/CS-II/2017/38) for the financial assistance.

References

Aggarwal AL, Raiyani CV, Patel PD, Shah PG, Chatterjee SK (1982) Assessment of exposure to benzo (a) pyrene in air for various population groups in Ahmedabad. Atmos Environ (1967) 16(4):867–870

Akyüz M, Çabuk H (2010) Gas–particle partitioning and seasonal variation of polycyclic aromatic hydrocarbons in the atmosphere of Zonguldak. Turkey Sci Tot Environ 408:5550–5558

Albinet A, Leoz-Garziandia E, Budzinski H, Villenave E (2007) Polycyclic aromatic hydrocarbons (PAHs), nitrated PAHs and oxygenated PAHs in ambient air of the Marseilles area (South of France): concentrations and sources. Sci Tot Environ 384:280–292

Bhargava A, Khanna RN, Bhargava SK, Kumar S (2004) Exposure risk to carcinogenic PAHs in indoor-air during biomass combustion whilst cooking in rural India. Atmos Environ 38(28):4761–4767

Bidleman TF (1988) Atmospheric processes: wet and dry deposition of organic compounds are controlled by their vapor-particle partitioning. Environ Sci Techno 22:361–367

Chirico R, Spezzano P, Cataldi D (2007) Gas–particle partitioning of polycyclic aromatic hydrocarbons during the spring and summer in a suburban site near major traffic arteries. Polycyclic Aromatic Compd 27:401–423

Cincinelli A, Bubba MD, Martellini T, Gambaro A, Lepri L (2007) Gas-particle concentration and distribution of n-alkanes and polycyclic aromatic hydrocarbons in the atmosphere of Prato (Italy) Chemo 68:472–478

Etchie TO, Sivanesan S, Etchie AT, Adewuyi GO, Krishnamurthi K, George KV, Rao PS (2018) The burden of disease attributable to ambient PM 2.5—bound PAHs exposure in Nagpur, India. Chemos

Gadi R, Singh DP, Saud T, Mandal TK, Saxena M (2012) Emission estimates of particulate PAHs from biomass fuels used in Delhi, India. Human Ecol Risk Assess Int J 18(4):871–887

Gaga EO, Ari A (2011) Gas–particle partitioning of polycyclic aromatic hydrocarbons (PAHs) in an urban traffic site in Eskisehir Turkey. Atmos Res 99(2):207–216

Harrison RM, Simith DJT, Luhana L (1996) Source apportionment of atmospheric polycyclic aromatic hydrocarbons collected from an urban location in Birmingham, U.K. Environ Sci Techno 30:825–832
Kumari KM, Lakhani A (2018) PAHs in gas and particulate phases: measurement and control. Environmental Contaminants (pp. 43–75). Springer, Singapore
Li J, Zhang G, Li XD, Qi SH, Liu GQ, Peng XZ (2006) Source seasonality of polycyclic aromatic hydrocarbons (PAHs) in a subtropical city, Guangzhou. South China. Sci Tot Environ 355:145–155
Lima ALC, Farrington JW, Reddy CM (2005) Combustion-derived polycyclic aromatic hydrocarbons in the environment—a review. Environ Forensic 6:109–131
Liu Y, Chen L, Huang QH, Li WY, Tang YJ, Zhao JF (2009) Source apportionment of polycyclic aromatic hydrocarbons (PAHs) in surface sediments of the Huangpu River, Shanghai China. Sci Total Environ 407(8):2931–2938
Miller JN, Miller JC (2010) Statistics and chemometrics for analytical chemistry. ISBN: 978-0-273-73042-2
Mohanraj R, Solaraj G, Dhanakumar S (2011a) Fine particulate phase PAHs in ambient atmosphere of Chennai metropolitan city India. Environ Sci Pollut Res 18:764–771
Mohanraj R, Solaraj G, Dhanakumar S (2011) PM 2.5 and PAH concentrations in urban atmosphere of Tiruchirappalli, India. Bull Environ Contam Toxicol 87:330–335
Mohanraj R, Dhanakumar S, Solaraj G (2012) Polycyclic aromatic hydrocarbons bound to PM 2.5 in urban Coimbatore, India with emphasis on source apportionment. Sci World J
Nisbet IC, Lagoy PK (1992) Toxic equivalency factors (TEFs) for polycyclic aromatic hydrocarbons (PAHs). Regul Toxicol Pharmacol 16(3):290–300
Odabasi M, Vardar N, Sofuoglu A, Tasdemir Y, Holsen TM (1999) Polycyclic aromatic hydrocarbons (PAHs) in Chicago air. Sci Tot Environ 227:57–67
Pandey P, Patel DK, Khan AH, Barman SC, Murthy RC, Kisku GC (2013) Temporal distribution of fine particulates (PM 2.5, PM 10), potentially toxic metals, PAHs and metal-bound carcinogenic risk in the population of Lucknow City, India. J Environ Sci Health Part A 48(7):730–745
Ringuet J, Albinet A, Leoz-Garziandia E, Budzinski H, Villenave E (2012) Reactivity of polycyclic aromatic compounds (PAHs, NPAHs and OPAHs) adsorbed on natural aerosol particles exposed to atmospheric oxidants. Atmos Environ 61:15–22
Rajput N, Lakhani A (2012) Particle associated polycyclic aromatic hydrocarbons (PAHs) in urban air of Agra. Polycyclic Aromat Compd 32(1):48–60
Schnelle-Kreis J, Sklorz M, Orasche J, Stölzel M, Peters A, Zimmermann R (2007) Semi volatile organic compounds in ambient PM 2.5. Seasonal trends and daily resolved source contributions. Environ Sci Techno 41:3821–3828
Singla V, Pachauri T, Satsangi A, Kumari KM, Lakhani A (2012) Characterization and mutagenicity assessment of PM 2.5 and PM 10 PAH at Agra, India. Polycyclic Aromat Compd **32**(2):199–220
Su Y, Lei YD, Wania F, Shoeib M, Harner T (2006) Regressing gas/particle partitioning data for polycyclic aromatic hydrocarbons. Environ Sci Technol 40(11):3558–3564
Tasdemir Y, Esen F (2007) Urban air PAHs: concentrations, temporal changes and gas/particle partitioning at a traffic site in Turkey. Atmos Res 84:1–12
Venkataraman C, Thomas S, Kulkarni P (1999) Size distributions of polycyclic aromatic hydrocarbons—gas/particle partitioning to urban aerosols. J Aerosol Sci 30(6):759–770
Verma PK, Sah D, Kumari KM, Lakhani A (2017) Atmospheric concentrations and gas–particle partitioning of polycyclic aromatic hydrocarbons (PAHs) and nitro-PAHs at indo-gangetic sites. Environ Sci Process Impact 19(8):1051–1060
Volckens J, Leith D (2003) Comparison of methods for measuring gas–particle partitioning of semi-volatile compounds. Atmos Environ 37:3177–3188
Wang XT, Miao Y, Zhang Y, Li YC, Wu MH, Yu G (2013) Polycyclic aromatic hydrocarbons (PAHs) in urban soils of the megacity Shanghai: occurrence, source apportionment and potential human health risk. Sci Total Environ 447:80–89

Wei C, Han Y, Bandowe BAM, Cao J, Huang R-J, Ni H, Tian J, Wilcke W (2015) Occurrence, gas/particle partitioning and carcinogenic risk of polycyclic aromatic hydrocarbons and their oxygen and nitrogen containing derivatives in Xi'an, central China. Sci Tot Environ 505:814–822

Yamasaki H, Kuwata K, Miyamoto H (1982) Effects of ambient temperature on aspects of airborne polycyclic aromatic hydrocarbons. Environ Sci Techno 16:189–194

Yang Y, Zhang N, Xue M, Tao S (2010) Impact of soil organic matter on the distribution of polycyclic aromatic hydrocarbons (PAHs) in soils. Environ Pollut 158(6):2170–2174

Zhou C, Zhu X, Wang Z, Ma X, Chen J, Ni Y, Li X (2013) Gas-particle partitioning of PAHs in the urban air of Dalian, China: measurements and assessments. Polycyclic Aromat Compd 33(1):31–51

Chapter 25
Estimation of Passenger Exposure to $PM_{2.5}$ on a Highway

Soma Sekhara Rao Kolluru and Aditya Kumar Patra

25.1 Introduction

The prominent sources that contribute to particulate matter (PM) pollution in urban areas include public roads (resuspended road dust), industries, transport, waste burning, and so forth. The primary concern for airborne PM is its adverse health effects on human being. PM can enter through the respiratory system to the lungs of the human being where it can react with lung cells and damage it, thus causing diminished lung function (Rundell et al. 2008). It can also cause triggering of inflammatory effects (Nel 2005) and lead to morbidity and mortality (Dockery et al. 1993). The adverse health effects associated with inhalation of PM include respiration difficulties, sore throat, cough, asthma, premature deaths, and cardiovascular diseases cardiac arrhythmias (Pope et al. 2004; Massey et al. 2009). The health effects primarily depend on the size of PM because the particle size determines the quantum of it that can be inhaled and the part of it that can penetrate to the deeper parts of the respiratory systems and reach lungs (Slezakova et al. 2007). PM, typically up to 10 μm, enters through the respiratory system. This is referred as respirable particulate matter (RPM/RSPM) or PM_{10}. Recent health studies indicate that finer fractions of the PM_{10} with an aerodynamic diameter up to 2.5 μm (referred to as $PM_{2.5}$) has greater potential for health damage. These particles have the ability to travel through the nasal canal and get deposited in the lungs and inner linings of the lungs (Anderson and Thundiyil 2012). $PM_{2.5}$ concentrations were measured in six eastern US cities

S. S. R. Kolluru
School of Environmental Science and Engineering, Indian Institute of Technology Kharagpur, Kharagpur 721302, India
e-mail: sekharkolluru09@gmail.com

A. K. Patra (✉)
Department of Mining Engineering, Indian Institute of Technology Kharagpur, Kharagpur 721302, India
e-mail: akpatra@mining.iitkgp.ac.in

S. M. Shiva Nagendra et al. (eds.), *Urban Air Quality Monitoring, Modelling and Human Exposure Assessment*, Springer Transactions in Civil and Environmental Engineering,
https://doi.org/10.1007/978-981-15-5511-4_25

and reported that the daily mortality increased by 1.5% when the two-day mean average of $PM_{2.5}$ increased by 10 μg m^{-3} (Schwartz et al. 1996). Out of the several sources of PM, transport is the main source. Whether it is less developed country, developing or a developed country, the transport sector is the dominant source for ambient PM (Srivastava et al. 2008). Millions of commuters spend a considerable amount of time in vehicles on a daily basis while they travel from their home to the workplace and back. It ranges from few minutes to several hours (Huang and Hsu 2009). On average, people spend nearly 100 min per day in commuting (Mcguckin and Srinivasan 2005). Exposure studies in Dublin, Ireland revealed that during the morning and evening peak hours, bus commuters were exposed to an average $PM_{2.5}$ concentration in the range of 103.81–128.16 μg m^{-3} (Mcnabola et al. 2008) which was nearly 4–5 times of World Health Organization (WHO) permissible limit. A study conducted in Beijing in 2010–2011 reported that the commuters in the bus are exposed to $PM_{2.5}$ concentrations of 42 μg m^{-3}, which exceeds the WHO limits (Huang et al. 2012).

In India, the road networks have increased from 3,99,942 km in 1951 to 54,72,144 km in 2015, at a compound annual growth rate (CAGR) of 4.2% (Ministry of Road transport and Highways 2014). Vehicles registrations have increased from 0.3 million in 1951 to 182.4 million in 2013 (Ministry of Road transport and Highways 2012). A study in Delhi found that commuters in auto rickshaw were exposed to fine PM concentrations of 190 μg m^{-3} (Apte et al. 2011). A study in Delhi concluded that unenclosed travel modes (walking, cycling, and two-wheeler) recorded up to 10–40% higher concentration than ambient location concentration (Goel et al. 2015). In cities, the traffic moves at a slower speed due to traffic density and frequently stops at traffic lights. Also, the big buildings in cities do not allow free flow of the air and recirculation of traffic emission takes place which further leads to build up of PM concentration in city road networks. On the other hand, in highways the traffic density is low, and therefore the traffic is free-flowing, it encounters no traffic lights (except at the junctions where one road diverts to the nearest city/town), and absence of high-rise buildings causes easy dispersion of the PM emitted. Thus, the characteristic of exposure in highways is different from exposure in cities. While studies have been conducted on commuter exposure from traffic in cities, limited studies (Lawrence et al. 2004; Pachauri et al. 2013) were available for commuters' exposure to respirable particles while they travel on highways. This chapter discusses commuter exposure to $PM_{2.5}$ on one of the busiest national highways in India.

25.2 Methodology

25.2.1 Study Route

A National Highway (NH) stretch of 210 km connecting Bhadrachalam in Telangana state and Vijayawada in Andhra Pradesh state was considered as study route. The road consists of 168 km in NH30 from Bhadrachalam to Ibrahimpatnam and 42 km in NH 65 from Ibrahimpatman to Vijayawada. This road passes through major towns such as Bhadrachalam-Kothagudem-Tiruvuru-Ibrahimpatnam-Vijayawada. It is one of the busiest national highways that are used by public buses, cars, and trucks (Fig. 25.1). Bhadrachalam (BCM) has a bus depot which is operated by Telangana State Road Transport Corporation (TSRTC). From here, buses travel to Vijayawada (VJA) along the study route. Indian Tobacco Corporation (ITC) operates one of its paper mills in this town. Kothagudem (KTDM) is a coal town having headquarters of Singareni Collieries Company Limited (SCCL) located here. SCCL operates four opencast mines around the town, namely Jalagam Vengala Rao (JVR) OC I expansion and OC II project, Gautamkani OC project (GKOC), and Kistaram OC project (KOC). Other industries in and around the town are Nava Bharat Ferro Alloy plant, Nava Bharat power, sponge iron plant and Kothagudem Thermal Power Station. A number of private vehicles like cars, trucks, minivans, and autos operate in the town. TSRTC is operating a bus depot in Kothagudem from where several buses travel to Vijayawada.

There are no notable industries in and around Tiruvuru (TVR). The sources of air pollution are vehicular traffic, road resuspension dust, and coal dust emission due to operations of trucks carrying coal from Sathupally coal mines to KTDM. Andhra Pradesh State Road Transport Corporation (APSRTC) operates a depot in TVR. APSRTC along with TSRTC operate many bus services via Tiruvuru from Bhadrachalam to Vijayawada.

Fig. 25.1 Study route from Bhadrachalam to Vijayawada. *Source* Google maps

Vijayawada (VJA) is a major city in the state of Andhra Pradesh. The city is one of the major trading and business centers of the state. As per 2011 census, the city had a population of 1048,240, making it the second largest city in the state in terms of population and it had an urban agglomeration population of 1491,202. The Pandit Nehru Bus Station (PNBS) in VJA is the fourth largest and busiest bus station in India. Many buses from other states like Chhattisgarh, Orissa, Karnataka, Tamilnadu, and Telangana arrive at this bus station daily. Andhra Pradesh Generation Corporation (APGENCO) operates one of its biggest power plants in this city.

25.2.2 Instruments

Passenger exposure was measured along the study route using Environmental Particulate Air monitor 5000 (EPAM 5000). EPAM 5000 is a portable air sampler with an optical sensor for the determination of PM concentration with simultaneous gravimetric collection on a 47 mm filter membrane (Fig. 25.2). It has the running time up to 24 h. It has three inlets for sampling PM of different sizes (PM_1, $PM_{2.5}$, and PM_{10}). During this study $PM_{2.5}$ exposure was measured. EPAM 5000 operates at a flow rate of 4.0 litre per minute.

Zero calibration was performed before each sampling session. The sampler was made to log data every 1 min. The battery power was properly checked before each sampling session and care was taken to prevent any obstacles while sampling.

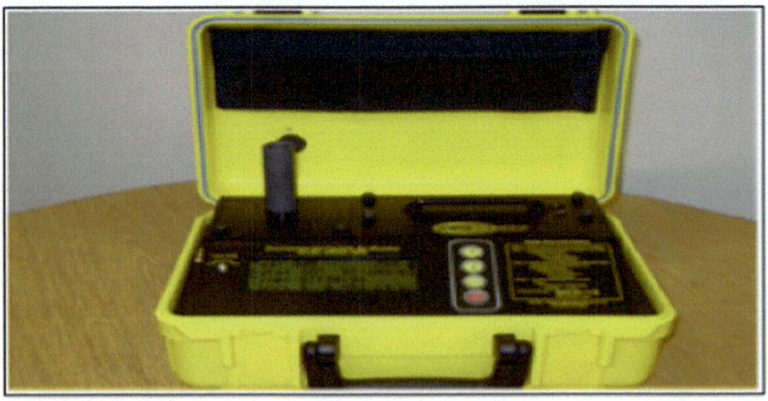

Fig. 25.2 EPAM 5000

25.2.3 Sampling Protocol

The study between BCM and VJA was carried out during July–August 2016. The modes considered for this study are: bus, car, and car (AC). For each mode, sampling was repeated for three trips. In bus, the sampler was placed on the lap and held near the windows in the middle parts of the bus on the driver side. The window was kept open during sampling. In rain event, the sampling continued but the windows were closed. In car, the sampler was placed at the backseat. All four windows of the car were kept open during the trips. Care was taken not to apply sudden brakes, as it may dislodge the equipment from its location. A Tata Indigo 2007 diesel model car was used in this study. The protocol which was followed for car was also followed for car (AC) mode. Additionally, the windows were closed and AC was operated in recirculation mode. A total of 9 trips were made in three travel modes. Each trip started at 8:00 h. Background $PM_{2.5}$ concentrations were measured at a location about 30 km from the study route (Fig. 25.1). The location is in a rural setting having no nearby pollution source from traffic and industry. Background sampling started at 9:00 a.m. and continued until 5:00 p.m.

25.3 Results

25.3.1 Bus

The opening of the windows enabled emissions to enter inside the bus cabin (Fig. 25.3). Average exposure obtained during the trips was 72.30 $\mu g\ m^{-3}$ ± 58.17 $\mu g\ m^{-3}$ (Mean ± SD). The overall $PM_{2.5}$ exposures between BCM and KTDM were 56.45 $\mu g\ m^{-3}$ ± 46.39 $\mu g\ m^{-3}$. The sampling started at BCM bus stand when bus started its trips. The density of traffic increased after 20 km from BCM. The sampling started at around ~8:00 a.m. during three trips. The exposure level of $PM_{2.5}$ increased as the bus approached KTDM. On July 17, 2016, due to higher traffic density, the bus could not overtake the concrete mixer lorry; instead, it followed the lorry during 9:00–9:10 h. Black exhaust from the truck could be easily visible from the bus which resulted in high $PM_{2.5}$ exposures during this period than remaining part of the section. The $PM_{2.5}$ exposures between KTDM and TVR were 74.16 $\mu g\ m^{-3}$ ± 57.42 $\mu g\ m^{-3}$. There were no notable pollution sources along this route. On July 17, 2016, the period between 10:06 a.m. to 10:21 h, the subject bus followed empty coal truck which traversed through the unpaved stretch of road. The concentration thus increased to 467 $\mu g\ m^{-3}$. The average concentration between TVR and VJA was 83.50 $\mu g\ m^{-3}$ ± 66.76 $\mu g\ m^{-3}$. Concentrations increased as the bus approached VJA, because of industrial emissions around the city, in addition to traffic exhaust and road resuspension. The sampling stopped when the bus reached VJA bus stand.

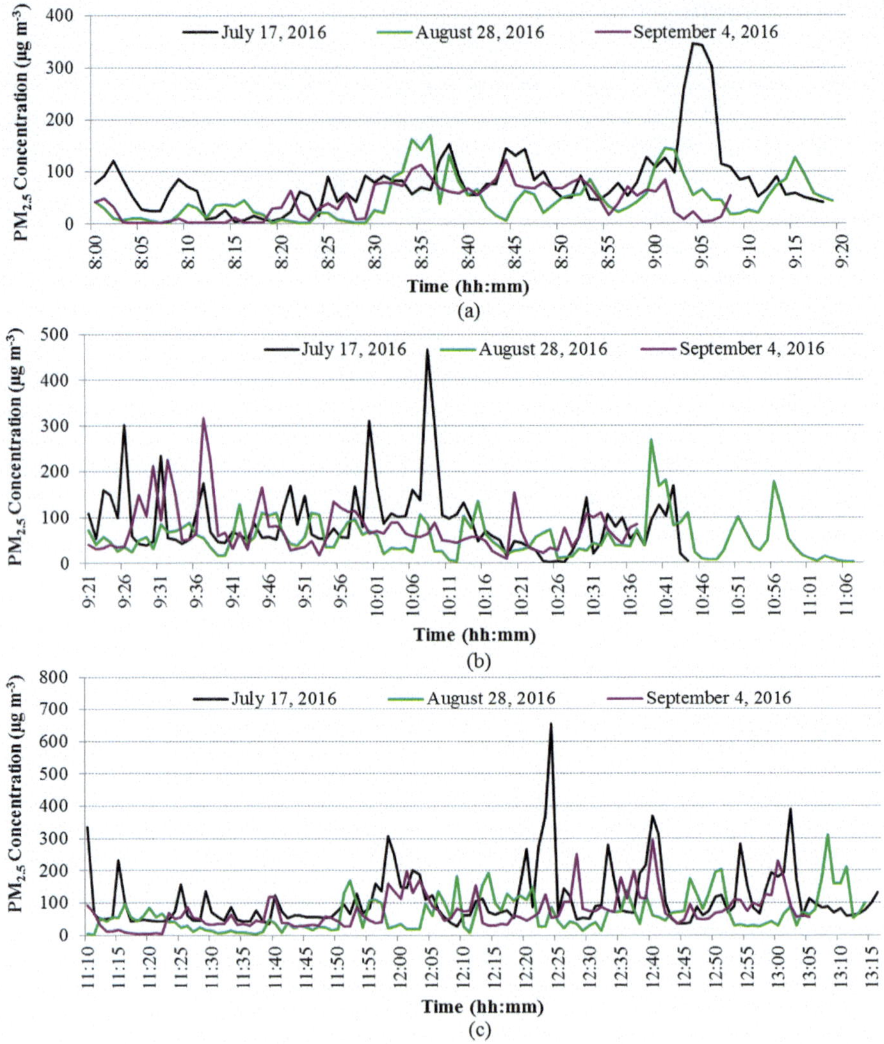

Fig. 25.3 Personal exposure levels in bus between **a** BCM and KTDM, **b** KTDM and TVR, and **c** TVR and VJA

25.3.2 *Car*

The mean exposure in car between BCM and VJA was 84.20 μg m^{-3} ± 68.38 μg m^{-3} (Fig. 25.4). For the section BCM and KTDM, sampling started at the bus stand, which is in the center of the town. The overall mean exposure between BCM and KTDM was 76.63 μg m^{-3} ± 59.3 μg m^{-3}. After crossing BCM town, the exposure levels increased due to increase in vehicle density and industrial emissions in KTDM.

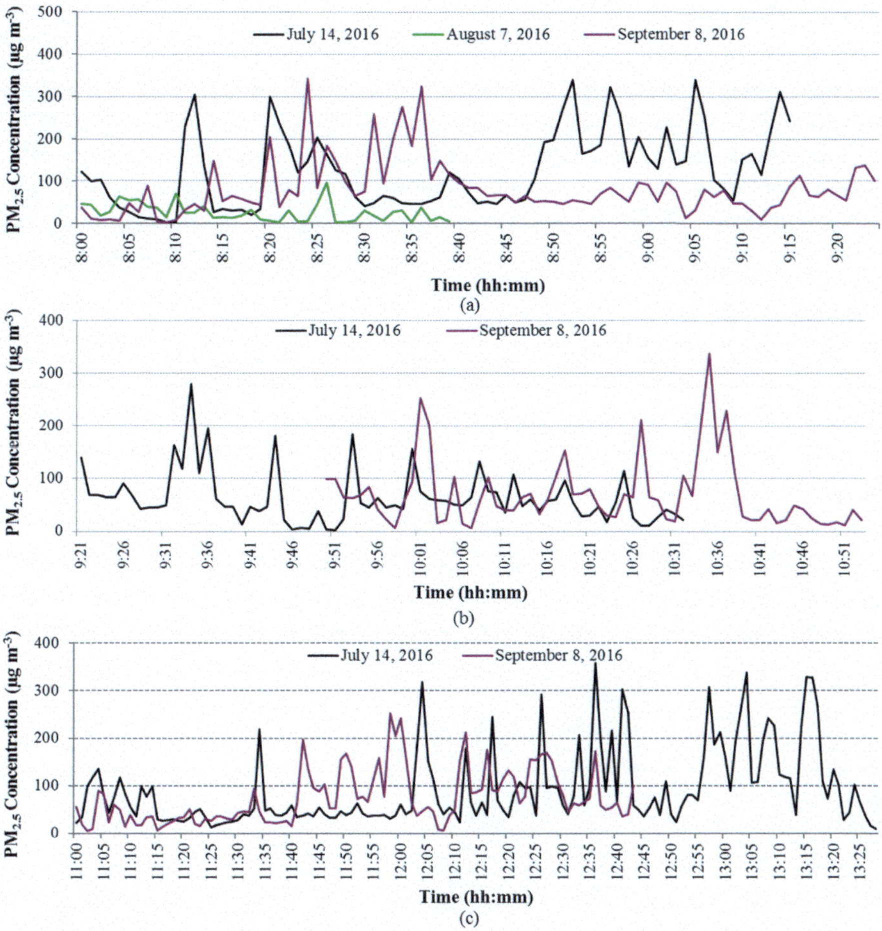

Fig. 25.4 Personal exposure levels in car between **a** BCM and KTDM, **b** KTDM and TVR, and **c** TVR and VJA

The overall mean exposure between KTDM and TVR section was 66.83 $\mu g\ m^{-3} \pm 58.00\ \mu g\ m^{-3}$. Tailpipe emissions of heavy vehicles directly entered into vehicle cabin due to low height of the car. Additionally, road resuspension dust also caused the higher exposure. Due to unexpected conditions on August 7, 2016, another route was taken in contrary to the original specified route. So, the samplers were stopped and concentrations were not recorded. The exposure profiles of two trips along this route were considered. From TVR to VJA, the exposure level increased as the car approached VJA. The overall average concentration of three trips was 70.03 $\mu g\ m^{-3} \pm 58.66\ \mu g\ m^{-3}$.

25.3.3 Car (AC)

Average exposure during the journey was 55.30 μg m^{-3} ± 34.13 μg m^{-3}, which is less than the exposure levels obtained during the trips by bus and car (Fig. 25.5). The average exposure between BCM and KTDM was 54.71 μg m^{-3} ± 35.33 μg m^{-3}. The exposure levels had an increasing trend when the car approached KTDM. The average exposure in KTDM-TVR section was 50.47 μg m^{-3} ± 25.22 μg m^{-3}. The average exposure in TVR-VJA section was 60.71 μg m^{-3} ± 41.85 μg m^{-3}. The exposure levels increased as the car approached VJA.

Background

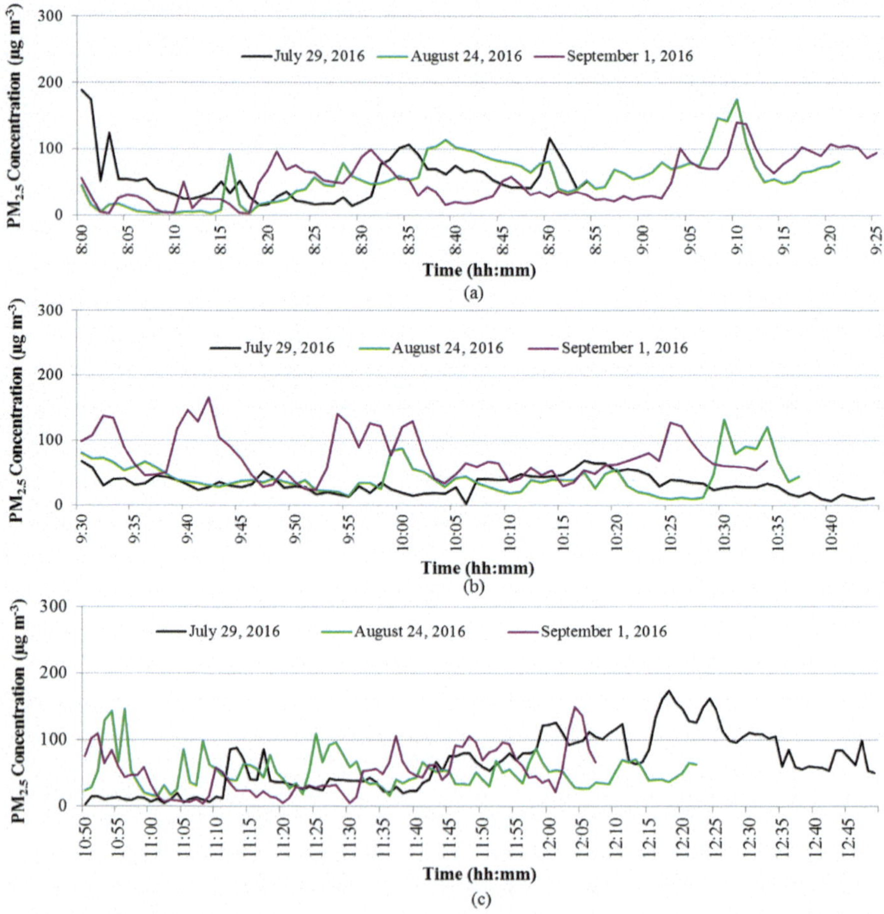

Fig. 25.5 Personal exposure levels in car (AC) between **a** BCM and KTDM, **b** KTDM and TVR, and **c** TVR and VJA

Fig. 25.6 Background exposure levels

Table 25.1 Summary $PM_{2.5}$ exposure levels in travel modes in three sections of the highway

Highway segment	$PM_{2.5}$ concentration ($\mu g\ m^{-3}$)			
	Bus	Car	Car (AC)	Background
BCM to KTDM	56.45 ± 46.39	76.63 ± 59.3	54.71 ± 35.33	24.39 ± 10.12
KTDM to TVR	74.16 ± 57.42	66.83 ± 58.00	50.47 ± 25.22	
TVR to VJA	83.50 ± 66.76	70.03 ± 58.66	60.71 ± 41.85	
BCM to VJA (entire route)	**72.30 ± 58.17**	**84.20 ± 68.38**	**55.30 ± 34.13**	

The background location was selected far away from known local pollution sources. The $PM_{2.5}$ concentrations during the sampling sessions were 24.39 $\mu g\ m^{-3}$ ± 10.12 $\mu g\ m^{-3}$ (Fig. 25.6).

A summary of the results was presented in Table 25.1. Exposure levels were highest for travel during TVR to VJA section for all travel modes. For all sections, the exposure levels were the least during travel by car (AC). For the entire route from BCM to VJA, passengers travelling in car experienced the highest exposure, followed by Bus and Car (AC).

25.4 Discussion

The commuters while traveling from BCM to VJA are exposed to a very high level of $PM_{2.5}$. Irrespective of the mode of transport, the exposure levels were higher than the limit values set by WHO (25 $\mu g\ m^{-3}$) (WHO: WHO Air quality guidelines for particulate matter, ozone, nitrogen dioxide and sulfur dioxide: Global update 2005) and USA (35 $\mu g\ m^{-3}$).[1] For only a few cases, the average exposure levels were less than the limit (60 $\mu g\ m^{-3}$) set through National Ambient Air Quality Standard in

[1] United States Environmental Protection Agency, National Ambient Air Quality Standard. Available at: https://www.epa.gov/criteria-airpollutants/naaqs-table.

India.[2] The exposure levels were 2.0–3.5 times of the background $PM_{2.5}$ level. The values across the bus, the car, and the car (AC) show that the exposure levels vary widely. Car with windows open resulted in higher exposure than other modes of transport. Exhaust pipes of the bus and truck are located at the lower level, almost at the cabin level of the car. Therefore, the vehicle exhaust, nearly in undiluted condition, reaches the cabin of the car and this might be the reason for high exposure levels recorded for the commuters in the car. The least commuter exposure was observed for car (AC) because car windows were closed. Therefore, only little exhaust could enter the car cabin through the AC system of the car and leaks through different parts of the car body. $PM_{2.5}$ exposure during travel in the segment TVR-VJA was usually higher than exposures in all other segments for all mode of transport. There could be three reasons for this high exposure level: (1) For the road segment near the VJA city, traffic becomes slower, and traffic density becomes higher, resulting in more exhaust emission and increase in exposure time, (2) emissions from several industries and other anthropogenic activities in VJA may contribute to a part of $PM_{2.5}$ exposures during travel from TVR-VJA, and (3) during the study period, the road relaying was taking place in this segment and resuspension of construction dust has contributed to the $PM_{2.5}$ exposure.

25.5 Conclusion

This chapter presents the findings of a passenger exposure study on a national highway, which is important because a large population regularly travel on national highways from one city to the other or from a small town to a city for their daily livelihood. Therefore, their exposure to respirable dust constitutes a health concern which needs to be assessed, but very few studies were conducted on national highways. Our study conducted on a national highway of 210 km connecting Bhadrachalam and Vijayawada revealed the following:

1. The average exposure levels were 72.30 $\mu g\ m^{-3} \pm 58.17\ \mu g\ m^{-3}$, 84.20 $\mu g\ m^{-3} \pm 68.38\ \mu g\ m^{-3}$, 55.30 $\mu g\ m^{-3} \pm 34.13\ \mu g\ m^{-3}$ for the bus, the car, and the car (AC), respectively.
2. Passengers were exposed to respirable PM as high as 2.0–3.5 times the WHO and US permissible limit.
3. $PM_{2.5}$ exposure levels in sections between TVR and VJA were higher than other sections, mainly due to pollution from several activities in VJA and road resuspension.
4. Commuters were exposed to the highest level of $PM_{2.5}$ while they travel in car with windows open.
5. The exposure is least for the commuters using car with ac when all the windows were closed.

---

[2]Central Pollution Control Board. National Ambient Air Quality Standard. Central Pollution Control Board, India. http://www.cpcb.nic.in/National_Ambient_Air_Quality_Standards.php.

Therefore, personal exposure studies on national highways are necessary for the prediction of safest modes of transport in terms of exposure levels and better implementation of the regulation measures and policies.

References

Anderson JO, Thundiyil JG (2012) Clearing the air: a review of the effects of particulate matter air pollution on human health. J Med Toxicol 8:166–175

Apte JS, Kirchstetter TW, Reich AH, Deshpande SJ, Kaushik G, Chel A, Marshall JD, Nazaroff WW (2011) Concentrations of fine, ultra-fine, and black carbon particles in auto-rickshaws in New Delhi. India Atmos Environ 45:4470–4480

Dockery DW, Pope CA, Xu X, Spengler JD, Ware JH, Fay ME, Ferris JBG, Speizer FE (1993) An association between air pollution and mortality in six U.S. cities. N Engl J Med 329: 1753–1759

Goel R, Shahzad G, Guttikunda SK, Wilson D, Tiwari G (2015) On-road $PM_{2.5}$ pollution exposure in multiple transport microenvironments in Delhi. Atmos Environ 123:129–138

Huang HL, Hsu DJ (2009) Exposure levels of particulate matter in long-distance buses in Taiwan. Ind Air 19:234–242

Huang J, Deng F, Wu S, Guo X (2012) Comparisons of personal exposure to $PM_{2.5}$ and CO by different commuting modes in Beijing, China. Sci Total Environ 425:52–59

Lawrence A, Masih A, Taneja A (2004) Indoor/outdoor relationships of carbon monoxide and oxides of nitrogen in domestic homes with roadside, urban and rural locations in a central Indian region. Ind Air 15:76–82

Massey D, Masih J, Kulshrestha A, Habil M, Taneja A (2009) Indoor/outdoor relationship of fine particles less than 2.5 ($PM_{2.5}$) in residential homes locations in central Indian Region Build Environ 44:2037–2045

Mcguckin N, Srinivasan N (2005) The journey-to-work in the context of daily travel. In: Census data for transportation planning Conference, pp 1–42

Mcnabola A, Broderick BM, Gill LW (2008) Relative exposure to fine particulate matter and VOCs between transport microenvironments in Dublin: Personal exposure and uptake. Atmos Environ 42:6496–6512

Ministry of Road transport and Highways (2012) Road transport year book (2012–2013). http://morth.nic.in/

Ministry of Road transport and Highways (2014) Basic road statistics of India. http://morth.nic.in/

Nel A (2005) Air pollution related illness: effects of particles. Science 308:804–806

Pachauri T, Satsangi A, Singla V, Lakhani A, Kumari KM (2013) Characteristics and sources of carbonaceous aerosols in $PM_{2.5}$ during winter time in Agra, India. Aerosol Air Qual Res 13:977–991

Pope CA, Burnett RT, Thurston GD, Thun MJ, Calle EE, Krewski D, Godleski JJ (2004) Cardiovascular mortality and long-term exposure to particulate air pollution. Epidemiological evidence general pathophysiological pathways of disease. Circulation 6:71–77

Rundell KW, Slee JB, Caviston R, Hollenbach AM (2008) Decreased lung function after inhalation of ultrafine and fine particulate matter during exercise is related to decreased total nitrate in exhaled breath condensate. Inhal Toxicol 20:1–9

Schwartz J, Dockery DW, Neas LM (1996) Is daily mortality associated specifically with fine particles? J Air Waste Manage Assoc 46:927–939

Slezakova K, Pereira MC, Reis MA, Alvim-Ferraz MC (2007) Influence of traffic emissions on the composition of atmospheric particles of different sizes—Part 1: concentrations and elemental characterization. J Atmos Chem 58:55–68

Srivastava A, Gupta S, Jain VK (2008) Source apportionment of total suspended particulate matter in coarse and fine size ranges over Delhi. Aerosol Air Qual Res 8:188–200

WHO: WHO Air quality guidelines for particulate matter, ozone, nitrogen dioxide and sulfur dioxide: Global update (2005) Report No. WHO/SDE/PHE/OEH/06.02. World Health Organization

Chapter 26
Assessment of Organic Markers in Fine Aerosol of Mumbai City

Abba Elizabeth Joseph, Seema Unnikrishnan, Rakesh Kumar, and S. Vivek Balachander

26.1 Introduction

Air pollution studies have documented that there is a dose–response relationship between particulate exposure and health effects like cancer, pulmonary and heart diseases (Valavanidis et al. 2008). Although concentrations of different types of PM are important indicators of pollutant levels and related health effects, in this paper, we trace the organic markers of fine particles which have an aerodynamic diameter of 2.5 μm or lesser. Fine particulate matters are primarily derived from direct emission from combustion processes (Pope C-III and Dockery 2006) which include various sources such as the vehicular use of petrol and diesel, emission from thermal power plants, industrial processes such as smelters, paper mills, biomass burning, meat cooking. Fine particles also consist of transformational products (Pope C-III and Dockery 2006) generated from the oxidation of the primary gases such as sulfur and nitrogen oxides into secondary particles.

Air pollution is a big concern in India and therefore Central Pollution Control Board (CPCB), India has introduced a new particulate standard of $PM_{2.5}$ as 60 $\mu g/m^3$

A. E. Joseph (✉) · S. Vivek Balachander
Xavier University, Bhubaneswar, Harirajpur 752050, India
e-mail: elizabeth@xsos.edu.in

S. Vivek Balachander
e-mail: us15023@stu.ximb.ac.in

S. Unnikrishnan
National Institutes of Industrial Engineering, Vihar Lake, Mumbai 400087, India
e-mail: seemaunnikrishnan@gmail.com

R. Kumar
National Environmental Engineering Research Institute,
Mumbai Zonal Laboratory, Mumbai 400018, India
e-mail: r_kumar@neeri.res.in

S. M. Shiva Nagendra et al. (eds.), *Urban Air Quality Monitoring, Modelling and Human Exposure Assessment*, Springer Transactions in Civil and Environmental Engineering,
https://doi.org/10.1007/978-981-15-5511-4_26

in the year 2009 (Central Pollution Control Board (CPCB) 2009). More detailed information regarding the sources of the fine PM can help us better understand and mitigate the malignant effects of air pollution in urban cities like Mumbai.

Ambient fine particles contain inorganic components as well as a carbonaceous fraction which is made up of hundreds of different organic structures (Herlekar et al. 2012). The organic chemical composition of the fine particles is useful for identifying their respective sources of emission. Organic complex mixtures contain molecular tracers that can be linked to a specific source or are by-products of atmospheric reactions.

26.2 Methodology

26.2.1 Study Area

Air quality monitoring was carried out during the year 2007–2008 at four different locations, representing control, kerb, residential, and industrial areas. The monitoring locations are depicted in Fig. 26.1. Colaba represented control site as the monitoring location is very close to the sea thereby getting impacted by the sea breeze. Continuous vehicular movement along with the commercial activities was observed at Dadar and hence it was designated as kerb site. Khar was identified as a residential site. It is located close to the western express highway and airport. This site is also characterized by number of hotels and bakeries. Due to the presence of major industries including petroleum refinery, thermal power plant, fertilizer and chemical manufacturing, Mahul was selected as an industrial site. This site is also characterized by continuous movement of heavy-duty diesel vehicles and tankers. Most of the roads in this area are unpaved. Along with this ambient sampling, indoor air quality monitoring was also undertaken in the houses located in the nearby areas. Several criteria like proper ventilation, absence of pets, and smokers at home were taken into consideration while selecting the sites. Indoor air at the control site was monitored in the space between the living room and residence of domestic help. The bedroom was chosen as the place for monitoring at kerb as well as industrial monitoring site, whereas at residential site, the indoor air quality monitoring was carried out in the living room.

26.2.2 Sample Collection

The seasonal air quality monitoring was carried out for 24 h for 15 days in summer (extending from March to May), post monsoon (in October and November), and also in winter (i.e., during December and January). $PM_{2.5}$ was collected on two types of filters, namely PTFE Teflon filter (46.2 mm diameter, 2.0 μm pore size) and

Fig. 26.1 Study area (Mumbai city)

quartz filters (47 mm diameter, Pallflex Tissue) using MiniVol samplers (Airmetrics, Eugene, OR, USA) operated at a flow rate of 5 L m^{-1}. As per the standard procedure, equilibration of Teflon filters was carried out for 24 h, before and after sampling, in a clean room chamber maintained at 20 °C and 40% RH. The gravimetric estimation of fine particulate matter was done using an electronic microbalance (Sartorius, Model ME5) with a sensitivity of 1 mg. As a part of quality assurance and quality control (QA/QC), the initial and final weight of each filter was recorded in duplicate on a pre-calibrated microbalance. The quartz filters were conditioned prior to sampling by heating at a temperature of 900 °C for 3 h by covering the silica crucible containing quartz filters with another crucible. This pre-conditioning is essential to remove the organics present, if any. After sampling, the evaporation of volatile compounds was prevented by storing all the filters at a temperature of −20 °C. Elemental and organic carbon content was estimated by analyzing it on thermal optical analyzer [Desert Research Institute (DRI), NV, USA]. After this, the concentration of organic molecular markers in the post-monsoon season samples was computed.

26.2.3 Analysis of Carbon Content and Molecular Markers

The quantification of molecular markers like polycyclic aromatic hydrocarbons (PAHs), n-alkanes and hopanes, steranes, alkenes, and other alkanes was done using injection port thermal desorption and subsequently through gas chromatography/mass spectrometry method at DRI. The details of the methodology and QA/QC followed are described elsewhere (Fraser et al. 2002). The concentration of levoglucosan was estimated by ion chromatography method. The detailed procedure and QA/QC for the same has been described elsewhere (Fraser et al. 2002).

26.3 Results and Discussion

The geometric means (GM) of the fine particles concentration at the control (*C*), kerb site (*K*), residential (*R*), and industrial (*I*) sites were 73 ± 27, 92 ± 39, 78 ± 40 and 92 ± 40 μg/m^3, respectively, for outdoors and 90 ± 26, 93 ± 44, 64 ± 27, and 72 ± 35 μg/m^3, respectively, for indoors. The geometric mean concentration of indoor fine particles was greater than its corresponding outdoor values at the control site, whereas the levels were less at indoors in residential and industrial areas. Table 26.1 presents the indoor and outdoor concentration of fine particles with OC and marker data. Equal indoor and outdoor fine particles concentration was observed at kerb site. The I/O ratios derived from geometric mean fine particle concentrations were 1.23 ± 0.12, 1.02 ± 0.14, 0.81 ± 0.12, and 0.78 ± 0.11 at *C*, *K*, *R*, and *I*, respectively.

The ratio at control sites (>1) indicates that there is a strong indoor source for fine particles. This household was a naval officer's quarters whereby domestic helper's quarter was also attached. In the officer's quarters used LPG as fuel for cooking,

Table 26.1 Concentration of organic markers in fine particles and data analysis of markers in Mumbai during 2007–2008

		Colaba	Dadar	Khar	Mahul
$PM_{2.5}$ (μg/m³)	Ambient	73 ± 27	92 ± 39	78 ± 40	92 ± 40
	Indoor	90 ± 26	93 ± 44	64 ± 27	72 ± 35
OC Avg (μg/m³)	Ambient	18.12 + 6.60	28.95 + 16.10	28.04 + 16.53	26.87 + 15.74
	Indoor	30.66 + 7.63	27.35 + 9.32	23.83 + 14.87	16.51 + 9.79
PAH Avg (ng/m³)	Ambient	0.82	1.023	4.185	2.27
	Indoor	1.78	1.93	3.19	1.17
Alkanes (ng/m³)	Ambient	8.399	7.611	22.804	17.840
	Indoor	8.860	13.131	14.561	10.630
Hopane (ng/m³)	Ambient	0.504	0.546	4.532	1.213
	Indoor	0.941	1.562	2.671	0.650
Sterane (ng/m³)	Ambient	0.304	0.261	2.224	0.761
	Indoor	0.753	0.699	1.188	0.323
Levoglucosan (ng/m³)	Ambient	425	460	1565	1550
	Indoor	686	1292	1972	1166
BaPE index	Ambient	3.5	3.8	18.2	7.9
	Indoor	6.5	9.7	13.6	4.5
CPI index	Ambient	1.24	1.14	1.19	1.18
	Indoor	1.22	1.17	1.23	1.12
Ratio of PAH/n-alkane	Ambient	0.22	0.24	0.41	0.26
	Indoor	0.45	0.31	0.53	0.20
The ratio I/O					
$PM_{2.5}$		1.2	1.0	0.81	0.78
PAH		2.17	1.89	0.76	0.52
Alkanes		1.05	1.73	0.64	0.60
Hopanes		1.87	2.86	0.59	0.54
Steranes		2.48	2.68	0.53	0.42
Levoglucosan		1.61	2.81	1.26	0.75

whereas in the domestic helper quarters, kerosene was used (500 ml/day) for cooking on a stove. Therefore, this household was unique in the sense that it had two kitchens and mixed fuel pattern. Fuel use and the cooking device could be the source of elevated pollutants levels indoor as compared to ambient air values. I/O ratio of (<1) at residential and industrial sites shows that outdoor fine particles are increasing in the residential and industrial area. In India, land use pattern is mixed. Roadside vehicles influence the residential site, while the industrial area has a power plant and refineries, which are a good source of fine particles. At kerb site, I/O ratio was 1.02 ± 0.14, close to 1 indicating equal indoor and outdoor levels. Indoor–outdoor ratio approaches unity if the penetration factor is much closer to 1 and if average exchange rates are higher (Lunden et al. 2008). The average I/O ratio of $PM_{2.5}$ in the present study at Mahul site is 0.78 ± 0.11, which is similar to an earlier study conducted at Chembur near industrial site estimated as 0.75 (Patil and Kumar 2002).

In the samples collected through the study, about 40 PAHs, 24 alkanes, 18 hopanes, 12 sterane, 2 methyl alkanes, 3 branched alkanes, 5 cycloalkanes, and 1 alkene were identified and quantified from each category.

PAH: PAHs are incomplete combustion products when organic material like coal, wood, petrol, and diesel is burnt. They are also emitted from domestic sources like tobacco smoke and area sources like forest fire, agricultural burning, and waste incineration (Harrison et al. 1996; Khalili et al. 1995).

In the present study, it was observed that the sum of the highest concentration of PAH was found to be at Khar (175.8 ng/m^3) and lowest was observed at the control site Colaba (35.3 ng/m^3). The sitewise variation of PAH and its distribution is presented in Fig. 26.2. The carcinogenic PAH compounds accounted for nearly 20% of total PAHs at all the sites. Presence of tracer picene was detected among the different PAHs; this indicates combustion of coal as one of the sources. Retene another tracer was detected at Mahul, and it indicates burning of wood as a source. Mahul has a lot of slums in its vicinity, who use wood for cooking.

The indoor concentration of PAHs ranges between 75 and 134 ng/m^3. The maximum concentration of PAH was observed for 7,12-dimethyl benz(a)anthracene at control and industrial and for cyclopenta(cd)pyrene at the kerb and residential. As indicated earlier that cyclopenta(cd)pyrene is a tracer for gasoline-fueled cars, the dominance of this PAH at the kerb and residential site indoors proves that indoor particles are derived from outdoor air sources.

Indoor and outdoor sites in Mumbai show a high concentration of PAHs. The maximum concentration is observed in the residential area followed by industrial site, respectively. Among 15 PAHs, the highest concentration was reported for benzo(a)pyrene followed by indeno[1,2,3-cd]pyrene. Control site has maximum I/O and outdoor air pollutants with respect to carbon content in $PM_{2.5}$ this city. This household was naval officer quarters whereby domestic helper quarter was also attached. Officer quarter was using LPG as fuel for cooking, whereas domestic helper using kerosene (500 ml/day) for cooking on a stove. Therefore, this household was unique in the sense that it had two kitchen and mixed fuel pattern. Fuel use and the cooking device could be the source of elevated pollutants levels indoor as compared to ambient air data. In general, it is believed that Indian homes are comparatively more influenced by outdoor air due to good air exchange. However, I/O ratios indicate the presence of indoor sources. Carbon and mass levels correlate well at kerb site corroborating the fact that the vehicular pollution has a direct impact on indoor households. Indoor and outdoor sites in Mumbai show a high concentration of PAHs as well. High indoor concentrations are likely to result in high individual exposures.

Benzo(a)pyrene BaPE is considered while estimating the carcinogenic potential of PAHs (Ji et al. 2007). World Health Organization suggested the unit risk of 0.00008 ng/m^3 for a lifetime of PAHs exposure, assuming one was exposed to the

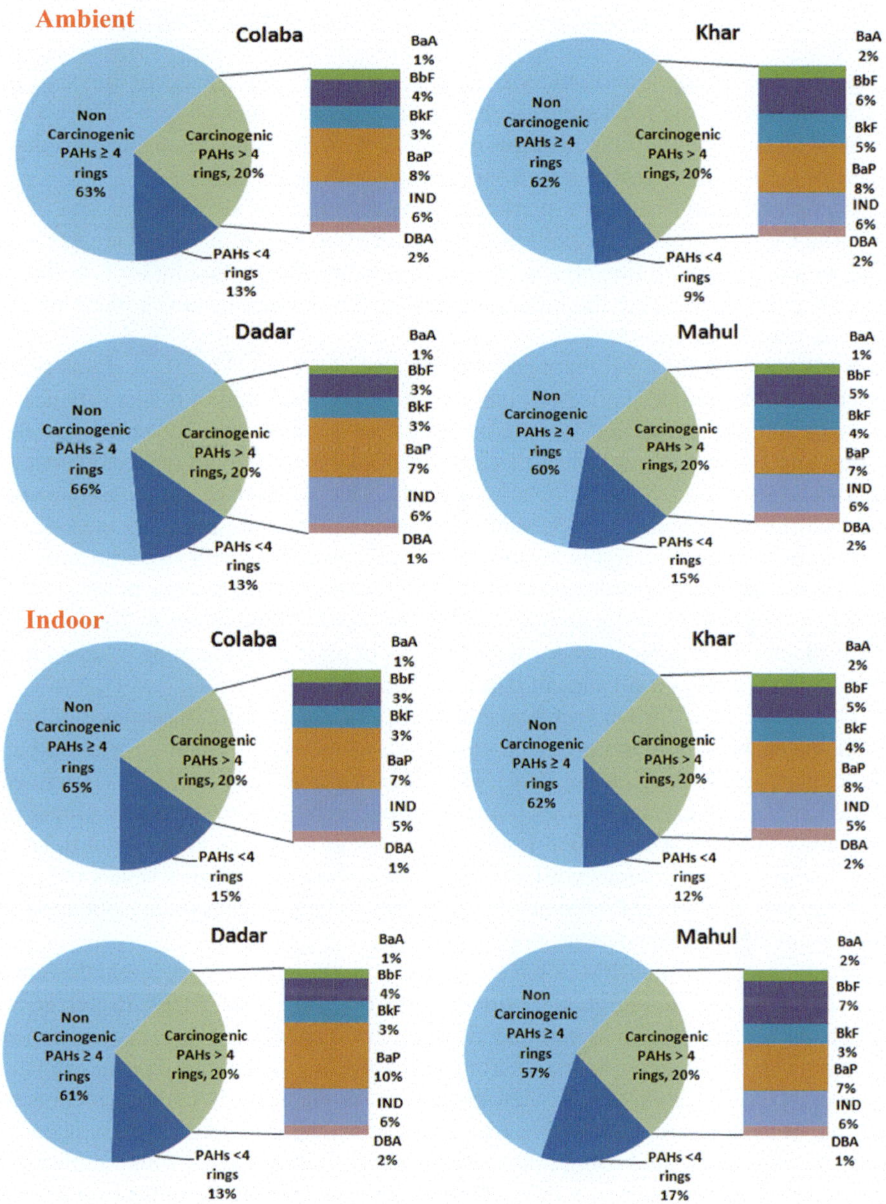

Fig. 26.2 Sitewise variation of PAHs (ambient and indoor environment)

average level of one-unit BaP concentration (1 ng/m^3) (World Health Organization 1987).

BaPE for the indoor and outdoor site is reported in Table 26.1. On the basis of BaPE calculation, the residential site has the maximum values of 18.8 and 13.6 for ambient and indoor sampling location, respectively. BaPE estimates at the residential site give an indication that in future cancer cases may arise.

Though I/O ratio of a fine particle is less than one at a residential site, it is important to note that toxic pollution characteristics of fine particles are greater. High indoor toxic pollution affects personal exposure. When outdoor air is polluted then infiltration can make the indoor air quality worse due to sources like a motor vehicle, waste burning, etc.

N-alkanes: N–alkane compounds ranging from C15 to C30 carbon atoms were observed in the samples collected at all the sites. The total sum of n-alkane concentration was highest at Khar (592.9 ng/m^3) and kerb site Dadar (197.9 ng/m^3) had the lowest totally average n-alkane concentration. Colaba and Mahul reported concentrations of 197.9 and 463.8 ng/m^3, respectively. The variation of n-alkanes sitewise and its different categories are presented in Fig. 26.3. The long-chain alkane (>C37) concentration was observed to be similar in all the four sites with slight differences in the composition of volatile and non-volatile n-alkanes. The presence of long-chain alkanes in ambient air is due to open burning of plastic (Simoneit et al. 2005); long-chain alkanes are organic markers of the burning of plastic. The burning of municipal wastes also produces n-alkanes in India (Fu et al. 2010).

The value of the Carbon Preference Index (CPI) indicates the importance of both fossil fuel combustion and biogenic emission as the primary source for the urban particulate and alkanes (Li et al. 2006). The alkane Carbon Preference Index is defined as the sum of odd carbon number homologs divided by sum of the even number of homologs. Alkanes with Carbon Preference Index close to one are thought to arise from anthropogenic sources while higher Carbon Preference Index values indicate biogenic origin. Carbon Preference Index values at indoor and outdoor obtained in the present study indicate anthropogenic sources.

Many sources of fine PM, such as gasoline- and diesel-powered vehicles emit both aromatic and non-aromatic hydrocarbons (Schauer et al. 1999; Rogge et al. 1993). Other sources such as natural gas (Rogge et al. 1993) and fuel oil combustion (Rogge et al. 1993) are enriched in aromatic hydrocarbons relative to non-aromatic hydrocarbons. The summed concentration of the aromatic hydrocarbons relative to the sum of the non-aromatic hydrocarbons has been determined in the present study. The ratio was calculated using polycyclic aromatic hydrocarbons with a molecular weight of 228 or greater to remove the influence of loss by volatilization during sampling. The ratios at sites, viz. *C*, *K*, and *I* have similar values indicating both diesel and gasoline vehicles. The ratio at '*R*' site and '*C*' was relatively more suggesting a higher proportion of aromatic HC which indicates the probability of sources like fuel oil or natural gas combustion. In the indoor control site, kerosene was used as fuel.

Hopanes and Steranes: In motor lubricating oils, engine oil has been identified as the source marker for vehicular emission of hopanes and steranes. Steranes are

Fig. 26.3 Sitewise variation of N-alkanes (ambient and indoor environment)

not found in sources such as petroleum and diesel, they are found in the atmosphere due to use of lubricant engine oil (Azevedo et al. 1999). The highest concentration of hopanes (81.6 ng/m^3) and steranes (26.68 ng/m^3) was found to be in the residential site Khar, which would be due to its close proximity with the western express highway and high traffic flow. The variation of hopanes sitewise is represented in Fig. 26.4

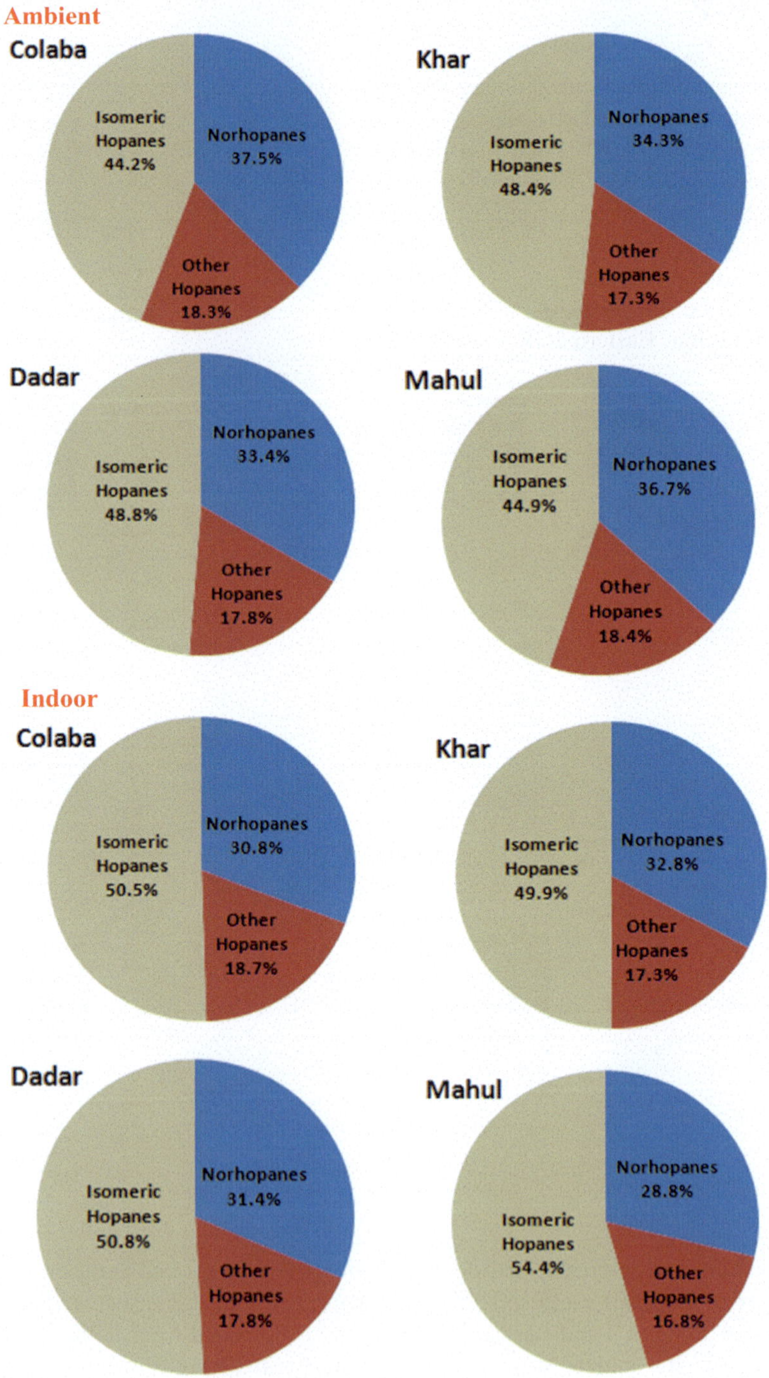

Fig. 26.4 Sitewise variation of hopanes (ambient and indoor environment)

and variation of steranes sitewise is represented in Fig. 26.5. The presence of hopane and sterane at all the four sites in Mumbai city confirms fossil fuel utilization and dominant vehicular traffic.

Higher hopanes and steranes in indoor atmosphere of Colaba and Dadar indicate significant infiltration of ambient air at these monitoring sites. Khar has the highest concentrations of both hopanes and steranes indicating the influence of vehicular exhaust (Simoneit 1984).

Khar had the highest total hopane concentration while Colaba and Dadar reported lower hopane concentration of 9.1 and 9.8 ng/m^3, respectively. Steranes concentration was relatively lower at Colaba (3.65 ng/m^3) and Dadar (3.13 ng/m^3). A previous study by (Chowdhury et al. 2007) in Mumbai indicated that the sum of hopanes and steranes at Worli a residential site was 23 $\pm$ 5 ng/m^3, which is a much lower figure than the one observed by the present study.

Levoglucosan: Biomass combustions release lots of gases and particles into the ambient air (Crutzen and Andreae 1990). Levoglucosan is a good tracer to detect burning of wood (Schauer and Cass 2000). Levoglucosan levels were high in Mahul and Khar; they were found to be 1550 and 1565 ng/m^3, respectively. Colaba and Dadar had relatively lower levoglucosan concentration of 425 and 460 ng/m^3, respectively. Sitewise percentage breakup of total levoglucosan concentration is presented in Fig. 26.6. The higher level of levoglucosan in Khar is indicative of garden waste burning since it is a residential area. Slums near sampling stations use wood for heating water and cooking.

In case of levoglucosan, except for Mahul site, indoor concentrations were higher than ambient concentrations indicating the prevalence of use of wood as fuel at Colaba, Dadar, and Khar.

26.4 Conclusion

The presence of hopanes, steranes, and levoglucosan indoor suggests that infiltration of outdoor fine particles to indoors. The present study was conducted without involving the air exchange rate. Further, research is required on the measurement of actual air exchange rate of different types of households for Indian conditions. More research is also needed to focus how the indoor air pollution affects personal exposure to varying urban population which resides very close to multiple outdoor sources. One season data is not enough to assess carcinogenic potential. Time series analysis needs to be done for PAHs and cancer in Mumbai city. Present BaPE can be a good indicator to understand future cancer cases. Highest BaPE was also observed at residential site indoors. Though I/O ratio is less than one at a residential site, toxic pollution characteristic of fine particles is greater and therefore, a cause for concern. High indoor toxic pollution affects personal exposure, when outdoor air is polluted then infiltration can make the indoor air quality worse due to sources like a motor vehicle, waste burning, etc.

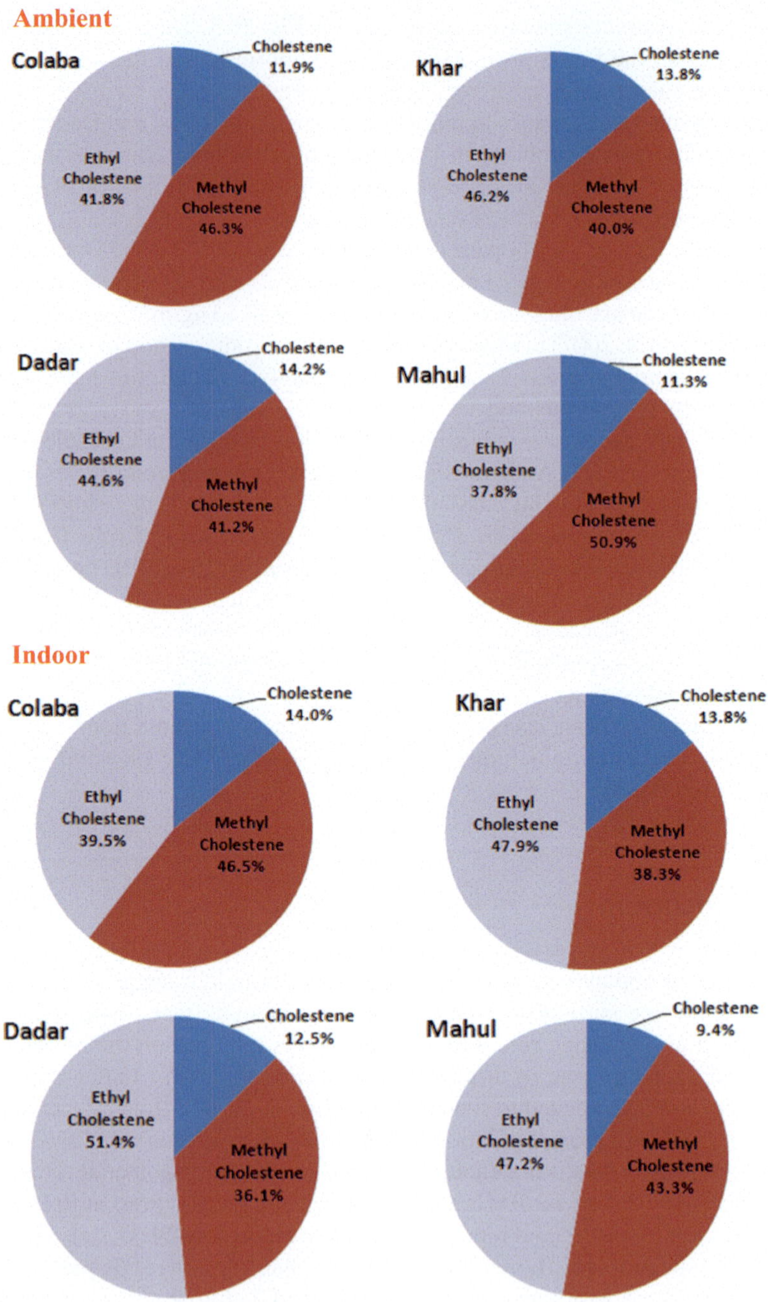

Fig. 26.5 Sitewise variation of steranes (ambient and indoor environment)

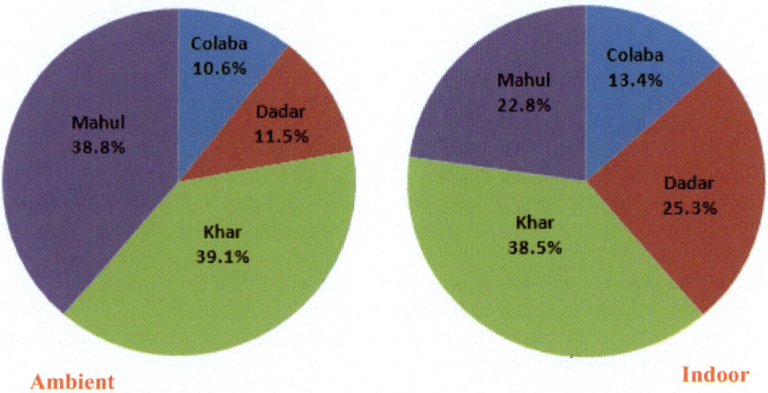

Fig. 26.6 Sitewise breakup of total levoglucosan concentration (ambient and indoor environment)

The movement of better knowledge generation should be continued with a focused goal for correlation of air quality monitoring results with the health status of the urban and rural population, especially for fine particles. Also, these results need to be used for better planning of air quality management. Though the percent contribution of markers to organic carbon ranges between 2 and 5%, it gives very important information to understand the sources. The present study attempted organic marker characterization in indoors and outdoors as they are a significant fraction of the fine particles in the urban sources because of their combustion origin.

The present study highlights that particle mass along with chemical speciation to understand toxic components of fine particles need to be considered before $PM_{2.5}$ regulation. The sampling and analysis of $PM_{2.5}$ will not only provide data w.r.t. compliance but also give information for future epidemiological and toxicological studies. The current research work expands spatial coverage of organic markers in Mumbai. Understanding the sources of fine particulates helps various stakeholders of the city or country to pinpoint emissions sources and identification of sources which are most harmful to health or any other environmental issues. This information can be used to prepare appropriate airshed management plans to protect human health and environment.

Acknowledgements The author is thankful to CSIR for providing fellowship to pursue PhD and acknowledges the support of CSIR NEERI, Delhi in carbon analysis. The authors would like to thank Desert Research Institute, NV, and United States for analyzing organic markers. The corresponding author would like to acknowledge the support of Shivaji Sawant and Mihir Herlekar in data analysis.

References

Azevedo DA, Moreira LS, Siqueira DS (1999) Composition of extractable organic matter in aerosols from urban areas of Rio de Janeiro City. Brazil Atmos Environ 33:4987–5001

Central Pollution Control Board (CPCB) (2009) National ambient air quality standards. http://www.cpcb.nic.in/National_Ambient_Air_Quality_Standards.php. Accessed 08 Jan 2015

Chowdhury Z, Zheng M, Schauer JJ, Sheesley RJ, Salmon L, Cass GR, Russell A (2007) Speciation of ambient fine organic carbon particles and source apportionment of PM2.5 in Indian Cities. J Geophys Res 112:D15303

Crutzen PJ, Andreae MO (1990) Biomass burning in the tropics: impact on atmospheric chemistry and biogeochemical cycles. Science 250(4988):1669–1678

Fraser MP, Yue ZW, Tropp RJ, Kohl SD, Chow JC (2002) Molecular composition of organic fine particulate matter in Houston, TX. Atmos Environ 36(38):5751–5758

Fu PQ, Kawamura K, Pavuluri CM, Swaminathan T, Chen J (2010) Molecular characterization of urban organic aerosol in tropical India: contributions of primary emissions and secondary photooxidation. Atmos Chem Phys 10:2663–2689

Harrison RM, Smith DJT, Luhana L (1996) Source appointment of atmospheric polycyclic aromatic hydrocarbons collected from an urban location in Birmingham U.K. Environ Sci Technol 30:825–832

Herlekar M Abba EJ, Kumar R, Gupta I (2012) Chemical speciation and source assignment of particulate (PM10) phase molecular markers in Mumbai. Aerosol Air Qual Res

Ji H, Zhang D, Shinohara R (2007) Size distribution and estimated carcinogenic potential of particulate polycyclic aromatic hydrocarbons collected at a downtown site in Kumamoto, Japan. Spring J Health Sci 53(6):700–707

Khalili NR, Scheff PA, Holsen TM (1995) PAH source fingerprints for coke oven, diesel and gasoline engines, highway tunnel and wood combustion emission. Atmos Environ 29:533–542

Li M, McDow SR, Tollerud DJ, Mazurek MA (2006) Seasonal abundance of organic molecular markers in urban particulate matter from Philadelphia, PA. Atmos Environ 40:2260–2273

Lunden MM, Kirchstetter TW, Thatcher TL, Hering SV, Brown NJ (2008) Factors affecting the indoor concentrations of carbonaceous aerosols of outdoor origin. Atmos Environ 42:5660–5671

Patil RS, Kumar VA (1 June, 2002) Modelling of indoor-outdoor relation of air pollutants in Mumbai City. In: 9th international conference on indoor air quality and climate, California, US

Pope C-III A, Dockery DW (2006) Health effects of fine particulate air pollution: lines that connect. J Air Waste Manage Assoc 56:709–710

Rogge WF, Hildemann LM, Mazurek MA, Cass GR, Simoneit BRT (1993) Sources of fine organic aerosol 2: noncatalyst and catalyst-equipped automobiles and heavy-duty diesel trucks. Env Sci Technol 27:636–651

Rogge WF, Hildemann LM, Mazurek MA, Cass GR, Simoneit BRT (1993) Sources of fine organic aerosol 8: boilers burning No. 2 distillate fuel oil. Env Sci Technol 31:2731–2737

Schauer JJ, Kleeman MJ, Cass GR, Simoneit BRT (1999) Measurement of emissions from air pollution sources 1: C-1 through C-29 organic compounds from meat charbroiling. Env Sci Technol 33:1566–1577

Schauer JJ, Cass GR (2000) Source apportionment of wintertime gas-phase and particle phase air pollutants using organic compounds as tracers. Environ Sci Technol 34(1):821–832

Simoneit BRT (1984) Application of molecular marker analysis to reconcile sources of carbonaceous particulates in tropospheric aerosols. Sci Total Env 36:61–72

Simoneit BRT, Medeiros PM, Didyk BM (2005) Combustion products of plastics as indicators for refuse burning in the atmosphere. Environ Sci Technol 39:6961–6970. Cited from Fu PQ, Kawamura K, Pavuluri CM, Swaminathan T, Chen J (2010) Molecular characterization of urban organic aerosol in tropical India: contributions of primary emissions and secondary photooxidation. Atmos Chem Phys 10:2663–2689

Valavanidis A, Fiotakis K, Vlachogianni T (2008) Airborne particulate matter and human health: toxicological assessment and importance of size and composition of particles for oxidative damage and carcinogenic mechanisms. J Environ Sci Health, Part C 26(4):339–362

World Health Organization (1987) Polycyclic aromatic hydrocarbons (PAH). Air quality guidelines for Europe, WHO regional publications, European series: No. 23, pp 105–107. World Health Organization, Geneva, Switzerland

Chapter 27
Distributions of n-Alkanes, Alkanoic Acids and Anhydrosugars in Wintertime Size-Segregated Aerosols Over Middle Indo-Gangetic Plain

Nandita Singh and Tirthankar Banerjee

27.1 Introduction

Aerosols have been recognized as an integral part of polluted air for decades if not centuries. Aerosols are complex mixture of solid and/or liquid inorganic and organic components of both primary and secondary origin. It's presence in the ambient environment continue to be a domain of high priority research. The persistence of airborne particulates and their long-term and short-term exposures have often associated with increased rate of mortality and morbidity, especially over the developing and least-developed world (Kumar et al. 2015; Xiao et al. 2014). Aerosols are the heterogeneous complex mixture and have the potential to influence the atmospheric processes from local to regional scale (Hodzic et al. 2015). They directly impact atmosphere's radiative budget by altering the solar radiations (through scattering and/or absorption of insolation), and indirectly by modifying cloud formation process (serving as cloud condensation nuclei) (Seinfeld and Pandis 2016). Aerosols originate from either naturally occurring processes (re-suspended dust, volcanic eruption, sea spray and biological materials) or anthropogenic activities (combustion of biomass and fossil fuels, secondary aerosols). Natural sources predominantly emit the coarser particles, whereas fine particles are mostly emitted from anthropogenic sources. Understanding the ways in which aerosols behave in ambient atmosphere, evolve and exert effects

N. Singh · T. Banerjee (✉)
Institute of Environment and Sustainable Development, Banaras Hindu University, Varanasi, India
e-mail: tb.iesd@bhu.ac.in

N. Singh
e-mail: nandita.singh@bhu.ac.in

T. Banerjee
DST-Mahamana Centre of Excellence in Climate Change Research, Banaras Hindu University, Varanasi, India

S. M. Shiva Nagendra et al. (eds.), *Urban Air Quality Monitoring, Modelling and Human Exposure Assessment*, Springer Transactions in Civil and Environmental Engineering,
https://doi.org/10.1007/978-981-15-5511-4_27

on environment requires extensive knowledge of its formation, transport, chemical transformation and removal processes.

The composition of aerosols is primarily the function of aerosol sources. In general, carbonaceous species contribute major fraction of particulate mass both in terms of elemental carbon (EC) and organic matter (OM). EC is different from OM in its light absorbing property, chemical inertness, extended aromatic ring, insolubility and refractory nature (Blanchard et al. 2014). The particulate-bound OM is commonly referred as the organic aerosol (OA) which accounts for significant fraction of airborne particulates mass and has potential to govern the physicochemical properties of aerosols. OA directly influence climate by modifying radiation budget and indirectly by modulating cloud microphysics, optical and hygroscopic properties of coexisting inorganic species (Zhang et al. 2011). OA causes adverse health effects such as cardiovascular diseases, respiratory illness like asthma, bronchitis, lung cancer and pre-mature mortality due to their carcinogenic, mutagenic and allergic behaviour (Ramírez et al. 2011; Lelieveld et al. 2015). The health and environmental effects of OA depend fundamentally on particle size, mass concentration, morphology and chemical composition. These parameters are also function of both emission sources and regional meteorology, which extends large spatial-temporal variation to aerosol properties. The ultrafine ($PM_{0.1}$, aerodynamic diameter $\leq 0.1\ \mu m$) and fine particles ($PM_{2.5} \leq 2.5\ \mu m$) are predominantly composed of OA, which can easily reach to bronchi, circulated in blood and responsible for several cardiovascular diseases (Nemmar et al. 2002). OA may have its origin directly from the sources (primary organic aerosol, POA) like from the burning of biomass and fuel, sea spray, volcanic eruption and plant emissions. However, the dominant proportion of OA is secondary (secondary organic aerosol, SOA) formed within the atmosphere via multiple gas-to-particle phase reactions and through successive oxidation and condensation of volatile organic compounds (VOCs) onto existing airborne particles (Hallquist et al. 2009). In coarser particles, OA mostly constitute from pollens, plant debris, sea-salt and soil dust.

OA are compounded by a number of organic molecules, primarily with aliphatic and aromatic hydrocarbons, anhydrosugars, alcohols, lipids, fatty acids, terpenes and steroids. The aliphatic hydrocarbons (n-alkanes, n-alkanoic acids) and anhydrosugars constitute significant fraction of OA. Due to the atmospheric stability and residence time, they retain their source signatures and serve as marker to particular emission sources (Bush and McInerney 2013). The high molecular weight hydrocarbon homologues are mostly biogenic in origin, derived from terrestrial plat wax, microorganism and vegetation debris, whereas the low molecular weight homologues have anthropogenic origin like cooking, biomass and fossil fuel burning (Kang et al. 2016). The anhydrosugars mostly emitted during the thermal degradation of cellulose, hemicellulose and lignin. The anhydrosugars like levoglucosan with its isomers mannosan and galactosan are the combustion product of cellulose and hemicellulose. Levoglucosan is widely used as a biomass burning tracer (Banerjee et al. 2015). Due to short atmospheric lifetime, the characteristic properties of OA are highly region specific. Therefore, in order to identify potential emission sources of OA, its environmental and human health impacts, and subsequently to develop emission abatement strategies, complete understanding of size distribution and chemical composition of OA is required.

Most of the Asian countries face huge burden of airborne particulates mainly due to the continuous rise in anthropogenic emissions and by the natural sources. The Indo-Gangetic Plain (IGP) in South Asia is one of the highly polluted and densely populated regions (Lelieveld et al. 2001). The entire IGP is well documented for having medium to high aerosol loading with considerable spatial and temporal variations (Sen et al. 2017; Singh et al. 2017). The large number of particulates emitted from post-harvest agricultural residue burning, vehicular and industrial emission, bio-fuel burning emissions and from long-range mineral dust transport eventually deteriorate air quality, reduced visibility and crop yield (Das and Jayaraman 2012). Prolonged exposure to these pollutants enhances the possibility of respiratory and cardiovascular diseases. Besides these regional impacts, the high aerosol burden over IGP has potential to influence regional climate. Being a sub-tropical region, the IGP usually encounters deep convective layer, which facilitate vertical movement of surface aerosols to a greater height resulting into increase in residence time. The aerosol mixed convective clouds carried apart from the source towards Africa and Mediterranean region by the influence of upper tropospheric wind (Lawrence and Lelieveld 2010). Further, a thick layer of brown cloud is reported over South Asia, which gradually rises upward and expanded horizontally towards west during summer monsoon, forming a giant umbrella over the Middle East, northern Africa and the eastern Mediterranean region (Vernier et al. 2011; Lawrence 2011). Additionally, during summer months, high organics and sulphates are also reported over Asian tropopause (Fig. 27.1a) (Yu et al. 2015).

The extreme variation in the meteorology and emission sources also causes seasonal changes in airborne particulates over IGP (Murari et al. 2017; Kumar et al. 2017). The transboundary transport of desert dust is frequent during summer months, whereas post-harvest agricultural residue burning dominates during post-monsoon (Singh et al. 2018). Severe fog with reduced visibility, extreme hazy condition, frequents events of biomass and coal burning are common characteristics during winter (Kumar et al. 2017). Beside the season-specific aerosol sources, fireworks during Diwali induce large particulate pollution over the entire region (Kumar et al. 2016). Recent studies reported that exposure to air pollution causes 1.66 million pre-mature deaths over India (Banerjee et al. 2017) and large pre-mature mortality primarily concentrated over IGP region (Fig. 27.1b) (Lelieveld et al. 2015). Studies have shown that due to high diffusion coefficient, fine and accumulation mode particles possess high deposition probability into respiratory ducts and lungs (Kheuhl et al. 2012). Therefore, it is absolutely necessary to identify chemical composition and size distributions of aerosols in order to establish aerosol-climate-human health effects. The contribution of various emission sources can be estimated by using organic molecular markers in conjunction with inorganic and trace metal species into the receptor models (Banerjee et al. 2015). During last decade, remarkable progress has been made in the molecular level characterization of OA, but such information is sparse over IGP, and only limited number of studies are available from very few

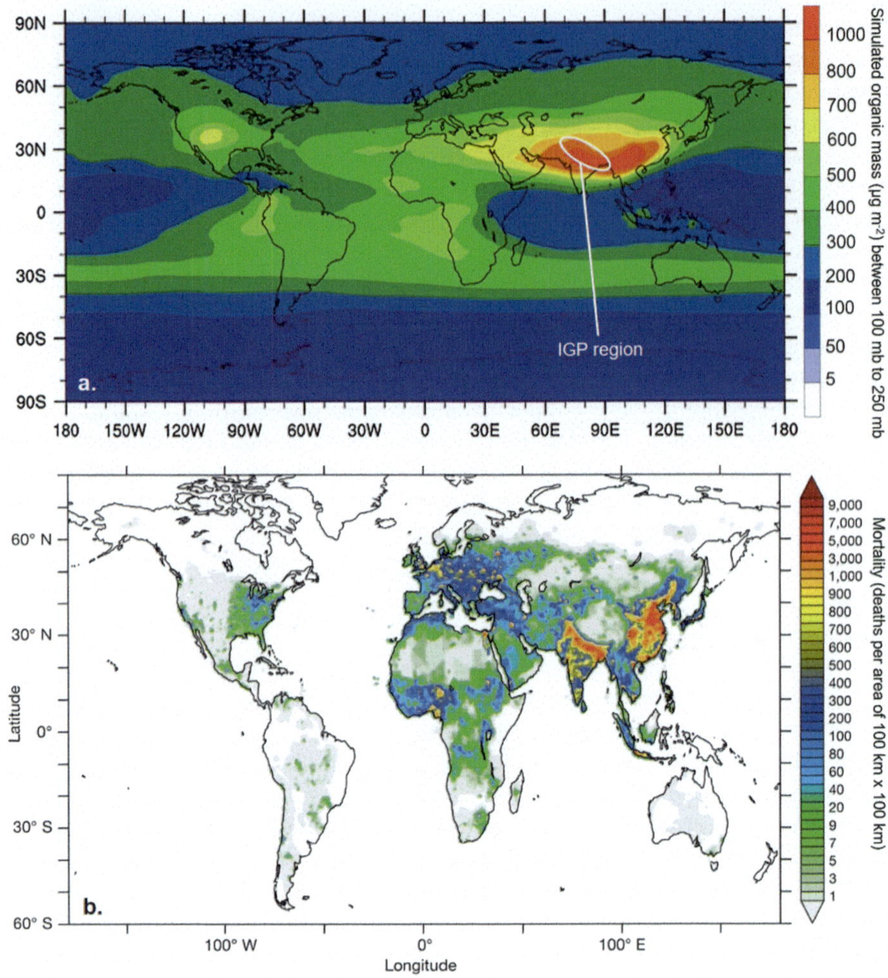

Fig. 27.1 Distribution of **a** simulated organic mass over tropopause layer. Modified from Yu et al. (2015), **b** pre-mature mortality related to $PM_{2.5}$ in 2010 (Lelieveld et al. 2015)

specific urban areas (Pant et al. 2015; Singh et al. 2013; Chowdhury et al. 2007). To overcome this knowledge gap, the characteristics of size-segregated aerosols in terms of mass concentration and organic constituents were studied at an urban site over middle GP.

27.2 Spatial Distribution and Emission Sources of OA Over IGP

The heterogeneity in emission sources and highly varied meteorology over IGP results into complicated aerosol chemistry with large uncertainty. Spatial distribution of aerosol optical depth (AOD) retrieved from Aqua Moderate-resolution Imaging Spectroradiometer (MODIS) atmosphere level-2 product over a span of five years (2011–2015) and varying composition of chemical species is shown in Fig. 27.2. It clearly indicates the presence of relatively high AOD (>0.6) over the entire IGP with considerable spatial differences. The AOD >0.8 particularly over the western IGP might be the result of transported dust (Kumar et al. 2018). In contrary to Western region, Eastern IGP showed the dominance of finer particles, which are

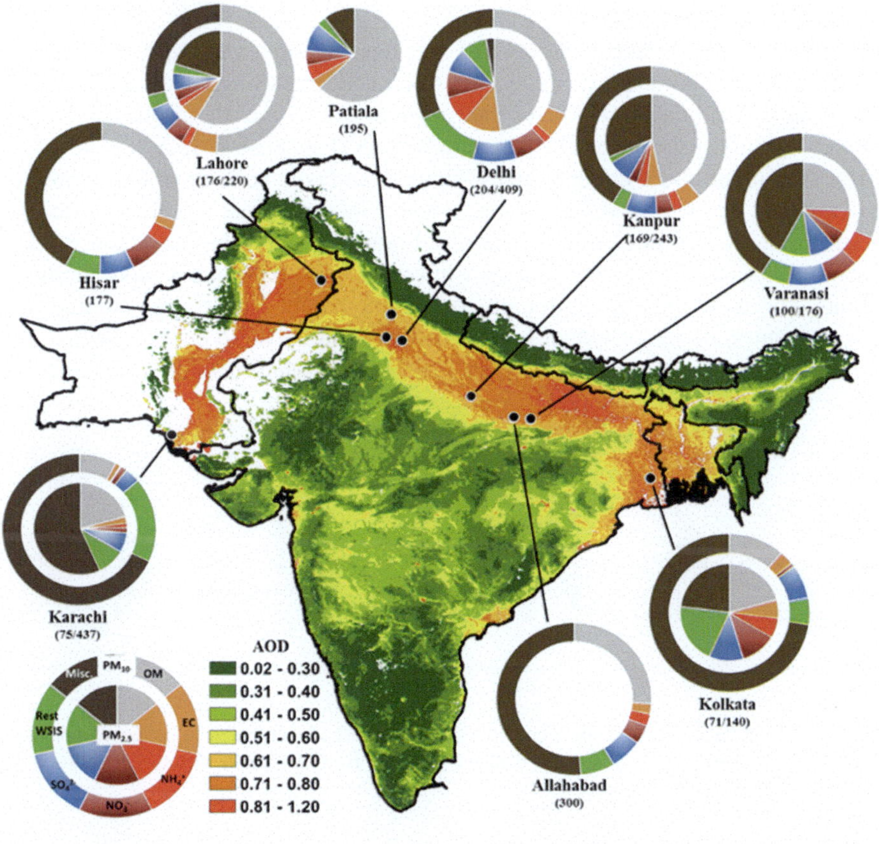

Fig. 27.2 Spatial variation of aerosol optical depth (AOD) across South Asia with relative contribution of major chemical species to surface-level total particulate mass (only for Hisar and Allahabad), and for $PM_{2.5}$, PM_{10} mass concentration (Singh et al. 2017). *Note* Data within the parenthesis indicate aerosol mass for $PM_{2.5}$ and PM_{10} or TSP (Singh et al. 2017, and references therein)

vowed to be influenced by regional meteorology (Kumar et al. 2018). Figure 27.2 also includes the information on ground-based particulate matter concentration and relative composition of aerosols, which is important to understand the population exposure-response relation. The chemical characterization of particulates in terms of carbonaceous species (OM & EC), secondary ionic species (NH_4^+, SO_4^{2-} and NO_3^-) and water-soluble ionic components (WSIS) are limited to only few urban locations. Over IGP, OA are predominantly emitted from various combustion processes such as agricultural residue burning, waste incineration, fossil fuel burning in vehicles and industries and contribute major portion (10–63%) of particulate mass. The relative contribution of OA is high in $PM_{2.5}$ than PM_{10}. Significant, diurnal variation in OA concentration is also reported with specific increase in night-time OA concentration, primarily influenced by biomass burning and stable meteorological conditions. Moreover, higher concentration of water-soluble organics and nitrate suggests the formation of SOA, especially in night which gets further enhanced during fogy condition (Rastogi et al. 2015).

Recently, a study conducted over IGP has reported high variability in chemical composition of airborne particulates with space and time (Singh et al. 2017, 2018). Source apportionment studies over the region conclude diesel, gasoline, coal and biomass burning as the prime OA emission sources (Pant et al. 2015,Villalobos et al. 2015). Among all these major sources, biomass and coal burning were reported to be dominated during post-monsoon and winter season, whereas diesel and gasoline-based emissions were dominated in summer. Figure 27.3 describes the spatial heterogeneity in OA emitting sources during winter. In Delhi, Agra and Kanpur, biomass burning contribute the highest in terms of particle-bound OA. In contrast, gasoline vehicles contribute major OA in Lahore and Kolkata, as these cities possess large number of decade-old vehicles in poor fuel ignition state, resulting into massive emission of particulate-bound OA (Stone et al. 2010).

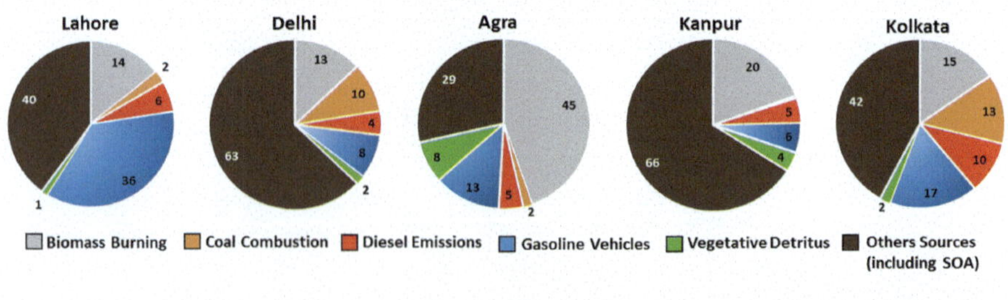

Fig. 27.3 Variation in organic carbon emitting sources during winter season at Lahore (Stone et al. 2010), Delhi (Chowdhury et al. 2007), Kolkata (Chowdhury et al. 2007), Agra (Villalobos et al. 2015) and Kanpur (Villalobos et al. 2015)

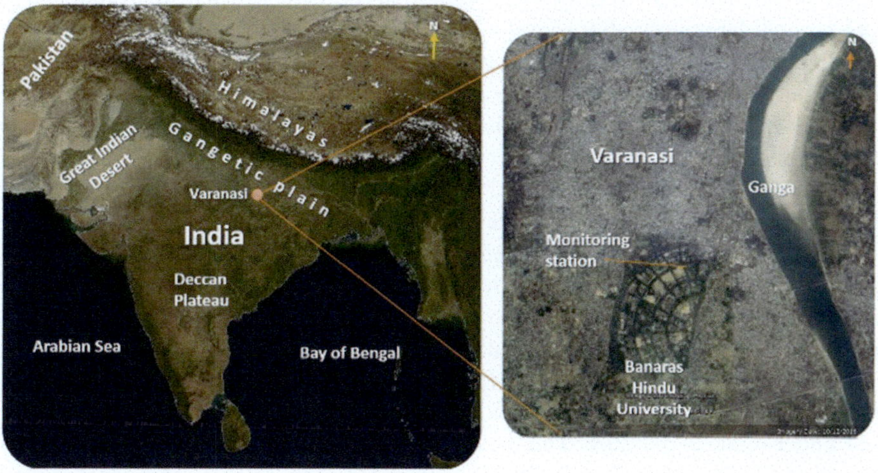

Fig. 27.4 Geographical location of particulate monitoring station. *Source* https://gibs.earthdata.nasa.gov

27.3 Site Description and Methodology

27.3.1 Study Area

Ground-based experiment was conducted in Varanasi located at 25°18′ N and 83°01′ E with an altitude of 82.2 m AMSL. Varanasi has an area of 225 km^2 and an estimated population of 3.4 million. The city is located on the bank of River Ganga (Fig. 27.4). Regional climate is typically humid sub-tropical and westerly predominate during winter. Different small-scale industries such as food processing; fabric printing and dyeing; paint manufacturing and manufacturing industries are located in and around the monitoring location.

27.3.2 Aerosols Monitoring and Extraction of OA

Monitoring for size-segregated particulate matter was continued during winter 2016 (January to February) with once in a week sampling frequency, continuously for 72-h. The particulate samples were collected with Anderson eight-stage cascade impactor (Tisch Environmental Inc., USA) having equivalent aerodynamic cut-off diameters at 50% efficiency are >9.0, 9.0–5.8, 5.8–4.7, 4.7–3.3, 3.3–2.1, 2.1–1.1, 1.1–0.65, 0.65–0.43 and <0.43 μm (backup filter). The flow rate was maintained at 28.3 L min^{-1} throughout study. The pre-combusted quartz filter papers were placed within the impactor's Petri dishes with plastic tweezers in a closed environment and were brought laboratory after sampling.

All the filters were desiccated for 24-h before and after the sampling. The mass of total particulates at each stage were determined gravimetrically using electronic balance (AY220, Shimadzu). The filter papers were grouped into 3 size fractions depending on the cut-off diameter: (1) coarse mode, particulates having aerodynamic cut-off diameter more than 2.1 μm, $PM_{>2.1}$; (2) fine mode, particles having aerodynamic cut-off diameter between 1.1 and 2.1 μm, $PM_{1.1-2.1}$; and (3) submicron mode, aerodynamic cut-off diameter less than 1.1 μm, $PM_{<1.1}$. A two-step extraction procedure was employed to extract maximum organic compounds from the filters. The filters were cut into small pieces and extracted by ultrasonication with 50 mL of each solvent mixture 1 (dichloromethane: hexane, 1:1, v/v) and mixture 2 (dichloromethane: methanol, 1:1, v/v). The combined extract was filtered with Teflon syringe filter and concentrated with vacuum rotatory evaporator under reduced pressure. The detailed discussions on extraction and analytical procedure of OA are included in the works of Hu et al. (2013). The concentrated extracts were derivatized using *N*,*O*-bis-(trimethylsilyl)-trifluoroacetamide plus 1% trimethylchlorosilane, and re-dissolved in 1 mL hexane.

27.3.3 Instrumentation

The final extracts after derivatization were analysed with gas chromatograph mass spectrometer (GCMS-QP2010 Ultra, Shimadzu, Japan). The samples (1 μl) were injected in splitless mode. Injector temperature was maintained at 260 °C. Initially, the column oven temperature was at 50 °C with isothermal hold of 2 min which linearly ramped up to 120 °C (30 °C min^{-1}) followed by 300 °C (6 °C min^{-1}) and hold for 11 min at 300 °C. Rxi-5MS fused silica capillary column 30 m × 0.25 mm id × 0.25 μm (Restek, Bellefonte, PA, USA) was used in the GCMS. Through electron impact ionization (70 eV) ions were produced and were further scanned for m/z 40–650 with a defined ion source (230 °C) and interface temperature (270 °C). Chromatograms peaks were analysed using fragmentation pattern and retention time of analytical standards (sigma).

27.4 Results and Discussion

27.4.1 Size Distribution of Atmospheric Particulates

The average concentration in a different particulate size fraction during study period (January–February, 2016) was summarized in Table 27.1 and Fig. 27.5. On average,

Table 27.1 Mass distribution of airborne particulate concentrations ($\mu g\ m^{-3}$)

Size	Mean	Max.	Min.	SD
$PM_{>2.1}$	209.1	366.8	116.8	76.0
$PM_{1.1-2.1}$	22.0	36.6	16.0	6.6
$PM_{<1.1}$	47.9	58.0	39.6	7.9

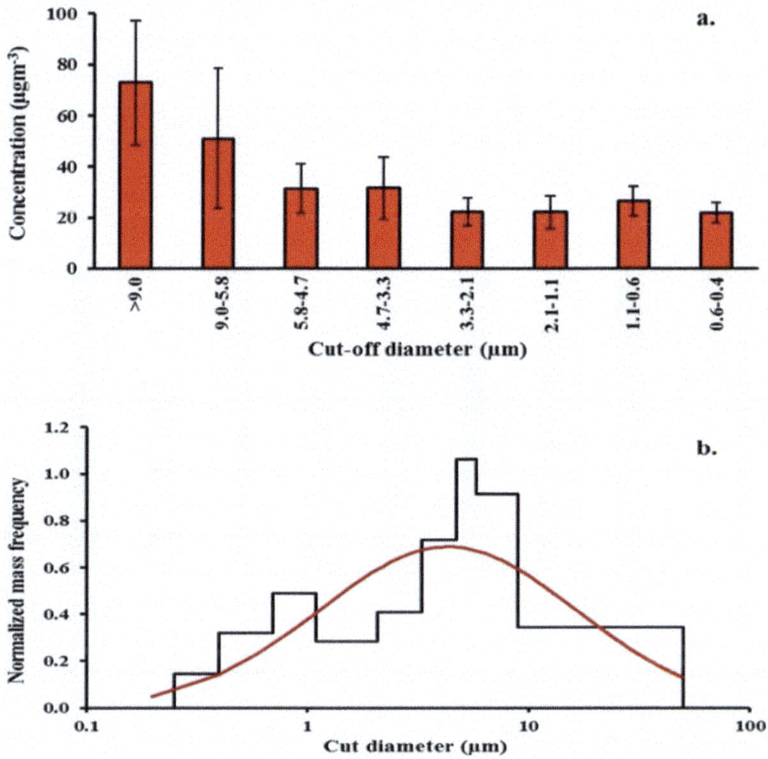

Fig. 27.5 Distribution of **a** size-segregated particulate mass concentration, **b** normalized mass frequency

74% mass of total particulate matter was contributed by coarse mode fraction (116.8–366.8 $\mu g\ m^{-3}$), and 26% by submicron (18%, 39.6–58.0 $\mu g\ m^{-3}$) and fine mode fractions (8%, 16.0–36.6 $\mu g\ m^{-3}$). Average mass concentration at each subsequent stage is shown in Fig. 27.5a which indicate highest fraction of mass loading occurred, especially at first two stages, having corresponding aerodynamic diameter $\geq$5.8 μm. The normalized mass frequency distribution (Fig. 27.5b) showed a bimodal distribution with max peak in coarse mode around 7–8 μm aerodynamic diameter and second peak in submicron region having aerodynamic diameter around 1 μm. For the entire duration, average mass median aerodynamic diameter (MMAD) was 4.4 μm and size-segregated aerosols geometric standard deviation (GSD) was 3.8.

The weekly variation of total aerosol in coarse, fine and submicron fractions were shown in Fig. 27.6a. The average mass concentration in coarse mode was found much higher (209 ± 76 μg m^{-3}) than fine (22 ± 7 μg m^{-3}) and submicron mode (48 ± 8 μg m^{-3}). On average, coarse mode contributes major fraction (74%) of total aerosol loading followed by submicron (18%) and fine (8%) mode particles. The mean particulate ratio ($PM_{2.1}/PM_{>2.1}$; 0.36 ± 0.12) indicates the dominance of coarser particle, whereas submicron to fine mode particle ratio ($PM_{<1.1}/PM_{1.1-2.1}$; 2.31 ± 0.64, Fig. 27.6b) showed dominance of submicron mode particles over fine mode. The relative dominance of submicron particles suggests the emission from fossil fuel combustion, waste incineration and residential wood burning.

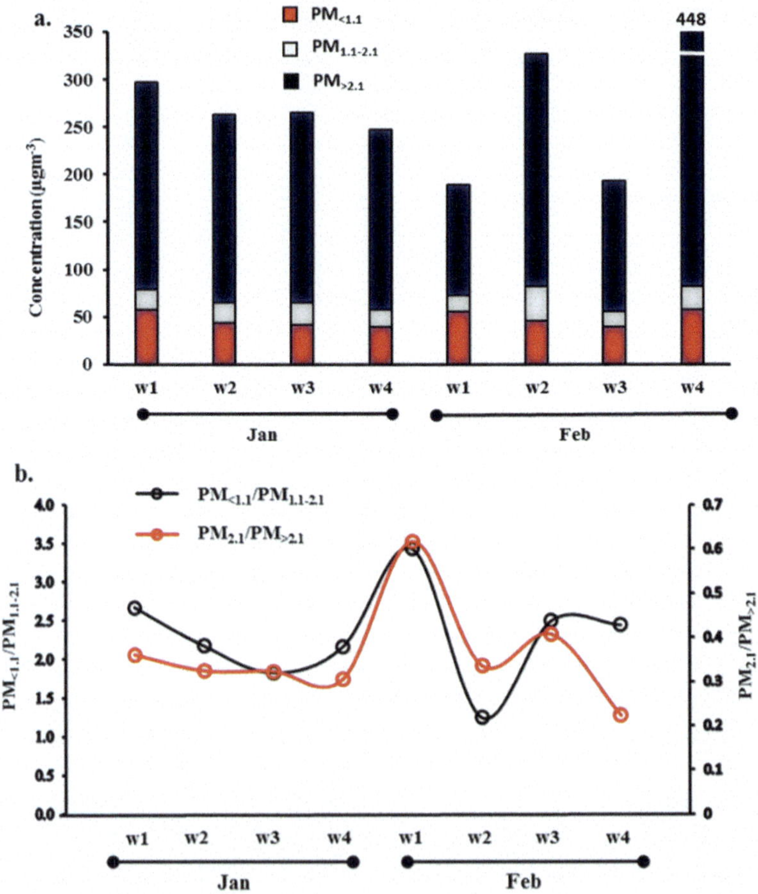

Fig. 27.6 Wintertime variation of **a** particulates mass concentration and **b** particle ratio

27.4.2 Particulate-Bound Organic Compounds

Entire IGP is facing a tremendous growth in industrialization and urbanization, resulting into serious air pollution concern. There are reports of massive increase in particle-bound OA over the cities across IGP, e.g. Delhi (Pant et al. 2015), Lahore (Stone et al. 2010), Kolkata (Chowdhury et al. 2007), Kanpur and Agra (Villalobos et al. 2015), often documented in constituting 20–90% of fine particulate mass (Rajput and Sarin 2014). Apart from negative health issues, the presence of OA potentially modifies the regional precipitation either by reducing the mean evaporation or by reducing the cloud formation processes (Riipinen et al. 2011). However, the effective characterization of OA across IGP is still lacking, and till the submission of the article, there were no published report on size-segregated distribution of organic aerosols and its chemical composition in Varanasi. Thus, we aimed to explore size distribution of particle-bound OA and to identify specific molecular markers which will be useful to estimate aerosol sources.

Particle-bound organic compounds (21 n-alkanes, 3 anhydrosugars and 11 n-alkanoic acids) in prementioned size fraction were quantified. Their average mass distribution as total with standard deviation is shown in Fig. 27.7. To understand and compare the contribution of various sources for specific particulate size, the specific source markers were analysed. The contribution from various biogenic, petrogenic and pyrogenic sources comprise homologous n-alkanes (C_{11}–C_{32}) and while wood burning markers include levoglucosan, galactopyranoside, mannopyranose and xylopyranose. Total identified organic compounds contributed 16% of total aerosols during winter season. Out of which, major portion was contributed by submicron (8%) and finer mode (7%) while least from coarse mode fraction (1%). Total

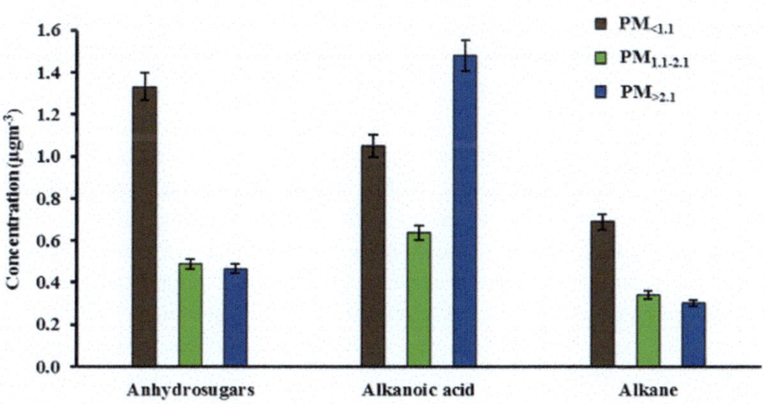

Fig. 27.7 Size distribution of organic species

concentration of levoglucosan, n-alkanes and n-alkanoic acid were found higher in comparison with Agra and Kanpur (Villalobos et al. 2015).

n-Alkanes

Figure 27.7 indicates the average concentration of n-alkanes in different aerosol size fractions with highest in submicron aerosols (mean ± SD: 672 ± 183 ng m^{-3}) in comparison with fine (352 ± 45 ng m^{-3}) and coarse mode fractions (293 ± 68 ng m^{-3}). The size distributions of individual n-alkane species were shown in Fig. 27.8a. The molecular distribution of n-alkanes (C_{11}–C_{32}) showed highly varied concentration (0–92 ng m^{-3}), having slight dominance of odd number with maxima at C_{29}. The high molecular weight n-alkanes homologues (C_{27}–C_{33}) mainly released from terrestrial higher plants (Stone et al. 2010). However, the prediction of exact source is ambiguous, i.e. whether it is coming from plant wax or from surface deposited plant litter with dust resuspension. The high concentration of C_{29} in submicron particles suggests its emission from the burning of wood which is used for residential heating, particularly in winter (Kang et al. 2016). It was also recognized that OA concentration was higher in submicron and finer fraction compared to coarse one. The high concentration in submicron size fraction suggests the dominance of vehicular and industrial sources (Singh et al. 2017). Further, this is also supported by the carbon prefix index (CPI), which varied between 0.8 and 1.2 (1.0 ± 0.2, mean ± SD) for all three size fractions. The lower CPI (close to unity) values are generally accepted as indication of anthropogenic emissions, whereas CPI >5 suggests the influence of biogenic sources like from plant wax (Simoneit et al. 2004).

Alkanoic Acids

The long chain (>C_8) alkanoic acids are fatty acids which is composed of long hydrocarbon (hydrophobic) chain and carboxylic group (polar) at the end. Due to their amphiphilic property, these fatty acids act as aerosol surfactants and can potentially change the aerosol characteristics (Tervahattu et al. 2005). In the present analysis, a homologous series (C_9 to C_{26}) of n-alkanoic acid were identified in different size fractions (Fig. 27.8b). Total fatty acids constitute highest fraction of solvent extractable organics with dominance in coarse (1486 ± 384 ng m^{-3}), submicron (1049 ± 374 ng m^{-3}) and finer mode (637 ± 124 ng m^{-3}). The larger CPI (2.5–3.4) shows the dominance of even fatty acids with maximum concentration of myristic acid (tetradecanoic, C_{14}) and steric acid (octadecanoic, C_{18}). Low molecular weight fatty acids ($C_{\leq 20}$) mostly emitted from fuel combustion, phytoplankton, cooking, small vascular plants and microbes, whereas high molecular weight fatty acids ($C_{>20}$) mainly originated from higher plant wax (Simoneit et al. 2004). The dominance of low molecular weight fatty acids suggests anthropogenic influence (motor vehicles, residential biomass burning and cooking). Moreover, the higher concentration of fatty acids (more oxidized organic compounds) suggests formation of secondary aerosols. The stable meteorological condition coupled with low boundary layers height and formation of fog (Singh et al. 2017; Kaul et al. 2011) during winter season leads to reduced mixing and increased lifetime for atmospheric aerosols.

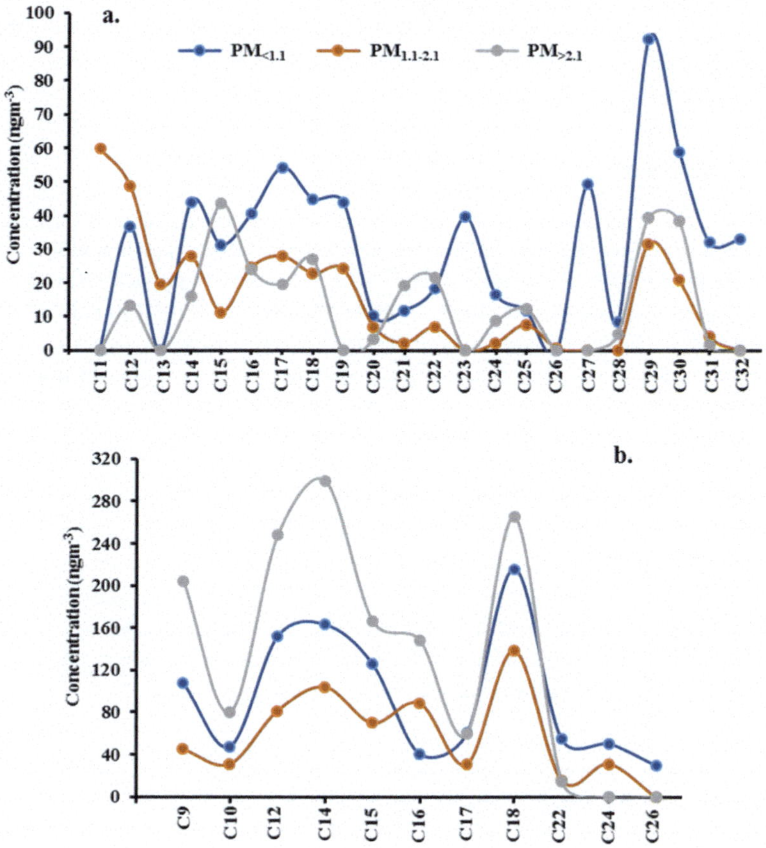

Fig. 27.8 Size distribution of individual **a** n-alkanes and **b** n-fatty acids during winter

Anhydrosugars

Anhydrosugars are produced from the pyrolysis of cellulose and used to identify the biomass burning source contribution. Among the anhydrosugars, levoglucosan is most frequently used molecular marker for biomass burning emissions. The mean anhydrosugars concentration in total aerosol was found 2283 ng m^{-3}. Anhydrosugars concentration showed large variations with maximum concentration noted in the submicron mode (1332 $\pm$ 133 ng m^{-3}) compared to fine (486 $\pm$ 76 ng m^{-3}) and coarse mode aerosols (465 $\pm$ 110 ng m^{-3}) (Fig. 27.7). The higher concentration during winter season is mainly due to the increased use of biomass fuel (charcoal, dung cake, agricultural residues and wood) for domestic heating (Banerjee et al. 2017). Burning of such biomass produces considerable amount of volatile compound along with soot, brown carbon and organic aerosols both in ultrafine and submicron mode (Liu et al. 2016; Srivastava et al. 2019). Compared to the present study, slightly higher concentrations of anhydrosugars were also reported in Lahore ($PM_{2.5} \approx 1350$ ng m^{-3})

(Stone et al. 2010) while relatively lower concentration is reported in Agra and Kanpur ($PM_{2.5} \approx 85–190$ n gm^{-3}) (Villalobos et al. 2015).

27.5 Conclusions

The manuscript discusses the size distribution of airborne particulates over a typical urban environment in middle IGP, with variation in particle-bound organic compounds. Size fractionated study showed that the 74% of total aerosol mass is contributed by coarse mode (>2.1 μm) particles followed by submicron (18%) and fine mode fraction (8%). The normalized mass frequency distribution follows bimodal distribution pattern having peaks in coarse and submicron range. The three major organic groups, n-alkane, n-alkanoic acid and anhydrosugars were also identified. The total analysed organic groups were found to constitute 16% of total particulates having higher concentration in submicron (8%) and fine fraction (7%). Dominance of OA in fine mode aerosols with low CPI (1.0 ± 0.2) for n-alkane indicate prevalence of anthropogenic emissions. We conclude that solid wood burning, residential cooking, re-suspended road dust, fossil fuel (mostly from vehicles since the city has no large industry) and biomass burning are the major OA sources during winter. However, understanding intra-seasonal variation in size-segregated aerosols and organic compounds requires further analysis over middle IGP.

Acknowledgements The research is financially supported by Department of Science and Technology, New Delhi (SR/FTP/ES-52/2014). N. Singh, duly acknowledges the financial support provided by DST Women Scientist Scheme (SR/WOS-A/EA-1012/2015). Authors also acknowledge the DBT grant (no. BT/PR3130/INF/22/139/2011) for using AIRF facility at JNU, Delhi as well as guidance and inspiration by Dean and Director, IESD-BHU.

References

Banerjee T, Murari M, Kumar M, Raju MP (2015) Source apportionment of airborne particulates through receptor modeling: Indian scenario. Atmos Res 164–165:167–187

Banerjee T, Kumar M, Mall RK, Singh RS (2017) Airing 'clean air' in clean India mission. Environ Sci Pollut Res 24(7):6399–6413

Blanchard CL, Chow JC, Edgerton ES, Watson JG, Hidy GM, Shaw S (2014) Organic aerosols in the southeastern United States: speciated particulate carbon measurements from the SEARCH network, 2006–2010. Atmos Environ 95:327–333

Bush RT, McInerney FA (2013) Leaf wax n-alkane distributions in and across modern plants: implications for paleoecology and chemotaxonomy. Geochim Cosmochim Acta 117:161–179

Chowdhury Z, Zheng M, Schauer JJ, Sheesley RJ, Salmon LG, Cass GR, Russell AG (2007) Speciation of ambient fine organic carbon particles and source apportionment of $PM_{2.5}$ in Indian cities. J Geophys Res 112:D15303

Das SK, Jayaraman A (2012) Long-range transportation of anthropogenic aerosols over eastern coastal region of India: Investigation of sources and impact on regional climate change. Atmos Res 118:68–83

Hallquist M et al (2009) The formation, properties and impact of secondary organic aerosol: current and emerging issues. Atmos Chem Phys 9:5155–5236

Hodzic A et al (2015) Organic photolysis reactions in tropospheric aerosols: effect on secondary organic aerosol formation and lifetime. Atmos Chem Phys 15:9253–9269

Hu Q-H, Xie Z-Q, Wang X-M, Kang H, Zhang PF (2013) Levoglucosan indicates high levels of biomass burning aerosols over oceans from the Arctic to Antarctic. Sci Rep 3:3119

Kang M, Fu P, Aggarwal SG, Kumar S, Zhao Y, Sun Y, Wang Z (2016) Size distributions of n-alkanes, fatty acids and fatty alcohols in springtime aerosols from New Delhi, India. Environ Pollut 219:957–966

Kaul DS, Gupta T, Tripathi SN, Tare V, Collett JL Jr (2011) Secondary organic aerosol: a comparison between foggy and nonfoggy days. Environ Sci Technol 45:7307–7313

Kheuhl PJ, Anderson TL, Candelaria G, Gershman B, Harlin K, Hesterman JY, Holmes Y, Hoppin J, Lackas C, Norenberg JP, Yu H, McDonald JD (2012) Regional particle size dependent deposition of inhaled aerosols in rats and mice. Inhal Toxicol 24(1):27–35

Kumar M, Singh RS, Banerjee T (2015) Associating airborne particulates and human health: exploring possibilities. Environ Int 84:201–202

Kumar M, Singh RK, Murari V, Singh AK, Singh RS, Banerjee T (2016) Fireworks induced particle pollution: a spatio-temporal analysis. Atmos Res 180:78–91

Kumar M, Raju MP, Singh RS, Banerjee T (2017a) Impact of drought and normal monsoon scenarios on aerosol induced radiative forcing and atmospheric heating in Varanasi over middle Indo-Gangetic Plain. J Aerosol Sci 113:95–107

Kumar M, Raju MP, Singh RK, Singh AK, Singh RS, Banerjee T (2017b) Wintertime characteristics of aerosols over middle Indo-Gangetic Plain: vertical profile, transport and radiative forcing. Atmos Res 183:268–282

Kumar M, Parmar KS, Kumar DB, Mhawish A, Broday DM, Mall RK, Banerjee T (2018) Long-term aerosol climatology over Indo-Gangetic Plain: Trend, prediction and potential source fields. Atmos Environ 180:37–50

Lawrence MG (2011) Asia under a high-level brown cloud. Nat Geosci 4:352–353

Lawrence MG, Lelieveld J (2010) Atmospheric pollutant outflow from southern Asia: a review. Atmos Chem Phys 10:11017–11096

Lelieveld J et al (2001) The Indian ocean experiment: widespread air pollution from South and Southeast Asia. Science 291(5506):1031–1036

Lelieveld J, Evans JS, Fnais M, Giannadaki D, Pozzer A (2015) The contribution of outdoor air pollution sources to premature mortality on a global scale. Nature 525:367–371

Liu C, Chung CE, Zhang F, Yin Y (2016) The colors of biomass burning aerosols in the atmosphere. Scientific reports. https://doi.org/10.1038/srep28267

Murari V, Kumar M, Mhawish A, Barman SC, Banerjee T (2017) Airborne particulate in Varanasi over middle Indo-Gangetic Plain: variation in particulate types and meteorological influences. Environ Monit Assess 189:157

Nemmar A, Hoet PHM, Vanquickenborne B, Dinsdale D, Thomeer M, Hoylaerts MF, Vanbilloen H, Mortelmans L, Nemery B (2002) Passage of inhaled particles into the blood circulation in humans. Circulation 105:411–414

Pant P, Shukla A, Kohl SD, Chow JC, Watson JG, Harrison R (2015) Characterization of ambient $PM_{2.5}$ at a pollution hotspot in New Delhi, India and inference of sources. Atmos Enviorn 109:178–189

Rajput P, Sarin MM (2014) Polar and non-polar organic aerosols from large-scale agricultural-waste burning emissions in Northern India: implications to organic mass-to-organic carbon ratio. Chemosphere 103:74–79

Ramírez N, Cuadras A, Rovira E, Marce RM, Borrull F (2011) Risk assessment related to atmospheric polycyclic aromatic hydrocarbons in gas and particle phases near industrial sites. Environ Health Perspect 119(8):1110–1116

Rastogi N, Patel A, Singh A, Singh D (2015) Diurnal variability in secondary organic aerosol formation over the Indo-Gangetic Plain during winter using online measurement of water-soluble organic carbon. Aerosol Air Qual Res 15:2225–2231

Riipinen I, Pierce JR, Yli-Juuti T, Nieminen T, Hakkinen S, Ehn M, Junninen H, Lehtipalo K, Petaja T, Slowik J, Chang R (2011) Organic condensation: a vital link connecting aerosol formation to cloud condensation nuclei (CCN) concentrations. Atmos Chem Phys 11:3865–3878

Seinfeld JH, Pandis SN (2016) Atmospheric chemistry and physics: from air pollution to climate change, 3rd edn. 978-1-118-94740-1

Sen A, Abdelmaksoud AS, Ahammed YN, Banerjee T, Bhat MA, Chatterjee A, Choudhuri AK, Das T, Dhir A, Dhyani PP, Gadi R (2017) Variations in particulate matter over Indo-Gangetic Plains and Indo-Himalayan Range during four field campaigns in winter monsoon and summer monsoon: role of pollution pathways. Atmos Environ 154:200–224

Simoneit BRT, Kobayashi M, Mochida M, Kawamura K, Lee M, Lim H-J, Turpin BJ, Komazaki Y (2004) Composition and major sources of organic compounds of aerosol particulate matter sampled during the ACE-Asia campaign. J Geophys Res Atmos 109

Singh DP, Gadi R, Mandal TK, Saud T, Saxena M, Sharma SK (2013) Emissions estimates of PAH from biomass fuels used in rural sector of Indo-Gangetic Plains of India. Atmos Environ 68:120–126

Singh N, Mhawish A, Deboudt K, Singh RS, Banerjee T (2017a) Organic aerosols over Indo-Gangetic Plain: sources, distributions and climatic implications. Atmos Environ 157:59–74

Singh N, Murari V, Kumar M, Barman SC, Banerjee T (2017b) Fine particulates over South Asia: review and meta-analysis of $PM_{2.5}$ source apportionment through receptor model. Environ Pollut 223:121–136

Singh N, Banerjee T, Raju MP, Deboudt K, Sorek-Hamer M, Singh RS, Mall RK (2018) Aerosol chemis-try, transport and climatic implications during extreme biomass burning emissions over Indo-Gangetic Plain. Atmos Chem Phys 18:14197–14215

Srivastava S, Kumar M, Singh RS, Rai BN, Mall RK, Banerjee T. Long-term observation of black carbon aerosols at an urban location over central Indo-Gangetic Plain, South Asia. Atmosfera. Accepted in press

Stone E, Schauer J, Quraishi TA, Mahmood A (2010) Chemical characterization and source apportionment of fine and coarse particulate matter in Lahore, Pakistan. Atmos Environ 44:1062–1070

Tervahattu V, Juhanoja J, Vaida V, Tuck AF, Niemi JV, Kupiainen K, Kulmala M, Vehkamäki H (2005) Fatty acids on continental sulfate aerosol particles. J Geophys Res 110:D06207

Vernier JP, Thomason LW, Pommereau JP, Bourassa A, Pelon J, Garnier A, Hauchecorne A, Blanot L, Trepte C, Degenstein D, Vargas F (2011) Major influence of tropical volcanic eruptions on the stratospheric aerosol layer during the last decade. Geophys Res Lett 38(12):L12807

Villalobos AM, Amonov MO, Shafer MM, Devi JJ, Gupta T, Tripathi SN, Rana KS, Mckenzie M, Bergin MH, Schauer JJ (2015) Source apportionment of carbonaceous fine particulate matter ($PM_{2.5}$) in two contrasting cities across the Indo-Gangetic Plain. Atmos Pollut Res 6:398–405

Xiao S, Wang QT, Cao JJ, Huang RJ, Chen WD, Han YM, Xu HM, Liu SX, Zhou YQ, Wang P, Zhang JQ, Zhan CL (2014) Longterm trends in visibility and impacts of aerosol composition on visibility impairment in Baoji, China. Atmos Res 149:88–95

Yu P, Toon OB, Neely RR, Martinsson BG, Brenninkmeijer CAM (2015) Composition and physical properties of the Asian Tropopause Aerosol Layer and the North American Tropospheric Aerosol Layer. Geophys Res Lett 42:2540–2546

Zhang Q, Jimenez JL, Canagaratna MR, Ulbrich IM, Ng NL, Worsnop DR, Sun Y (2011) Understanding atmospheric organic aerosols via factor analysis of aerosol mass spectrometry: a review. Anal Bioanal Chem 401:3045–3067

Chapter 28
Exposure of $PM_{2.5}$ and Carbonaceous Matter Amongst Rural Inhabitants in and Around Durg District of Chhattisgarh, India

Shailendra Kumar Kushawaha, Yasmeen F. Pervez, Sumita Nair, and Shamsh Pervez

28.1 Introduction

Air pollution has been identified as a global environmental problem which is affecting human health (Pant et al. 2016). The size of fine particulate matter (PM) in aerodynamic diameter is 2.5 μm ($PM_{2.5}$) is generally considered as fine particulate and has shown adverse effect on human health. In a broader sense, exposure is used in qualitative and quantitative calculation of pollutant and it is explained as description of the pollutant's concentration and the quality and quantity of time individual spends in touch with pollutants through breathing (Pant et al. 2016; Moschandreas et al. 2002; Morawska et al. 2013). PM is a blend of several chemical species and notable particulates components are organic fractions, such as organic carbon (OC) and elemental carbon (EC). Carbonaceous matter comprises of organic and elemental carbon where OC acts as a major component and its contribution is 90% in carbonaceous matter. The main pollutant discharged or by-product, which emits after all the process of burning industrial emission, traffic, outdoor fires and domestic biomass, is elemental carbon (EC), also known as soot and black carbon (BC), and contributes 5–10% which is relatively low (Ram et al. 2008). EC is a primary pollutant, emitted as by-product of all combustion processes such as industrial emissions, traffic, outdoor fires, and household biomass fuels. EC plays a main role in global climatic change

S. K. Kushawaha · Y. F. Pervez (✉)
Department of Chemistry, Chhatrapati Shivaji Institute of Technology, Durg, C.G 491001, India
e-mail: dr.ypervez@gmail.com

S. Nair
Department of Applied Chemistry, Bhilai Institute of Technology, Durg, C.G 491001, India

S. Pervez
School of Studies in Chemistry, Pt. Ravishankar Shukla University, Raipur, C.G 492010, India

S. M. Shiva Nagendra et al. (eds.), *Urban Air Quality Monitoring, Modelling and Human Exposure Assessment*, Springer Transactions in Civil and Environmental Engineering,
https://doi.org/10.1007/978-981-15-5511-4_28

as it possesses a powerful capability of absorbing solar radiation and causes a positive radiative forcing (Hansen et al. 2000; Jacobson 2001; Pervez et al. 2016). The molecule refers OC contains carbon-hydrogen bond and constitutes a large diversity of organic compound. OC is a crucial component which straight discharge or formed in situ the atmosphere and helps in scattering light and chilling the atmosphere along with health problem (Pervez et al. 2016). Thus, carbonaceous particulates studies needed more attention and research, thus, measurement of OC and EC is important (Pipal et al. 2014). The study has been conducted in rural environment of Durg District, Chhattisgarh. This study was carried out to assess the mass concentration level of $PM_{2.5}$ and exposure level to rural inhabitants and characterize fine particulate matter in terms of carbonaceous species.

28.2 Materials and Methodology

28.2.1 *Sampling Site*

The study was conducted at Anjora (AN), Achhoti (AC), Chandkhuri (CH) rural areas situated in Durg district, Chhattisgarh in Central India (Fig. 28.1). Durg stretches across 20°54′ to 21°32′ N and 81°10′ to 81°36′ E. The height of about 317 m above the level of the sea. Durg has population of 3,343,079 and 61.59% population lives in rural area (Census 2011). The climate of study area is tropical, wet, and dry. In summer, the average temperature varies between 25 and 43 °C while in winter the average temperature varies between 21 and 17 °C. The selection of sampling sites was households, based on surrounding environment of rural residents and the types of solid biomass fuel for cooking (Matawle et al. 2016). Most households in the study region were using fuel wood, crops straw, and dung cakes as cooking fuels (Matawle et al. 2016). AN is directly affected by heavy traffic emission on NH-6. This site is located near major national highway (NH-6) which is one of the busiest highways, connecting Mumbai to Kolkata. CH is situated on Durg–Balod road and the number of small-scale aluminium wares industries is present in and around areas. These industries are burning bulk amount of coal to mould aluminium for making aluminium wares. AC is located away from the main road site but predominated by heavy household biomass burning.

28.2.2 *Sampling and Analysis*

28.2.2.1 Sampling

Low volume samplers (MiniVol, Air Metrics, USA) were used in sampling sites to collect integrated 24-h $PM_{2.5}$ sample between two seasons: (1) winter (October

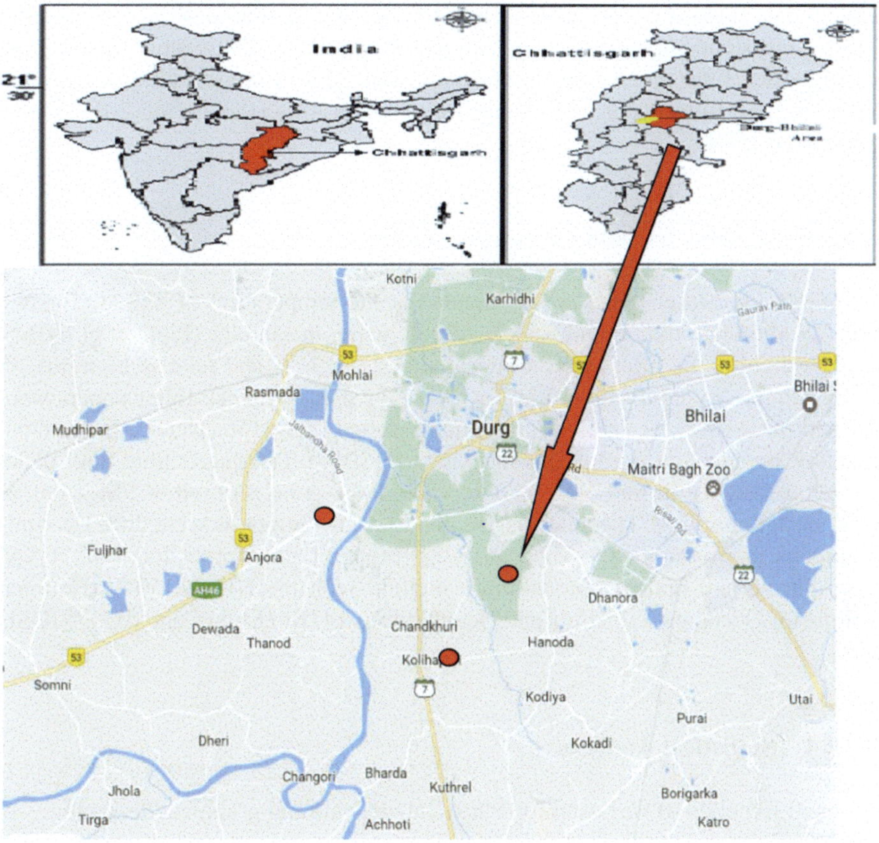

Fig. 28.1 Map of Durg district shows sampling sites *Source* mapdata©2014

2014–February 2015), (2) summer (March 2015–June 2015) (Matawle et al. 2016). A total of 210 samples were collected during entire period. In winter and summer with same no.: Anjora (AN) $n = 48$, Achhoti (AC) $n = 30$, Chandkhuri (CH) $n = 27$. To measure exposure level of $PM_{2.5}$ concentration, samplers were installed at the height of 1.5 metres from the ground level, inside the house (Matawle et al. 2016). Samples were collected on pre-fired quartz fibre filters (QFF) with the flow rate of 5 L per minute (Matawle et al. 2016). The used filters were stored in filter cassettes at 4 °C to its chemical analysis was made.

28.2.2.2 Quality Control

The quality control in monitoring was followed, flow rate calculation checked on daily basis and the fluctuation in flow rate was maintained within range (Pant et al. 2016). The impactor was used to classify particulate depending on their size of less than 2.5 μm. Sample was regularly clean to keep it dust-free to avoid the counting of dust mass concentration of sample (Pipal et al. 2014). Chow and Watson 1998 had suggested filter change in the impactor after 72 h of the sampling, but in this study, the used filter paper was changed after every 48 h (Pipal et al. 2014; Chow and Watson 1998).

28.2.2.3 Gravimetric and Chemical Analysis

The quartz fibre filter paper was preheated at the temperature of 450 °C for two hours, to eliminate the organic species before using in sampler (Pipal et al. 2014). Before weighing the filters were desiccated at 27 ± 1 °C and relative humidity 35 ± 1% in humidity-controlled room for 24 h (Pipal et al. 2014). Filter papers were weighed three times in an electronic microbalance (TB-2150, Denver Instrument, Germany, precision of ±10 μg) (Matawle et al. 2016). The quartz fibre filter paper was weighed twice to find out gravimetrically mass concentration of $PM_{2.5}$ before and after the sampling. The analysis OC and EC concentrations were done by semi-continuous thermal/optical carbon analyser (Sunset Laboratory, Model 4L, USA) using the thermal optical transmittance method with the NIOSH 5040 (National Institute for Occupational Safety and Health) protocol (Birch and Cary 1996; NIOSH 1999).

28.2.2.4 Statistical Analysis

Microsoft Excel 2010 was used for statistical analysis and graphical designing.

28.3 Results and Discussion

28.3.1 Concentration of $PM_{2.5}$

The average seasonal variations of $PM_{2.5}$ concentrations for winter and summer seasons for all sampling locations were as follows: Anjora (AN); 524.36 ± 99.42 μg m^{-3} and 317 ± 59.77 μg m^{-3}, Achhoti (AC); 533.13 ± 98.5 μg m^{-3} and 380 ± 74.19 μg m^{-3}, Chandkhuri (CH) 611.37 ± 112.57 μg m^{-3} and 428.74 ± 80.64 μg m^{-3}, respectively, are shown in Table 28.1; Fig. 28.2. It was found that the concentration of $PM_{2.5}$ was higher than the annual standards value of National Ambient Air Quality Standard (NAAQS) ($PM_{2.5}$ = 40 μg m^{-3}) and by the United State Environmental Protection Agency (USEPA) ($PM_{2.5}$ = 15 μg m^{-3}) (Pervez et al. 2016). The highest average mass concentration of $PM_{2.5}$ was highest at Chandkhuri (CH) which was 14.6% and 16.5% higher than Achhoti (AC) and Anjora (AN) in winter and 12.8% and 35.2% higher than Achhoti (AC) and Anjora (AN) in summer,

Table 28.1 Seasonal concentration (geometric mean ± standard derivation) ($\mu g\ m^{-3}$) of $PM_{2.5}$ and carbonaceous species in sampling site

Site	$PM_{2.5}$	OC	EC	TC	SOC
Winter					
AN	524.36 ± 99.42	127.30 ± 23.29	11.19 ± 1.96	138.53 ± 25.00	104.80 ± 20.00
AC	533.13 ± 98.5	115.66 ± 25.71	9.70 ± 2.36	125.46 ± 27.67	96 ± 22.04
CH	611.37 ± 112.57	123.56 ± 31.06	17.22 ± 6.21	141.35 ± 35.72	87.27 ± 23.94
Summer					
AN	317 ± 59.77	67.05 ± 12.54	8.82 ± 2.40	75.98 ± 14.19	49.10 ± 9.87
AC	380 ± 74.19	61.11 ± 13.13	6.06 ± 1.78	68.70 ± 13.29	43.70 ± 13.50
CH	428.74 ± 80.64	63.82 ± 9.16	10.20 ± 2.19	74.13 ± 10.87	43.04 ± 6.64

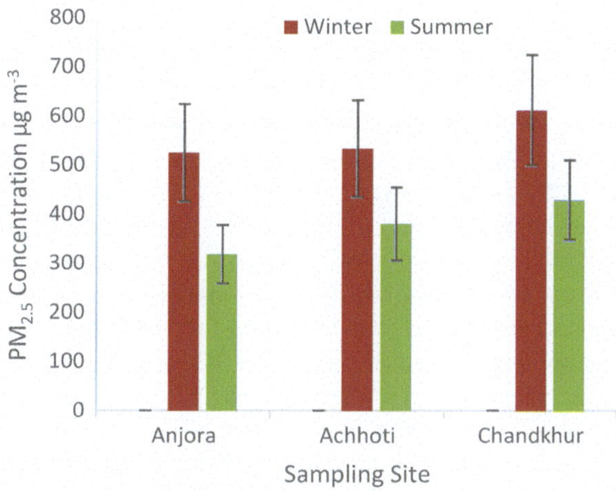

Fig. 28.2 Concentration of $PM_{2.5}$ in sampling site in winter and summer

respectively. In winter, it was observed that the concentration of $PM_{2.5}$ observed highest in all the sampling sites because of slow wind speed, lower mixing height, and less atmospheric temperature (Pipal et al. 2014). The particles could not scatter in stable and cold condition and settled at low height in atmosphere (Pipal et al. 2014). In Chandkhuri (CH), the $PM_{2.5}$ mass concentration was found high as compare to other sites in all seasons due to traffic emissions, massive biomass burning, and huge combustion of fossil fuels in small-scale industries in the region. Some studies reported that the average mass concentration of $PM_{2.5}$ was 590 and 741 $\mu g\ m^{-3}$ in kitchens due to using wood and cow dung in rural areas (Matawle et al. 2016; Balakrishanan et al. 2013) and 197.5 ± 84.3 $\mu g\ m^{-3}$ due to household solid burning fuels in a different rural area in Pune, India (Satsangi et al. 2014).

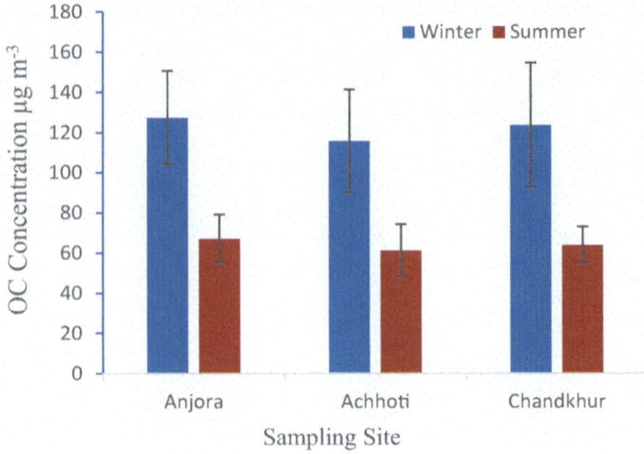

Fig. 28.3 Concentration of OC in sampling site in winter and summer

28.3.2 Concentration of Carbonaceous Species

Table 28.1 represents the average concentrations of OC, EC, and TC. Highest average OC concentration was observed in (AN) in winter and summer; 127.30 ± 23.29 µg m^{-3} and 65.59 ± 12.54 µg m^{-3}, respectively, followed by (AC) 115.66 ± 25.71 µg m^{-3} and 61.11 ± 13.13 µg m^{-3} than at (CH);123.56 ± 31.06 µg m^{-3} and 63.82 ± 9.16 µg m^{-3}. Highest OC concentration level in Anjora (AN) in both seasons may be attributed due to more biomass burning for cooking purpose in rural houses. Highest EC was observed in Chandkhuri (CH) in both seasons, i.e. winter and summer; 17.22 ± 6.21 and 10.20 ± 2.19 µg m^{-3}, respectively. Heavy traffic load and massive coal burning contributes to high EC as carbonaceous fraction. The order of EC occurrence in winter and summer at (AN);11.19 ± 1.96 µg m^{-3} and 8.82 ± 2.40 µg m^{-3}, at Achhoti (AC); 9.70 ± 2.36 µg m^{-3} and 6.06 ± 1.78 µg m^{-3}, Chandkhuri (CH) 17.22 ± 6.21 µg m^{-3} and 10.20 ± 2.19 µg m^{-3}.

The total carbon (TC) variation for both the seasons (winter and summer) in all three locations were as follows: Chandkhuri (CH); 141.35 ± 35.72 µg m^{-3} and 74.13 ± 10.87 µg m^{-3} > Anjora (AN);138 ± 25.00 µg m^{-3} and 75.98 ± 14.19 µg m^{-3} > Achhoti (AC); 125.46 ± 27.67 µg m^{-3} and 68.70 ± 13.29 µg m^{-3} (Figs. 28.3 and 28.4).

28.3.3 OC/EC Ratio

OC/EC ratio is useful tool to get information about the source of origin (Turpin et al. 1991; Chow et al. 1996; Lin and Tai 2001; Zeng and Wang 2011). If the value of OC/EC ratio is found more than 2, then it indicates the development of SOC a part from primary emission source (Pipal et al. 2014). Different studies have found a different ratio of OC/EC at a different level for different types of emission source. OC/EC value in between (1) 1–4.2 indicates diesel combustion and gasoline

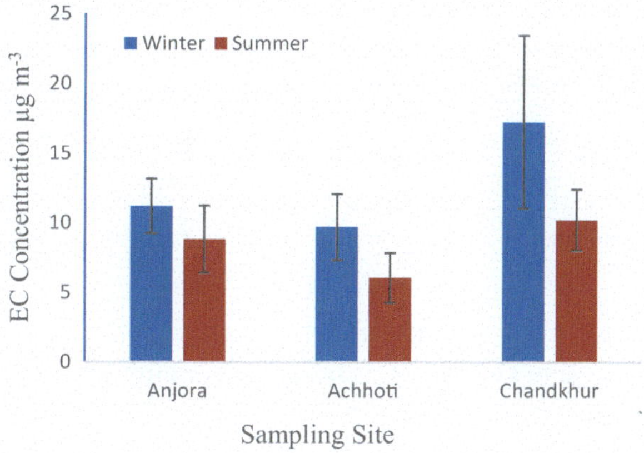

Fig. 28.4 Concentration of EC in sampling site in winter and summer

Table 28.2 OC/EC ratio of the present study

Location	Period	OC/EC
Anjora (Durg)	October 2014–February 2015	11.3
	March 2015–June 2015	7.7
Achhoti (Durg)	October 2014–February 2015	12.5
	March 2015–June 2015	10.09
Chandkhuri (Durg)	October 2014–February 2015	7.4
	March 2015–June 2015	6.25

combustion vehicles (Pipal et al. 2014), (2) 16.8–40.0 indicates wood combustion (Schauer et al. 2001), (3) 2.5–10.5 indicates residential coal smoke (Chen et al. 2006), (4) 7.7 indicates biomass burning (Zhang et al. 2007) and (5) 32.9–81.6 indicates kitchen emissions (He et al. 2004).

The OC/EC ratios in this study were found as follows (winter and summer, respectively) (Table 28.2):

Anjora; 11.3 and 7.72, Achhoti; 12.5 and 10.09, Chandkhuri; 7.4 and 6.25. Higher OC/EC ratio for winter compared to summer might be due to higher moisture content in burning fuels which result in longer episode of smouldering phases, known to produce higher OC (Chakrabarty et al. 2013; Matawle et al. 2014; Bano et al. 2018). Findings of OC/EC in the present study were compared to average OC/EC ratio of 6.9 which was reported due to biomass burning emission in Agra (Pachauri et al. 2013), 2.1 was reported for vehicular emission in Mumbai (Venkataraman et al. 2002), 8.9 and 7.2 were reported for biomass burning emission in Hisar and Allahabad, respectively (Ram et al. 2012), 2.4 was reported for fossil fuel burning in Pune (Safi et al. 2014).

Positive correlation between OC and EC at Anjora, Achhoti, and Chandkhuri attributed to similar source origin for both species (Table 28.3). Strong OC-EC correlations ($R^2 = 0.67$–0.74) at Anjora and Achhoti in winter indicates similar sources of emission. Moderate OC-EC correlation ($R^2 = 0.54$–0.50) in Chandrakhuri indicates

Table 28.3 Correlation of OC and EC in selected sampling site [M = slope (μg m^{-3}) and C = intercept (μg m^{-3})]

Sampling site	Winter			Summer		
	M	C	R^2	M	C	R^2
Anjora	10.24	12.59	0.75	4.16	30.72	0.52
Achhoti	8.9	29.48	0.67	4.23	36.37	0.33
Chandkhuri	3.54	65.22	0.50	3.07	32.31	0.54

that only a part of the OC in the house-indoors is generated in indoor while cooking and rest of OC and EC is from outdoor infiltration (Smith et al. 1996).

28.3.4 Estimation of Secondary Organic Carbon (SOC)

Organic carbon has sources of primary and secondary emission. Biomass and fossil fuel burning have emitted primary organic carbon (POC). SOC is not measured by direct method. The concentration of SOC is calculated by EC tracer method (Cabada et al. 2004). SOC concentration was found to be 104.8 ± 20.00 μg m^{-3} in winter and 49.10 ± 9.87 μg m^{-3} in summer at Anjora (AN), 96.00 ± 22.04 μg m^{-3} in winter and 43.70 ± 13.50 μg m^{-3} in summer at Achhoti (AC), 87.27 ± 23.94 μg m^{-3} in winter and 43.04 ± 6.64 μg m^{-3} in summer at Chandkhuri (CH). The % values of SOC of total OC are as follows (winter and summer, respectively): AN; 82.5% and 73.2%, AC; 83% in and 71.5%, CH; 70.6% and 67.4%. High concentration of SOC was reported in Anjora in both seasons as OC concentration was very high in this region (Fig. 28.5).

28.4 Conclusion

This study has demonstrated seasonal variations in concentration of $PM_{2.5}$, OC, and EC. Concentration of $PM_{2.5}$ was highest in Chandkhuri > Anjora > Achhoti and approximately 15 folds higher in winter and 11 folds higher in summer season than National Ambient Air Quality Standards (NAAQS) which is 40 μg m^{-3}. The contribution of TC, OC, and EC to $PM_{2.5}$ was estimated as follows: In AN; 26.4, 24.2, and 2.13% in winter, 23.9, 21.5, and 2.7% in summer, In AC; 23.5, 21.6, and 1.8% in winter and 18.07, 16.08, and 1.59% in summer, in CH; 23.12, 20.2, and 2.8% in winter and 17.29, 14.88, and 2.37% in summer. In all sampling sites, the observed concentrations of OC and EC were higher in winter season, as compared to summer. The OC/EC ratios in sampling sites were observed in the range of 6.25–12.5, which suggested the major contribution from biomass burning to indoor $PM_{2.5}$ and

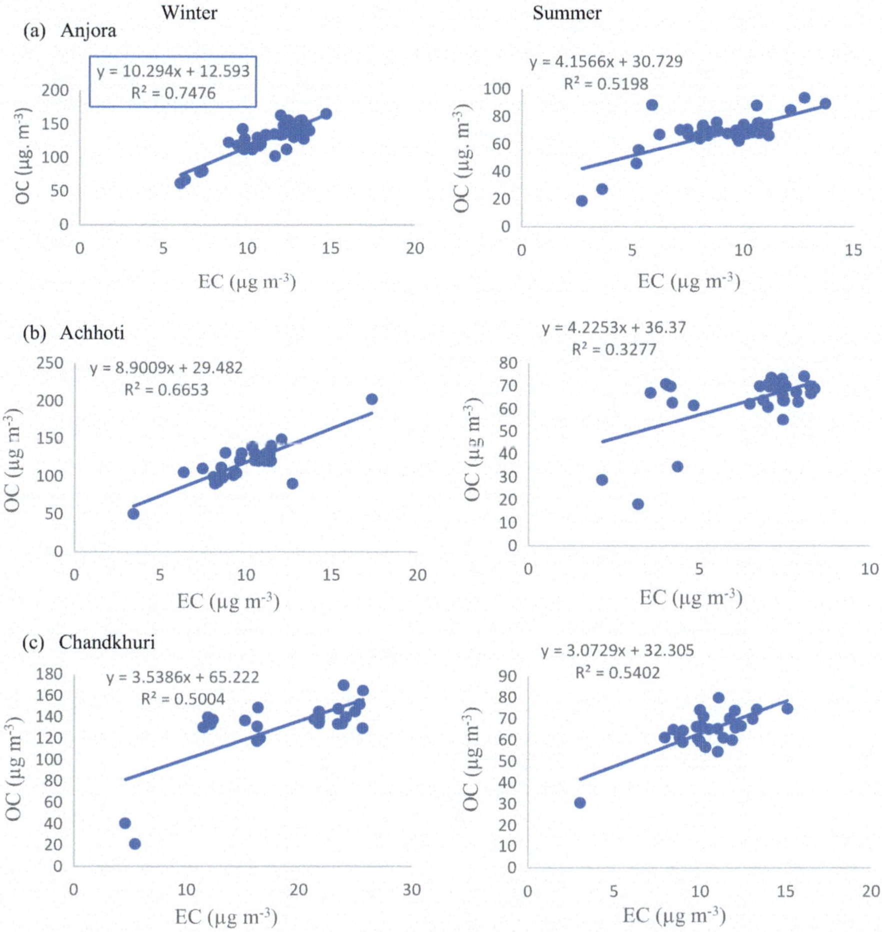

Fig. 28.5 Correlation scattered plot of OC with EC in selected sampling sites in winter and summer

carbonaceous fractions. SOC were considered to be formed by condensations of semi-volatile organic carbons (SVOCs) and atmospheric transformation of biogenic and aromatic hydrocarbons (Pipal and Satsangi 2015). Strong and moderate correlation between OC and EC indicated the source of organic carbon from primary emissions as well as secondary emissions, respectively, in sampling sites. The study of exposure level and the seasonal variability of fine particulates and carbonaceous fractions suggested the need to evoke awareness amongst rural inhabitants about the health implications of biomass burning. It is a need of time for an important and instant step in reduction of the emissions of carbonaceous matter in the rural household environment of Durg region of Chhattisgarh State.

Acknowledgements One of the authors (SK) thanks to the Department of Chemistry, Chhatrapati Shivaji Institute of Technology (CSIT), Durg and School of Studies in Chemistry, Pt. Ravishankar Shukla University, Raipur, for giving all necessary like library and laboratory facilities. Author is also very grateful to Indian Institute of Tropical Meteorology (IITM), New Delhi for extending its helping hand for analysis and equipment support.

References

Balakrishanan K, Ghosh S, Ganguli B, Sambandam S, Bruce N, Barnes DF (2013) State and National household concentrations of $PM_{2.5}$ from solid cook fuel use: Results from measurements and modelling in India for estimation of the global burden of disease. Environ Health 12:77

Bano S, Pervez S, Chow JC, Matawle JL, Watson JG, Sahu RK, Srivastava A, Tiwari S, Pervez YF, Deb MK (2018) Coarse particle ($PM_{10-2.5}$) source profile for emissions from domestic cooking and industrial process in Central India. Sci Total Environ 627:1137–1145

Birch ME, Cary RA (1996) Elemental carbon-based method for monitoring occupational exposures to particulate diesel exhaust. Aerosol Sci Technol 25:221–241

Cabada JC, Pandis SN, Subramanian R, Robinson AL, Polidori A, Turpin B (2004) Estimating the secondary organic aerosol contribution to $PM_{2.5}$ using the EC tracer method. Aerosol Sci. Tech. 38:140–155

Chakrabarty RK, Pervez S, Chow JC, Watson JG, Dewangan S, Robles J, Tian G (2013) Funeral pyres in south Asia: brown carbon aerosol emissions and climate impacts. Environ Sci Tech 44–48

Chen Y, Zhi G, Feng Y, Fu J, Feng J, Sheng G, Simoneit BRT (2006) Measurement of emission factors for primary carbonaceous particles from residential raw coal combustion in china. Geophys Res Lett 33. http://dx.doi.org/10.1029/2006GL026966

Chow JC, Watson JG (1998) Guideline on speculated particulate monitoring. Prepared for US Environmental Protection Agency, Research Triangle Park, NC , by Desert Research Institute, Reno, NV. http://epa.gov/ttn/amtic/files/ambient/pm25/spec/drispec.pdf.

Chow JC, Watson JG, Lu Z, Lowentha DH, Frazier CA, Solomon PA, Thuillier RH, Magliano KL (1996) Descriptive analysis of $PM_{2.5}$ and PM_{10} at regionally representative locations during SJVAQS/AUSPEX. Atmos Environ 30:2079–2112

Hansen J, Sato M, Ruedy R, Lacis A, Oinas V (2000) Global warming in the twenty-first century: an alternative scenario. Proc Natl Acad Sci USA 97:9875–9880

He Z, Kim YJ, Ogunjobi KO, Kim JE, Ryu SY (2004) Carbonaceous aerosols characteristics of $PM_{2.5}$ particles in North eastern Asia in summer 2002. Atmos Enov 38:1795–1800

Jacobson MZ (2001) Strong radiative heating due to the mixing state of blackcarbon in atmospheric aerosols. Nature 409:695–697

Lin JJ, Tai HS (2001) Concentrations and distributions of carbonaceous species in ambient particles in Kaohsiung City, Taiwan. Atmos Environ 35:2627–2636

Matawle J, Pervez S, Dewangan S, Tiwari S, Bisht DS, Pervez YF (2014) $PM_{2.5}$ chemical source profile of emissions resulting from industrial and domestic burning activities in India. Aerosol Air Qual Res 14:2051–2066

Matawle JL, Pervez S, Shrivastava A, Tiwari S, Pant P, Deb MK, Bisht DS, Pervez YF (2016) $PM_{2.5}$ pollution from household solid fuel burning practices in central India: 1. Impact on indoor air quality and associated health risks. Environmental Geochemistry and Health 39(5):1045–1058

Morawska L, Afshari A, Bae GN, Buonanno G, Chao CYH, Hänninen O, Hofmann W, Isaxon C, Jayaratne ER, Pasanen P, Salthammer T, Waring M, Weirzbicka A (2013) Indoor aerosols: from personal exposure to risk assessment. Indoor Air 23:462–487

Moschandreas DJ, Watson J, D'Aberton P, Scire J, Zhu T, Klein W, Saksana S (2002) Chapter three: methodology of exposure modelling. Chemosphere 49:923–946

NIOSH (1999) In manual of analytical methods. National Institute for Occupation Safety and Health, Cincinnati, OH

Pachauri T, Singla V, Satsangi A, Lakhani A, Kumari KM (2013) Characterization of carbonaceous aerosols with special reference to episodic events at Agra, India. Atmos Res 123:98–110

Pant P, Guttikunda SK, Peltier RE (2016) Exposure to particulate matter in India: a synthesis of findings and future directions. Environ Res 147:480–496

Pervez YF, Kushawaha SK, Nair S, Pervez P (2016) Status of atmospheric carbonaceous matter in various locations of India. Asian J Chem 28(6):1261–1266

Pipal AS, Satsangi PG (2015) Study of carbonaceous species, morphology and source of fine ($PM_{2.5}$) and coarse (PM_{10}) particles along with their climatic nature in India. Atmos Res 154:103–115

Pipal AS, Tiwari S, Satsangi PG, Taneja A, Bisht DS, Srivastava AK, Srivastava MK (2014) Sources and characteristics of carbonaceous aerosols at Agra "World heritage site" and Delhi "capital city of India, Environmental Science and Pollution Research 21(14):8678–8691

Ram K, Sarin MM, Hegde P (2008) Atmospheric abundances of primary and secondary carbonaceous species at two high—altitude sites in India: Sources and temporal variability. Atmos Environ 42:6785–6796

Ram K, Sarin MM, Sudheer AK, Rengarajan R (2012) Carbonaccous and sccondary aerosols during winter time fog and haze over urban sites in the Indo Gangetic Plain. Aerosol Air Qual Res 12:359–370

Safi PD, Raju MP, Rao PSP, Pandithurai G (2014) Characterization of carbonaceous aerosols over the urban tropical location and a new approach to evaluate their climatic importance. Atmos Environ 92:493–500

Satsangi PG, Yadav S, Pipal AS, Kumbhar N (2014) Characteristics of trace metals in fine ($PM_{2.5}$) and inhalable (PM_{10}) particles and health risks assessment along with in silico approach in indoor environment of India. Atmos Environ 92:384–393

Schauer JJ, Kleeman MJ, Cass GR, Simoneit BRT (2001) Measurement of emission from air pollution sources. 3. C1–C32 organic compound from fireplace combustion of wood. Environ Sci Tech 35:1716–1728

Smith DJT, Harrison RM, Luhana L, Pio CA, Castro LM, Tariq MN, Harat S, Quraishi T (1996) Concentration of particulate airbone polycyclic aromatic hydrocarbons and metal collected in Lahore Pakistan. Atmos Environ 30:4031–4040

Turpin BJ, Huntzicker JJ, Larson SM, Cass GR (1991) Las Angeles summer middy particulate carbon: primary and secondary aerosol. Environ Sci Tech 25:1788–1793

Venkataraman C, Reddy CK, Josson S, Reddy MS (2002) Aerosol size and chemical characteristics at Mumbai, India, during the INDOEXIFP (1999). Atmos Environ 36:1979–1991

Zeng T, Wang Y (2011) Nationwide summer peaks of OC/EC ratios in the contiguous united states. Atmos Environ 45:578–586

Zhang Y, Shao M, Zhang Y, Zeng L, He L, Zhu B, Wei B, Wei Y, Zhu X (2007) Source profile of particulate organic matters from cereal straw burnings. Environ Sci 19:167–175

Chapter 29
Assessment of Biomass Burning Emissions from India—A Comprehensive Study

N. Manojkumar and B. Srimuruganandam

29.1 Introduction

Biomass is a collective term given to the sectors of forest, agriculture crop residue, municipal solid waste, and residential firewood (Permadi and Kim Oanh 2013; Ozgen et al. 2014). Biomass sector accounts for 14% in the world's energy consumption next to coal, oil, and natural gas (Bai et al. 2013). Prediction on the future global energy supply shows that the contribution of biomass will reach 400 EJ yr^{-1} in 2050 (Berndes et al. 2003; Williams et al. 2012). In this scenario, annual biomass burned area between 1997 and 2011 is estimated as 348 million hectares (Mha) yr^{-1} (Giglio et al. 2013) and on account of small fires the burned area increased to 464 Mha yr^{-1} (Randerson et al. 2012). The term biomass burning (BB) refers to the burning of various biomass materials, viz. forest area, agricultural crop residues, grass, and solid waste. An average of 2 PgC yr^{-1} is emitted from BB during 1997--2009 with reasonable interannual variability. In BB, 44% of carbon emissions are from grasslands and savannas burning followed by fires in woodland (16%), forest (15%), peat (3%), and crop residue burning (3%) (Van Der Werf et al. 2010). Another recent study estimated CO_2 emissions from BB to be 9950 Tg yr^{-1} (Mieville et al. 2010). Burning is subject to high interannual variability (Koppmann et al. 2005; Van Der Werf et al. 2006) with different emissions of greenhouse gases (GHG), chemically active gases and volatile organic compounds (VOC) (Chan et al. 2003; Wang et al. 2014). BB contribution to the world annual emission of carbon monoxide (CO), oxides of nitrogen, and methane (CH_4) is about 25%, 18%, and 6%, respectively, (Watson et al. 2001). In addition to this, BB also emits toxic gases (polychlorinated

N. Manojkumar · B. Srimuruganandam (✉)
School of Civil Engineering, Vellore Institute of Technology (VIT), Vellore, TN, India
e-mail: bsrimuruganandam@vit.ac.in

N. Manojkumar
e-mail: manoj.ebet@gmail.com

S. M. Shiva Nagendra et al. (eds.), *Urban Air Quality Monitoring, Modelling and Human Exposure Assessment*, Springer Transactions in Civil and Environmental Engineering,
https://doi.org/10.1007/978-981-15-5511-4_29

dibenzo compounds) and halogenated compounds (methyl bromide and methyl chloride). Hydroxyl radical, a crucial in the atmosphere is found predominantly in the BB area (Crutzen and Andreae 1990; Akagi et al. 2011). All these BB emissions will modify the chemical composition of atmosphere and subsequently impact the air quality (Koppmann et al. 2005).

Apart from these gaseous emissions, BB produces both solid and liquid aerosol particles (Reid et al. 2005). Smoke from BB comprises numerous components like aqueous aerosols, inorganic carbonaceous soot and organic carbonaceous soot, viz. black carbon (BC) and organic carbon (OC) (Williams et al. 2012). The physical properties (size and number) and chemical constituents of the particulate matter may vary depending upon nucleation, coagulation, and adsorption/desorption processes (Calvo et al. 2013). BC has the property to absorb solar radiation in the atmosphere, and so became significant global warming pollutant next to CO_2 (Ramanathan and Carmichael 2008; Caserini et al. 2013). BC in atmosphere is a good indicator of anthropogenic air pollutants. OC has the property of weak absorbing or complete scattering of sunlight radiation, and it causes a cooling effect in the affected region (Bond et al. 1999; Bond 2001; Solomon 2007). Particulates from BB not only affect atmospheric radiative transfer but also indirectly modify the cloud properties through the formation of cloud condensation nuclei (Keywood et al. 2013). Thus, understanding the influence of BB emissions in global climate change requires a clear-cut knowledge of the interdependence of both anthropogenic and natural BB emissions and its influence in climate system along with their emission characteristics in both temporal and spatial scales. Knowing the importance of BB emissions and their impacts, this study aims to assess BB emissions caused by forest fires in India using satellite data and their potential impacts on global climate change.

29.2 Study Region

The region selected for the present study is Indian subcontinent. India has land area of 3,287,263 km^2 and is situated between 8°4$''$ and 37°6$''$ N and 68°7$''$ and 97°25$''$ E (Fig. 29.1). India has tropical monsoon climate and includes four seasons, viz. summer (March to May), monsoon (southwest: June to September; northeast: October and November), and winter (December to February). It experiences both hot and cold climates over various parts throughout the year. Southwest monsoon provides 80% of total precipitation in India. The Indian subcontinent has 2.4% of the world's land area that comprises 11% flora and 6.5% fauna of the world. India has ten biogeography zones and 26 biogeography provinces (given in Table 29.1), which are contributors to significant ecosystems in the world.

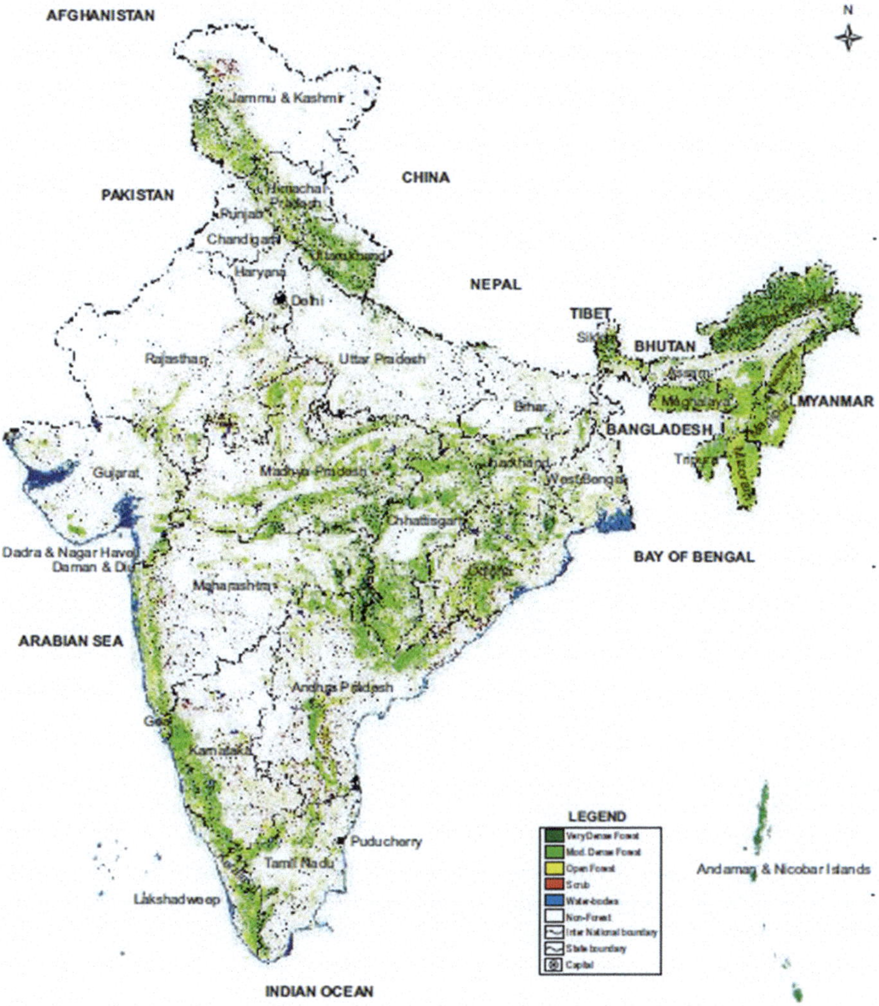

Fig. 29.1 Map showing the forest covers in India. *Source* Forest Survey of India (2012)

29.2.1 India's Forest Cover

India's total forest cover as of 2012 is 69.79 Mha, and it occupies 21.23% of the geographical area. India's forest cover is periodically estimated by Forest Survey of India (FSI), and it defines the forest cover having ≥10% tree canopy with at least one-hectare area. Figure 29.1 shows the detailed map of forest cover in India. Different classes of the Indian forests concerning canopy density are given in Table 29.2. A land which does not satisfy any of the classes comes under the non-forest category. In total forest cover, the scrub class is not included (Forest Survey of India 2012). In

Table 29.1 Indian biogeographic zones and provinces (Rodgers et al. 2000)

Biogeographic Zones	Biogeography provinces
Trans-Himalayan zone	Ladakh mountains and Tibetan plateau
Himalayan zone	Himalaya (four zonal: Northwest, West, Central, and East)
Desert zone	Thar and Katchchh
Semi-arid zone	Punjab plains and Gujarat
Western Ghats zone	Malabar plains and the Western Ghats mountains
Deccan plateau zone	Highlands (Central and Eastern), Chotta Nagpur, Central Plateau, and South Deccan
Gangetic plain zone	Upper Gangetic plain and Lower Gangetic plain
Northeast zone	Brahmaputra Valley and Northeast hills
Coastal Zone	West Coast, East Coast, and Lakshadweep
Islands	Andaman Nicobar

Table 29.2 Indian forest classes

Class	Canopy density percentage
Very dense forest	70% and above
Moderately dense forest	40% and above and less than 70%
Open forest	10% and above and less than 40%
Scrub	Less than 10%

Source Forest Survey of India (2012)

India, moderately dense forest account for 9.7% followed by open forest (8.99%), very dense forest (2.54%), and scrub (1.26%). The non-forest area is about 77.51% of the geographical area. In India, tropical forest types contribute to 80.69% of total forest area. In tropical type, tropical dry deciduous forest (41.87%) and tropical moist deciduous forest (19.73%) contribute a higher percentage. Various forest species and their percentages are presented in Table 29.3.

Table 29.3 Forest types and its percentage

Forest type	% Of forest area
Tropical forest	80.69
Subtropical forest	5.35
Himalayan forest	4.96
Alpine forest	2.55
Plantation	5.07
Montane wet temperate forest	0.69
Littoral and swamp forest	0.69
Total	100

Source Forest Survey of India (2012)

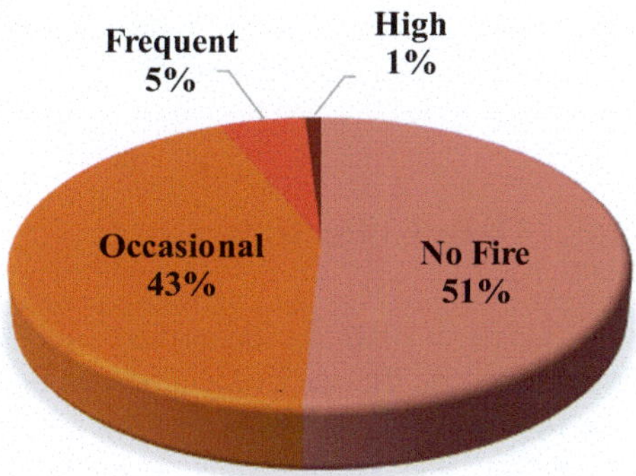

Fig. 29.2 Percentage of fire-prone areas in India. *Source* Forest Survey of India (2012)

29.2.2 Forest Fires in India

Three types of fire occur in the forest: ground fire, surface fire, and crown fire. Ground fire occurs below the litter surface with no flames. In a surface fire, most of the litter in the forest burn and in case of crown fire, whole part of trees are susceptible to burn. Crown fires are highly destructible when compared to surface fires. In India of any given year, about 55% of forest area is prone to fires. A forest fire can be very high, high, frequent, and occasional. Percentage of fire-prone areas in India is shown in Fig. 29.2.

In India, 90% of forest fires are human-induced. Accidental fires, shifting cultivation for agriculture, controlled burning, firewood burning, and agriculture residue burning are significant contributors to BB in India (Prasad et al. 2001).

29.2.3 Agriculture Residue Burning

After harvesting, the biomass left in the field is said to be agricultural residues in the form of straws, stalks, and leaves. These crop residues are utilized for feeding animals, roof thatching, fuel in residential cooking and industries. Despite its various useful applications, a massive amount of residues are left in the fields. In a recent estimate, a total of 620 Mt of the residue is generated during 2008--2009. Of 620 Mt, straw from rice and wheat contributed 62% followed by sugarcane residue (20%). Disposing of such quantities is a major challenge for farmers. Burning the crops is the cost-effective and quickest way to prepare the field for next plantation. In north India, field burning is most commonly practiced during the post-harvesting period. In the Indian scenario, crop residue burning emits 406.7 Tg yr^{-1} of air pollutants into the atmosphere. CO_2 and CO contributed 97.7% of total emissions. This higher contribution is because CO_2 and CO are the crucial products in residue burning. Harvest seasons and burning of crop residue vary within India, with peak emissions in April and October. In India, significant agriculture residue burning activities occur in the states of Punjab, Uttar Pradesh, Haryana, and Himachal Pradesh (Sahai 2007; Venkataraman et al. 2006; Jain 2014).

29.3 Data Collection

Emissions from BB are quantified using the following parameters: burned area, emission factor, the mass of available fuel for burning, and mass of fuel burnt in the area. For burned area, quantification several satellite products such as MCD45A1, MCD64A1, and Along-Track Scanning Radiometer (ATSR) are available. Each product locates the burned area based on an algorithm. MCD45A1 has bidirectional reflectance model approach whereas MCD64A1 and ATSR have hybrid algorithm using active fire hotspots and dynamic threshold. ATSR has a dataset with 1 km × 1 km spatial resolution. Apart from biomass burning, the ATSR data also include the fire events during volcanic activity, flares from oil and gas production, and crude oil production. MCD64A1 generates the product with 500 m resolution, but it has errors in detecting the small and fragmented burned area (Tsela et al. 2014; Fornacca et al. 2017; Sahul et al. 2015). Hence, in the present study, the burned area is estimated from MCD45A1, a Moderate Resolution Imaging Spectroradiometer (*MODIS*) dataset using *ArcGIS software*. *MODIS* imaging instrument is operating on both Terra and Aqua satellites. It images the surface of the earth in every 1–2 days. *MODIS* capture data with 36 spectral bands in different spatial resolution. Bands 1 and 2 have a spatial resolution of 250 m and bands from 3 to 7 have 500 m spatial resolution. Remaining bands have 1 km spatial resolution. Bands 1–19 bandwidth ranges from 620 to 965 nm and for bands 20–36 bandwidth ranges from 1.36 to 14.385 μm. Out of these bands, the 21st band having 3.929–3.989 μm is meant for forest fires and volcanoes. Likewise, the surface temperature is captured by 22, 23, 29, 31, and 32 bands. *MODIS* data are available in daily, 8 days, 16 days, 32 days, and monthly.

MCD45A1 used in this study is downloaded from the *MODIS5.1* collection. This collection contains monthly data on level 3 gridded 500 m product. For each month, two science datasets are available based on the burn date and confidence of detection. Burned dataset provides the approximate day of burning in the given year with codes for discriminating burned and unburned areas. A dataset with the confidence of detection provides code from 1 to 5 indicating from highest to least confidence. All data are downloaded in *GeoTIFF* subsets for the months from January 2013 to October 2014. *GeoTIFF* subsets cover the world in 24 windows and our study area India lies in 18th and 19th windows. In this study, the *GeoTIFF* subset with the confidence of detection for window 18 and 19 is downloaded.

Using *ArcGIS* version 9.3 burned areas are mapped and calculated with confidence ranging from 1 to 3. Remaining confidence levels 4 and 5 are not considered based on its low accuracy and instructions from the user guide. After mapping, burned area is estimated from pixels count using *ArcGIS*. BB estimation in this study mainly focuses on forest fire emissions. The possible reason is there is no sign of agriculture residue burning because we had taken the confidence interval from 1 to 3. Hence, the estimated burned area and emissions are of high confidence. The mapped forest burned area is shown in Figs. 29.3 and 29.4 for the year 2013 and 2014 separately. After mapping, burned area is estimated from pixels count using *ArcGIS*. About 1.03 and 0.8 Mha of forest area are burned from forest fires of India in the years 2013 and 2014, respectively.

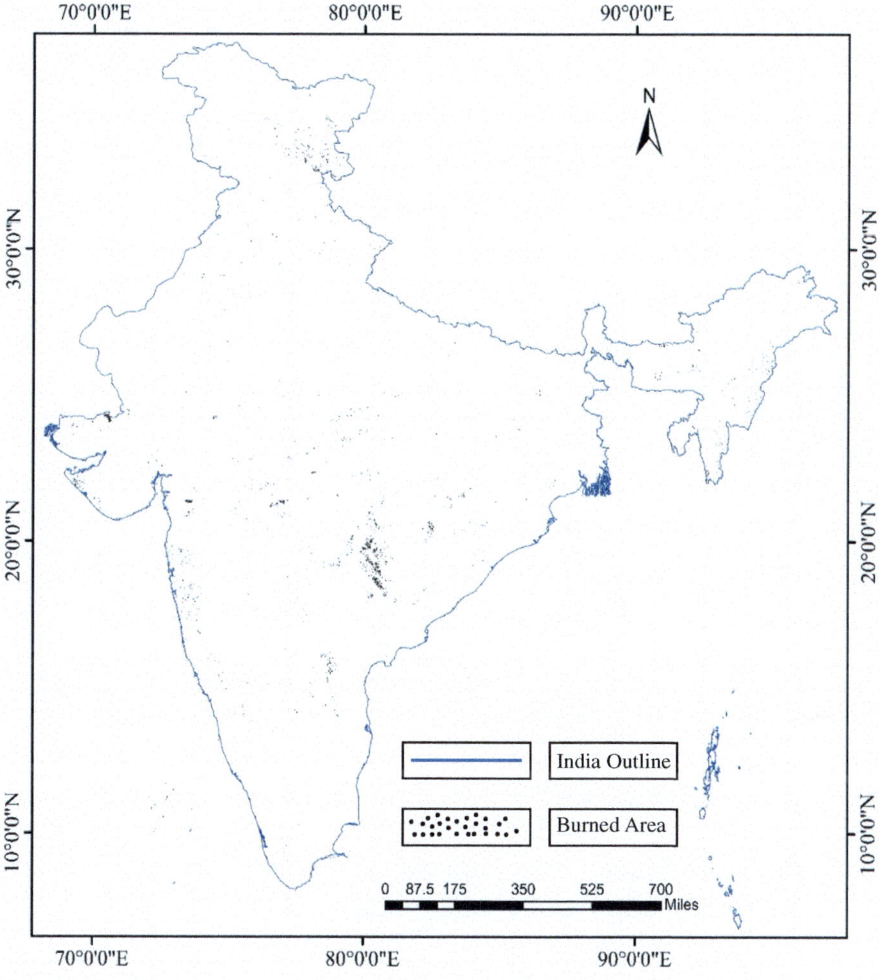

Fig. 29.3 Forest burned area map of India during the year 2013

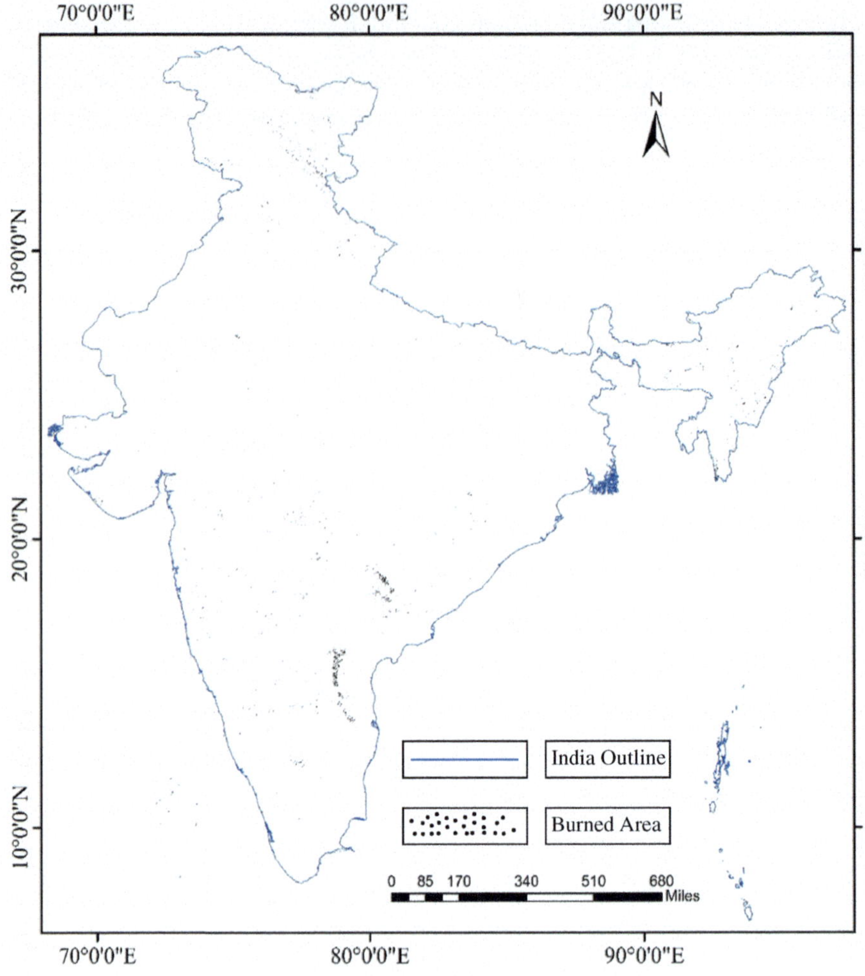

Fig. 29.4 Forest burned area map of India during the year 2014

29.4 Emission Factor, Mass of Fuel Available for Combustion and Combustion Factor

For quantifying emissions from a point, area or line source, emission factor (EF) is used. EF is a constant value that gives the amount of pollutant in interest released into the atmosphere when the associated activity takes place. EF is usually expressed by the ratio of the weight of pollutant to unit weight. For BB, the EF of various species is given by Akagi.et al. (2011) and Andreae and Merlet (2001). In BB, EF represents gaseous, or particulates or ions emitted per kg or ton of fuel burned. The EF for gases and particulates is a constant value, and it varied according to vegetation burned, viz. tropical, Savannah, grasslands, etc., and burning phases, i.e., flaming and combustion. It is expressed in g kg^{-1} or kg $tons^{-1}$. EF is calculated as per the following Eq. (29.1) (Andreae and Merlet 2001).

$$E.F \cong \frac{[x]}{\sum\left(\left[C_{CO_2} + C_{CO} + C_{CH_4} + C_{VOC} + C_{aerosol}\right]\right)}[C]_{biomass} \tag{29.1}$$

where $[x]$ = 'x' species concentration in the smoke, C_x = various carbon species and $[C]_{biomass}$ = concentration of carbon in the burned biomass. In this study, EF for tropical forest given in Akagi et al. (2011) and global fire emission database (GFED) are used. GFED estimates the emissions from BB at 0.25° spatial resolution globally.

The mass of fuel available for combustion in the burning area includes litter, biomass, and dead organic matter. The mass of fuel available for combustion is affected by fuel density and type of fire. Fuel density varies for each type of forest, its age, and available litter. For example, in tropical deciduous forest, litter fuel will increase during the winter season. When fire intensity is low, the fuel will be limited to dead organic matter and ground litter, and in the case of high fire intensity, the fuel will be litter, live, and dead organic matter. Combustion factor is the actual fuel combusted in the fire, and it varies with size and moisture content. The mass of fuel available for combustion and combustion factor is available in Chap. 2 of IPCC Guidelines (2006) (Venkataraman et al. 2006; Haripriya 2000; Chhabra et al. 2002). For the best estimate, values related to Indian forest are used in this study. The values used for the mass of fuel available for combustion and combustion factor are 9.48 kg m^{-2} and 0.46, respectively.

29.5 BB Emission Calculations

The total emission from BB is estimated using the IPCC recommendations (Guidelines 2006) as in Eq. (29.2).

$$E = A * B * C * \mathrm{EF} \tag{29.2}$$

where E = emission in grams; A = area burned in hectares; B = mass of fuel available for combustion in tons per hectare; C = combustion factor, and EF = emission factor in g per kg. The estimated emissions are tabulated and compared with previous findings in Table 29.4.

Table 29.4 Calculated emissions of gaseous, particulates, and volatile organic compounds

Compound	Previous estimates		Emission factors used (g kg^{-1})	This study estimate (Gg yr^{-1})	
	Crop residue emissions	Forest emissions		2013	2014
Particulates					
TPM	1280.61	–	13	592	497
$PM_{2.5}$	572–2393	279–924	9.1	414	348
TC	–	–	5.24	234	192
OC	211–970	188–559	4.71	214	180
BC	86–372	16–38	0.52	23.7	19.9
Gaseous compounds					
Carbon dioxide (Tg yr^{-1})	149–539	49–100	1643	74.8	62.8
Carbon monoxide (Tg yr^{-1})	10–74	3–7	93	4.23	3.55
Methane	313–1164	108–322	5.07	231	194
Hydrogen	–	–	3.36	153	128
Oxides of nitrogen	289–1290	103–255	2.55	116	97.4
Ammonia	151–560	38–100	1.33	60.5	50.8
Hydrogen cyanide	–	–	0.42	19.1	16
Sulfur dioxide	46–172	20–65	0.4	18.2	15.3
Nitrous oxide	–	–	0.2	9.1	7.64

Compound	Emission factors used (g kg^{-1})	This study estimate (Gg yr^{-1})	
		2013	2014
Volatile organic compounds			
Methanol	2.43	111	92.8
Formaldehyde	1.73	78.7	66.1
Acetic acid	1.55	70.5	59.2
Ethylene	1.06	48.5	40.7
Formic acid	0.79	36	30.2
Ethane	0.71	32.3	27.1
Propylene	0.64	29.1	24.4
Acetone	0.63	28.7	24.1
Acetylene	0.44	20	16.8
Benzene	0.39	17.8	14.9
Higher alkenes	0.277	12.6	10.6
Toluene	0.26	11.8	9.93

(continued)

Table 29.4 (continued)

Compound	Emission factors used (g kg^{-1})	This study estimate (Gg yr^{-1})	
		2013	2014
Terpene	0.15	6.83	5.73
Isoprene	0.13	5.92	4.97
Propane	0.126	5.73	4.81
Xylene	0.11	5.01	4.2
Higher alkanes	0.083	3.78	3.17
Ethanol	0.037	1.68	1.41
Dimethyl Sulfide	0.001	0.061	0.051

Note All emission estimates are in Gg yr^{-1} unless otherwise specified

29.6 Radiative Forcing Calculations

Impacts in the climate system by GHGs are represented by radiative forcing (RF). RF can be positive or negative. Positive RF indicates the tendency of increasing global temperature by trapping and reflecting radiation toward earth and vice versa.

Simplified expressions from Myhre et al. (1998) are used for calculating RF by CO_2, CH_4, and N_2O and are given in Eqs. (29.3)–(29.5).

CO_2 Forcing

$$\Delta F = \alpha . \ln \frac{c}{c0} \tag{29.3}$$

where ΔF is the RF by CO_2 gas; α is the constant (5.35); c is the current CO_2 concentration in 2013 is 0.009 ppmv and in 2014 is 0.08 ppmv; and c_0 is the pre-industrial CO_2 concentration. The RF values of CO_2 in 2013 and 2014 are found to be 1.88 W m^{-2} and 1.90 W m^{-2}, respectively.

CH_4 Forcing

$$\Delta F = \alpha\left(\left(\sqrt{M} - \sqrt{M_0}\right) - \left(f(M, N_0) - f(M_0 - N_0)\right)\right) \tag{29.4}$$

where ΔF is the *RF* by CH_4 gas; α is the constant (0.036); M is the present CH_4 concentration; M_0 is the pre-industrial CH_4 concentration; N is the present N_2O concentration in 2013 is 0.08 ppbv and in 2014 is 0.06 ppbv; N_0 is the pre-industrial N_2O concentration. The *RF* values of CH_4 in 2013 and 2014 are found to be 0.496 W m^{-2} and 0.499 W m^{-2}, respectively.

N_2O Forcing

$$\Delta F = \alpha\left(\left(\sqrt{N} - \sqrt{N_0}\right) - (f(M_0, N) - f(M_0 - N_0))\right) \quad (29.5)$$

where ΔF is RF by CH_4 gas; α is constant (0.12); M is present CH_4 concentration; M_0 is pre-industrial CH_4 concentration; N is present N_2O concentration in 2013 is 0.0018 ppbv and in 2014 is 0.0015 ppbv; N_0 is pre-industrial N_2O concentration. The RF values of N_2O in 2013 and 2014 are found to be 0.183 W m^{-2} and 0.187 W m^{-2}, respectively.

29.7 Conclusion

This study presents a comprehensive assessment of BB emissions and its influence on the global climate change regarding regional/continental scale from forest fires in India. The main findings are highlighted here.

- BB emissions are assessed using 500 m resolution MODIS dataset from GFED with a confidence level of 1–3 (forest area specific) for determining burned area which resulted in highly accurate estimations.
- High interannual variability in forest fires, burned area, and BB emissions are found between the years 2013 and 2014.
- Most of the forest fire is mapped in the central and northeastern states of India.
- Forest burned area is estimated as 1.03 ha and 0.8 ha in 2013 and 2014, respectively.
- CO_2 and CO are recorded with high emissions throughout the study period as main combustion products in BB followed by TPM, $PM_{2.5}$, TC, OC, and CH_4.
- RF of CO_2 (1.9 W m^{-2}), CH_4 (0.499 W m^{-2}) and N_2O (0.187 W m^{-2}) are found to be high in 2014 than the year 2013.

References

Akagi SK et al (2011) Emission factors for open and domestic biomass burning for use in atmospheric models. Atmos Chem Phys 11:4039–4072

Andreae MO, Merlet P (2001) Emission of trace gases and aerosols from biomass burning. Glob Biogeochem Cycles 15:955–966

Bai J, Sun X, Zhang C, Xu Y, Qi C (2013) The OH-initiated atmospheric reaction mechanism and kinetics for levoglucosan emitted in biomass burning. Chemosphere 93:2004–2010

Berndes G, Hoogwijk M, van den Broek R (2003) The contribution of biomass in the future global energy supply: a review of 17 studies. Biomass Bioenerg 25:1–28

Bond TC (2001) Spectral dependence of visible light absorption by carbonaceous particles emitted from coal combustion. Geophys Res Lett 28:4075–4078

Bond TC et al (1999) Light absorption by primary particle emissions from a lignite burning plant. Environ Sci Technol 33:3887–3891

Calvo AI et al (2013) Research on aerosol sources and chemical composition: past, current and emerging issues. Atmos Res 120–121:1–28

Caserini S et al (2013) A methodology for elemental and organic carbon emission inventory and results for Lombardy region, Italy. Sci Total Environ 450–451:22–30

Chan CY et al (2003) Characteristics of biomass burning emission sources, transport, and chemical speciation in enhanced springtime tropospheric ozone profile over Hong Kong. J Geophys Res 108:1–13

Chhabra A, Palria S, Dadhwal VK (2002) Growing stock-based forest biomass estimate for India. Biomass Bioenergy 22:187–194

Crutzen PJ, Andreae MO (1990) Biomass burning in the tropics: impact on atmospheric chemistry and biogeochemical cycles. Science 250:1669–1678

Forest Survey of India (2012) State forest reports (2005, 2009, 2011 & 2013)

Fornacca D, Ren G, Xiao W (2017) Performance of three MODIS fire products (MCD45A1, MCD64A1, MCD14ML), and ESA Fire_CCI in a mountainous area of Northwest Yunnan, China, characterized by frequent small fires. Remote Sens 9:1–20

Giglio L, Randerson JT, Van Der Werf GR (2013) Analysis of daily, monthly, and annual burned area using the fourth-generation global fire emissions database (GFED 4). J Geophys Res Biogeosci 118:317–328

IPCC Guidelines (2006) IPCC Guidelines for national greenhouse gas inventories, vol 4: agriculture, forestry and other land use, Chapter 2: generic methodologies applicable to multiple land-use categories, pp 1–59

Haripriya GS (2000) Estimates of biomass in Indian forests. Biomass Bioenergy 19:245–258

Jain N (2014) Emission of air pollutants from crop residue burning in India. Aerosol Air Qual Res 14:422–430

Keywood M et al (2013) Fire in the air: biomass burning impacts in a changing climate. Crit Rev Environ Sci Technol 43:40–83

Koppmann R, Czapiewski KV, Reid JS (2005) A review of biomass burning emissions, part I: gaseous emissions of carbon monoxide, methane, volatile organic compounds, and nitrogen containing compounds. Atmos Chem Phys Discuss 5:10455–10516

Mieville A et al (2010) Emissions of gases and particles from biomass burning during the 20th century using satellite data and an historical reconstruction. Atmos Environ 44:1469–1477

Myhre G, Highwood Eleanor J, Shine Keith P (1998) New estimates of radiative forcing due to well-mixed greenhouse gases. Geophys Res Lett 25:2715–2718

Ozgen S et al (2014) Emission factors from small scale appliances burning wood and pellets. Atmos Environ 94:144–153

Permadi DA, Kim Oanh NT (2013) Assessment of biomass open burning emissions in Indonesia and potential climate forcing impact. Atmos Environ 78:250–258

Prasad VK et al (2001) Biomass and combustion characteristics of secondary mixed deciduous forests in Eastern Ghats of India. Atmos Environ 35:3085–3095

Ramanathan V, Carmichael G (2008) Global and regional climate changes due to black carbon. Nat Geosci 1:221–227

Randerson JT, Chen Y, Van Der Werf GR, Rogers BM, Morton DC (2012) Global burned area and biomass burning emissions from small fires. J Geophys Res 117:G04012

Reid JS, Koppmann R, Eck TF, Eleuterio DP (2005) A review of biomass burning emissions part II : intensive physical properties of biomass burning particles. Atmos Chem Phys 5:799–825

Rodgers WA, Panwar HS, Mathur VB (2000) Wildlife protected area network in india: a review: executive Summary. Wildlife Institute of India

Sahai S et al (2007) A study for development of emission factors for trace gases and carbonaceous particulate species from in situ burning of wheat straw in agricultural fields in India. Atmos Environ 41:9173–9186

Sahul LK et al (2015) Regional biomass burning trends in India: analysis of satellite fire data Regional biomass burning trends in India. J Earth Syst Sci 91:1–25

Solomon S et al (2007) Climate change 2007: the physical science basis. In: Contribution of working group I to the fourth assessment report of the Intergovernmental Panel on Climate Change. Cambridge University Cambridge, UK and New York, USA

Tsela P et al (2014) Validation of the two standard MODIS satellite burned-area products and an empirically-derived merged product in South Africa. Remote Sens 6:1275–1293
Van Der Werf GR et al (2006) Interannual variability of global biomass burning emissions from 1997 to 2004. Atmos Chem Phys Discuss 6:3175–3226
Van Der Werf GR et al (2010) Global fire emissions and the contribution of deforestation, savanna, forest, agricultural, and peat fires (1997–2009). Atmos Chem Phys 10:11707–11735
Venkataraman C et al (2006) Emissions from open biomass burning in India: Integrating the inventory approach with high-resolution Moderate Resolution Imaging Spectroradiometer (MODIS) active-fire and land cover data. Glob Biogeochem Cycles 20:1–12
Wang H et al (2014) Source profiles of volatile organic compounds from biomass burning in yangtze river delta, China. Aerosol Air Qual Res 14:818–828
Watson RT et al (2001) Climate change 2001: synthesis report, contribution of working groups I, II, and III to the third assessment report of the Intergovernmental Panel on Climate Change. Cambridge University, Cambridge, UK and New York, USA
Williams A, Jones JM, Ma L, Pourkashanian M (2012) Pollutants from the combustion of solid biomass fuels. Prog Energy Combust Sci 38:113–137

Chapter 30
Correlation Assessment of Indoor Air Pollutants Emitted by Household Fuels and Its Health Impacts

B. S. Shilpa and K. S. Lokesh

30.1 Introduction and Literature

Present day air pollutants emitted due to combustion of solid fuels, in households of developing countries, are considered as a major source of health risks to the exposed populations. According to WHO report, biomass or coal burning for cooking and heating exposes about 2 billion people worldwide to high concentrations of particulate matter and gases (Smith et al. 2003). The concentrations of these pollutants are 10--20 times higher than health guidelines for typical urban outdoor concentrations (Balakrishnan et al. 2004; Khare et al. 2000; Saiyed et al. 2001). Usually, the stove is located at floor level, adding to the risk of accident and leading to unhygienic condition. In addition, these are often not fitted with a chimney to remove the pollutants. The children, aged people, and women are group that spends more time in indoors of such households, who are vulnerable to get affected by these pollutants (Colbeck et al. 2007; Kulshreshtha and Khare 2011).

India has the second highest per capita burden of respiratory illness in the world, which accounts for about 1/9th of National Burden of Disease (NBD) (Balakrishnan et al. 2004; Sussan et al. 2014). Approximately about 6% of the national burden of disease in terms of disability adjusted life years (DALYs) and includes 500,000 excess deaths annual, attributed to domestic fuel use (Balakrishnan et al. 2004; Jindal 2007; Kang et al. 2009; Mishra et al. 1999). In rural India, nearly 90% of primary energy used is accounted for biomass (wood, 56%; crop residues, 16%; dung, 21%)

B. S. Shilpa (✉)
Department of Civil Engineering, Vidyavardhaka College of Engineering, Mysuru, Karnataka 570002, India
e-mail: shilpa.bs@vvce.ac.in

K. S. Lokesh
Department of Environmental Engineering, Sri Jayachamarajendra College of Engineering, Mysuru, Karnataka 570006, India
e-mail: lokeshkaggere@yahoo.com

S. M. Shiva Nagendra et al. (eds.), *Urban Air Quality Monitoring, Modelling and Human Exposure Assessment*, Springer Transactions in Civil and Environmental Engineering,
https://doi.org/10.1007/978-981-15-5511-4_30

(TEDDY 1998; Venkataraman et al. 2010). According to many researchers, air pollutants in ambient environment act as surrogates for personal exposure. This exposure shows the contribution of PM to various health risks such as cardiovascular mortality, total mortality, lung cancer, and various other respiratory illnesses (Balakrishnan et al. 2004; Colbeck et al. 2007; Crist et al. 2008; Andersen et al. 2005). However, these researches show a weak association between personal and ambient concentrations, most likely because people spend most of their time indoors (Andersen et al. 2005).

From the preceding context, it is clear that indoor air pollution associated with household fuels use in India is a significant public health concern. As India has second highest per capita burden of respiratory illness in the world (Habre et al. 2014; Balakrishnan et al. 2004), where standards for IAQ are not available for Indian condition, the objective of present research is to assess the various domestic fuels emission in the indoors, morphological characteristics of RSPM, and to develop a relationship for associated health risks.

30.2 Methodology

In the present work, western part of Mysuru city was considered as the study area. Mysuru is the second-largest city in the state of Karnataka, India. Located at the base of the Chamundi hills about 146 km southwest of the state capital Bengaluru, it is spread across an area of 128.42 km^2 Mean Sea Level (altitude) of Mysuru city is 765 m, Latitude is 12.3024° N and Longitude is 76.6386° E. The sampling stations were considered in and around parts of Mysuru city for five different fuels, i.e., kerosene, biomass, cowdung cakes, coal, and LPG in different households as shown in Fig. 30.1.

Indoor and ambient sampling of PM were carried out simultaneously for a time period of 2 h before cooking, during cooking, and after cooking in indoor locations of different fuels and 8 h sampling for ambient condition, respectively. The equipments used for the indoor air sampling are Personal Air Sampler (APS2), Handy sampler (APM 821), and Gastec precision gas detecting system (GV 100S). While ambient air samples are collected using High Volume Air Sampler (APM 415), Respirable Dust Sampler (APM 451), and Gastec precision gas detecting system (GV 100S), respectively.

The air pollutants such as SPM, RSPM, NO_2, SO_2, and CO are monitored both in indoors and outdoors of study locations and analyzed using Standard analytical techniques. RSPM is subjected to morphological analysis using Scanning Electron Microscope.

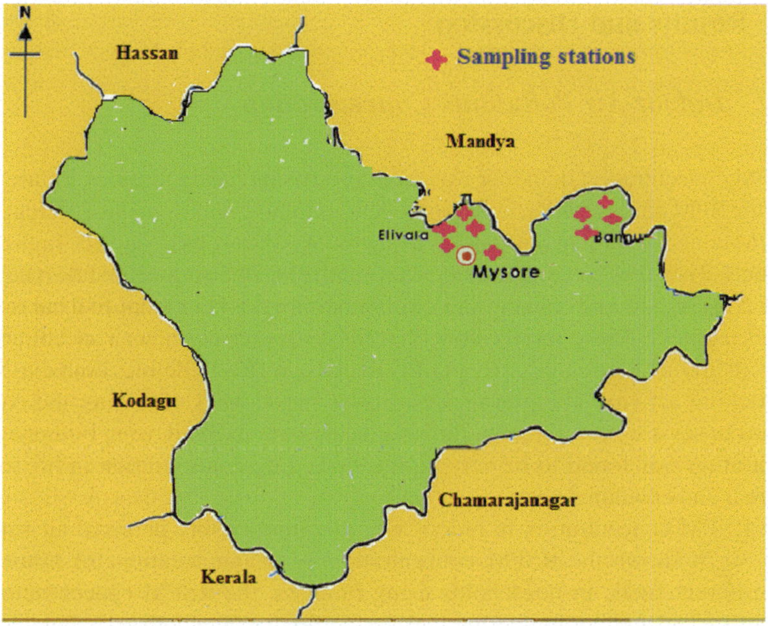

Fig. 30.1 Location of sampling stations

30.2.1 Scanning Electron Microscopy

To characterize the morphology of RSPM, filter paper was examined using Scanning Electron Microscopy (LEO 435-VP, England, UK) at Central Laboratory Facility, CFTRI, Mysore. Filter materials were mounted on metallic stub which is gold coated (~100 Å) with sputter coater (Polaron Sputter Coat System, Model 5001, England) and viewed under SEM 435 VP (Leo 40 Electron Microscopy Ltd. Cambridge, UK) at 10 kV. Electronic beam scans the sample; electrons are emitted from the surface during scanning. The brightness of the image on the monitor depends on the number of electrons emitted from the sample. These emitted electrons were recorded by a detector, which enable to produce an element contrast picture due to high sensitivity of electrons.

The monitored data has been subjected to statistical analysis using SPSS version 16. The statistical procedure is carried for factor analysis to determine relation with household air pollutants, fuel, and its health impacts.

30.3 Results and Discussion

30.3.1 Indoor Air Pollutants Concentration

The SPM concentration in indoor was monitored using handy sampler. Figure 30.2a shows the SPM concentration in indoor locations of households using different fuels. The SPM concentration in households using biomass was observed to be higher than other fuels, 2 h before cooking activity was found to be 74.56 μg m^{-3} while it has risen to 1438.5 μg m^{-3} during cooking and 2 h after burning has been stopped it has reduced to 605.7 μg m^{-3}. These results show that SPM prevails at higher concentration in indoors of all fuel households except LPG than the WHO guidelines both during and after cooking activity. The prevalence of these particulates even after the cooking activities poses a serious threat to the occupants. In household using biomass, SPM concentration was found to be very high followed by concentration in households using coal and cowdung cakes.

The RSPM concentration in indoor was monitored using personal air sampler. Figure 30.2b shows the RSPM concentration in indoor locations of households using different fuels. In households using biomass, the RSPM concentration 2 h before cooking activity was found to be 67.06 μg m^{-3} while it has risen to 1484.9 μg m^{-3} during cooking, and 2 h after burning has been stopped, it has reduced to 486.76 μg m^{-3}. These results show that RSPM prevails at higher concentration in indoors of all fuel households than the WHO guidelines. Biomass fuel emits higher concentration RSPM when compared to coal and cowdung cakes.

The NO_2 concentration in indoor was monitored using handy sampler by modified Jacob method. Figure 30.3a shows the NO_2 concentration in indoor locations of households using different fuels. Highest NO_2 concentration in household using coal was observed than other fuels households; 2 h before cooking activity is found

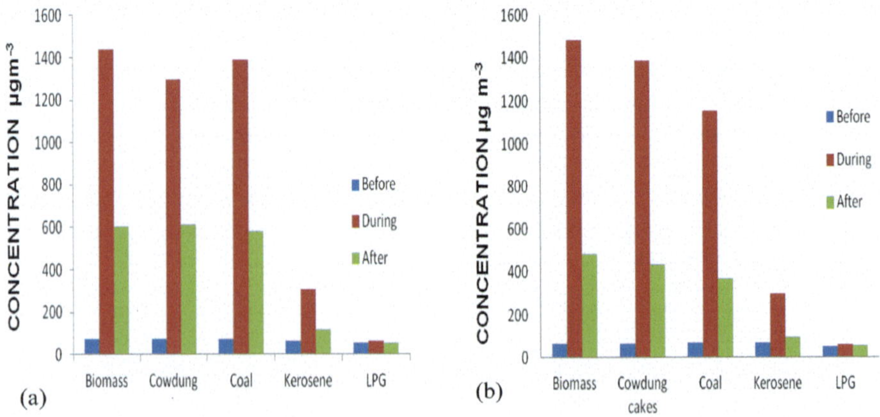

Fig. 30.2 **a** SPM, **b** RSPM concentration in indoors at various time interval for different fuels

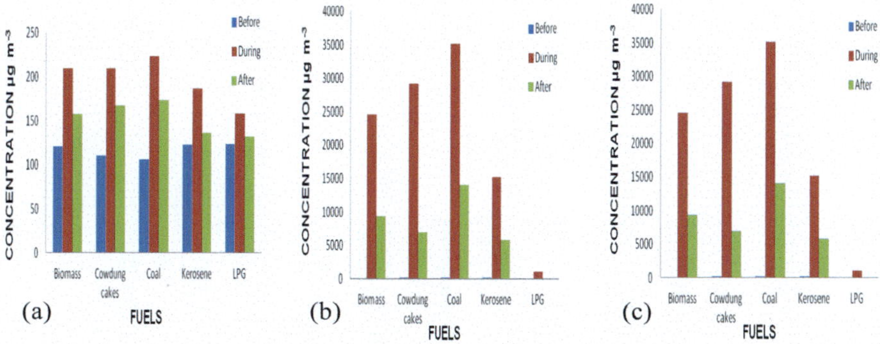

Fig. 30.3 Concentration **a** NO_2, **b** SO_2, **c** CO in indoors at various time interval for different fuels

to be 106 µg m^{-3} while it has risen to 223.45 µg m^{-3} during cooking, and 2 h after dousing of fire, it has reduced to 173.74 µg m^{-3}. These results show that NO_2 prevails at higher concentration in indoors of all fuel households except LPG than the WHO Guidelines both during and after cooking activity. In coal household, NO_2 was found to be in higher concentration than in households using biomass and cowdung cakes.

The SO_2 concentration in indoor was monitored using handy sampler by Improved West and Gaeke method. Figure 30.3b shows the SO_2 concentration in indoor locations of households using different fuels. In household using coal, the SO_2 concentration was observed to be highest, 2 h before cooking activity is found to be 80.51 µg m^{-3} while it has risen to 169.37 µg m^{-3} during cooking and 2 h after burning has been stopped, it has reduced to 112.62 µg m^{-3}. These results show that SO_2 prevails at higher concentration in indoors of all fuel households than the WHO guidelines both during and after cooking activity. SO_2 concentration was found to be higher in households using kerosene, biomass, cowdung cakes, and LPG.

The CO concentration in indoor was monitored using Gastec precision gas detector system. Figure 30.3c shows the CO concentration in indoor locations of households using different fuels. In household using coal, the CO concentration 2 h before cooking activity was found to be 234 µg m^{-3} while it has risen to 35,100 µg m^{-3} during cooking and 2 h after burning has been stopped, it has reduced to 14,040 µg m^{-3}.These results show that CO prevails at slightly at high concentration (100 µg m^{-3}) as per WHO guidelines in indoors of all fuel households except for LPG fuel. High concentration is observed in household using coal followed by cowdung cakes and biomass.

30.3.2 Ambient Pollutants Concentration

Figure 30.4 shows the pollutants ambient concentration at households using different fuels. At biomass household, Ambient NO_2 of 115.32 µg m^{-3}, CO concentration of

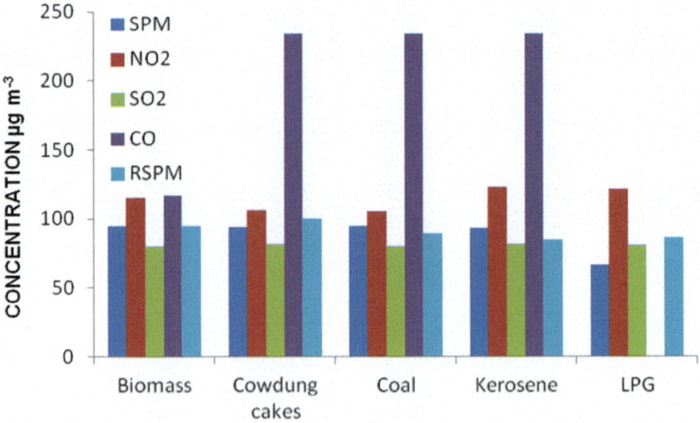

Fig. 30.4 Ambient concentration of pollutants at different fuel households

117 µg m^{-3}, and RSPM concentration of 94.85 µg m^{-3} were observed to be high. At cowdung cakes household, coal household and kerosene, Ambient CO concentration of 234 µg m^{-3} was observed to be high in concentration than other pollutants. At LPG household, Ambient NO_2 of 121.56 µg m^{-3} was observed to be high than other IAP. As most of these sampling stations are located in the vicinity of traffic junctions, CO and NO_2 are found to be in concentration.

30.3.3 *Morphological Characterization of RSPM*

Morphological study reveals that particles are mainly from anthropogenic activities (e.g., combustion). This activity releases ultra-fine particulates (UFPs) within the PM_{10} metric are considered to be 'nanoparticles' (NPs; particles less than 100 nm in at least one dimension) but the anthropogenic NPs are further defined as having an equivalent spherical diameter of less than 100 nm. Figure 30.5 shows the comparison of control filter paper where glass fibers are clear, with filter paper having RSPM emitted by biomass, coal, cowdung cakes, and kerosene attached to glass fibers.

From these SEM images, it is observed that particles have a tendency to form flocculated aggregates, referred to as '**spherules**' less than 100 nm in equivalent spherical diameter (ESD), called as nano-sized PM, where as individual particles or '**spherulites**,' are less than 10 nm in ESD. As there is greater pulmonary deposition efficiency of NPs, with their large surface area (SA) leads to cardiopulmonary toxicity (Berbe et al. 2008). These PM may then be able to cross the air barrier into the respiratory tract, the accumulation tends to reduce the epithelial cells activity, in turn leads to health effects such as inflammation, cardiac disruption, and even may lead to death.

Typical images generated with SEM shown in Figs. 30.6 and 30.8 reveals that high

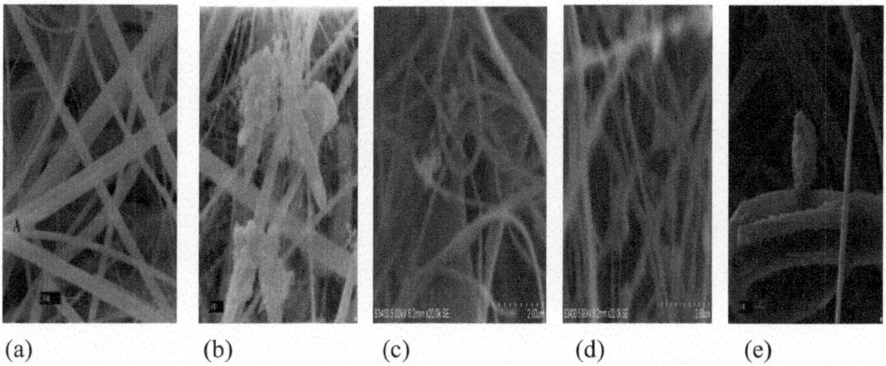

Fig. 30.5 Comparison of control filter paper (A) and filter paper having RSPM emitted during cooking from different household fuels. **a** Control, **b** biomass, **c** coal, **d** cowdung cakes, **e** kerosene

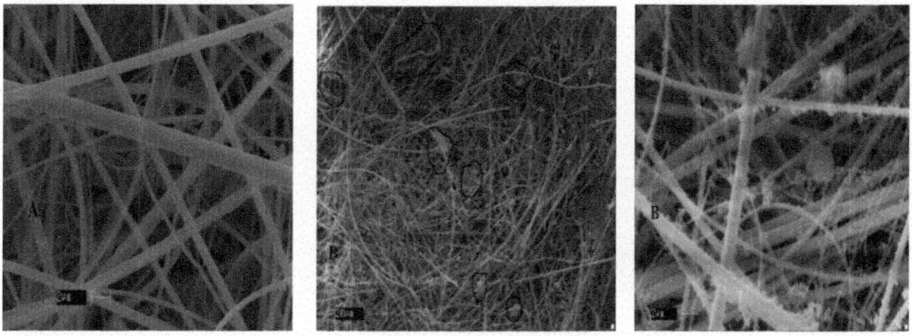

Fig. 30.6 SEM images at various magnification showing spherules emitted by biomass during cooking

RSPM load is observed with biomass than cowdung cakes used household. It also reveals that the concentration of particulates smaller than 1 μm is more. Figure 30.7 shows the morphology of RSPM, and particulates emitted by biomass are irregular shaped when compared with other fuels generated. Figures 30.9 and 30.10 show the morphology of RSPM emitted by coal and kerosene. RSPM emitted by kerosene is spherical shaped and it has higher concentration of spherules.

30.3.4 Indoor to Outdoor Ratio of Pollutants

Indoor/outdoor (I/O) ratio describes the significance of emission contributions by indoor sources compared to outdoor sources. The indoor sources contribution will be significant if the ratio value is greater than one, while ratio value is lower than one, it indicates that ventilation of that indoor environment is good and also indoor

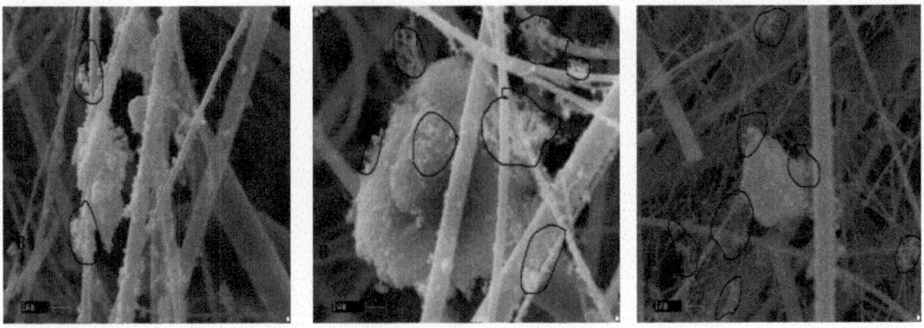

Fig. 30.7 SEM images reveal about concentration of RSPM and morphology of particles emitted by biomass

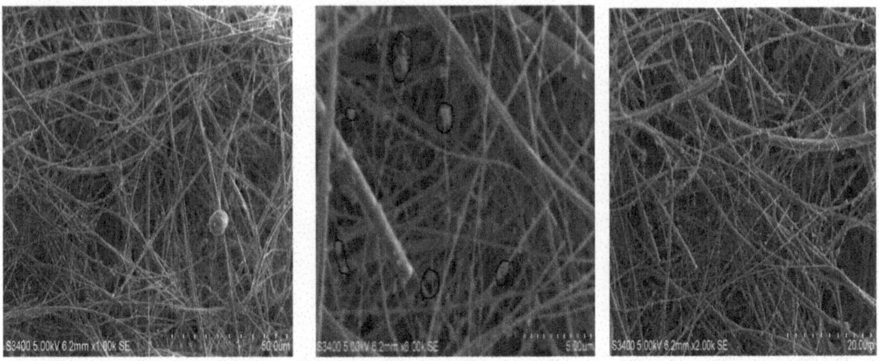

Fig. 30.8 SEM images of spherules emitted by cowdung cakes

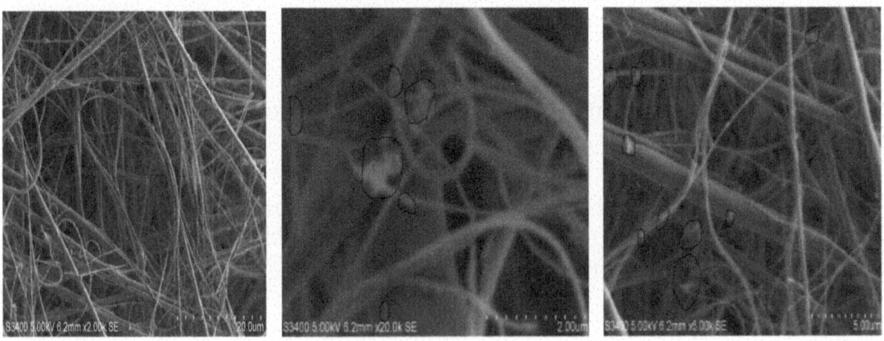

Fig. 30.9 SEM images of spherules emitted by coal

Fig. 30.10 SEM images at various magnification showing spherules emitted from kerosene during cooking

emission sources are weak. The following sections discuss about indoor/outdoor ratio for pollutants considered in households using different fuels.

- Indoor/outdoor ratio of SPM in different households

The I/O ratio of SPM was found to be very high during cooking activity in all households except LPG used household. Figure 30.11a shows the I/O ratio for SPM in different households. I/O ratio for SPM has value higher than 1 (1.26–15.16) except for LPG where observed ratio is below 1. A significant observation has been made with respect to I/O ratio after two hours of dousing the fire. An average of over 65% reduction has been recorded for fuels like biomass, coal, cowdung cakes, and kerosene. This indicates that either SPMs are getting settled or they might have got escaped from within the household.

- Indoor/outdoor Ratio of Respirable Suspended Particulate Matter

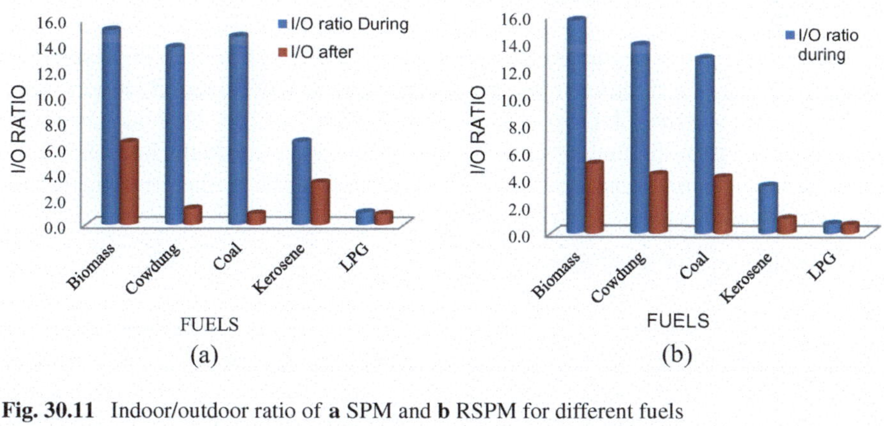

Fig. 30.11 Indoor/outdoor ratio of **a** SPM and **b** RSPM for different fuels

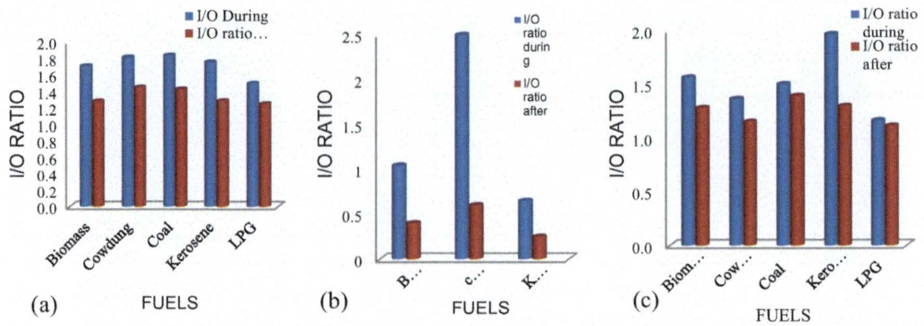

Fig. 30.12 Indoor/outdoor ratio of **a** NO_2, **b** SO_2 and **c** CO for different fuels

With respect to RSPM, I/O ratio has shown higher than 1 (1.1–15.7) except for LPG where observed ratio is below 1. A significant observation has been made w.r.t I/O ratio after two hours of using the fuel. An average of over 30% reduction has been recorded for fuels like biomass, coal, cowdung, and kerosene. This indicates that either RSPM is getting settled or escape from within households. The I/O ratio of RSPM was found to be more during cooking activity in all households as shown in Fig. 30.11b.

- Indoor/outdoor Ratio of Nitrogen Dioxide

After careful evaluation and analysis of obtained ratio, it has been observed that there is not much of a significant increase in I/O ratio. The observed I/O ratio values of NO_2 are ranging from 1.25 to 1.84 as shown in Fig. 30.12a. This indicates that ratio is just little over 1, justifying the presence of NO_2 within and outside the households in almost equivalent concentrations. However, this needs to be further investigated to find out the exact reason.

- Indoor/outdoor Ratio of Sulfur Dioxide

The I/O ratio of SO_2 is evaluated as shown in Fig. 30.12. It has been observed that there is not much of a significant increase in I/O ratio. The observed I/O ratio values of NO_2 are ranging from 1.12 to 1.97 as shown in Fig. 30.12b. This indicates that ratio is over 1, justifying the presence of SO_2 within and outside the households in almost same concentrations. However, this also needs to be further investigated to find out the exact reason.

- Indoor/outdoor Ratio of Carbon Monoxide

The monitored values of CO have shown ratio of over 2.5 in the households using cowdung cakes, while other fuels using households have shown less ratio (<1) as shown in Fig. 30.12c. CO concentrations contributed by cowdung cakes seem to be significant while CO emissions are not at all significant in the households sing other types of fuels. These may lead to severe impact on occupant's health. The reason for higher ratio of CO might be improper burning of cowdung cakes. However, before cooking, CO-monitored values were found to be BDL indicating the absence of CO in the ambient air.

The reviewed publications have not revealed such ratios particularly for different households for various fuels. However, there are few reports publishing the results of penetration of traffic-originated pollution into indoors. It is observed from the results reported by Massey et al. (2009) that the $PM_{2.5}$ in rural area I/O ratio was found to be 1.15 in households using wood and kerosene, nearly one in cowdung-cake used household and less than one in LPG used household. The available literature reports only for particulates, while for gaseous pollutants, the literature is either very scanty or no literature.

From these studies, the I/O ratio for biomass burning households was found to be significantly high (>15) followed by cowdung indicating the higher concentration of indoor pollution. This would lead to severe health implications on the inmates of households.

30.3.5 Assessment of Health Impacts Through Factor Analysis

A varimax-rotated factor analysis is carried out for indoor data sets to identify the possible pollutants emitted by household fuels causing health implications. Factor analysis was carried out using varimax with Kaiser normalization as rotation method. To limit the number of extracted factors, the Eigen threshold value is set to 1.0. Factor loadings more than 0.3 are considered to be significant. Information on health implications caused by pollutants are collected through literature. Tables 30.1, 30.2, 30.3, 30.4, and 30.5 show the results of factor analysis (FA) for different household

Table 30.1 Factor analysis for indoor air quality data---Fuel: biomass

Biomass		*F*1	*F*2
Rotated component score	RSPM	0.774	0.663
	CO	0.722	0.692
	SPM	0.719	0.695
	SO_2	0.635	0.773
	NO_2	0.683	0.729
Eigen value		4.977	0.021
% Of variance		99.546	0.424
Cumulative (%)		99.546	99.970
Possible pollutant responsible for health impairment		RSPM	NO_2
Possible health impairment		COPD, exacerbation of asthma	ARI

Table 30.2 Factor analysis for indoor air quality data---Fuel: coal

Coal		$F1$	$F2$
Rotated component score	RSPM	0.784	0.585
	SPM	0.721	0.660
	CO	0.713	0.668
	SO_2	0.694	0.641
	NO_2	0.593	0.776
Eigen value		4.951	0.038
% Of variance		99.016	0.755
Cumulative (%)		99.016	99.771
Possible pollutant responsible for health impairment		RSPM	NO_2
Possible health implications		Asthma COPD	ARI

Table 30.3 Factor analysis for indoor air quality data---Fuel: cowdung cakes

Cowdung cakes		$F1$	$F2$
Rotated component score	CO	0.807	0.591
	RSPM	0.788	0.615
	SO_2	0.719	0.694
	SPM	0.692	0.722
	NO_2	0.587	0.810
Eigen value		4.938	0.061
% Of variance		98.761	1.225
Cumulative (%)		98.761	99.986
Possible pollutant responsible for health impairment		CO	NO_2
Possible health implications		Cardiovascular diseases	ARI

fuels and its health implications.

Table 30.1 shows the varimax-rotated FA results for household using biomass with possible health impairment. Two factors were extracted from indoor data which explained about the total variance, the first factor accounting for 99.546% and the second factor 0.424%. First factor shows almost equal loading for all pollutants with range from 0.774 to 0.683, with high loading for RSPM, followed by CO and least loading for NO_2. The exposure for RSPM will definitely affect the lung tissues leading to malfunctioning of lungs. It can be argued that $F1$ probably represents a chronic obstructive pulmonary diseases and asthmatic problems. Second factor shows highest loading for SO_2 (0.773) followed by NO_2 (0.729) and least loading for RSPM (0.663). This indicates second component mainly leads to lung infections as these pollutants affect the lung tissues.

Table 30.4 Factor analysis for indoor air quality data---Fuel: kerosene

Kerosene		*F*1	*F*2
Rotated component score	SPM	0.789	0.614
	NO_2	0.752	0.659
	SO_2	0.709	0.705
	CO	0.614	0.789
	RSPM	0.658	0.753
Eigen value		4.958	0.039
% Of variance		99.159	0.788
Cumulative (%)		99.159	99.947
Possible pollutant responsible for health impairment		SPM	CO
Possible health implications		Exacerbation of asthma	ARI

Table 30.5 Factor analysis for indoor air quality data---Fuel: LPG

LPG		*F*1	*F*2
Rotated component score	NO_2	0.999	−0.005
	SO_2	0.991	−0.005
	CO	0.983	0.101
	RSPM	−0.246	0.969
	SPM	0.529	0.848
Eigen value		3.343	1.615
% Of variance		66.857	32.303
Cumulative (%)		66.857	99.159
Possible pollutant responsible for health impairment		NO_2	RSPM
Possible health implications		Cardiovascular diseases, COPD	ARI

Table 30.2 shows the varimax-rotated FA results for household using coal as fuel with possible health implications. Two factors were considered from indoor data which explained about the total variance, the first factor accounting for 99.016% and the second factor 0.755%. First factor shows almost equal loading for all pollutants ranging from 0.784 to 0.593, with high loading for RSPM, followed by CO and least loading for NO_2. The exposure for RSPM will definitely affect the lung tissues leading to malfunctioning of lungs. It can be argued that first component probably represents a chronic obstructive pulmonary diseases and asthmatic problems. Second factor shows highest loading for NO_2 (0.776) followed by CO (0.668) and least loading for RSPM (0.585). This indicates that second component mainly leads to lung infections as these pollutants affect the lung tissues.

The varimax-rotated FA results for household using cowdung cakes as fuel with possible health impairment are shown in Table 30.3. Two factors were extracted from indoor data which explained about the total variance, the first factor accounting for 98.761% and the second factor 1.225%. First factor shows almost equal loading for all pollutants ranging from 0.807 to 0.587, with high loading for CO (0.807), followed by RSPM (0.788) and least loading for NO_2 (0.587). The exposure for CO will definitely affect the heart tissues leading to ischemic heart diseases. It can be argued that first component probably represents cardiovascular diseases. Second factor shows highest loading for NO_2 (0.810) followed by SPM (0.722) and least loading for CO (0.591). This shows that second components mainly lead to lung infections as these pollutants affects the lung tissues causing ARI.

Table 30.4 shows the varimax-rotated FA results for household using kerosene as fuel with possible health impairment. Two factors were extracted from indoor data which explained about the total variance, the first factor accounting for 99.159% and the second factor 0.788%. First factor shows almost equal loading for all pollutants ranging from 0.789 to 0.614, with high loading for SPM (0.789) followed by NO_2 (0.752) and least loading for CO (0.614). The exposure for SPM will affect the tissues of respiratory. It can be argued that first component probably represents COPD and asthma. Second factor shows highest loading for CO (0.789) followed by RSPM (0.753) and least loading for SPM (0.614). This shows that second components mainly contribute toward lung infections as these pollutants affect the lung tissues causing ARI.

Table 30.5 shows the varimax-rotated FA results for LPG household with possible health impairment. Two factors were extracted from indoor data which explained about the total variance, the first factor accounting for 66.857% and the second factor 32.303%. First factor shows high loading for NO_2, SO_2, CO, and RSPM when compared to other pollutants. The exposure for all the above pollutants will definitely affect the lung and heart tissues leading to malfunctioning of lungs and ischemic heart diseases. It can be argued that $F1$ probably represents a chronic obstructive pulmonary diseases and cardiovascular diseases. Second factor shows highest loading for RSPM (0.969) followed by SPM (0.848) and least loading for CO (0.101). This indicates $F2$ mainly leads to breathing problems or ARI due to high loading of RSPM, SPM and CO which affects the lung tissues.

30.4 Conclusions

From this work, conclusions drawn are as,

- The average NO_2 indoor concentrations during cooking were found to be high in households where coal is used as fuel whereas minimum concentration was found to be in households using LPG. SO_2 was found to be high in coal-used households while low concentration was recorded for LPG. SPM and RSPM were found to be high in households using biomass. High CO concentration was found during cooking period in coal-used households. Ambient air quality parameters were within NAAQ Standards except for SPM and RSPM concentrations.

- The emitted PM from biomass fuel and kerosene fuel has been subjected to morphological study. Morphological studies have been conducted using Scanning Electron Microscopy. From our investigation it can be concluded that, the concentration of RSPM having aerodynamic size less than 1 μm can be spherulites. These spherulites were more in concentration during combustion of biomass when compared to other fuels. The results are indicated the presence of spherules. These spherules will have severe health impact on people.
- Higher Indoor/outdoor ratio for SPM, NO_2 and RSPM was found (value of 15.160, 1.710 and 15.655, respectively) for biomass fuel. Kerosene has shown higher value of I/O ratio for SO_2 (value of 1.968), while CO has recorded an I/O ratio of 2.5 in cowdung cakes used households. From this research results, out of all the fuels studied, cowdung has shown higher concetrations of emissions while LPG has recorded least emissions.
- Factor analysis results have confirmed that the RSPM emitted from biomass and coal has severe impacts on occupants reporting COPD, asthma, and exacerbation of asthma. While, CO is dominating in causing severe health impacts in the households where cowdung cakes were used. It has the potential of causing cardiovascular diseases.

Acknowledgements Authors extend heartfelt thanks to the occupants of the households for their kind cooperation. Authors acknowledge Mr. Srinivasa Reddy, Proprietor, M/s SHE Techno Laboratories, Bengaluru, and Prof. Ananthu K. M., Assoc. Professor, Department of Civil Engineering, PESCE, Mandya, for providing indoor air monitoring equipments.

References

Andersen PR, Ramachandran G, Pai P, Maynard A (2005) Women's personal and indoor exposures to $PM_{2.5}$ in Mysore, India: Impact of domestic fuel usage. Atmos Environ 39:5500–5508

Balakrishnan K, Mehta S, Kumar P, Ramaswamy P, Sambandam S, Kumar KS, Smith KR (2004) Indoor air pollution associated with household fuel use in India an exposure assessment and modeling exercise in rural districts of Andhra Pradesh, India. The World Bank project report

Berube K, Jones T, Jones C (2008) Environmental particles: a breath of fresh air? An article published in "INFOCUS", Issue 9, pp 7–18

Colbeck I, Nazir ZA, Hasnain S, Sultan S (2008) Indoor air quality at rural and urban sites in Pakistan. J Water Air Soil Pollut 8:61–69

Crist KC, Liu B, Kim M, Deshpande SR, John K (2008) Characterization of fine particulate matter: indoor, outdoor and personal exposures. Environ Res 106:62–71

Ezzathi M (2005) Indoor air pollution and Health in developing countries. The Lancet 366:104–106

Habre R, Moshier E, Castro W, Nath A, Grunin A, Rohr A, Godbold J, Schachter N, Kattan M, Coull B, Koutrakis P (2014) Effects of $PM_{2.5}$ and its components from indoor and outdoor sources on cough and wheeze symptoms in asthmatic children. J Expos Sci Environ Epidemiol 24:380–387

Jindal SK (2007) A health Perspective of indoor air pollution. Ed Ind J Med Res 126:409–411

Kang S, Chuaoli L, Wang F, Zhang Q, Cong Z (2009) Total suspended particulate matter and toxic elements indoors during cooking with Yakdung cakes. 43:4343–4346

Khare M, Gupta S (2000) Indoor air pollution. J IAEM 27:103–110

Kulshreshtha P, Khare M (2011) Indoor exploratory analysis of gaseous pollutants and respirable particulate matter at residential homes of Delhi, India. Atmos Pollut Res 2:337–350

Massey D, Masih J, Kulshrestha A, Habil M, Taneja A (2009) Indoor/outdoor relationship of fine particles less than 2.5 μm (PM2.5) in residential homes locations in central Indian region . Build Env 44:2037–2045

Mishra VK, Rutherford RD, Smith KR (1999) Cooking with biomass fuels increases the risk of blindness. National Family Health Survey Bulletin of International Institute for Population Sciences, No. 14, ISSN 1083–8678

Saiyed HN, Patel TS, Gokani VN (2001) Indoor air pollution in India—A major environmental and public health concern. ICMR Bull 31(5):1–9. ISSN 0377-4910

Smith KR, Mehta S, Maeusezahl-Feuz M (2003) Indoor air pollution from household use of solid fuels. In: Comparative quantification of health risk—Chapter 18, pp 1435–1493

Sussan TE, Ingole V, Kim JH, McCormick S, Negherbon J, Fallica J, Akulian J (2014) Source of biomass cooking fuel determines pulmonary response to household air pollution. Am J Respir Cell Mol Biol 50(3):538–548

TEDDY (1998) Tata energy data directory yearbook. Tata Energy Research Institute, New Delhi

Venkataraman C, Sagar AD, Habib G, Lam N, Smith KR (2010) The Indian national initiative for advanced biomass cookstoves: the benefits of clean combustion. Energy Sustain Dev 14:63–72

The manufacturer's authorised representative in the EU is Springer Nature Customer Service Centre GmbH, Europaplatz 3, 69115 Heidelberg, Germany. If you have any concerns regarding our products, please contact ProductSafety@springernature.com

Printed and bound by CPI Group (UK) Ltd, Croydon, CR0 4YY
30/09/2026
02225328-0011